Association of Dietary Intake with Chronic Disease and Human Health

Association of Dietary Intake with Chronic Disease and Human Health

Guest Editors

Domenico Tafuri
Francesca Latino

MDPI

Basel • Beijing • Wuhan • Barcelona • Belgrade • Novi Sad • Cluj • Manchester

Guest Editors
Domenico Tafuri
Department of Medical,
Motor and Wellness Sciences
University of Naples
"Parthenope"
Naples
Italy

Francesca Latino
Department of Education and
Sport Sciences
Pegaso University
Naples
Italy

Editorial Office
MDPI AG
Grosspeteranlage 5
4052 Basel, Switzerland

This is a reprint of the Special Issue, published open access by the journal *Nutrients* (ISSN 2072-6643), freely accessible at: https://www.mdpi.com/journal/nutrients/special_issues/34Y8NHGRN4.

For citation purposes, cite each article independently as indicated on the article page online and as indicated below:

Lastname, A.A.; Lastname, B.B. Article Title. *Journal Name* **Year**, *Volume Number*, Page Range.

ISBN 978-3-7258-3657-4 (Hbk)
ISBN 978-3-7258-3658-1 (PDF)
https://doi.org/10.3390/books978-3-7258-3658-1

Contents

About the Editors

Tafuri Domenico

Tafuri Domenico is a full professor at the University of Naples "Parthenope" and a director of the Department of Medical, Motor, and Wellness Sciences. He also serves as the university's referent for students with disabilities. His research interests span human health, sports medicine, and physiology, with a particular focus on the physiological and biomechanical adaptations induced by physical activity. He has authored numerous publications in national and international journals, contributing to advancements in exercise science and performance optimization. Additionally, he collaborates on interdisciplinary projects aimed at integrating medical and motor sciences to enhance well-being and prevent lifestyle-related diseases.

Francesca Latino

Francesca Latino is a researcher in the Department of Education and Sport Sciences at the "Pegaso" University of Naples. Her research focuses on the physiology of exercise across different age groups, with particular attention to its effects on muscular, cardiovascular, and neurohumoral adaptations. She has authored publications in national and international journals, contributing to the understanding of exercise-induced physiological changes. Additionally, her work explores inclusive methodologies and intervention strategies aimed at enhancing learning, preventing sedentary behavior, and promoting motor skill acquisition.

nutrients

MDPI

Editorial

Association of Dietary Intake with Chronic Disease and Human Health

Domenico Tafuri [1] and Francesca Latino [2,*]

1 Department of Medical, Motor and Wellness Sciences, University of Naples "Parthenope", via Medina 40, 80133 Naples, Italy; domenico.tafuri@uniparthenope.it
2 Department of Education and Sport Sciences, Pegaso University, 80100 Naples, Italy
* Correspondence: francesca.latino@unipegaso.it

1. Introduction

Eating habits are among the major determinants of health, and in most countries in the world part of the population suffers from one or more problems related to malnutrition, by default (due to an acute or chronic lack of micronutrients) or by excess (energy, sugars, and fats, and deficiencies in important micronutrients) [1]. Extensive scientific literature has long confirmed that diet has a strong impact on the health status of a population. The most relevant factors and diseases related to poor nutrition are obesity, diabetes, cardiovascular diseases, tumors, non-alcoholic fatty liver disease, chronic intestinal diseases, neuropsychiatric diseases, and neurodegenerative diseases [2,3].

The spread of unhealthy eating habits has caused an increase in overweight rates for almost 60 percent of the world's population, a third of which suffer from obesity. These problems are associated with diseases such as cardiovascular disease, type 2 diabetes, and some forms of cancer. Major social and economic changes, such as trends towards globalization and urbanization, challenge the ability to eat healthy diets [4]. Products with low nutritional value challenge food cultures and traditions. Sedentary lifestyles and a lack of physical activity increase progressively with age and affect more than one in three adults, compounding the problem of unhealthy eating patterns [5]. According to the Atlas of Heart Disease and Stroke, recently published by the World Health Organization (WHO) [6], the combination of an unhealthy diet and a sedentary lifestyle is the second most responsible cause of the 17 million deaths from cardiovascular and brain diseases, preceded only by smoking, and of cancer. A varied and balanced diet reduces the risk of cancer [7].

Healthy eating and good nutrition are biological acts necessary for meeting the energy needs of individuals and are fundamental for their health [8]. The busyness that very often characterizes our lives compels the selection of convenience foods and eating too quickly, which challenge healthy eating patterns. Healthy diets are a very powerful tool in the prevention of pathological conditions that can compromise quality of life and even life expectancy. It is, at this point, conceivable how non-communicable diseases represent a threat to the goals of the 2030 Agenda [9], in terms of both affecting sustainable development and their high social costs. In fact, the global burden of non-communicable diseases, with the threat they pose, constitutes a serious public health problem, hampering social and economic development worldwide.

Received: 21 November 2024
Revised: 29 November 2024
Accepted: 23 January 2025
Published: 26 January 2025

Citation: Tafuri, D.; Latino, F. Association of Dietary Intake with Chronic Disease and Human Health. *Nutrients* **2025**, *17*, 446. https://doi.org/10.3390/nu17030446

2. Nutrition as a Key Modifiable Factor for the Prevention of Chronic Disease and the Promotion of Human Health

2.1. Prevention of Chronic Disease by Means of Diet and Lifestyle Changes

Several studies have confirmed the significant role of foods in the development of certain forms of cancer and, conversely, the ability of other foods to prevent their formation [10,11]. According to the American Institute for Cancer Research (AICR), more than 30% of cancers are directly attributable to a poor diet, understood in both quantitative and qualitative terms. In this Special Issue, Wang et al. [12] explore the potential role of selenium in the developmental process of esophageal cancer. To explore the relationship between selenium-related factors and esophageal cancer through bioinformatic analysis, a case–control study was conducted to verify the results. The findings showed lower glutathione peroxidase 3 (GPx3) expression in tumor tissues compared to normal tissues. The findings did not establish a direct association between selenium itself and esophageal squamous cancer; however, the significant difference observed between GPx3 and core protein fatty acid binding protein 1 (FABP1) in the two groups suggested the potential significance of selenium in esophageal cancer prevention.

In addition, particular behaviors and habits can significantly affect individual cardiovascular risk: often the simple correction of these situations substantially modifies the incidence of coronary heart disease and therefore represents an important measure in cardiovascular prevention [13,14]. These aspects are highlighted by Sikand and Severson [15], who elucidate that suboptimal dietary quality has eclipsed all other mortality risk determinants, culminating in 11 million fatalities and constituting half of global cardiovascular disease mortality. In their investigation, they assert that adults are encouraged to adhere to a cardioprotective diet that prioritizes plant-based alimentary sources, including vegetables, fruits, legumes, nuts, and whole grains, alongside lean protein sources and fish. They are advised to curtail the consumption of foods replete in saturated fats and dietary cholesterol while minimizing sodium intake. Furthermore, the avoidance of trans-fats, processed meats, refined carbohydrates, and sugar-sweetened consumables is recommended. The incorporation of dietary adjuncts, such as viscous fiber, plant sterols/stanols, and long-chain omega-3 fatty acids, is also advocated. They concluded that the implementation of medical nutrition therapy resulted in enhancements in lipid profiles, weight management, HbA1c levels, and hypertension, alongside fiscal savings. Empirical studies indicate that nutrition education and engagement in physical activity can equip students with the ability to discern the influence of a healthful lifestyle on psychosomatic well-being. Additionally, insufficient physical activity coupled with a substandard diet can adversely affect children's weight status, metabolic function, and cardiovascular health. In light of the immediate and prolonged ramifications of children's health behaviors, there exists an urgent imperative to facilitate the adoption and sustenance of healthy lifestyles among children. Consequently, the enhancement of children's physical activity and nutritional practices are recognized as two paramount public health initiatives.

Chen et al. [16] conducted an investigation into the impact of consistent nutritional consultations on the mitigation of risk factors pertinent to cardiovascular diseases, which include body mass index, body composition, blood pressure, lipid profiles, glucose-related biomarkers, and inflammatory mediators. The data were amassed from a sample of participants (n = 129) who underwent eight dietary consultations, subsequently categorized into two distinct groups based on the frequency of consultations: an irregular group (experiencing non-uniform consultation intervals; n = 39) and a regular group (who received consultations every three weeks; n = 90). In comparison to the irregular group, the regular group exhibited more pronounced declines in cardiovascular disease risk factors, encompassing body mass index, body fat percentage, triglyceride levels, total cholesterol, low-density lipoprotein cholesterol, and insulin concentrations. Furthermore, participants with a body mass index of $\geq$27 kg/m^2 demonstrated significantly noteworthy enhancements in cardiovascular risk parameters, including body weight, body mass index, visceral fat mass, and levels of triglycerides, total cholesterol, low-density lipoprotein cholesterol, glycated hemoglobin, and insulin. The authors concluded that there exists substantial evidence supporting the efficacy of regular nutritional consultations for adults exhibiting cardiovascular disease risk factors, particularly among those classified as obese. Similarly, Capra and colleagues [17] underscore the significance of nutritional intervention as a preliminary therapeutic approach for individuals exhibiting elevated cardiovascular risk, with particular emphasis on its critical relevance for pediatric and adolescent populations. They scrutinize the predominant dietary patterns prevalent among children and adolescents, aiming to assess their impact on cardiovascular risk factors and the prevention of cardiovascular diseases. Their findings indicated that the Mediterranean diet, DASH diet, Nordic diet, and certain plant-based diets appear to be the most promising dietary frameworks concerning cardiovascular health during developmental stages. The analysis of the role of nutrition in the management of diabetes was another topic of investigation in this Special Issue. The fundamental role of diet in the management of diabetes and its self-control is undisputed; it is for all intents and purposes a therapy. Gal et al. [18] postulate that dietary habits and consumption behaviors are integral to both the prevention and management of type 2 diabetes mellitus (T2DM). Willey and associates [19] investigated dietary habits among a variety of ethnic, cultural, and regional groups. The researchers posited that individuals exhibiting a shared clinical phenotype (age, body mass index, duration since T2DM diagnosis, and insufficient glycemic regulation) would reveal analogous high-calorie "western" dietary patterns and suboptimal dietary quality, notwithstanding the variations in geographic locales and cultural backgrounds. Dietary information was obtained from a cohort of 611 adults across Argentina, Germany, Poland, Serbia, Slovakia, Slovenia, Spain, Turkey, and the USA through three 24 h dietary recalls. The authors inferred that the dietary pattern data indicate the persistence of influences from more traditional region-specific diets. Nevertheless, the overall quality of the diet was found to be inadequate, which may exacerbate insufficient glycemic control, potentially as a consequence of the excessive consumption of high-calorie, nutrient-poor foods, likely linked to the global transitions in the food supply that are currently underway. Finally, within this Special Issue, the crucial role of nutrition in the management of inflammatory bowel disease has been extensively discussed. Chronic inflammatory bowel diseases, characterized by a chronic recurrent inflammatory process affecting one or more intestinal segments, are strongly influenced by the intake of different foods. A proper diet cannot cure inflammatory bowel disease, but some foods can aggravate symptoms, while others can relieve them and promote healing. Palamenghi, Usta, Leone, and Graffigna [20] emphasize the importance of food intake in the management of inflammatory bowel disease (IBD) by offering valuable insights into the complex motivations that drive food choices among individuals with IBD. Based on their findings, food engagement had a significant impact on participants' food preferences, with higher levels of food involvement associated with a propensity to seek out pleasurable food; however, individuals with high levels of health involvement reflected a better emotional state that resulted in a lower need for food. Sasson et al. [21] propose that dietary factors may exacerbate intestinal inflammation through the dysregulation of the immune response, modifications in intestinal permeability and the mucosal barrier, the promotion of microbial dysbiosis, and various other pathways. Consequently, modifications in dietary practices may be integrated into therapeutic approaches for inflammatory bowel diseases (IBDs).

2.2. Improving Nutrition Education at Schools: Challenges and Opportunities

Prevention is a fundamental activity for safeguarding health [22,23]. In this vein, schools, as institutions spread throughout the world, are the main place in which to exercise and strengthen food education processes as a preventive measure [24]. This could guide new generations to develop adequate sensitivity to issues of personal and collective well-being, sustainability, and the adoption of correct lifestyles. Concepts such as the value of food are important, through which one can rediscover a sense of belonging to a community in which food is always part of its history and place [25]. These values are also accompanied by multiple virtuous effects, such as the protection of production territories, supply chains, and food systems, the dissemination of the culture of responsible consumption, and the containment of food waste. The principle of prevention also makes health and welfare systems more sustainable, and a deep-rooted food culture helps entire populations make informed choices with a view to healthiness and sustainability [26]. These themes are the basis of the articles that represent this Special Issue. Moscatelli and colleagues [27] have highlighted how the years spent at university represent a critical period that can affect both quality of life and the eating habits of later adulthood, as well as, in the long term, the health of an individual. The aim of this study was to investigate the lifestyles of university students living far from home. Their study showed that it is possible to improve lifestyle through educational events aimed at making students aware of the health risks of unhealthy lifestyles.

Latino, Tafuri, Saraiello, and Tafuri [28] suggested that since youths spend the greater part of the day at school, this is considered an ideal setting for fostering active and healthy living. The authors investigated the connection between physical activity and academic achievement in normal-weight and overweight adolescents. One hundred students (aged 14–15) from a public high school in the south of Italy participated in either a 12-week classroom-based physical activity break program performed during science classes (60′/2 days per week) in which a nutritional educational program was carried out or in regular science lessons (60′/2 days per week). The conclusions of this research underpin the notion that classroom-based physical activity breaks are a successful approach for enhancing child students' physical well-being.

2.3. The Potential of Nutrition to Promote Psychological and Behavioral Well-Being

Eating is one of the primary needs of all living beings, but it often represents much more than a tool to satisfy a physiological need of the body, and there are many psychological implications that are hidden behind every meal. Every meal, starting with the three main ones, is a very important social moment for human beings, who often like to eat in company [29].

Sometimes food becomes an outlet and something to cling to in fighting stress, anxiety or depression. It then happens that we eat too much or badly, favoring fatty or sweet foods, often very caloric, which also satisfy the need to fill a "void" from a chemical point of view. Other times, however, it may happen that we eat irregularly, ingesting everything possible and also giving in to real binge eating, which is very harmful in the long run for health and the maintenance of a normal weight.

In this perspective, Latino et al. [28] concluded that the association of a nutritional educational program with physical activity breaks is a successful approach for enhancing students' psycho-physical well-being, as well as academic achievement.

Gheisary et al. [30] aim to provide a comprehensive list of associations between lifestyle traits and the general health and well-being status of individuals with arthritis in the Canadian population, using binary logistic regression analysis on data from the Canadian Community Health Survey, which includes 104,359 respondents. Their study highlights the need for personalized, holistic approaches that can include a combination of dietary interventions, sleep therapies, and smoking cessation for better prevention and lifestyle improvement.

Li and colleagues [31], on the other hand, underline in their study the importance of nutritional changes and, in particular, the intake of phytanic acid in the management of the symptoms and psychosocial implications that follow in patients with Refsum disease.

Shawon and colleagues [32] conducted an investigation into the sex-specific prevalence of psychological distress and maladaptive eating behaviors among adolescents across various nations and regions, with the aim of examining their potential correlations. The study utilized data obtained from the Global School-Based Health Survey (GSHS), encompassing 61 nations. Their results reveal a significant disparity in the prevalence of psychological distress and maladaptive dietary practices among distinct regions. Furthermore, adolescents who reported experiencing psychological distress were found to be at an increased likelihood of engaging in unhealthy eating behaviors.

2.4. Nutrition and Physical Activity in the Prevention and Treatment of Chronic Disease

In addition to what we eat, it is essential to always keep the body moving. The habit of physical activity, together with its frequency and intensity, contributes to reducing or increasing our energy needs [33].

With advancing age, body fat increases, especially visceral fat, water in the tissues decreases, and bone density as well as muscle mass are reduced, causing a drop in basic metabolism. This explains the importance of physical activity for the reduction in and control of body weight, now recommended for all healthy people (regardless of age). To preserve mental and physical well-being, the WHO [34] recommends that we do at least 150 min of moderate physical activity per week. It has been calculated, in fact, that 30 min a day is enough to prevent weight gain in a person who follows a varied and balanced diet. On the other hand, the commitment must increase (60–90 min a day) if the person has to lose weight or maintain the weight achieved by following a diet (the recommendation is all the more significant the greater the weight loss).

Sedentary behaviors, such as spending too much time in front of the television, are not only a harmful habit for young people, but also affect adults. In children and adolescents, multimodal interventions aim at the systematic reduction in sedentary lifestyles and the increase in habitual AF, which together with proper nutrition education could reduce the growth of overweight and obesity and ensure better levels of health in populations in the future (EU Action Plan on Childhood Obesity 2014–2020), as highlighted by the studies of Latino and colleagues in addition to Moscatelli and colleagues [27,28]. The interventions can be combined with a training approach, as the practice of AF activates in young maturation processes, promoting personal and social development [35].

Physical exercise is also an effective health intervention in the treatment of numerous chronic diseases and, in terms of reducing mortality, provides benefits similar to those obtainable from drugs in the secondary prevention of coronary heart disease, post-stroke rehabilitation, heart failure, and diabetes prevention; in less severe pathologies (back pain, osteoarthritis), it brings significant benefits to symptoms and quality of life [36].

According to this line of thought, Janiczak and colleagues [37] highlight the role of proper nutrition in the management of athlete recovery.

2.5. Environmental and Policy Approaches in the Prevention of Chronic Diseases: The Role of Governments and Health Systems

The role of the health system in the implementation of preventive programs should be to disseminate greater scientific knowledge of the role of nutrition in the public health and economic sustainability plans of governments [38–43]. Through primary prevention, an attempt is made to improve lifestyles in the population to reduce the burden of diseases. Lifestyle has a significant impact on health, as it acts on the main risk factors. Eating habits have a powerful influence on health, favoring it if body weight is kept within optimal limits.

The emergence of the importance of managing dietary habits in combating chronic disabling diseases has placed an urgent need on governments to develop sustainable and effective public intervention models [44–47]. So far, the effectiveness and convenience of certain types of preventive programs have been confirmed: those implemented in hospitals, workplaces, and those aimed at the elderly population. On the other hand, there is still insufficient evidence to understand which types of interventions are most effective in producing a significant and lasting increase in dietary changes, and which are most suitable for extending the benefits to large segments of the population.

This line of research is extensively explored worldwide. In their study, Downer and colleagues [48] critically assess contemporary initiatives aimed at integrating food and nutrition into the prevention, management, and therapeutic approaches for diet-related diseases within healthcare frameworks. The authors underscore the significance of interventions that are being implemented with increasing frequency in the United States and have been trialed to a certain degree in various other nations, including medically tailored meals, medically tailored grocery provisions, and produce prescription initiatives.

3. Conclusions

Overall, the fifteen studies enclosed within this Special Issue encompass both experimental and descriptive methodologies pertinent to design (for instance, correlational, survey, and developmental inquiries). It is significant to note that all of the research disclosed highly favorable outcomes regarding the association between nutrition, chronic diseases, and human health across diverse populations. These inquiries contribute to the existing body of knowledge in this domain. We aspire for these articles to be regarded as intellectually stimulating and academically engaging. Furthermore, we anticipate the emergence of additional studies employing innovative methodologies and advanced techniques to further explore this critical research domain and to advocate for healthier lifestyles.

Author Contributions: This editorial is the result of a study designed and shared between the authors. All authors have contributed equally to this work. They contributed intellectually to the manuscript, read the manuscript, and approved the submission. All authors have read and agreed to the published version of the manuscript.

Conflicts of Interest: The authors declare no conflict of interest.

References

1. Noce, A.; Romani, A.; Bernini, R. Dietary intake and chronic disease prevention. *Nutrients* **2021**, *13*, 1358. [CrossRef] [PubMed]
2. Neuhouser, M.L. The importance of healthy dietary patterns in chronic disease prevention. *Nutr. Res.* **2019**, *70*, 3–6. [CrossRef] [PubMed]
3. Billingsley, H.E.; Heiston, E.M.; Bellissimo, M.P.; Lavie, C.J.; Carbone, S. Nutritional Aspects to Cardiovascular Diseases and Type 2 Diabetes Mellitus. *Curr. Cardiol. Rep.* **2024**, *26*, 73–81. [CrossRef] [PubMed]
4. Srour, B.; Kordahi, M.C.; Bonazzi, E.; Deschasaux-Tanguy, M.; Touvier, M.; Chassaing, B. Ultra-processed foods and human health: From epidemiological evidence to mechanistic insights. *Lancet Gastroenterol. Hepatol.* **2022**, *7*, 1128–1140. [CrossRef]
5. Samuel, P.O.; Edo, G.I.; Emakpor, O.L.; Oloni, G.O.; Ezekiel, G.O.; Essaghah, A.E.A.; Agoh, E.; Agbo, J.J. Lifestyle modifications for preventing and managing cardiovascular diseases. *Sport Sci. Health* **2024**, *20*, 23–36. [CrossRef]
6. World Health Organization. *WHO Manual on Sugar-Sweetened Beverage Taxation Policies to Promote Healthy Diets*; World Health Organization: Geneva, Switzerland, 2022.
7. Jayedi, A.; Soltani, S.; Abdolshahi, A.; Shab-Bidar, S. Healthy and unhealthy dietary patterns and the risk of chronic disease: An umbrella review of meta-analyses of prospective cohort studies. *Br. J. Nutr.* **2020**, *124*, 1133–1144. [CrossRef]

8. Beslay, M.; Srour, B.; Méjean, C.; Allès, B.; Fiolet, T.; Debras, C.; Chazelas, E.; Deschasaux, M.; Wendeu-Foyet, M.G.; Hercberg, S.; et al. Ultra-processed food intake in association with BMI change and risk of overweight and obesity: A prospective analysis of the French NutriNet-Santé cohort. *PLoS Med.* **2020**, *17*, e1003256. [CrossRef]
9. Petkoska, A.T.; Trajkovska-Broach, A. Sustainable food systems and healthy diets: The case of mediterranean diet. *Acta Hortic. Regiotect.* **2021**, *24*, 110–118. [CrossRef]
10. Naghshi, S.; Sadeghi, O.; Willett, W.C.; Esmaillzadeh, A. Dietary intake of total, animal, and plant proteins and risk of all cause, cardiovascular, and cancer mortality: Systematic review and dose-response meta-analysis of prospective cohort studies. *BMJ* **2020**, *370*, m2412. [CrossRef]
11. Kurniawan, A.W.; Wiguno, L.T.H.; Setyawan, H.; Shidiq, A.A.P.; García-Jiménez, J.V.; Eken, Ö.; Latino, F.; Tafuri, F.; Pranoto, N.W.; Rahmatullah, M.I.; et al. I-Spring Assisted Development of a Basketball Shooting Technique Program. *Retos Nuevas Perspect. Educ. Física Deporte Recreación* **2024**, *55*, 874–881.
12. Wang, N.; Pan, D.; Zhu, X.; Ren, X.; Jin, X.; Chen, X.; Wang, Y.; Su, M.; Sun, G.; Wang, S. Selenium May Be Involved in Esophageal Squamous Cancer Prevention by Affecting GPx3 and FABP1 Expression: A Case-Control Study Based on Bioinformatic Analysis. *Nutrients* **2024**, *16*, 1322. [CrossRef] [PubMed]
13. Wang, Y.J.; Yeh, T.L.; Shih, M.C.; Tu, Y.K.; Chien, K.L. Dietary sodium intake and risk of cardiovascular disease: A systematic review and dose-response meta-analysis. *Nutrients* **2020**, *12*, 2934. [CrossRef] [PubMed]
14. Mozaffarian, D.; Aspry, K.E.; Garfield, K.; Kris-Etherton, P.; Seligman, H.; Velarde, G.P.; Williams, K.; Yang, E.; ACC Prevention of Cardiovascular Disease Section Nutrition and Lifestyle Working Group and Disparities of Care Working Group. "Food is medicine" strategies for nutrition security and cardiometabolic health equity: JACC state-of-the-art review. *J. Am. Coll. Cardiol.* **2024**, *83*, 843–864. [CrossRef] [PubMed]
15. Sikand, G.; Severson, T. Top 10 dietary strategies for atherosclerotic cardiovascular risk reduction. *Am. J. Prev. Cardiol.* **2020**, *4*, 100106. [CrossRef] [PubMed]
16. Chen, Y.; Lin, J.; Chen, J.; Chen, Y.; Yang, S. Regular nutrition consultations reduced risk factors for cardiovascular diseases in adults. *Nutrition* **2024**, *118*, 112259. [CrossRef] [PubMed]
17. Capra, M.E.; Monopoli, D.; Decarolis, N.M.; Giudice, A.; Stanyevic, B.; Esposito, S.; Biasucci, G. Dietary Models and Cardiovascular Risk Prevention in Pediatric Patients. *Nutrients* **2023**, *15*, 3664. [CrossRef]
18. Gal, A.M.; Iatcu, C.O.; Popa, A.D.; Arhire, L.I.; Mihalache, L.; Gherasim, A.; Nita, O.; Soimaru, R.M.; Gheorghita, R.; Graur, M.; et al. Comprendere l'interazione tra assunzione dietetica e comportamento alimentare nel diabete di tipo 2. *Nutrients* **2024**, *16*, 771. [CrossRef]
19. Willey, J.; Wakefield, M.; Silver, H.J. Exploring the Diets of Adults with Obesity and Type II Diabetes from Nine Diverse Countries: Dietary Intakes, Patterns, and Quality. *Nutrients* **2020**, *12*, 2027. [CrossRef]
20. Palamenghi, L.; Usta, D.; Leone, S.; Graffigna, G. Food-Related Behavioral Patterns in Patients with Inflammatory Bowel Diseases: The Role of Food Involvement and Health Engagement. *Nutrients* **2024**, *16*, 1185. [CrossRef]
21. Sasson, A.N.; Ananthakrishnan, A.N.; Raman, M. Diet in Treatment of Inflammatory Bowel Diseases. *Clin. Gastroenterol. Hepatol.* **2021**, *19*, 425–435.e3. [CrossRef] [PubMed]
22. Alves, J.G.B.; Alves, L.V. Early-life nutrition and adult-life outcomes. *J. Pediatr.* **2024**, *100* (Suppl. 1), S4–S9. [CrossRef] [PubMed]
23. Danao, K.; Rokde, V.; Nagar, A.; Bendale, A.R.; Karande, N. Advances in food and nutrition research. In *Nutrigenomics and Nutraceuticals*; Apple Academic Press: Palm Bay, FL, USA, 2024; pp. 65–87.
24. Saha, S.; Dawson, J.; Murimi, M.; Dodd, S.; Oldewage-Theron, W. Effects of a nutrition education intervention on fruit and vegetable consumption-related dietary behavioural factors among elementary school children. *Health Educ. J.* **2020**, *79*, 963–973. [CrossRef]
25. Love, P.; Booth, A.; Margerison, C.; Nowson, C.; Grimes, C. Food and nutrition education opportunities within Australian primary schools. *Health Promot. Int.* **2020**, *35*, 1291–1301. [CrossRef] [PubMed]
26. Hargreaves, D.; Mates, E.; Menon, P.; Alderman, H.; Devakumar, D.; Fawzi, W.; Greenfield, G.; Hammoudeh, W.; He, S.; Lahiri, A.; et al. Strategies and interventions for healthy adolescent growth, nutrition, and development. *Lancet* **2022**, *399*, 198–210. [CrossRef]
27. Moscatelli, F.; De Maria, A.; Marinaccio, L.A.; Monda, V.; Messina, A.; Monacis, D.; Toto, G.; Limone, P.; Monda, M.; Messina, G.; et al. Assessment of Lifestyle, Eating Habits and the Effect of Nutritional Education among Undergraduate Students in Southern Italy. *Nutrients* **2023**, *15*, 2894. [CrossRef]
28. Latino, F.; Tafuri, F.; Saraiello, E.; Tafuri, D. Classroom-Based Physical Activity as a Means to Improve Self-Efficacy and Academic Achievement among Normal-Weight and Overweight Youth. *Nutrients* **2023**, *15*, 2061. [CrossRef] [PubMed]
29. Mozaffarian, D.; Blanck, H.M.; Garfield, K.M.; Wassung, A.; Petersen, R. A Food is Medicine approach to achieve nutrition security and improve health. *Nat. Med.* **2022**, *28*, 2238–2240. [CrossRef]

30. Gheisary, Z.; Hoja, I.; Liu, J.; Papagerakis, P.; Weber, L.P.; Fenton, M.; Katselis, G.S.; Lieffers, J.R.L.; Papagerakis, S. Association of Sleep Quality and General, Mental, and Oral Health with Lifestyle Traits (Dietary Intake, Smoking Status) in Arthritis: A Cross-Sectional Study from the Canadian Community Health Survey (CCHS). *Nutrients* **2024**, *16*, 2091. [CrossRef] [PubMed]
31. Li, J.J.; Kim, J.J.; Nausheen, F. Phytanic Acid Intake and Lifestyle Modifications on Quality of Life in Individuals with Adult Refsum Disease: A Retrospective Survey Analysis. *Nutrients* **2023**, *15*, 2551. [CrossRef] [PubMed]
32. Shawon, M.S.R.; Rouf, R.R.; Jahan, E.; Hossain, F.B.; Mahmood, S.; Das Gupta, R.; Islam, I.; Al Kibria, G.M.; Islam, S. Il peso del disagio psicologico e dei comportamenti alimentari non sani tra 222.401 adolescenti in età scolare provenienti da 61 paesi. *Sci. Rep.* **2023**, *13*, 21894. [CrossRef]
33. Chen, C.C.; Walker, A.; Collins, K.; Heydari, H.; Lim, S.; Kulinna, P. The Feasibility of University Adapted Physical Activity and Nutrition Education for Young Adults with Intellectual Disabilities. *J. Health Phys. Lit.* **2023**, *2*, 14–25.
34. DiPietro, L.; Al-Ansari, S.S.; Biddle, S.J.; Borodulin, K.; Bull, F.C.; Buman, M.P.; Cardon, G.; Carty, C.; Chaput, J.P.; Chastin, S.; et al. Advancing the global physical activity agenda: Recommendations for future research by the 2020 WHO physical activity and sedentary behavior guidelines development group. *Int. J. Behav. Nutr. Phys. Act.* **2020**, *17*, 143. [CrossRef]
35. O'Connell, M.L.; Coppinger, T.; McCarthy, A.L. The role of nutrition and physical activity in frailty: A review. *Clin. Nutr. ESPEN* **2020**, *35*, 1–11. [CrossRef] [PubMed]
36. Abu-Baker, N.N.; Eyadat, A.M.; Khamaiseh, A.M. The impact of nutrition education on knowledge, attitude, and practice regarding iron deficiency anemia among female adolescent students in Jordan. *Heliyon* **2021**, *7*, e06348. [CrossRef] [PubMed]
37. Janiczak, A.; Alcock, R.; Forsyth, A.; Trakman, G.L. A systematic review of interventions targeting modifiable factors that impact dietary intake in athletes. *Br. J. Nutrition.* **2024**, *131*, 229–247. [CrossRef] [PubMed]
38. Ross, A.C.; Caballero, B.; Cousins, R.J.; Tucker, K.L. *Modern Nutrition in Health and Disease*; Jones & Bartlett Learning: Burlington, MA, USA, 2020.
39. Lawrence, M.; Burlingame, B.; Caraher, M.; Holdsworth, M.; Neff, R.; Timotijevic, L. Public health nutrition and sustainability. *Public Health Nutr.* **2015**, *18*, 2287–2292. [CrossRef]
40. Carlisle, S.; Hanlon, P. Connecting food, well-being and environmental sustainability: Towards an integrative public health nutrition. *Crit. Public Health* **2014**, *24*, 405–417. [CrossRef]
41. Ridgway, E.M.; Lawrence, M.A.; Woods, J. Integrating environmental sustainability considerations into food and nutrition policies: Insights from Australia's National Food Plan. *Front. Nutr.* **2015**, *2*, 29. [CrossRef] [PubMed]
42. Fanzo, J. Food policies' roles on nutrition goals and outcomes: Connecting of food and public health systems. In *International Food Law and Policy*; Springer: Cham, Switzerland, 2016; pp. 213–251.
43. Herforth, A.; Frongillo, E.A.; Sassi, F.; Mclean, M.S.; Arabi, M.; Tirado, C.; Remans, R.; Mantilla, G.; Thomson, M.; Pingali, P. Toward an integrated approach to nutritional quality, environmental sustainability, and economic viability: Research and measurement gaps. *Ann. N. Y. Acad. Sci.* **2014**, *1332*, 1–21. [CrossRef] [PubMed]
44. Bodirsky, B.L.; Dietrich, J.P.; Martinelli, E.; Stenstad, A.; Pradhan, P.; Gabrysch, S.; Mishra, A.; Weindl, I.; Le Mouël, C.; Rolinski, S.; et al. The ongoing nutrition transition thwarts long-term targets for food security, public health and environmental protection. *Sci. Rep.* **2020**, *10*, 19778. [CrossRef]
45. Grosso, G.; Mateo, A.; Rangelov, N.; Buzeti, T.; Birt, C. Nutrition in the context of the Sustainable Development Goals. *Eur. J. Public Health* **2020**, *30* (Suppl. 1), i19–i23. [CrossRef]
46. Carey, R.; Caraher, M.; Lawrence, M.; Friel, S. Opportunities and challenges in developing a whole-of-government national food and nutrition policy: Lessons from Australia's National Food Plan. *Public Health Nutr.* **2016**, *19*, 3–14. [CrossRef] [PubMed]
47. Slater, J.; Yeudall, F. Sustainable livelihoods for food and nutrition security in Canada: A conceptual framework for public health research, policy, and practice. *J. Hunger. Environ. Nutr.* **2015**, *10*, 1–21. [CrossRef]
48. Downer, S.; Harlan, T.S. Food is medicine: Actions to integrate food and nutrition into healthcare. *BMJ* **2020**, *369*, m2482. [CrossRef]

nutrients

MDPI

Article

Association between Antibiotic Exposure and Type 2 Diabetes Mellitus in Middle-Aged and Older Adults

Lei Chu [1], Deqi Su [1], Hexing Wang [2], Dilihumaer Aili [1], Bahegu Yimingniyazi [1], Qingwu Jiang [2] and Jianghong Dai [1,*]

1 School of Public Health, Xinjiang Medical University, 567 Shangde North Road, Urumqi 830000, China
2 Key Laboratory of Public Health Safety of Ministry of Education, School of Public Health, Fudan University, Shanghai 200032, China
* Correspondence: epi102@sina.com

Abstract: Background: Although previous studies have shown an association between clinically used antibiotics and type 2 diabetes, the relationship between antibiotic exposure from food and drinking water and type 2 diabetes in middle-aged and older adults is unclear. Objective: This study was aimed at exploring the relationship between antibiotic exposures from different sources and type 2 diabetes in middle-aged and older people, through urinary antibiotic biomonitoring. Methods: A total of 525 adults who were 45–75 years of age were recruited from Xinjiang in 2019. The total urinary concentrations of 18 antibiotics in five classes (tetracyclines, fluoroquinolones, macrolides, sulfonamides and chloramphenicol) commonly used in daily life were measured via isotope dilution ultraperformance liquid chromatography coupled with high-resolution quadrupole time-of-flight mass spectrometry. The antibiotics included four human antibiotics, four veterinary antibiotics and ten preferred veterinary antibiotics. The hazard quotient (HQ) of each antibiotic and the hazard index (HI) based on the mode of antibiotic use and effect endpoint classification were also calculated. Type 2 diabetes was defined on the basis of international levels. Results: The overall detection rate of the 18 antibiotics in middle-aged and older adults was 51.0%. The concentration, daily exposure dose, HQ, and HI were relatively high in participants with type 2 diabetes. After model adjustment for covariates, participants with HI > 1 for microbial effects (OR = 3.442, 95%CI: 1.423–8.327), HI > 1 for preferred veterinary antibiotic use (OR = 3.348, 95%CI: 1.386–8.083), HQ > 1 for norfloxacin (OR = 10.511, 96%CI: 1.571–70.344) and HQ > 1 for ciprofloxacin (OR = 6.565, 95%CI: 1.676–25.715) had a higher risk of developing type 2 diabetes mellitus. Conclusions: Certain antibiotic exposures, mainly those from sources associated with food and drinking water, generate health risks and are associated with type 2 diabetes in middle-aged and older adults. Because of this study's cross-sectional design, additional prospective studies and experimental studies are needed to validate these findings.

Keywords: antibiotic; type 2 diabetes mellitus; hazard quotient; hazard index

Citation: Chu, L.; Su, D.; Wang, H.; Aili, D.; Yimingniyazi, B.; Jiang, Q.; Dai, J. Association between Antibiotic Exposure and Type 2 Diabetes Mellitus in Middle-Aged and Older Adults. *Nutrients* **2023**, *15*, 1290. https://doi.org/10.3390/nu15051290

Academic Editors: Domenico Tafuri and Francesca Latino

Received: 10 February 2023
Revised: 2 March 2023
Accepted: 3 March 2023
Published: 5 March 2023

1. Introduction

Antibiotics have been widely used for the treatment of bacterial infections in humans and animals and for the growth of animals since Alexander Fleming discovered that penicillin can be used to treat bacterial infections in 1928. Aureomycin was subsequently found to promote growth in animals in the 1940s [1–3]. The overuse of antibiotics in humans and particularly in animals, in which they are used in large quantities, has led to the presence of antibiotic residues in animals. A substantial proportion of antibiotics (30–90%) is excreted in an unchanged form or as active metabolites through urine or feces in animals [4]. Our previous studies have shown that antibiotic residues in animals persist in processed foods (meat foods, livestock and poultry products, aquatic products, milk, etc.) and enter the human diet [5]. Moreover, antibiotic residues in excreted urine and feces, known as metabolite contaminants, can enter the environment (in surface water and sediment) and then human drinking water [6]. Existing studies have detected antibiotic residues in food and water environments [7].

The relationship between antibiotic exposure and type 2 diabetes mellitus (T2DM) has received increasing attention, given the adverse effects of excessive antibiotic use in humans and animals, and the similar effects of antibiotic exposure and use in middle-aged and older adults [8,9]. Some epidemiological evidence suggests that T2DM is associated with alterations in microbiota composition and function [10,11]. Specific antibiotics have been associated with perturbed glucose homeostasis in patients with T2DM [12]. However, these studies have examined the relationships between clinically administered antibiotics and T2DM, whereas the relationship between daily exposure to antibiotics from one's diet or drinking water and T2DM remains unclear [13,14].

Many indicators are available for assessing antibiotic exposure, such as the concentration, detection rate, daily exposure dose (DED), hazard quotient (HQ) and hazard index (HI). The HI is an index for the quantitative and systematic evaluation of antibiotic exposure levels, which can provide a comprehensive indication of antibiotic exposure levels [6]. This study was conducted to assess the relationship between antibiotic exposure and the risk of T2DM in middle-aged and older people in Xinjiang, by testing urinary antibiotic exposure and then calculating the risk index through a biomonitoring method. Notably, unlike previous clinical studies based solely on questionnaire surveys, this study did not overlook the endogenous antibiotic exposure pathway.

2. Materials and Methods

2.1. Study Population

Participants were from the National Key Research and Development Program "Xinjiang Multi-Ethnic Natural Population Cohort Construction and Health Follow-Up Study," which has been conducted by our group since 2018 [15]. In the 2019 baseline survey, subsamples of participants were recruited randomly from three townships in Huocheng County in the Yili region (Langgan Township, Sarbulak Township and Luchaogou Township) according to the main local ethnic groups. The inclusion criteria were as follows: middle-aged and older people who were 45–75 years of age; residents who had lived in the areas for more than 3 years; and those who had a non-acute onset state. The exclusion criteria were severe liver or kidney disease, or mental illness. A total of 659 middle-aged or older adults were enrolled in this study; 134 had incomplete questionnaires or lacked urine or blood glucose information. Thus, 525 middle-aged or older people were included in this study, including 145 Han, 136 Hui, 138 Uyghur and 106 Kazakh individuals. These comprised 345 middle-aged people and 180 older people, of whom 264 were men and 261 were women. All participants signed an informed consent form, and the study was approved by the Ethics Committee of the Xinjiang Uygur Autonomous Region Hospital of Traditional Chinese Medicine (2018XE0108).

2.2. Urine Testing and Medical Information Collection

For the field investigation, the study participants were instructed to fast before the physical examination. An amount of 12 mL of morning urine was collected from the participants, and the urine samples were stored in the dark and frozen in a −40 °C refrigerator on site after collection. Urine samples (1 mL) were purified with an Oasis HLB 96-well solid phase extraction plate and analyzed via ultraperformance liquid chromatography coupled with high-resolution quadrupole time-of-flight mass spectrometry with an isotopic internal standard in 1 mL of urine, hydrolyzed by β-glucuronidase. All antibiotics were analyzed with an HSST3 chromatography column for separation. The three phenols were separated with acetonitrile and an aqueous mobile phase in the negative ionization mode, and the other antibiotics were separated with methanol and an aqueous mobile phase with 0.1% formic acid in the positive ionization mode. Each batch was based on 96-well solid phase extraction plates. A total of 96 samples were analyzed: 92 real urine samples, 2 solvent blank samples and 2 spiked urine samples of 10 ng/mL. The solvent blank samples and the urine samples spiked with the standard were analyzed together with the real urine samples. Solvent blanks were used to monitor background interference, and spiked urine samples were used to monitor precision and accuracy. The limits of detection and limits of quantification were defined as signal-to-noise ratios of 3 and 10, respectively. The limits of detection and limits of quantification for all antibiotics ranged from 0.04 to 1.31 ng/mL and from 0.13 to 4.37 ng/mL, respectively.

Physical examinations were performed by trained nurses or physicians. A 20 mL blood sample was collected for each participant with a vacuum blood collection device with an intravenous anticoagulant. Biochemical and routine blood tests were performed with 4 mL blood samples. These tests were performed at the township health center nearest to the survey site. Whole blood samples (3 mL) were transferred to three cryotubes immediately after blood sample collection. The blood samples used to separate plasma and leukocytes were centrifuged (4 °C, 3000 rpm, 10 min) within 2 h after blood sample collection.

The questionnaires were administered by medical students who had received professional training. The baseline questionnaire was mainly the China Kadoorie Biobank study baseline questionnaire, with minor modifications based on the comments of experts from Northwestern Medical College, China [16]. The questionnaire collected information on sociodemographic information; tea and coffee consumption; alcohol intake; smoking status; dietary status (the consumption frequency of pork, beef, mutton, fried food, vegetables and fruits was referred to in the food frequency method questionnaire) [17]; passive smoking and indoor air pollution; personal and family medical history; physical activity; mental health; and female reproductive history.

2.3. *Antibiotic Health Risk Assessment*

The daily exposure dose (DED) is an indicator used to assess the daily antibiotic exposure of an organism according to the following formula for the DED of an environmental contaminant [18,19]:

$DED = \frac{CsV}{MbP}$ (in µg/kg/day), where Cs is the antibiotic concentration in µg/L; V is the daily urine volume in L/day (daily urine volume was set at 1.70 L/day and 1.60 L/day for men and women, respectively); Mb is the body weight in kg; and P is the proportion of antibiotic excreted in urine, from the human pharmacokinetic data in an unchanged form and in a glucuronide bound form.

The acceptable daily intake (ADI) is an indicator of exposure applicable to long-term low doses of antibiotics, rather than the short-term high doses used in clinical settings. The current common standard for the ADI, established by the World Health Organization and the Food and Agriculture Organization of the United Nations in 1957, aims to establish a safe limit value to minimize the hazards threatening population health, which in turn is used to assess the risk of exposure [20]. We quantified the ADI of antibiotics by referring to the international ADI standard. With the exception of the three sulfonamides, the ADIs were established on the basis of the microbiological effects of common bacteria in the human gut microbiota. The ADIs for the three sulfonamides were established on the basis of toxicological effects.

The HQ is used to assess the health risk of each antibiotic and is defined as the ratio of daily exposure, i.e., the ratio of DED to the daily available dose ADI [21]: $HQ = \frac{DED}{ADI}$; $HI = \sum HQ$.

The HI is used to assess the cumulative health risk of combined antibiotic exposures and is based on a dose-additive concept of similar effects induced by multiple antibiotics, defined as the sum of HQs with similar effect endpoints. It is a new index for the quantitative and systematic evaluation of antibiotic exposure levels that provides a comprehensive indication of the level of antibiotic exposure. The HI is currently applied in evaluating the health risk of the long-term low-dose cumulative intake of antibiotics in humans. Because adults can be exposed to multiple antibiotics simultaneously, the HI was determined for 3 sulfonamides according to toxicological effects, 11 antibiotics according to microbial effects, 4 antibiotics for veterinary use and 10 antibiotics for preferred veterinary use, on the basis of different effect endpoints. An HQ or HI $\geq$1 indicates the presence of a potential health risk.

2.4. Statistical Analysis

The 18 antibiotics were grouped according to their antimicrobial mechanism or use, and new variables were generated via the summation of antibiotic mass concentrations in urine. The three new variables were human antibiotics (HAs), veterinary antibiotics (VAs) and preferred veterinary antibiotics (PVAs). The five new variables for antimicrobial mechanisms were tetracyclines, fluoroquinolones, macrolides, sulfonamides and phenicols. The same categories of HI were summed to generate new variables, which were divided into HI for microbiological effects, HI for toxicological effects, HI for veterinary use and HI for preferred veterinary use. Descriptive analysis provided the frequency of antibiotic testing or percentage of selected concentrations (P95, P99). The rank-sum test was used to analyze differences in blood glucose or antibiotic concentrations and other demographic variables, such as sex, age, education, income level, physical activity, diet, smoking and alcohol consumption.

The HI was classified into two groups according to hazard to human health: the HI $\leq$ 1 group and the HI > 1 group. Participants with fasting blood glucose values $\geq$7.0 mmol/L and self-reported T2DM were classified as the T2DM group, and those with fasting blood glucose values <7.0 mmol/L were classified as the non-T2DM group [22]. Variables were included in Model A as covariates if the p-value for the demographic variable of the total glucose or total antibiotic concentration derived from the t-test, ANOVA or rank test (Kruskal–Wallis test) was <0.05. Model A included the following covariates: age and income level. If the p value of the demographic variables of the total blood glucose or total antibiotic concentration obtained from the t-test, ANOVA or rank test (Kruskal–Wallis test) was <0.2, these variables were included in Model B as covariates. Model B included the variables from Model A as covariates plus sex, education, frequency of eating pork, frequency of eating mutton, frequency of eating fried food, vegetable consumption and smoking. The covariates associated with the relationship between antibiotic use and T2DM that had been studied in previous clinical settings were included in Model C. Model C included the variables in Model B as covariates plus physical activity, frequency of eating beef, fruit consumption and drinking (i.e., the covariates adjusted for in Model C were the risk factors for type 2 diabetes and all the demographic characteristics we focused on). Finally, the three models were statistically analyzed with binary logistic regression models to assess the relationship between HI and T2DM. All statistical analyses were performed in the statistical software SPSS 26. p-values <0.05 were considered statistically significant.

3. Results

All 18 antibiotics were detected in urine, with an overall detection rate of 51.0% and detection rates of individual antibiotics ranging from 0.2% to 20.0%. Urinary antibiotic concentrations ranged from below the limit of detection to above 1000 ng/mL (Table 1). Two or more antibiotics were detected in urine in 20.9% of individuals. Almost all non-T2DM populations had lower concentrations of antibiotics at the 95th percentile and 99th percentile than did the T2DM populations, except for tetracycline, norfloxacin, ofloxacin, thiamphenicol and sulfamethoxazole. The composition ratios of antibiotic detection rates were also lower in almost all non-T2DM populations than they were in the T2DM populations, except for chlortetracycline, doxycycline, clarithromycin, sulfadiazine and sulfamethoxazole.

Table 1. Frequency and concentration distribution of antibiotics detected in the urine of middle-aged and older people with and without type 2 diabetes in Xinjiang (n = 525).

Antibiotic	Usage	All (n = 525) n (%) [a]	Non-T2DM (n = 461)				T2DM (n = 64)			
			n (%) [a]	Percentiles		Maximum	n (%) [a]	Percentiles		Maximum
				95th	99th			95th	99th	
All antibiotics [b]		268 (51.0)	228 (49.5)	78.09	20,991.06	44,588.26	40 (62.5)	2363.48	24,282.26	24,282.26
HAs		54 (10.3)	43 (9.3)	0.87	77.06	4615.54	11 (17.2)	41.30	2989.35	2989.35
VAs		60 (11.4)	48 (10.4)	2.02	148.17	36,112.85	12 (18.8)	4.48	2090.43	2090.43
PVAs		224 (42.7)	190 (41.2)	44.97	7343.72	44,418.14	34 (53.1)	272.99	24,282.26	24,282.26
Tetracyclines [c]		79 (15.0)	64 (13.9)	13.35	20,991.06	44,565.07	15 (23.4)	42.71	2362.84	2362.84
Chlortetracycline	VA	4 (0.8)	4 (0.9)	—	—	119.02	0 (0)	—	—	—
Tetracycline	PVA	60 (11.4)	50 (10.8)	7.41	626.39	44,416.90	10 (15.6)	12.48	272.42	272.42
Doxycycline	PVA	3 (0.6)	3 (0.7)	—	—	11,315.50	0 (0)	—	—	—
Oxytetracycline	VA	38 (7.2)	30 (6.5)	1.43	75.75	36,112.85	8 (12.5)	4.48	2090.43	2090.43
Fluoroquinolones [c]		170 (32.4)	142 (30.8)	7.09	726.97	2876.96	28 (43.8)	67.69	24,282.26	24,282.26
Enrofloxacin	VA	11 (2.1)	7 (1.5)	—	0.21	9.40	4 (6.3)	0.20	0.96	0.96
Norfloxacin	PVA	66 (12.6)	55 (11.9)	1.28	18.14	1012.23	11 (17.2)	1.11	24,282.26	24,282.26
Ciprofloxacin	PVA	35 (6.7)	29 (6.3)	0.57	6.24	928.11	6 (9.4)	3.86	8.66	8.66
Ofloxacin	PVA	105 (20.0)	92 (20.0)	1.63	39.86	2876.96	13 (20.3)	1.08	66.89	66.89
Macrolides [c]		29 (5.5)	24 (5.2)	0.87	77.06	4615.54	5 (7.8)	41.30	2989.35	2989.35
Azithromycin	HA	15 (2.9)	13 (2.8)	—	5.11	479.12	2 (3.1)	—	248.70	248.70
Clarithromycin	HA	3 (0.6)	3 (0.7)	—	—	4615.54	0 (0)	—	—	—
Roxithromycin	HA	15 (2.9)	12 (2.6)	—	23.10	132.94	3 (4.7)	—	2989.35	2989.35
Sulfonamides [c]		47 (9.0)	40 (8.7)	0.25	1.81	7232.59	7 (10.9)	0.21	2.74	2.74
Sulfamethazine	PVA	16 (3.0)	13 (2.8)	—	0.28	11.66	3 (4.7)	—	0.28	0.28
Sulfadiazine	PVA	1 (0.2)	1 (0.2)	—	—	10.51	0 (0)	—	—	—
Sulfamethoxazole	PVA	9 (1.7)	9 (2.0)	—	0.69	5546.27	0 (0)	—	—	—
Trimethoprim	PVA	33 (6.3)	29 (6.3)	0.12	1.12	1686.32	4 (6.3)	0.06	2.74	2.74
Phenicols [c]		46 (8.8)	37 (8.0)	0.05	0.63	30.93	9 (14.1)	0.24	1.04	1.04
Chloramphenicol	HA	30 (5.7)	23 (5.0)	—	0.12	9.98	7 (10.9)	0.05	0.78	0.78
Florfenicol	VA	18 (3.4)	15 (3.3)	—	0.43	9.86	3 (4.7)	—	0.43	0.43
Thiamphenicol	PVA	3 (0.6)	2 (0.4)	—	—	11.10	1 (1.6)	—	0.08	0.08

HA: human antibiotic; VA: veterinary antibiotic; PVA: preferred veterinary antibiotic. [a] Positive detection (detection frequency, %). [b] Positive detection of one or more studied antibiotics. [c] Positive detection of one or more antibiotics in the corresponding category. —Urinary concentration of antibiotics is below the limit of detection.

Daily doses of antibiotic exposure were mostly lower in the 95th and 99th percentiles of the non-T2DM population than they were in the T2DM population, except for tetracycline, ofloxacin, sulfamethoxazole and florfenicol (Table 2).

Table 2. Estimated daily intake of antibiotics (μg/kg/day), on the basis of urinary concentrations of antibiotics in middle-aged and older people in Xinjiang (n = 525).

	Non-T2DM (n = 461) Percentiles			T2DM (n = 64) Percentiles		
Antibiotic	**95th**	**99th**	**Maximum**	**95th**	**99th**	**Maximum**
Tetracyclines [a]	0.44	569.61	1619.25	1.27	91.25	91.25
Chlortetracycline	0.00	0.00	5.44	0.00	0.00	0.00
Tetracycline	0.32	25.64	1612.89	0.45	11.91	11.91
Doxycycline	0.00	0.00	336.85	0.00	0.00	0.00
Oxytetracycline	0.06	3.11	1443.79	0.16	79.34	79.34
Fluoroquinolones [a]	0.22	37.14	57.11	2.13	873.98	873.98
Enrofloxacin	0.00	0.03	0.89	0.03	0.11	0.11
Norfloxacin	0.05	0.76	45.57	0.04	873.98	873.98
Ciprofloxacin	0.03	0.24	51.21	0.17	0.37	0.37
Ofloxacin	0.04	0.96	57.11	0.02	2.09	2.09
Macrolides [a]	0.04	4.59	383.10	1.01	108.08	108.08
Azithromycin	0.00	1.57	167.41	0.00	108.08	108.08
Clarithromycin	0.00	0.00	383.10	0.00	0.00	0.00
Roxithromycin	0.00	0.83	3.85	0.00	94.88	94.88
Sulfonamides [a]	0.01	0.12	610.71	0.01	0.10	0.10
Sulfamethazine	0.00	0.01	0.35	0.00	0.01	0.01
Sulfadiazine	0.00	0.00	0.40	0.00	0.00	0.00
Sulfamethoxazole	0.00	0.07	536.56	0.00	0.00	0.00
Trimethoprim	0.00	0.05	74.15	0.00	0.10	0.10
Phenicols [a]	0.00	0.03	1.02	0.01	0.03	0.03
Chloramphenicol	0.00	0.00	0.21	0.00	0.02	0.02
Florfenicol	0.00	0.03	0.41	0.00	0.02	0.02
Thiamphenicol	0.00	0.00	0.40	0.00	0.00	0.00
All antibiotics [b]	3.82	615.82	1619.98	95.29	873.98	873.98

[a] Sum of estimated daily intakes of antibiotics in the corresponding category. [b] Sum of estimated daily intake of all antibiotics.

Among 525 middle-aged and older people in Xinjiang, 40 individuals (7.7%) had an HI value ≥ 1 and a health risk. The values of the HQ were mostly higher for the T2DM than they were in the non-T2DM population, except for individual antibiotics such as tetracycline, norfloxacin, ofloxacin, florfenicol and sulfamethoxazole. Notably, the 99th percentile HQ values of oxytetracycline and norfloxacin were relatively high in the T2DM population, at 26.448 and 62.427, respectively, thus indicating that a high oxytetracycline and norfloxacin intake was associated with a high risk of susceptibility to human health effects (Table 3).

Table 4 shows that, in older adults who were 60–74 years of age, compared with middle-aged adults aged 45–59 years, blood glucose values and total antibiotic exposure concentrations increased with age. This trend was also seen with an increasing frequency of vegetable consumption; participants who ate vegetables daily had higher blood glucose values and total antibiotic exposure concentrations than did those who did not eat vegetables daily. Moreover, occasional pork eaters had higher blood glucose values but lower total antibiotic concentrations than did those who did not eat pork, and occasional alcohol drinkers had higher blood glucose values but lower total antibiotic concentrations than did non-drinkers.

Table 3. Hazard quotient (HQ) of veterinary antibiotics and human/veterinary antibiotics among middle-aged and older people in Xinjiang (n = 525).

Antibiotics	Effect End Point	ADI (μg/kg/day)	Overall *N* (%)	Non-T2DM (n = 461)			T2DM (n = 64)		
				Percentiles		Maximum	Percentiles		Maximum
				95th	99th		95th	99th	
Hazard index [a]									
Chlortetracycline	Microbiological	3	1 (0.2)	0.000	0.000	1.813	0.000	0.000	0.000
Tetracycline	Microbiological	3	12 (2.3)	0.107	8.547	537.630	0.150	3.970	3.970
Doxycycline	Microbiological	3	1 (0.2)	0.000	0.000	112.284	0.000	0.000	0.000
Oxytetracycline	Microbiological	3	6 (1.1)	0.019	1.038	481.264	0.052	26.448	26.448
Enrofloxacin	Microbiological	6.2	0 (0)	0.000	0.005	0.143	0.004	0.017	0.017
Norfloxacin	Microbiological	14	5 (1.0)	0.004	0.054	3.255	0.003	62.427	62.427
Ciprofloxacin	Microbiological	0.15	11 (2.1)	0.202	1.589	341.393	1.129	2.464	2.464
Ofloxacin	Microbiological	3.2	2 (0.4)	0.014	0.299	17.847	0.008	0.653	0.653
Trimethoprim	Microbiological	4.2	1 (0.2)	0.001	0.012	17.656	0.001	0.025	0.025
Florfenicol	Microbiological	3	0 (0)	0.000	0.010	0.137	0.000	0.007	0.007
Thiamphenicol	Microbiological	2.5	0 (0)	0.000	0.000	0.159	0.000	0.002	0.002
Hazard index [b]									
Sulfamethazine	Toxicological	1.6	0 (0)	0.000	0.006	0.220	0.000	0.008	0.008
Sulfadiazine	Toxicological	20	0 (0)	0.000	0.000	0.020	0.000	0.000	0.000
Sulfamethoxazole	Toxicological	130	1 (0.2)	0.000	0.001	4.127	0.000	0.000	0.000

[a] Sum of HQ based on microbiological effect. [b] Sum of HQ based on toxicological effect; N: value of HQ $\geq$ 1.

Table 4. Blood glucose and all antibiotics in relation to demographic variables.

Variable	Blood Glucose (mmol/L) [a]	p-Value [d]	All Antibiotics (ng/mL) [c]	p-Value [b]
Gender		0.698		0.090
Male (n = 264)	5.90 ± 1.68		0 (1.78)	
Female (n = 261)	5.84 ± 1.79		0.11 (2.45)	
Age		0.373		0.041
45–59 (n = 432)	5.73 ± 1.43		0 (1.65)	
60–74 (n = 227)	6.15 ± 2.18		0.17 (2.27)	
Education		0.457		0.101
≤Primary (n = 383)	5.89 ± 1.86		0.08 (1.78)	
Secondary (n = 112)	5.92 ± 1.46		0 (1.82)	
≥High school (n = 30)	5.49 ± 0.48		0 (0.36)	
Income level		0.022		0.821
<10,000 RMB/year (n = 76)	6.27 ± 2.40		0.16 (1.63)	
10,000–35,000 RMB/year (n = 332)	5.89 ± 1.71		0.03 (1.69)	
≥35,000 RMB/year (n = 116)	5.57 ± 1.14		0 (2.37)	
Physical activity		0.824		0.267
Never (n = 427)	5.87 ± 1.78		0.04 (2.09)	
Occasionally (n = 98)	5.90 ± 1.52		0 (0.82)	
Pork		0.101		0.094
Do not eat (n = 444)	5.82 ± 1.73		0.05 (2.03)	
Occasionally eat (n = 78)	6.17 ± 1.77		0 (0.77)	
Beef		0.974		0.478
Do not eat (n = 155)	5.88 ± 1.63		0.07 (1.69)	
Occasionally eat (n = 281)	5.86 ± 1.88		0 (1.45)	
Eat every day (n = 87)	5.91 ± 1.41		0.17 (2.27)	
Mutton		0.057		0.138
Do not eat (n = 128)	6.19 ± 2.30		0.03 (1.72)	
Occasionally eat (n = 205)	5.82 ± 1.58		0 (0.94)	
Eat every day (n = 189)	5.73 ± 1.41		0.08 (3.07)	
Fried food		0.457		0.130
Do not eat (n = 328)	5.83 ± 1.64		0.07 (2.03)	
Occasionally eat (n = 197)	5.95 ± 1.88		0 (1.41)	
Vegetables		0.274		0.166
Not every day (n = 28)	5.52 ± 1.34		0 (0.68)	
Every day (n = 497)	5.89 ± 1.75		0.04 (1.78)	
Fruits		0.562		0.997
Not every day (n = 384)	5.85 ± 1.68		0.04 (1.69)	
Every day (n = 141)	5.95 ± 1.88		0.02 (2.70)	
Smoking		0.539		0.059
Never (n = 397)	5.92 ± 1.84		0.10 (2.05)	
Occasionally (n = 16)	5.88 ± 1.26		0.14 (4.36)	
Every day (n = 112)	5.71 ± 1.39		0 (0.96)	
Drinking		0.408		0.453
Never (n = 440)	5.85 ± 1.76		0.04 (1.75)	
Occasionally (n = 85)	6.02 ± 1.58		0 (1.39)	

[a] Mean ± standard deviation. [b] Rank test. [c] Median (interquartile range). [d] t-test or ANOVA.

The binary logistic regression results indicate that, after adjustment for covariates in Model A, the HI for microbiological effects, HI for PVAs, HQ for norfloxacin and HQ for ciprofloxacin were associated with a higher risk of T2DM in participants with higher exposure risk (HI > 1 or HQ > 1), and this association persisted after adjustment for more covariates in Models B and C (Table 5). Specifically, the risk of T2DM in the group with a microbiological effect HI > 1 was 2.948 times greater than the risk of T2DM in the group with microbiological effect HI ≤ 1 after adjustment for covariates in Model A, with a 95% *CI* and an OR of 2.948 (1.287–6.756). The confidence intervals were 3.391 (1.417–8.120) and 3.442 (1.423–8.327) after adjustment for covariates in Model B and Model C, respectively. After adjustment for covariates in Model A, the risk of T2DM was 2.928 times higher in those with an HI > 1 for PVAs than it was in the group with an HI ≤ 1 for PVAs, with a confidence interval of 2.928 (1.279–6.706). After adjustment for covariates in Models B and C, the confidence intervals were 3.271 (1.371–7.802) and 3.348 (1.386–8.083). After adjustment for covariates in Model A, those in the HQ > 1 group for norfloxacin had 10.075 times the T2DM risk of those in the HQ ≤ 1 group for norfloxacin, with a credible interval of 10.075 (1.612–62.952), and they still had a higher risk after adjustment for covariates in Models B and C, with credible intervals of 11.243 (1.728–73.142) and 10.511 (1.571–70.344). The risk of T2DM was 4.789 times greater in the ciprofloxacin HQ > 1 group than it was in the ciprofloxacin HQ ≤ 1 group after adjustment for covariates in Model A, with a confidence interval of 4.789 (1.336–17.175), and also after adjustment for covariates in Model B and Model C, with confidence intervals of 5.241 (1.377–19.946) and 6.565 (1.676–25.715).

Table 5. Associations of urinary antibiotic risk index, risk quotient and diabetes risk in middle-aged and older people, on the basis of binary logistic regression.

	Model A		Model B		Model C	
Variables	**Tertile 1**	**Tertile 2**	**Tertile 1**	**Tertile 2**	**Tertile 1**	**Tertile 2**
Group based on hazard index						
HI for microbiological effects	Ref.	2.948 (1.287–6.756) *	Ref.	3.391 (1.417–8.120) *	Ref.	3.442 (1.423–8.327) *
HI for toxicological effects	Ref.	—	Ref.	—	Ref.	—
HI for veterinary antibiotics	Ref.	0.962 (0.109–8.472)	Ref.	1.004 (0.105–9.607)	Ref.	0.952 (0.098–9.243)
HI for preferred veterinary antibiotics	Ref.	2.928 (1.279–6.706) *	Ref.	3.271 (1.371–7.802) *	Ref.	3.348 (1.386–8.083) *
Group based on hazard quotient						
HQ for tetracycline	Ref.	1.164 (0.245–5.534)	Ref.	1.195 (0.241–5.926)	Ref.	1.090 (0.218–5.456)
HQ for oxytetracycline	Ref.	1.195 (0.129–11.052)	Ref.	1.398 (0.136–14.372)	Ref.	1.238 (0.118–12.994)
HQ for norfloxacin	Ref.	10.075 (1.612–62.952) *	Ref.	11.243 (1.728–73.142) *	Ref.	10.511 (1.571–70.344) *
HQ for ciprofloxacin	Ref.	4.789 (1.336–17.175) *	Ref.	5.241 (1.377–19.946) *	Ref.	6.565 (1.676–25.715) *

* p-value < 0.05. Model A adjusted for age, pork and vegetable consumption, and drinking. Model B adjusted for sex, age, education and income level; pork, mutton, fried food and vegetable consumption; and smoking and drinking. Model C adjusted for sex, age, education, income level and physical activity; pork, beef, mutton, fried food, vegetable and fruit consumption; and smoking and drinking. —: Odds ratios are not provided because the proportion of participants with diabetes with a toxicological effect HQ ≥ 1 was too low.

4. Discussion

In this study, we measured the concentrations of 18 common antibiotics in urine and then calculated the DED, HQ and HI, thus indicating the relationship between T2DM and antibiotic risk in middle-aged and older people. The HI for microbiological effects >1, HI for PVA use >1, HQ for norfloxacin >1 and HQ for ciprofloxacin >1 in middle-aged and older people were positively associated with T2DM. These associations persisted after adjustment for several confounding factors known to be associated with T2DM. To our knowledge, this study is the first to use urinary antibiotic biomonitoring to explore different sources of antibiotic exposure, and to assess the health risks and consequent effects on T2DM.

Studies based on the biomonitoring of urinary antibiotic concentrations are lacking. Several studies have been performed in China and Korea [23–27]. Few studies have evaluated the HQ and HI of urinary antibiotics, all of which have been in China (Figure 1) [6,20,28–31]. Figure 1 shows the comparison of HQs in this study and six previous studies. The HQ levels in Xinjiang middle-aged and older adults were generally exceeded for 11 antibiotics, thus indicating a health risk, particularly for tetracycline (2.3%) and ciprofloxacin (2.1%). Throughout the western and eastern regions of China, the HI for ciprofloxacin was above the human health level threshold in all six studies. The detection rate of a ciprofloxacin HQ > 1 in middle-aged and older adults in Xinjiang was lower than that in several other regions (2.1%). Shanghai school-age children had the highest (8.07%), and the levels of ciprofloxacin HQ > 1 were comparable in Shanghai children and adults, both at 5.6%, and were higher than those in pregnant women in Eastern China (3.7%) and Anhui (3.8%). However, the levels of human antibiotic health risk in each of these regions varied, likely because of differences in locations, lifestyle habits and dietary and drinking habits among regions [32–34].

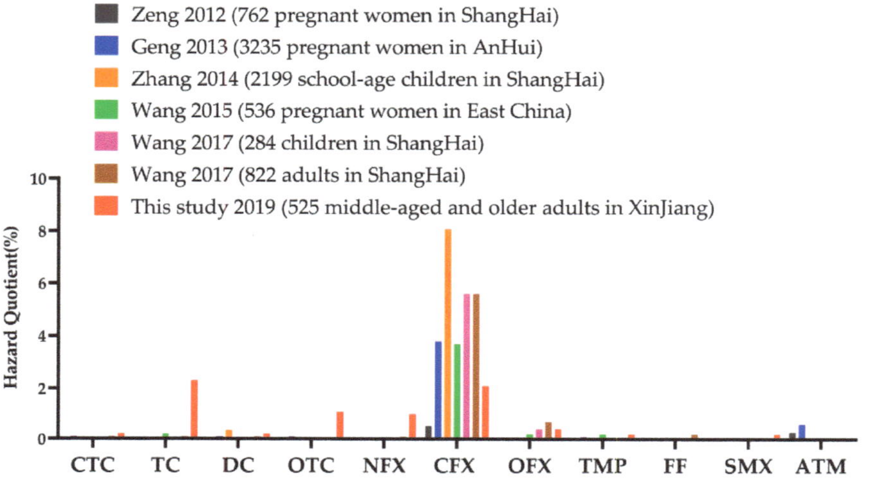

Figure 1. Comparison of hazard quotients of 11 antibiotics in urine between this study and six previous studies conducted by Zeng et al. [28], Geng et al. [29], Zhang et al. [30], Wang et al. [20], Wang et al. [31] and Wang et al. [6]. (CTC: Chlortetracycline; TC: Tetracycline; DC: Doxycycline; OTC: Oxytetracycline; NFX: Norfloxacin; CFX: Ciprofloxacin; OFX: Ofloxacin; TMP: Trimethoprim; FF: Florfenicol; SMX: Sulfamethoxazole: ATM: Azithromycin).

All 17 antibiotics have short half-lives of <15 h, except for azithromycin, which has a half-life of approximately 40 h [35]. Previous studies have hypothesized that exposure levels to antibiotics based on a single sampling test may represent previous long-term exposure levels, because antibiotic exposure depends on antibiotic use, and human exposure to antibiotics in daily life occurs primarily through contaminated food and drinking water (HAs, VAs and PVAs). Therefore, the concentration of urinary antibiotics in a single sample should reflect long-term antibiotic exposure to some extent, but not the intensity of antibiotic use or health risk [23]. This hypothesis is supported by our findings, which suggest that people with T2DM have higher biomonitoring urinary antibiotic concentrations, detection rates, daily exposure doses and HQ levels than do people without T2DM, and that people with diabetes have relatively higher HQ values for oxytetracycline and norfloxacin and have a higher health risk. Because the gut microbiome differed between the T2DM population and non-T2DM population, and the gut microbiome components are closely associated with glucose metabolism (e.g., insulin secretion, insulin sensitivity, etc.), the levels of antibiotic exposure should differ between these groups [36–40]. Because oxytetracycline and norfloxacin are common therapeutic agents in diabetes treatment, high and prolonged doses of oxytetracycline or norfloxacin may cause health risks in patients with diabetes [41].

A growing body of evidence from experimental studies suggests a causal relationship between the gut microbiota and diabetes [42]. Age factors, particularly aging, have been shown to be predisposing factors for abnormal glucose metabolism and abnormal regulation [43]. The diversity of the gut microbiota is diminished and is less stable in adults who are older than middle-age [44]. Antibiotics have multiple effects on host physiology, particularly on glucose homeostasis, by disrupting the intestinal microbiota. These effects are consistent with our findings, indicating an increase in antibiotics and glucose with age in people who are older than middle-age. In contrast, different antibiotic exposures arising from different diets, and the recognition of short-chain fatty acids and secondary bile acids by enteroendocrine cells, the vagus nerve and enteric neurons, among other aspects, result in different effects on glucose homeostasis [45,46]. Eating vegetables and pork and drinking alcohol with different frequencies in our study were associated with different trends in blood glucose levels and antibiotic concentrations. In this study, we observed higher blood glucose values in people who ate pork occasionally than those in people who did not eat pork; in people who ate vegetables daily than those in people who did not eat vegetables daily; and in people who drank alcohol occasionally than those in people who did not drink alcohol. Because pork consumption is a known risk factor for developing T2DM [47,48], and pigs are raised with antibiotics that promote growth and sterilization [49], people with T2DM are more likely to be tested for antibiotic exposure and to experience health risks from antibiotic exposure. A low intake of vegetables increases the risk of developing T2DM [50]. Vegetables are grown with added antibiotics [51], thus potentially increasing antibiotic exposure and risk in people with T2DM. Similarly, alcohol consumption is strongly associated with the development of T2DM, and excessive alcohol consumption can even trigger the risk of cancer in people with T2DM [52,53]. In contrast, alcohol production is prone to contamination with phytotoxins and low pH contaminants, such as antibiotics [54]. Therefore, antibiotics are often detected in patients with T2DM.

Although clinical studies have reported a relationship between antibiotics and T2DM [8,9,12–14,55,56], no studies have assessed the association between antibiotic exposure in daily life and the risk of T2DM. Previous findings from a clinical perspective are consistent with the results of this study, in that people who used antibiotics for longer periods of time had a higher risk of developing T2DM than did people who did not use antibiotics; moreover, those who used multiple antibiotics had a higher risk of developing T2DM than did those who used a single antibiotic. A significant dose-dependent relationship has been reported between antibiotic exposure and the incidence of T2DM [8,9,13]. Antibiotic exposure significantly increases the risk of T2DM in people above 50 years of age [14]. Moreover, fluoroquinolones are associated with an increased risk of diabetes [8]. However, results have varied among studies; for example, several studies have reported that single antibiotic use is not associated with the risk of developing T2DM [8,9]. Participants receiving more antibiotics had no increased risk of diabetes [57]. The reason for the discrepancy in the results may be due to differences in the doses and frequencies of medication and the variety of medications used for T2DM [58,59]. HI > 1 for microbiological effects in older people, HI > 1 for preferred veterinary use, HQ > 1 for norfloxacin (a PVA) and HQ > 1 for ciprofloxacin (a PVA) were significantly associated with T2DM and were risk factors for T2DM. These findings suggest that the cumulative health risk triggered by long-term exposure to low doses of antibiotics (contaminated food or drinking water) in daily life can affect T2DM development in middle-aged and older people, and that PVAs should be used in a regulated manner.

This study performed biomonitoring to assess the health risks of middle-aged and older people in Xinjiang. However, it has several limitations. First, this study was a cross-sectional study and thus could not demonstrate a causal relationship between antibiotics and T2DM. Antibiotic exposure might potentially be a proxy for some conditions associated with T2DM. Participants with T2DM might have had different levels of exposure to antibiotics in food or drinking water, and thus different levels of assessed health risks, depending on their dietary structure and intake. Second, because people are intermittently exposed to antibiotics in their daily lives, urinary concentrations in people may vary throughout the day. Antibiotics have short half-lives, and we collected morning urine from participants, thus potentially underestimating the exposure and the health risk assessment after multiple exposures to antibiotics. Single urine samples may not reflect the levels of long-term antibiotic exposure, thus potentially weakening the association between the risk of antibiotic exposure and T2DM in middle-aged and older adults.

5. Conclusions

We used a biomonitoring approach to assess antibiotic exposure and thus the risk of antibiotic exposure, and we found that exposure to antibiotics in daily life (mainly from food or drinking water contaminated with antibiotics), particularly with an HI > 1 for microbiological effects, HI > 1 for PVAs, HQ > 1 for norfloxacin and HQ > 1 for ciprofloxacin, may be risk factors associated with T2DM in middle-aged and older people. Because cross-sectional studies and primary urine sampling reflect long-term antibiotic exposure, more prospective studies and experimental studies are needed to further explore these associations and their underlying mechanisms.

Author Contributions: Conceptualization, L.C. and J.D.; Data curation, D.A. and Q.J.; Formal analysis, D.A.; Funding acquisition, J.D.; Investigation, H.W. and Q.J.; Methodology, D.S., H.W., B.Y. and J.D.; Project administration, J.D.; Resources, H.W., D.A. and Q.J.; Software, B.Y. and Q.J.; Supervision, J.D.; Visualization, B.Y.; Writing—original draft preparation, L.C.; Writing—review and editing, D.S. All authors have read and agreed to the published version of the manuscript.

Funding: This work was supported by a project of the National Key Research and Development Project of China (grant No. SQ2017YFSF090013), a subproject of the National Key Research and Development Project of China (grant No.2017YFC0907203), the National Natural Science Foundation of China (grant No.82160640) and Public Health and Preventive Medicine-Specialties of Higher Education Institutions in Xinjiang Uygur Autonomous Region in the 14th Five-Year Plan (grant No. New Education Letter [2022] No. 112). The sponsors did not play any role in the study design; in the collection, analysis and interpretation of data; in the writing of this report; or in the decision to submit the article for publication.

Institutional Review Board Statement: This study was conducted in accordance with the Declaration of Helsinki and was approved by the Institutional Review Board of the Xinjiang Uygur Autonomous Region Hospital of Traditional Chinese Medicine (2018XE0108).

Informed Consent Statement: Informed consent was obtained from all subjects involved in the study.

Data Availability Statement: Research data will be submitted as required.

Acknowledgments: We thank all participants and staff involved in the study, as well as Qingwu Jiang and Hexing Wang from Fudan University.

Conflicts of Interest: The authors declare no conflict of interest.

References

1. Castanon, J.I. History of the use of antibiotic as growth promoters in European poultry feeds. *Poult. Sci.* **2007**, *86*, 2466–2471. [CrossRef]
2. Dibner, J.J.; Richards, J.D. Antibiotic growth promoters in agriculture: History and mode of action. *Poult. Sci.* **2005**, *84*, 634–643. [CrossRef]
3. Roberts, P.J. Penicillin man: Alexander Fleming and the antibiotic revolution. *J. Hosp. Infect.* **2006**, *62*, 393. [CrossRef]
4. Carvalho, I.T.; Santos, L. Antibiotics in the aquatic environments: A review of the European scenario. *Environ. Int.* **2016**, *94*, 736–757. [PubMed]
5. Chu, L.; Wang, H.X.; Su, D.Q.; Zhang, H.W.; Yimingniyazi, B.H.; Aili, D.L.; Luo, T.; Zhang, Z.W.; Dai, J.H.; Jiang, Q.W. Urinary Antibiotics and Dietary Determinants in Adults in Xinjiang, West China. *Nutrients* **2022**, *14*, 4748. [CrossRef]
6. Wang, H.; Yang, J.; Yu, X.; Zhao, G.; Zhao, Q.; Wang, N.; Jiang, Y.; Jiang, F.; He, G.; Chen, Y.; et al. Exposure of adults to antibiotics in a Shanghai Suburban Area and Health Risk Assessment: A biomonitoring-based study. *Environ. Sci. Technol.* **2018**, *52*, 13942–13950. [CrossRef] [PubMed]
7. Wang, H.; Ren, L.; Yu, X.; Hu, J.; Chen, Y.; He, G.; Jiang, Q. Antibiotic residues in meat, milk and aquatic products in Shanghai and human exposure assessment. *Food Control* **2017**, *80*, 217–225. [CrossRef]
8. Boursi, B.; Mamtani, R.; Haynes, K.; Yang, Y.X. The effect of past antibiotic exposure on diabetes risk. *Eur. J. Endocrinol.* **2015**, *172*, 639–648. [CrossRef]

9. Mikkelsen, K.H.; Knop, F.K.; Frost, M.; Hallas, J.; Pottegard, A. Use of antibiotics and risk of type 2 diabetes: A population-based case-control study. *J. Clin. Endocrinol. Metab.* **2015**, *100*, 3633–3640. [CrossRef] [PubMed]
10. Omori, M.; Kato-Kogoe, N.; Sakaguchi, S.; Kamiya, K.; Fukui, N.; Gu, Y.H.; Nakamura, S.; Nakano, T.; Hoshiga, M.; Imagawa, A.; et al. Characterization of salivary microbiota in elderly patients with type 2 diabetes mellitus: A matched case-control study. *Clin. Oral Investig.* **2022**, *26*, 493–504. [CrossRef]
11. Qin, J.; Li, Y.; Cai, Z.; Li, S.; Zhu, J.; Zhang, F.; Liang, S.; Zhang, W.; Guan, Y.; Shen, D.; et al. A metagenome-wide association study of gut microbiota in type 2 diabetes. *Nature* **2012**, *490*, 55–60. [CrossRef]
12. Chou, H.W.; Wang, J.L.; Chang, C.H.; Lee, J.J.; Shau, W.Y.; Lai, M.S. Risk of severe dysglycemia among diabetic patients receiving levofloxacin, ciprofloxacin, or moxifloxacin in Taiwan. *Clin. Infect. Dis.* **2013**, *57*, 971–980. [CrossRef]
13. Park, S.J.; Park, Y.J.; Chang, J.; Choi, S.; Lee, G.; Son, J.S.; Kim, K.H.; Oh, Y.H.; Park, S.M. Association between antibiotics use and diabetes incidence in a nationally representative retrospective cohort among Koreans. *Sci. Rep.* **2021**, *11*, 21681. [CrossRef] [PubMed]
14. Zhou, J.; Lin, Y.; Liu, Y.; Chen, K. Antibiotic exposure and risk of type 2 diabetes mellitus: A systematic review and meta-analysis. *Environ. Sci. Pollut. Res. Int.* **2021**, *28*, 65052–65061. [CrossRef]
15. Tao, L.; Tian, T.; Liu, L.; Zhang, Z.; Sun, Q.; Sun, G.; Dai, J.; Yan, H. Cohort profile: The Xinjiang Multiethnic Cohort (XMC) study. *BMJ Open* **2022**, *12*, e048242. [CrossRef]
16. Chen, Z.; Chen, J.; Collins, R.; Guo, Y.; Peto, R.; Wu, F.; Li, L. China Kadoorie Biobank collaborative, g. China Kadoorie Biobank of 0.5 million people: Survey methods, baseline characteristics and long-term follow-up. *Int. J. Epidemiol.* **2011**, *40*, 1652–1666. [CrossRef] [PubMed]
17. Saravia, L.; Miguel-Berges, M.L.; Iglesia, I.; Nascimento-Ferreira, M.V.; Perdomo, G.; Bove, I.; Slater, B.; Moreno, L.A. Relative validity of FFQ to assess food items, energy, macronutrient and micronutrient intake in children and adolescents: A systematic review with meta-analysis. *Br. J. Nutr.* **2021**, *125*, 792–818. [CrossRef] [PubMed]
18. Gao, H.; Zhu, B.B.; Tao, X.Y.; Zhu, Y.D.; Tao, X.G.; Tao, F.B. Temporal variability of cumulative risk assessment on phthalates in Chinese pregnant women: Repeated measurement analysis. *Environ. Sci. Technol.* **2018**, *52*, 6585–6591. [CrossRef]
19. Guo, Y.; Alomirah, H.; Cho, H.S.; Minh, T.B.; Mohd, M.A.; Nakata, H.; Kannan, K. Occurrence of phthalate metabolites in human urine from several Asian countries. *Environ. Sci. Technol.* **2011**, *45*, 3138–3144. [CrossRef] [PubMed]
20. Wang, H.; Wang, N.; Qian, J.; Hu, L.; Huang, P.; Su, M.; Yu, X.; Fu, C.; Jiang, F.; Zhao, Q.; et al. Urinary antibiotics of pregnant women in Eastern China and cumulative health risk assessment. *Environ. Sci. Technol.* **2017**, *51*, 3518–3525. [CrossRef]
21. Soeborg, T.; Frederiksen, H.; Andersson, A.M. Cumulative risk assessment of phthalate exposure of Danish children and adolescents using the hazard index approach. *Int. J. Androl.* **2012**, *35*, 245–252. [CrossRef] [PubMed]
22. Zhu, B.; Wang, J.; Chen, K.; Wang, J.; Chen, K.; Yan, W.; Wang, A.; Wang, W.; Gao, Z.; Tang, X.; et al. A high triglyceride glucose index is more closely associated with hypertension than lipid or glycemic parameters in elderly individuals: A cross-sectional survey from the Reaction Study. *Cardiovasc. Diabetol.* **2020**, *19*, 112. [CrossRef]
23. Wang, H.; Wang, B.; Zhao, Q.; Zhao, Y.; Fu, C.; Feng, X.; Wang, N.; Su, M.; Tang, C.; Jiang, F.; et al. Antibiotic body burden of Chinese school children: A multisite biomonitoring-based study. *Environ. Sci. Technol.* **2015**, *49*, 5070–5079. [CrossRef] [PubMed]
24. Ji, K.; Kho, Y.; Park, C.; Paek, D.; Ryu, P.; Paek, D.; Kim, M.; Kim, P.; Choi, K. Influence of water and food consumption on inadvertent antibiotics intake among general population. *Environ. Res.* **2010**, *110*, 641–649. [CrossRef]
25. Li, N.; Ho, K.W.K.; Ying, G.G.; Deng, W.J. Veterinary antibiotics in food, drinking water, and the urine of preschool children in Hong Kong. *Environ. Int.* **2017**, *108*, 246–252. [CrossRef] [PubMed]
26. Wang, H.; Wang, N.; Wang, B.; Fang, H.; Fu, C.; Tang, C.; Jiang, F.; Zhou, Y.; He, G.; Zhao, Q.; et al. Antibiotics detected in urines and adipogenesis in school children. *Environ. Int.* **2016**, *89–90*, 204–211. [CrossRef] [PubMed]
27. Zhou, Y.J.; Zhu, F.; Zheng, D.Y.; Gao, M.M.; Guo, B.F.; Zhang, N.; Meng, Y.; Wu, G.L.; Zhou, Y.L.; Huo, X. Detection of antibiotics in the urine of children and pregnant women in Jiangsu, China. *Environ. Res.* **2021**, *196*, 110945. [CrossRef]
28. Zeng, X.; Zhang, L.; Chen, Q.; Yu, K.; Zhao, S.; Zhang, L.; Zhang, J.; Zhang, W.; Huang, L. Maternal antibiotic concentrations in pregnant women in Shanghai and their determinants: A biomonitoring-based prospective study. *Environ. Int.* **2020**, *2138*, 105638. [CrossRef]
29. Geng, M.; Liu, K.; Huang, K.; Zhu, Y.; Ding, P.; Zhang, J.; Wang, B.; Liu, W.; Han, Y.; Gao, H.; et al. Urinary antibiotic exposure across pregnancy from Chinese pregnant women and health risk assessment: Repeated measures analysis. *Environ. Int.* **2020**, *145*, 106164. [CrossRef]
30. Zhang, Y.; Tang, W.; Wang, Y.; Nian, M.; Jiang, F.; Zhang, J.; Chen, Q. Environmental antibiotics exposure in school-age children in Shanghai and health risk assessment: A population-based representative investigation. *Sci. Total Environ.* **2022**, *824*, 153859. [CrossRef]
31. Wang, H.; Tang, C.; Yang, J.; Wang, N.; Jiang, F.; Xia, Q.; He, G.; Chen, Y.; Jiang, Q. Predictors of urinary antibiotics in children of Shanghai and health risk assessment. *Environ. Int.* **2018**, *121*, 507–514. [CrossRef]

32. Feng, L.; Cheng, Y.; Zhang, Y.; Li, Z.; Yu, Y.; Feng, L.; Zhang, S.; Xu, L. Distribution and human health risk assessment of antibiotic residues in large-scale drinking water sources in Chongqing area of the Yangtze River. *Environ. Res.* **2020**, *185*, 109386. [CrossRef] [PubMed]
33. Guo, X.Y.; Hao, L.J.; Qiu, P.Z.; Chen, R.; Xu, J.; Kong, X.J.; Shan, Z.J.; Wang, N. Pollution characteristics of 23 veterinary antibiotics in livestock manure and manure-amended soils in Jiangsu province, China. *J. Environ. Sci. Health B* **2016**, *51*, 383–392. [CrossRef] [PubMed]
34. Qian, M.; Wu, H.; Wang, J.; Zhang, H.; Zhang, Z.; Zhang, Y.; Lin, H.; Ma, J. Occurrence of trace elements and antibiotics in manure-based fertilizers from the Zhejiang Province of China. *Sci. Total Environ.* **2016**, *559*, 174–181. [CrossRef] [PubMed]
35. Cunha, B.A. *Antibiotic Essentials*; Jones & Bartlett: Burlington, MA, USA, 2013.
36. Alvarez-Silva, C.; Kashani, A.; Hansen, T.H.; Pinna, N.K.; Anjana, R.M.; Dutta, A.; Saxena, S.; Støy, J.; Kampmann, U.; Nielsen, T.; et al. Trans-ethnic gut microbiota signatures of type 2 diabetes in Denmark and India. *Genome Med.* **2021**, *13*, 37. [CrossRef]
37. Doumatey, A.P.; Adeyemo, A.; Zhou, J.; Lei, L.; Adebamowo, S.N.; Adebamowo, C.; Rotimi, C.N. Gut microbiome profiles are associated with type 2 diabetes in Urban Africans. *Front. Cell Infect. Microbiol.* **2020**, *10*, 63. [CrossRef]
38. Shi, A.; Li, T.; Zheng, Y.; Song, Y.; Wang, H.; Wang, N.; Dong, L.; Shi, H. Chlorogenic acid improves NAFLD by regulating gut microbiota and GLP-1. *Front. Pharmacol.* **2021**, *12*, 693048. [CrossRef]
39. Zhang, X.; Zhao, A.; Sandhu, A.K.; Edirisinghe, I.; Burton-Freeman, B.M. Functional Deficits in gut microbiome of young and middle-aged adults with prediabetes apparent in metabolizing bioactive (poly)phenols. *Nutrients* **2020**, *12*, 3595. [CrossRef] [PubMed]
40. Zhang, J.; Ni, Y.; Qian, L.; Fang, Q.; Zheng, T.; Zhang, M.; Gao, Q.; Zhang, Y.; Ni, J.; Hou, X.; et al. Decreased abundance of Akkermansia muciniphila leads to the impairment of insulin secretion and glucose homeostasis in lean type 2 diabetes. *Adv. Sci.* **2021**, *8*, e2100536. [CrossRef]
41. Herrera, C.R.; Vidal, G.X. Cardiovascular events and mortality among type 2 diabetes mellitus patients newly prescribed first-line blood glucose-lowering drugs monotherapies: A population-based cohort study in the Catalan electronic medical record database, SIDIAP, 2010–2015. *Prim. Care Diabetes* **2021**, *15*, 323–331. [CrossRef]
42. Blaser, M.J. Antibiotic use and its consequences for the normal microbiome. *Science* **2016**, *352*, 544–545. [CrossRef]
43. Yu, F.; Han, W.; Zhan, G.; Li, S.; Jiang, X.; Wang, L.; Xiang, S.; Zhu, B.; Yang, L.; Luo, A.; et al. Abnormal gut microbiota composition contributes to the development of type 2 diabetes mellitus in db/db mice. *Aging* **2019**, *11*, 10454–10467. [CrossRef]
44. Cryan, J.F.; O'Riordan, K.J.; Cowan, C.S.M.; Sandhu, K.V.; Bastiaanssen, T.F.S.; Boehme, M.; Codagnone, M.G.; Cussotto, S.; Fulling, C.; Golubeva, A.V.; et al. The microbiota-gut-brain axis. *Physiol. Rev.* **2019**, *99*, 1877–2013. [CrossRef]
45. Grasset, E.; Burcelin, R. The gut microbiota to the brain axis in the metabolic control. *Rev. Endocr. Metab. Disord.* **2019**, *20*, 427–438. [CrossRef]
46. Muller, P.A.; Matheis, F.; Schneeberger, M.; Kerner, Z.; Jové, V.; Mucida, D. Microbiota-modulated CART+ enteric neurons autonomously regulate blood glucose. *Science* **2020**, *370*, 314–321. [CrossRef] [PubMed]
47. Hu, F.B. Globalization of diabetes: The role of diet, lifestyle, and genes. *Diabetes Care* **2011**, *34*, 1249–1257. [CrossRef] [PubMed]
48. Lalrohlui, F.; Ghatak, S.; Zohmingthanga, J.; Hruaii, V.; Kumar, N.S. Fermented pork fat (Sa-um) and lifestyle risk factors as potential indicators for type 2 diabetes among the Mizo population, Northeast India. *J. Health Popul. Nutr.* **2021**, *40*, 32. [CrossRef] [PubMed]
49. Patyra, E.; Kwiatek, K. HPLC-DAD analysis of florfenicol and thiamphenicol in medicated feedingstuffs. *Food Addit. Contam. Part A Chem. Anal. Control. Expo. Risk Assess* **2019**, *36*, 1184–1190. [CrossRef]
50. Ahmed, A.; Lager, A.; Fredlund, P.; Elinder, L.S. Consumption of fruit and vegetables and the risk of type 2 diabetes: A 4-year longitudinal study among Swedish adults. *J. Nutr. Sci.* **2020**, *9*, e14. [CrossRef]
51. Tadic, D.; Bleda Hernandez, M.J.; Cerqueira, F.; Matamoros, V.; Pina, B.; Bayona, J.M. Occurrence and human health risk assessment of antibiotics and their metabolites in vegetables grown in field-scale agricultural systems. *J. Hazard Mater.* **2021**, *401*, 123424. [CrossRef]
52. Lam, B.Q.; Srivastava, R.; Morvant, J.; Shankar, S.; Srivastava, R.K. Association of diabetes mellitus and alcohol abuse with cancer: Molecular mechanisms and clinical significance. *Cells* **2021**, *10*, 3077. [CrossRef]
53. Mallet, V.; Parlati, L.; Martinino, A.; Scarano Pereira, J.P.; Jimenez, C.N.; Sakka, M.; Bouam, S.; Retbi, A.; Krasteva, D.; Meritet, J.F.; et al. Demosthenes research, g. Burden of liver disease progression in hospitalized patients with type 2 diabetes mellitus. *J. Hepatol.* **2022**, *76*, 265–274. [CrossRef]
54. Moschona, A.; Liakopoulou-Kyriakides, M. Encapsulation of biological active phenolic compounds extracted from wine wastes in alginate-chitosan microbeads. *J. Microencapsul.* **2018**, *35*, 229–240. [CrossRef] [PubMed]
55. Davis, P.J.; Liu, M.; Alemi, F.; Jensen, A.; Avramovic, S.; Levy, E.; Hayes, R.B.; Schwartz, M.D. Prior antibiotic exposure and risk of type 2 diabetes among Veterans. Prim. *Care Diabetes* **2019**, *13*, 49–56. [CrossRef]
56. Zhou, J.; Lin, Y.; Liu, Y.; Chen, K. Response to Antibiotic exposure and risk of type 2 diabetes mellitus: A systematic review and meta-analysis by Maryam et al. (2021) by J Zhou and K Chen (2021). *Environ. Sci. Pollut. Res. Int.* **2022**, *29*, 18298–18299. [CrossRef] [PubMed]
57. Ye, M.; Robson, P.J.; Eurich, D.T.; Vena, J.E.; Xu, J.-Y.; Johnson, J.A. Systemic use of antibiotics and risk of diabetes in adults: A nested case-control study of Alberta's Tomorrow Project. *Diabetes Obes Metab.* **2018**, *20*, 849–857. [CrossRef] [PubMed]

58. Baik, S.H.; McDonald, C.J. Independent effects of 15 commonly prescribed drugs on all-cause mortality among US elderly patients with type 2 diabetes mellitus. *BMJ Open Diabetes Res. Care* **2020**, *8*, e000940. [CrossRef] [PubMed]
59. Ida, S.; Kaneko, R.; Imataka, K.; Okubo, K.; Shirakura, Y.; Azuma, K.; Fujiwara, R.; Takahashi, H.; Murata, K. Effects of oral antidiabetic drugs and glucagon-like peptide-1 receptor agonists on left ventricular diastolic function in patients with type 2 diabetes mellitus: A systematic review and network meta-analysis. *Heart Fail. Rev.* **2021**, *26*, 1151–1158. [CrossRef] [PubMed]

nutrients

MDPI

Article

Classroom-Based Physical Activity as a Means to Improve Self-Efficacy and Academic Achievement among Normal-Weight and Overweight Youth

Francesca Latino [1], Francesco Tafuri [2], Emma Saraiello [3] and Domenico Tafuri [3,*]

1 Faculty of Human Sciences, Pegaso University, 80100 Napoli, Italy
2 Heracle Lab Research in Educational Neuroscience, Niccolò Cusano University, 00166 Roma, Italy
3 Department of Movement Sciences and Wellness, University of Napoli "Parthenope", 80100 Napoli, Italy
* Correspondence: domenico.tafuri@uniparthenope.it

Abstract: Although physical activity has positive physical and mental health outcomes, particularly among adolescents, a significant percentage of young people maintain a largely sedentary lifestyle. Considering that the youths spend the greater part of the day at school, this is considered an ideal setting to foster active and healthy living. Consequently, this study is intended to investigate the connection between physical activity, self-efficacy and academic achievement in normal-weight and overweight adolescents. In total, 100 students (aged 14–15) from a public high school placed in the south of Italy were enrolled. They participated either in a 12-week classroom-based physical activity break program performed during science classes (60′/2 days per week) in which a nutritional educational program was carried out or in regular science lessons (60′/2 days per week). At the beginning and end of the intervention programs, a set of standardized motor evaluation tests (standing long jump test, Harvard step test, push up, sit and reach test), the scholastic self-efficacy test and the Amos 8-15 were administered. As a result, a meaningful Time × Group interaction for the self-efficacy variable and Amos 8-15 was observed in the intervention group. Specifically, they reported significant improvement in study skills, motivational factors, concentration and self-efficacy, as well as a decrease in anxiety and BMI ($p < 0.001$). No significant change was observed in the control group. The conclusions of this research underpin the notion that classroom-based physical activity break is a successful approach for enhancing students' psycho-physical well-being, as well as academic achievement.

Keywords: academic performance; weight status; exercise; scholastic self-efficacy; cognitive function; lifestyle

Citation: Latino, F.; Tafuri, F.; Saraiello, E.; Tafuri, D. Classroom-Based Physical Activity as a Means to Improve Self-Efficacy and Academic Achievement among Normal-Weight and Overweight Youth. *Nutrients* **2023**, *15*, 2061. https://doi.org/10.3390/nu15092061

Academic Editor: Pascale Duche

Received: 23 February 2023
Revised: 18 April 2023
Accepted: 21 April 2023
Published: 25 April 2023

1. Introduction

The scholastic readiness and well-being of the youth are major concerns in the 21st century [1]. The school is seen as a perfect environment for promoting an active and healthy lifestyle, in order to improve the youngers' well-being. Hence, the school should focus on being tasked with preventing obesity as well as the need to provide a high education and academic achievement for all students [2]. In fact, children and adolescents spend the greater part of their time attending school. Nevertheless, it should be highlighted that traditionally schools are mainly focused on increasing literacy and numeracy proficiency to the detriment of physically active behaviors [3]. In this regard, thanks to the fact that children and adolescents are accessible in this setting, many studies tried to focus on childhood obesity by developing interventions that may be used within the educational system [4].

Obesity, an excessive body fat storage state, is caused mostly by improper lifestyle [5,6]. It is the result of both an unhealthy diet and a lack of physical activity [7]. As World Health Organization (WHO) claimed, in recent times, the percentage of obesity has globally more than doubled among children and tripled among adolescents [8]. Moreover, they are not following the 2020 WHO physical activity guidelines, which urges young people from 5 to 17 years to reach at least 60 min of moderate-to-vigorous physical activity (MVPA) per day [9].

The maintenance of a desirable level of physical activity plays a key role in the improvement of cognitive, motor, and social skills. It is widely acknowledged that decreased levels of physical activity represent a crucial risk factor for the dramatical rising levels of obesity [10]. Furthermore, fitness seems related to better cognitive function and academic performance, as well as to the own psycho-social behavior [11–15].

Several researchers suggest that physical activity can be associated with cognition over the course of development [16]. They argued that young people participating in an increased amount of physical activity better execute attentional tasks which involve a higher level of cognitive control, such as planning, organization, problem solving, working memory, motor control and inhibitory control [17,18].

To explain the positive influence of physical activity on academic performance, we made the assumption that the related neurophysiological changes in the brain are due to the increase in brain blood flow, better neuroelectric function, and solicitation of the brain-derived neurotrophic factor (BDNF) release. These are all factors that encourage learning and keep cognitive functions by enhancing synaptic plasticity [19,20].

Moreover, the combination between weight status, eating habits and academic achievement is affected by individuals' psychosocial variables, such as self-efficacy [21,22]. Self-efficacy concerns the conviction that one can effectively participate in a certain behavior [23].

Perceived self-efficacy is the result of a self-referential and self-regulated system that guides and directs the behavior, orients the individual's relationship with the environment and creates the conditions for the development of new experiences and abilities [24].

Individuals who have low levels of self-efficacy often present closing and avoidance behaviors and poor performance or failure, while people with a higher level of perceived self-efficacy have a good chance of obtaining satisfactory results. Thus, those who believe they will be able to accomplish a task will obtain higher performance than those who assess it negatively [25]. This behavior deeply affects both weight control and academic performance.

Concerning that, a growing body of literature has documented that fatness and academic failure seem to be related to poor self-efficacy [26–29] and that physical activity is able to solicit significant behavior changes [30,31]. Physical activity has the potential to boost self-efficacy beliefs that are associated with changes in the person's mood when exercise ends [32,33]. This happens thanks to the fact that the experience of physical activity encourages the ability to leverage one's mental and social attitude to gain an extensive repertory of motor skills which can be expressed in various environments and in all daily activities [34].

In light of this theoretical framework, it is possible to indicate that perceived self-efficacy improves thanks to physical activity and that physical activity plays a crucial role in mediating self-efficacy and the impact of the interventions on dietary behavior and academic performance. Consequently, an intervention that combines physical activity, nutritional education and academic achievement could provide an unrepeatable occasion to intervene and propose a program that both enhances health and academic performance. In the school context, an alternative way to practice physical activity is through an "active classroom", where students incorporate movement into learning. It represents an opportunity to help them retain knowledge in a meaningful way, as well as to acquire an active lifestyle.

Several studies suggest that classroom-based PA performed for 10–15 min, 2 or 3 days per week, has a positive impact on increasing PA levels, classroom and learning behavior, as well as decreased weight status [35,36]. Concerning that, in a 2019 Italian study, Calella et al. [37] conducted a classroom-based active break (CABs) intervention that integrates PA during school time and assessed its potential effect on reducing inactivity and overweight in primary school children. As they argued, the CAB program is a safe tool to reduce inactivity and overweight and increase moderate/vigorous PA. According to Masini et al. [38], active breaks (CABs) led by teachers inside the classroom represent a good strategy for promoting PA. In their 2020 study, the authors supported the feasibility and effectiveness in terms of increased PA level of a CAB program in children attending primary school. Similarly, Di Maglie and colleagues [39] observed that the school-based intervention program represents an effective strategy for decreasing the number of children with obesity. It is also interesting to mention a study by Fiori et al. [40], in which the authors compared the quality rather than the duration of time spent in the gym during school hours. They indicated that quality PA was effective in improving children's physical fitness, significantly decreasing their body weight.

Therefore, the purpose of this work was to analyze the link between physical activity, body mass index (BMI), self-efficacy and academic achievement in adolescents. Moreover, this work aimed to establish if there are meaningful differences in perceived self-efficacy and school performance between overweight and non-overweight high school students measured through standardized assessment tests. We assumed that a quantifiable disparity of results would be found among overweight and non-overweight students at the end of a classroom-based physical activity program.

2. Methods

2.1. Study Design

The research is regarded as a controlled study to analyze the mediating role of a school-based physical activity program on overweight adolescents' self-efficacy and academic achievement. The research was carried out in a high school. The participants were involved in a scholastic intervention that combined lessons in nutritional education with physical activities carried out during science classes. Thus, the intervention program involved 24 lessons of moderate-to-vigorous aerobic (MVPA) exercises for the intervention group and a regular science lesson for the control group, during which no practice was carried out. The interventions consisted of 2 weekly lessons of sciences of the duration of 60 min in which a classroom-based physical activity lasting 20 min was integrated. The science classes consisted of nutritional educational lessons. The evaluations were performed before and at the end of the intervention programs (Figure 1).

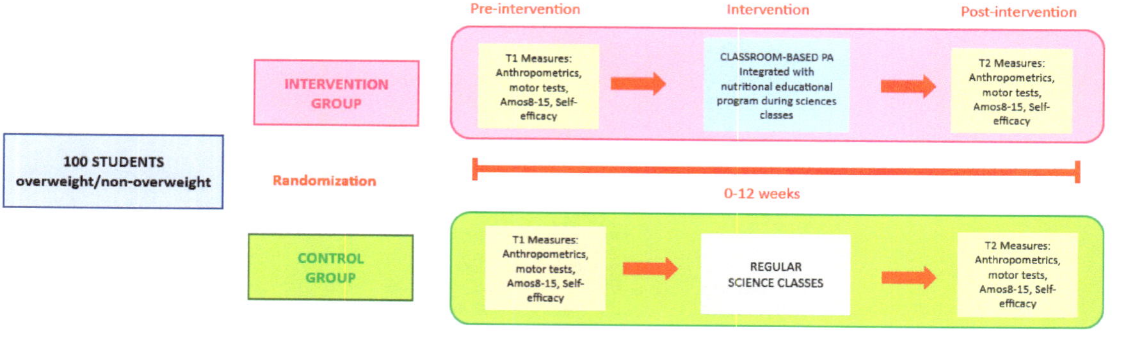

Figure 1. Study design graph.

2.2. Participants

The students enrolled were from a public high school placed in the south of Italy. One hundred students, as a convenient sample, were recruited with an age range of 14–15 years (M age = 14.64, *SD* = ±0.47). Participation in this trial was on a voluntary basis, and all the first- and second-year students were invited to participate. Of these students, 119 accepted the invite. To be included in this study, participants had to be between 14 and 15 years of age, be capable of finishing a moderate-to-vigorous intensity aerobic exercise session and attending the selected school, and be ready to avoid all physical activity beyond those indicated by the protocol. Exclusion criteria included acute illness at the time of testing, injuries that prevent performing exercises, and those unable to avoid physical activity beyond those indicated by the protocol. In total, 109 students met the inclusion criteria and were asked to participate in the trial. Of those recruited, 100 (56 males and 44 females) accepted to be involved in this research and rounded out the evaluations at baseline and post, while 9 of them refused because of personal issues.

As a result, the final sample consisted of 100 participants, of which 41 were overweight and 59 were non-overweight (41% overweight and 59% non-overweight), who completed the assessments at baseline and post-intervention. We matched participants randomly to one of two treatment conditions, and the percentage of overweight children was 46% for the experimental group (EG) and 36% for the control group (CG) (EG: overweight n = 23, non-overweight n = 27; CG: overweight n = 18, non-overweight n = 32).

The participants were sent an email to be advised in advance of their inclusion in the study. Following that, a second email was sent to the parents of all students, urging them to read a briefing that highlighted the goals of the trial and produce written consent for the participation of their kids in the research. Consent forms were signed by all parents prior to the start of the study. They contained the purpose of the study, the selection criteria, the procedures, the benefits and risks, alternatives, confidentiality, withdrawal and an injury disclaimer. Participants were assigned in a random way to one of two programs (EG n = 50; CG n = 50).

Using G*Power (version 3.1.9.6), an a priori power analysis was completed and showed that a sample size of 74 would provide a statistical power ($\alpha = 0.05$, $1\text{-}\beta = 0.80$) to identify a medium effect size ($f = 0.25$ or 0.4), given a coefficient of correlation $p = 0.80$ with 95% power and $\alpha = 0.05$, using a within-between mixed design. To prevent experimental mortality due to participants leaving, 100 subjects were enrolled.

The anonymity of all participants was ensured by the researchers. The study was conducted from October 2022 to December 2022. The research was conducted based on the Declaration of Helsinki.

2.3. Procedures

Initially (one week before), all participants were invited to participate in a briefing especially conceived to provide information concerning the program. Moreover, at that time, the motivation of participants was verified.

Participants were assessed for anthropometric measurements, including BMI, in order to place each student into the non-overweight or overweight category. After the BMI assessment, the researchers assigned participants to one of the two program conditions in a random way. In addition, to establish the beginning level of each subject's physical fitness, standardized motor evaluation tests were utilized. Moreover, two cognitive tests were administered to examine the student's self-efficacy perception and academic performance.

The students were tested individually and undertook each test in the same order, at the same time of the day, and under similar trial conditions. They completed the evaluations two days before and immediately after the trial to enable the impact assessment of the physical activity program. All evaluations and both physical activity programs were outlined, supervised and carried out by 2 certified physical education teachers.

2.4. Measures

2.4.1. Anthropometric Measures

A standard protocol was followed, and standardized instruments were used to carry out the anthropometric evaluations [35]. Height and weight measurements were performed in a room where the participant's privacy was ensured. They carried out the measurements three times for each participant, and the verification was always performed by the same experienced researcher in order to avoid incurring errors. Participants were weighed using a portable medical floor scale with a precision of 0.1 kg (Charder's MS6121R, Charder

Electronic Co., Taichung City, Taiwan). A wall-mounted digital stadiometer was used to measure height to the nearest 0.1 cm (Charder's HM200D, Charder Electronic Co., Taichung City, Taiwan). The height and weight measurements were used to calculate the body mass index (BMI) (kg/m^2) [36]. BMI was converted to BMI percentile, and the international BMI cutoff points were utilized to classify overweight and non-overweight students [37]. BMI evaluation was performed before and after the intervention programs.

2.4.2. Motor Tests

The measurements of physical fitness involved 4 evaluation tests:

1. Standing long jump test, a measure used to evaluate lower-body horizontal explosiveness [38].
2. Harvard Step test, utilized to evaluate aerobic fitness [39].
3. Push-up test, employed to measure upper-body strength and endurance [40].
4. Sit and reach, which measures the extensibility of the hamstring muscles and lower back [41].

These tests were chosen because they are easier to undertake, take little time and require simple equipment [42]. For all these reasons, they are perfect for school use. These tests were used before and after the intervention programs.

2.4.3. Amos 8-15 Questionnaire

The Amos 8-15 [43] is a set of tests designed for the Italian educational environment. It is meant to evaluate study skills and different motivational factors of students aged 8 to 15. It allows users to recognize students' weaknesses and strengths in order to launch targeted activities aimed at promoting effective study methods and motivational strategies related to the process of learning.

It includes easy-to-administer tools that help analyze different aspects involved in learning activities, such as approaches to the study, uses of strategies for the study, beliefs about oneself as a student, and accidental attributions about both successful or failed events. This battery test comprises: (i) study approach questionnaire (QAS); (ii) study strategies questionnaire (QS1 e QS2); (iii) convictions questionnaire (QC1I, QC2F, QC3O); (iv) attributions questionnaire (QCA); (v) objective study tests.

The researcher has the possibility to decide to employ all the existing tests or some of them. The QAS and the Objective study tests were chosen by the authors to conduct the research. These tests were carried out at the start and following the experimental trial.

2.4.4. Study Approach Questionnaire (QAS)

It evaluates various variables of the student's study attitude. The QAS consists of 49 items classified into 7 macro-areas: each area includes 7 items; 5 are positive, and 2 are negative. Anxiety is an exception because it consists of 2 positive and 5 negative items. These macro-areas include: (i) motivation; (ii) organization; (iii) didactic material development; (iv) study flexibility; (v) concentration; (vi) anxiety; (vii) attitude towards school.

The QAS uses a 3-point Likert-type scale, requiring responses ranging from 1 "disagree" to 3 "strongly agree". This questionnaire needs about 10 to 20 min to be completed (including the instruction and practice phase). If the respondent's QAS increases, this suggests they are experiencing an inappropriate approach to the study.

2.4.5. Objective Study Tests

The objective study test is an evaluation that allows checking the ability of the student to comprehend and memorize. It provides a text to be studied for 30 min subjectively. Afterward, on completion of a 15 min break, participants perform 3 different tests:

1. Choice of titles wherein students choose 3 most significant titles from a list of 8 titles. For each valid title, it is assigned 1 point.

2. Open questions in which the subject responds to 6 questions concerning the text previously studied. To assess the accuracy of the response, it is assigned from 1 to 3 points.
3. True/False questions ask the student to respond true or false to 12 questions. In addition, 1 point is given for each right response, 0 point is awarded for answers not given, and −1 point for each wrong response.

The sum of the scores obtained determines the total scoring system. This questionnaire required about 75 to 90 min to be completed.

2.4.6. Scholastic Self-Efficacy Scale

The scholastic self-efficacy scale [44] is a questionnaire thought to evaluate self-beliefs in one's competence to keep doing well in study and the self-regulation of learning. It is a 19-item questionnaire using a 5-point Likert-type scale. The evaluation requires responses ranging from 1 "not at all capable" to 5 "highly capable". The evaluation requires about 15 min (comprising the instruction and practice phase). The score ranges from 19 to 95. An increase in self-efficacy is identified with higher scores. This questionnaire was carried out at the start and following the experimental trial.

2.4.7. Physical Activity Intervention

The exercise training intervention for the experimental group was planned in accordance with the social-cognitive theory [45]. It sees self-efficacy perceptions as feelings based on 4 main lines of evidence: past performances, vicarious experiences, verbal persuasion and physiological state [23]. Thus, to enhance the sense of self-efficacy among overweight and non-overweight students, key elements of the intervention program based upon principles of the social cognitive theory were summarized as follows:

1. It was an enjoyable and appealing activity designed specifically to help overweight students to achieve success (i.e., utilizing activities easier to perform and using students' favorite music);
2. It created occasions for students to positively influence each other (i.e., experience a team mentality);
3. It verbally encouraged students to overcome their limits (i.e., 'you can do it');
4. It reduced any kind of anxiety related to exercise, avoiding competition.

Moreover, it was planned as follows: The first part focused on flexibility (3 min), followed by a core part of moderate-to-vigorous aerobic exercise (15 min), and lastly, a cool-down session (2 min) to preserve a safe heart rate. Warm-up parts involved the following exercises: marching in place, walking jacks, knee to chest, heel digs, arm circles, shoulder rolls, knee lifts, butt kicks, lunges, side steps and high knees. Moreover, cool-down parts included static exercises, such as neck stretch, behind-head tricep stretch, standing hip rotation, hamstring stretch, hip flexor stretch, side stretch and butterfly stretch.

Each active break was planned to be performed between and within science classes, during which a program of nutritional education was carried out.

A typical session of an active classroom intervention included:

1. Active breaks between and within learning activities;
2. Learning activities which involve movement;
3. Working at benches, standing desks, on the floor, or in combination to create movement between work areas;
4. Learning outdoors.

The intensity of each training session was monitored through an OMNI scale to respect exertion in the MVPA range of $5 < RPE < 8$ and to prevent any differences between training sessions [46].

2.4.8. Nutritional Education Intervention

Nutritional education consisted of a series of interventions that provided visual and verbal information and directions to attendees.

The goals of this intervention were to:

1. Provide a good approach toward healthy food and physical activity and stimulate reasoning for enhanced eating habits and active lifestyles in order to promote well-being for students.
2. Solicit students to think critically about nutrition and healthy lifestyles so that they can make suitable choices about what to eat, excluding junk food.
3. Help the students to critically recognize and analyze appropriate resources of information.

The framework for this nutrition education activity was represented by the *Dietary Guidelines* drawn up by the Italian Association of Dietitians. They comprised information concerning normal-weight maintenance, active lifestyle habits, healthy food and alcohol consumption. Therefore, this program provided information regarding:

1. The recommended food intake as needed to maintain daily nutritional needs;
2. Knowledge about the physiology of the human body;
3. Behavioral practices, including the factors that affect students' eating and food preparation habits;
4. Different ways to acquire a healthy and active lifestyle;
5. The importance of a healthy diet and physical activity in preventing diseases.

2.4.9. Statistical Analysis

The IBM SPSS version 25.0 (IBM, Armonk, NY, USA) was used to conduct the statistical analyses. Data were introduced as group mean (*M*) values and standard deviations (*SD*). In addition, they were checked for assumptions of normality through the Shapiro–Wilk test and homogeneity of variances by using Levene's test. To evaluate group differences at baseline, an independent sample *t*-test was used. To explore the impact of the physical activity program on all dependent variables, a two-way ANOVA (group (experimental/control) × time (pre-/post-intervention) with repeated measures on the time dimension was carried out. When 'Group × Time' interactions showed significance, a paired *t*-test was performed to outline the significant comparisons. Lastly, to analyze the magnitude of the significant 'Time × Group' interaction, a partial eta squared ($\eta^2{}_p$) value was used. It was interpreted as follows: small ($\eta^2{}_p < 0.06$), medium ($0.06 \leq \eta^2{}_p < 0.14$), and large ($\eta^2{}_p \geq 0.14$). Moreover, Cohen's *d* was used to determine the effect sizes for the pairwise comparisons. It was interpreted as small ($0.20 \leq d < 0.50$), moderate ($0.50 \leq d < 0.79$) and large ($d \geq 0.80$) [47]. Statistical significance was set at $p < 0.05$.

3. Results

All participants received the treatment conditions as allocated, and no participants reported injuries during the course of the trial. Students involved in the research did not differ in age, sex, anthropometric characteristics and psychological measures, as well as in socioeconomic status ($p > 0.05$). Data results for all dependent measures are shown in Table 1.

3.1. Motor Tests

Using a two-factor repeated measures ANOVA, we found a significant 'Time × Group' interaction for all four Motor tests carried out: standing long jump test ($F_{1,98} = 23.98$, $p < 0.001$, $\eta^2{}_p = 0.19$, large effect size), Harvard step test ($F_{1,98} = 374.01$, $p < 0.001$, $\eta^2{}_p = 0.79$, large effect size), push up tests ($F_{1,98} = 363.39$, $p < 0.001$, $\eta^2{}_p = 0.78$, large effect size) and sit and reach test ($F_{1,98} = 386.83$, $p < 0.001$, $\eta^2{}_p = 0.79$, large effect size). Moreover, post hoc analysis showed that, from pre- to post-test, the experimental group made a meaningful increase in standing long jump test ($t = 9.86$, $p < 0.001$, $d = 1.39$ large effect size), Harvard step test (t = 25.78, $p < 0.001$, d = 3.64, large effect size), push up test (t = 20.42, $p < 0.001$, d = 2.88, large effect size) and sit and reach test (t = 19.04, $p < 0.001$, d = 2.69, large effect size). The control group did not report any significant changes ($p > 0.05$).

Table 1. Changes in physical fitness, self-efficacy, study abilities and BMI after a 12-week active break program.

	Experimental Group (n = 50)			Control Group (n = 50)		
	Baseline	Post-Test	Δ	Baseline	Post-Test	Δ
Motor Tests						
Standing long jump test	1.51 (0.06)	1.54 (0.05) †*	0.03 (0.02)	1.49 (0.60)	1.49 (0.06)	0.05 (0.39)
Harvard step test	35.08 (12.98)	39.98 (12.81) †*	4.90 (1.34)	36.72 (13.3)	34.27 (13.41)	−2.45 (2.33)
Push up test	6.36 (1.85)	13.62 (3.50) †*	7.26 (2.51)	5.80 (1.87)	5.64 (1.89)	−0.15 (1.17)
Sit and Reach test	4.82 (2.32)	9.28 (3.10) †*	4.46 (3.18)	4.98 (2.23)	3.56 (2.60)	−1.41 (1.32)
Amos 8-15—QASQuestionnaire						
Motivation	13.76 (2.09)	15.54 (2.42) †*	1.78 (1.23)	14.24 (2.34)	12.86 (2.10)	−1.38 (1.24)
Organisation	14.96 (1.91)	17.02 (1.96) †*	2.06 (1.47)	15.04 (2.64)	13.72 (2.24)	−1.32 (0.93)
Didactic material development	15.22 (1.50)	14.94 (1.69)	−0.28 (0.75)	14.52 (1.47)	13.52 (1.48) 2.25	−1 (1.06)
Study flexibility	15.66 (1.89)	17.24 (2.02) †*	1.58 (0.83)	15.72 (1.60)	14.18 (1.82)	−1.54 (0.88)
Concentration	16.94 (2.03)	18.26 (1.87) †*	1.32 (0.91)	17.18 (2.08)	15.56 (2.18)	−1.62 (1.17)
Anxiety	17.10 (1.66)	15.02 (1.67) †*	−2.08 (1.20)	15.38 (2.35)	16.50 (2.13)	1.12 (1.11)
Attitude towards school	15.72 (2.44)	15.64 (2.51)	−0.08 (0.56)	16.62 (1.79)	15.12 (1.98)	−1.50 (0.70)
Amos 8-15—Objective Study Tests	18.76 (2.97)	22.56 (3.84) †*	3.80 (2.61)	18.98 (2.89)	16.98 (2.99)	−2 (0.88)
Scholastic Self-efficacy	46.76 (2.72)	54.12 (2.52)†*	7.36 (1.28)	46.86 (2.80)	44.92 (3.07)	−1.94 (1.66)
BMI percentile median (QR) $^{\Omega}$	91.65 (4.01)	88.79 (6.17) †*	−2.84 (2.54)	91.82 (4.03)	91.73 (4.01)	−0.09 (0.36)

Note: values are presented as mean (±SD); Δ: pre- to post-training changes; † Significant 'Group × Time' interaction: significant effect of the intervention ($p < 0.001$). * Significantly different from pre-test ($p < 0.001$); $^{\Omega}$ BMI percentile indicates the relative position of the child's BMI number among children of the same sex and age.

3.1.1. BMI

A meaningful 'Time×Group' interaction was also reached for BMI ($F_{1,98}$ = 59.30, $p < 0.001$, $\eta^2{}_p$ = 0.37, large effect size). Clearly, this case post hoc analysis also showed that the experimental group showed an important decrease in BMI from pre- to post-test (t = 7.97, $p < 0.001$, d = 1.12 large effect size). After the intervention program, the percentage of overweight children decreased from 46% to 28%. The control group did not report any significant changes ($p > 0.05$).

3.1.2. Study Approach Questionnaire QAS

A substantial 'Time × Group' interaction was obtained for motivation ($F_{1,98}$ = 162.70, $p < 0.001$, $\eta^2{}_p$ = 0.62, large effect size), organization ($F_{1,98}$ = 186.97, $p < 0.001$, $\eta^2{}_p$ = 0.65, large effect size), study flexibility ($F_{1,98}$ = 328.50, $p < 0.001$, $\eta^2{}_p$ = 0.77, large effect size), concentration ($F_{1,98}$ = 194.89, $p < 0.001$, $\eta^2{}_p$ = 0.66, large effect size) and anxiety ($F_{1,28}$ = 188.68, $p < 0.001$, $\eta^2{}_p$ = 0.65, large effect size). When the post hoc analysis was performed, it showed that the experimental group reached an important increase in motivation (t = 10.20, $p < 0.001$, d = 1.44, large effect size), organization (t = 9.86, $p < 0.001$, d = 1.39, large effect size), study flexibility (t = 13.37, $p < 0.001$, d = 1.89, large effect size) and concentration (t = 10.21, $p < 0.001$, d = 1.44, large effect size). In addition, a significant decrease in anxiety scores was obtained by the EG (t = −12.16, $p < 0.001$, d = 1.72, large effect size). Lastly, no significant "Time × Group" interaction was reached in didactic material development and attitude toward school ($p > 0.05$). The control group did not report any significant changes ($p > 0.05$).

3.1.3. Objective Study Tests

The results of a two-factor repeated measure ANOVA showed meaningful "Time × Group" interaction for objective study tests ($F_{1,98} = 220.36$, $p < 0.001$, $\eta^2_p = 0.69$, large effect size). Carrying out the post hoc analysis, it was found that EG made a significant increase in the score for objective study tests ($t = 10.26$, $p < 0.001$, $d = 1.45$, large effect size). The control group did not report any significant changes ($p > 0.05$).

3.1.4. Scholastic Self-Efficacy

The results of a two-factor repeated measure ANOVA showed a meaningful "Time × Group" interaction for self-efficacy tests ($F_{1,99} = 981.36$, $p < 0.001$, $\eta^2_p = 0.90$, large effect size). When the post hoc analysis was performed, it showed that the experimental group reached an important increase in self-efficacy tests ($t = 40.34$, $p < 0.001$, $d = 5.70$, large effect size). The control group did not report any significant changes ($p > 0.05$).

4. Discussion

The aim of this work was to analyze the relationship between a 12-week physical activity classroom break intervention, the perceived sense of self-efficacy and academic achievement among overweight and non-overweight high school students, grounded on the posit that physical activity plays a crucial role as a mediating factor between self-efficacy and the impact of programs on eating behavior and academic achievement.

In the frame of this work, the findings show that the 12-week active break performed during a program of nutritional education was most effective in promoting the highest academic performance and perceived sense of self-efficacy in overweight and non-overweight high school students. On the contrary, the regular lessons where no practice was performed were less efficient in achieving implications that matched the objectives set out.

The first result of the present study was the significant difference between overweight and non-overweight students with respect to self-efficacy. This difference is most likely attributable to the fact that participants categorized as overweight had poor self-confidence and were considerably less trusting in trying to find opportunities to engage in physical activity and select physically active tasks instead of those sedentary ones than their non-overweight counterparts. Our findings are consistent with previous findings which showed that lower baseline self-efficacy was linked to a higher weight status [48–51]. Consequently, the sense of perceived self-efficacy was associated with physical activity in overweight individuals [52]. Therefore, the interventions targeting it represent the most effective means to increase physical activity in this population [53]. Specifically, interventions focused on increasing self-efficacy were shown to lead to higher levels of physical activity in overweight adolescents.

The present findings also showed that as self-efficacy and levels of physical activity increased, the BMI decreased. It was coherent with earlier scientific literature which demonstrated that increased self-efficacy was related to greater weight loss. In fact, it is possible that physical activity programs specifically designed with the aim to enhance students' self-efficacy could be proven to be effective for intervention outcomes. Otherwise, enhancements in self-efficacy could be the result of weight loss [54–56]. Therefore, according to what was discussed, self-efficacy does not work in a direct way on weight, but it affects weight control behaviors. This happens even more if physical activity program is associated with a nutritional education program. It is in accordance with the social cognitive theory whereby a result obtained in a certain domain is the outgrowth of behavior change. At the same time, behavior changes depend on positive experiences and expectations [23,57]. In fact, health behavior theories argued that improvements in self-efficacy over the course of time will result in behavior change. In this frame, self-efficacy seems to be linked with the choice to follow several health behaviors, such as changes in weight and weight-related behaviors [58]. Therefore, it might be useful to promote certain behavioral strategies of self-efficacy (i.e., nutritional education program and physical activity) specially developed to improve the perceived sense of self-efficacy. These strategies will allow encouraging enhancements in eating intake, physical activity, and weight loss [59].

Perhaps, the major result of the present research was proof of the effectiveness of this active break program to enhance the learning skills of participating students. In line with previous research, it seemed that this finding was a consequence of lower anxiety, greater motivation toward learning, improved working memory, concentration, attention skills [60–62], and last but not least, greater ability to organize the study and decrease off-task behaviors [63,64]. This result stemmed from the advantages provided by the school-based active break. Several pieces of evidence posit that physical activity may enhance students' academic performance owing to the fact that they develop faster-responding skills with regard to different cognitive tasks (i.e., executive function skills and on-task behavior) at the end of a training session [65–67]. Moreover, their responses become more accurate [68,69]. Indeed, physical activity is a visible mediator of the enhancements in academic achievement since it is able to allow the allocation of neural resources underlying performance on working memory tasks [70,71]. A large body of research has demonstrated that individuals who perform high levels of physical activity show greater problem-solving and decision-making capabilities [72,73]. In addition, this supports the idea that the introduction of an active break in the daily routine may solicit arousal and decrease boredom [74,75]. Dishman and colleagues (2019) [76] argued that boosting physical activity could lead to an improvement in self-efficacy and consequently improved classroom behavior, as well as academic outcomes. Regarding the anxiety-reducing status found at the end of the intervention, it demonstrates that physical activity may decrease symptoms of stress and anxiety-related depression [77], leaving a greater amount of working memory that can be utilized for cognitive tasks. On the contrary, negative feelings stemming from anxiety take up working memory, decreasing the students' resources required for learning [78]. In fact, it seemed that lower indices of self-concept and higher levels of BMI could interfere with scholastic functioning [79,80]. They could impact school achievement and participation in school physical activity.

Students with an increase in the sense of perceived self-efficacy and a decrease in weight as a consequence of an active and healthy lifestyle acquired demonstrated greater motivation toward learning, better concentration, reduction of anxiety and higher capacity to organize study tasks relative to sedentary peers [5]. A growing body of scientific research has also revealed a negative relationship between academic achievement and body mass index through the use of self-reported measures of performance [81,82]. Our study offers added value in expanding the scientific evidence thanks to the fact that it reports objective measures of student performance. Moreover, it explores variables that, to the best of our knowledge, had not yet been fully discussed previously, such as the capacity to organize study tasks and be more flexible toward studying. They provide a much better picture of student performance.

Clearly, a meaningful connection between BMI and active lifestyle was reported in this work [83]. Students who enhanced their BMI were also those who reported the best results in the learning ability tests, at the same time improving their health. On the contrary, adolescents that experienced a significant reduction in the sense of perceived self-efficacy and a sedentary lifestyle showed poor academic performance. Earlier research about physical activity among younger demonstrated that lower levels of physical activity appear to be a significant determining factor in the occurrence of negative psychosocial determinants, as well as academic failure [84]. Physical activity is, therefore, an important contributing factor in the development and/or maintenance of self-efficacy and academic achievement.

Lastly, it does not come as a surprise that students who were able to decrease their weight seem to feel better athletically than when they were overweight. This finding is in accordance with the assumption that a lack of physical activity and sedentarism are determinants in the storage of adolescent fatness. Overweight situations in adolescents cause negative long-term implications, given that they have more limited chances to finish school on time, achieve high income, and live a wealthy life as adults than their non-overweight peers [85].

But what is surprising was to find that both overweight and non-overweight students participated in school-based sports teams. There were no differences in sports participation by weight status. This unexpected finding could be a consequence of the "no-cut" policy for sports and physical activity engagement of the selected school. This suggests that organized sports are a viable physical activity option for overweight adolescents.

According to what was discussed, these findings were in line with earlier studies confirming the evidence of a significant association between an active lifestyle, self-efficacy and academic achievement [86–91]. Physical activity appears to have several positive effects on academic outcomes. They occur via different direct and indirect physiological, emotional, cognitive and learning processes [92–95]. For these reasons, schools should provide occasions for students to keep a steady pace of physical activity [96]. Masini et al. (2022) [97] argued that the integration of school-based active breaks during school daily routine is the most profitable way to develop and enhance both cognitive functions and learning performance.

Although this work provides support concerning the positive relationship between physical activity, self-efficacy, weight loss and school outcomes, some limitations of this study warrant consideration. First of all, the present study is restricted by the inclusion of pupils who attended the same school. Therefore, the findings may not be extended to pupils at other schools or with different backgrounds. Moreover, the small sample size (N = 100) spawned the difficulties in engaging participants is another limitation. Lastly, another limitation is the lack of evaluation of the long-term effects of physical activity on cognitive functions. Furthermore, it did not explore a wide age range, and data were collected from a single window of time. Therefore, it is recommended that subsequent research analyzes similar factors on a wider and varied sample which involves pupils in the primary, secondary and high school. Nevertheless, the findings achieved may give significant directions for further studies. Therefore, the strengths of this research were provided by this effective strategy that allow increasing physical fitness and academic performance, as well as psychosocial factors.

5. Conclusions

This study suggests that an active lifestyle represents an effective way to maintain psycho-physical health and well-being, particularly among adolescents. In fact, it is of paramount importance that the recommended physical activity levels are achieved and maintained among all age groups. Nonetheless, a lack of physical activity is booming significantly among the youngest generations. This phenomenon is mainly the result of the increase in sedentary behaviors taken during the school day.

According to evidence-based recommendations to promote physical activity in youth, it will be important to integrate more physical exercise into daily school routines and to encourage students whenever they are physically active. Future research should target further approaches to tackle the consequences of too many sedentary activities and to effectively increase physical activity in youth.

Therefore, it is important that the school pays more attention to physical activity during the daily school routine in order to increase self-efficacy and academic performance, as well as to reduce overweight populations.

Author Contributions: F.L. designed the study, conducted the research, collected data, carried out the statistical analysis, was involved in the interpretation of data, and wrote and revised the manuscript. F.T. collected data, was involved in the interpretation of data, and revised the manuscript. E.S. was involved in data collection and helped supervise the intervention program. D.T. coordinated the study, interpreted the data, and revised the manuscript. All authors contributed intellectually to the manuscript. All authors have read and agreed to the published version of the manuscript.

Funding: No sources of funding were used to assist the preparation of this manuscript.

Institutional Review Board Statement: The study was conducted in accordance with the Declaration of Helsinki.

Informed Consent Statement: Written informed consent has been obtained from the patient(s) to publish this paper.

Data Availability Statement: Not applicable.

Conflicts of Interest: The authors declare no conflict of interest.

References

1. Braveman, P.; Gottlieb, L. The Social Determinants of Health: It's Time to Consider the Causes of the Causes. *Public Health Rep.* **2014**, *129* (Suppl. S2), 19–31. [CrossRef]
2. Martin, A.; Booth, J.N.; Laird, Y.; Sproule, J.; Reilly, J.J.; Saunders, D.H. Physical activity, diet and other behavioural interventions for improving cognition and school achievement in children and adolescents with obesity or overweight. *Cochrane Database Syst. Rev.* **2018**, *3*, CD009728. [CrossRef]
3. Montesano, P.; Mazzeo, F. Sports Activities in Obese Teenagers Improve Social Inclusion and Health. *Sport Mont* **2019**, *17*, 55–60. [CrossRef]
4. Donnelly, J.E.; Lambourne, K. Classroom-based physical activity, cognition, and academic achievement. *Prev. Med.* **2011**, *52* (Suppl. S1), S36–S42. [CrossRef]
5. Latino, F.; Fischetti, F.; Cataldi, S.; Monacis, D.; Colella, D. The Impact of an 8-Weeks At-Home Physical Activity Plan on Academic Achievement at the Time of COVID-19 Lock-Down in Italian School. *Sustainability* **2021**, *13*, 5812. [CrossRef]
6. Mazzeo, F.; Santamara, S.; Monda, V.; Tafuri, D.; Dalia, C.; Varriale, L.; De Blasio, S.; Esposito, V.; Messina, G.; Monda, M. Dietary supplements use in competitive and non-competitive boxer: An exploratory study. *Biol. Med.* **2016**, *8*, 1000294.
7. Jakicic, J.M.; Davis, K.K. Obesity and Physical Activity. *Psychiatr. Clin. N. Am.* **2011**, *34*, 829–840. [CrossRef]
8. Piccinno, A.; Colella, D. Differences in cardiovascular fitness of Italian high-school adolescents according to different physical activity levels assessed by IPAQ-A: A cross-sectional study. *Sport Sci. Health* **2016**, *13*, 149–155. [CrossRef]
9. World Health Organization. *World Health Organization (WHO) Guidelines on Physical Activity and Sedentary Behavior*; World Health Organization: Geneva, Switzerland, 2020.
10. Pate, R.R.; Wang, C.Y.; Dowda, M.; Farrell, S.W.; O'Neill, J.R. Cardiorespiratory fitness levels among US youth 12 to 19 years of age: Findings from the 1999–2002 National Health and Nutrition Examination Survey. *Arch. Pediatr. Adolesc. Med.* **2006**, *160*, 1005–1012. [CrossRef]
11. Datar, A.; Sturm, R.; Magnabosco, J.L. Childhood overweight and academic performance: National study of kin-dergartners and first-graders. *Obes. Res.* **2004**, *12*, 58–68. [CrossRef]
12. Li, Y.; Dai, Q.; Jackson, J.C.; Zhang, J. Overweight Is Associated with Decreased Cognitive Functioning among School-Age Children and Adolescents. *Obesity* **2008**, *16*, 1809–1815. [CrossRef]
13. Roberts, C.K.; Freed, B.; McCarthy, W.J. Low Aerobic Fitness and Obesity Are Associated with Lower Standardized Test Scores in Children. *J. Pediatr.* **2010**, *156*, 711–718.e1. [CrossRef]
14. Yu, Z.B.; Han, S.P.; Cao, X.G.; Guo, X.R. Intelligence in relation to obesity: A systematic review and meta-analysis. *Obes. Rev.* **2010**, *11*, 656–670. [CrossRef]
15. Shore, S.M.; Sachs, M.L.; Lidicker, J.R.; Brett, S.N.; Wright, A.R.; Libonati, J.R. Decreased Scholastic Achievement in Overweight Middle School Students. *Obesity* **2008**, *16*, 1535–1538. [CrossRef]
16. Sibley, B.A.; Etnier, J.L. The Relationship between Physical Activity and Cognition in Children: A Meta-Analysis. *Pediatr. Exerc. Sci.* **2003**, *15*, 243–256. [CrossRef]
17. Trudeau, F.; Shephard, R.J. Physical education, school physical activity, school sports and academic performance. *Int. J. Behav. Nutr. Phys. Act.* **2008**, *5*, 5–10. [CrossRef]
18. Tomporowski, P.D.; Davis, C.L.; Miller, P.H.; Naglieri, J.A. Exercise and children's intelligence, cognition, and aca-demic achievement. *Educ. Psychol. Rev.* **2008**, *20*, 111–131. [CrossRef]
19. Hillman, C.H.; Erickson, K.I.; Kramer, A.F. Be smart, exercise your heart: Exercise effects on brain and cognition. *Nat. Rev. Neurosci.* **2008**, *9*, 58–65. [CrossRef]
20. Monda, M.; Viggiano, A.; Vicidomini, C.; Iannaccone, T.; Tafuri, D.; De Luca, B. Expresso coffee increases parasympathetic activity in young, healthy people. *Nutr. Neurosci.* **2009**, *12*, 43–48. [CrossRef]
21. Hopkins, C.M.; Bennett, G.G. Weight-Related Terms Differentially Affect Self-Efficacy and Perception of Obesity. *Obesity* **2018**, *26*, 1405–1411. [CrossRef]
22. Bonavolontà, V.; Cataldi, S.; Latino, F.; Carvutto, R.; De Candia, M.; Mastrorilli, G.; Messina, G.; Patti, A.; Fischetti, F. The Role of Parental Involvement in Youth Sport Experience: Perceived and Desired Behavior by Male Soccer Players. *Int. J. Environ. Res. Public Health* **2021**, *18*, 8698. [CrossRef] [PubMed]
23. Bandura, A. *Self-Efficacy: The Exercise of Control*; W H Freeman/Times Books/Henry Holt & Co.: New York, NY, USA, 1997.
24. Di Palma, D.; Tafuri, D. Special needs and inclusion in sport management: A specific literature review | Posebne potrebe i ukljucenje u sportskom menadzmentu: Specifican pregled literature. *Sport Sci.* **2016**, *9*, 24–31.

25. Salles, A. Self-Efficacy as a Measure of Confidence. *JAMA Surg.* **2017**, *152*, 506–507. [CrossRef]
26. Bagherniya, M.; Darani, F.M.; Sharma, M.; Maracy, M.R.; Birgani, R.A.; Ranjbar, G.; Taghipour, A.; Safraian, M.; Keshavarz, S.A. Assessment of the Efficacy of Physical Activity Level and Lifestyle Behavior Interventions Applying Social Cognitive Theory for Overweight and Obese Girl Adolescents. *J. Res. Health Sci.* **2018**, *18*, e00409. [PubMed]
27. Schmidt, G.; Pichler, S. General Self-Efficacy and Body Weight: The Role of Race and Gender. *Psychol. Rep.* **2020**, *124*, 2476–2500. [CrossRef] [PubMed]
28. Pont, S.J.; Puhl, R.; Cook, S.R.; Slusser, W.; Section on Obesity, & Obesity Society. Stigma Experienced by Children and Adolescents with Obesity. *Pediatrics* **2017**, *140*, e20173034. [CrossRef]
29. Rachmah, Q.; Setyaningtyas, S.W.; Rifqi, M.A.; Indriani, D.; Nindya, T.S.; Megatsari, H.; Mahmudiono, T.; Kriengsinyos, W. Self-Efficacy to Engage in Physical Activity and Overcome Barriers, Sedentary Behavior, and Their Relation to Body Mass Index Among Elderly Indonesians with Diabetes. *J. Prev. Med. Public Health* **2019**, *52*, 242–249. [CrossRef]
30. Greco, G.; Fischetti, F.; Cataldi, S.; Latino, F. Effects of Shotokan Karate on resilience to bullying in adolescents. *J. Hum. Sport Exerc.* **2019**, *14*, 890–899. [CrossRef]
31. Kumar, B.; Robinson, R.; Till, S. Physical activity and health in adolescence. *Clin. Med.* **2015**, *15*, 267–272. [CrossRef]
32. Jerome, G.J.; Marquez, D.X.; McAuley, E.; Canaklisova, S.; Snook, E.; Vickers, M. Self-efficacy effects on feeling states in women. *Int. J. Behav. Med.* **2002**, *9*, 139–154. [CrossRef]
33. McAuley, E.; Talbot, H.M.; Martinez, S. Manipulating self-efficacy in the exercise environment in women: Influences on affective responses. *Health Psychol.* **1999**, *18*, 288–294. [CrossRef] [PubMed]
34. Robinson, L.E.; Stodden, D.F.; Barnett, L.M.; Lopes, V.P.; Logan, S.W.; Rodrigues, L.P.; D'Hondt, E. Motor Competence and Its Effect on Positive Developmental Trajectories of Health. *Sports Med.* **2015**, *45*, 1273–1284. [CrossRef] [PubMed]
35. Martínez-López, E.J.; Ruiz-Ariza, A.; de la Torre-Cruz, M.; Suárez-Manzano, S. Alternatives of Physical Activity within School Times and Effects on Cognition. A Systematic Review and Educational Practical Guide. *Psicol. Educ.* **2021**, *27*, 37–50. [CrossRef]
36. Resaland, G.K.; Moe, V.F.; Aadland, E.; Steene-Johannessen, J.; Glosvik, Ø.; Andersen, J.R.; Kvalheim, O.M.; A McKay, H.; A Anderssen, S. Active Smarter Kids (ASK): Rationale and design of a cluster-randomized controlled trial investigating the effects of daily physical activity on children's academic performance and risk factors for non-communicable diseases. *BMC Public Health* **2015**, *15*, 709. [CrossRef]
37. Calella, P.; Mancusi, C.; Pecoraro, P.; Sensi, S.; Sorrentino, C.; Imoletti, M.; Franzese, A.; Gallè, F.; Liguori, G.; Valerio, G. Classroom active breaks: A feasibility study in Southern Italy. *Health Promot. Int.* **2019**, *35*, 373–380. [CrossRef] [PubMed]
38. Masini, A.; Marini, S.; Leoni, E.; Lorusso, G.; Toselli, S.; Tessari, A.; Ceciliani, A.; Dallolio, L. Active Breaks: A Pilot and Feasibility Study to Evaluate the Effectiveness of Physical Activity Levels in a School Based Intervention in an Italian Primary School. *Int. J. Environ. Res. Public Health* **2020**, *17*, 4351. [CrossRef] [PubMed]
39. Di Maglie, A.; Marsigliante, S.; My, G.; Colazzo, S.; Muscella, A. Effects of a physical activity intervention on schoolchildren fitness. *Physiol. Rep.* **2022**, *10*, e15115. [CrossRef]
40. Fiori, F.; Bravo, G.; Parpinel, M.; Messina, G.; Malavolta, R.; Lazzer, S. A 3-year school-based intervention improved physical fitness and reduced the prevalence of overweight and obesity in Italian prepubertal children. *J. Sports Med. Phys. Fit.* **2021**, *61*, 1682–1689. [CrossRef]
41. Weiner, J.S.; Lourie, J.A. (Eds.) *Practical Human Biology*; Academic Press: London, UK, 1981.
42. Cole, T.J.; Bellizzi, M.C.; Flegal, K.M.; Dietz, W.H. Establishing a standard definition for child overweight and obesity worldwide: International survey. *BMJ* **2000**, *320*, 1240–1243. [CrossRef]
43. Cole, T.J.; Lobstein, T. Extended international (IOTF) body mass index cut-offs for thinness, overweight and obesity. *Pediatr. Obes.* **2012**, *7*, 284–294. [CrossRef]
44. Eurofit. *Eurofit Tests of Physical Fitness*, 2nd ed.; Sports Division Strasbourg, Council of Europe Publishing and Documentation Service; Sports Division Strasbourg, Council of Europe Publishing and Documentation Service: Strasbourg, France, 1993.
45. Brouha, L.; Health, C.W.; Graybiel, A. Step test simple method of measuring physical fitness for hard muscular work in adult men. *Rev. Can. Biol.* **1943**, *2*, 86.
46. Rozenek, R.; Byrne, J.J.; Crussemeyer, J.; Garhammer, J. Male-Female Differences in Push-up Test Performance at Various Cadences. *J. Strength Cond. Res.* **2022**, *36*, 3324–3329. [CrossRef]
47. Mayorga-Vega, D.; Merino-Marban, R.; Viciana, J. Criterion-Related Validity of Sit-and-Reach Tests for Estimating Hamstring and Lumbar Extensibility: A Meta-Analysis. *J. Sports Sci. Med.* **2014**, *13*, 1–14.
48. Krishnan, A.; Sharma, D.; Bhatt, M.; Dixit, A.; Pradeep, P. Comparison between Standing Broad Jump test and Wingate test for assessing lower limb anaerobic power in elite sportsmen. *Med. J. Armed Forces India* **2017**, *73*, 140–145. [CrossRef] [PubMed]
49. Cornoldi, C.; De Beni, R.; Zamperlin, C.; Meneghetti, C. *Amos 8-15. Strumenti Di Valutazione Di Abilità E Motivazione Allo Studio per Studenti Dagli 8 AI 14 Anni*; Erickson: Trento, Italy, 2005.
50. Pastorelli, C.; Picconi, L. Scala di Autoefficacia Scolastica Percepita. In *La Valutazione Dell'autoefficacia*; Caprara, G.V., Ed.; Erickson: Trento, Italy, 2001.
51. Bandura, A. *Social Foundations of Thought and Action. A Social Cognitive Theory*; Prentice-Hall: Englewood Cliffs, NJ, USA, 1986.
52. Utter, A.C.; Robertson, R.J.; Nieman, D.C.; Kang, J. Children's OMNI Scale of Perceived Exertion: Walking/Running Evaluation. *Med. Sci. Sports Exerc.* **2002**, *34*, 139–144. [CrossRef] [PubMed]
53. Cohen, J. A power primer. *Psychol. Bull.* **1992**, *112*, 155–159. [CrossRef]

54. Jeffery, R.W.; Bjornson-Benson, W.M.; Rosenthal, B.S.; Lindquist, R.A.; Kurth, C.L.; Johnson, S.L. Correlates of weight loss and its maintenance over two years of follow-up among middle-aged men. *Prev. Med.* **1984**, *13*, 155–168. [CrossRef]
55. A Linde, J.; Jeffery, R.W.; Levy, R.L.; E Sherwood, N.; Utter, J.; Pronk, N.P.; Boyle, R.G. Binge eating disorder, weight control self-efficacy, and depression in overweight men and women. *Int. J. Obes. Relat. Metab. Disord* **2004**, *28*, 418–425. [CrossRef]
56. Teixeira, P.J.; Going, S.B.; Houtkooper, L.B.; Cussler, E.C.; Metcalfe, L.L.; Blew, R.M.; Sardinha, L.; Lohman, T.G. Pretreatment predictors of attrition and successful weight management in women. *Int. J. Obes. Relat. Metab. Disord.* **2004**, *28*, 1124–1133. [CrossRef]
57. Wamsteker, E.W.; Geenen, R.; Iestra, J.; Larsen, J.K.; Zelissen, P.M.; van Staveren, W.A. Obesity-related beliefs predict weight loss after an 8-week low-calorie diet. *J. Am. Diet. Assoc.* **2005**, *105*, 441–444. [CrossRef]
58. Olander, E.K.; Fletcher, H.; Williams, S.; Atkinson, L.; Turner, A.; French, D.P. What are the most effective techniques in changing obese individuals' physical activity self-efficacy and behaviour: A systematic review and meta-analysis. *Int. J. Behav. Nutr. Phys. Act.* **2013**, *10*, 29. [CrossRef] [PubMed]
59. Buckley, J. Exercise self-efficacy intervention in overweight and obese women. *J. Heath Psychol.* **2014**, *21*, 1074–1084. [CrossRef] [PubMed]
60. Prochaska, J.O.; Norcross, J.C.; Fowler, J.L.; Follick, M.J.; Abrams, D.B. Attendance and outcome in a work site weight control program: Processes and stages of change as process and predictor variables. *Addict. Behav.* **1992**, *17*, 35–45. [CrossRef]
61. Fontaine, K.R.; Cheskin, L.J. Self-efficacy, attendance, and weight loss in obesity treatment. *Addict. Behav.* **1997**, *22*, 567–570. [CrossRef] [PubMed]
62. Richman, R.; Loughnan, G.; Droulers, A.; Steinbeck, K.; Caterson, I. Self-efficacy in relation to eating behaviour among obese and non-obese women. *Int. J. Obes.* **2001**, *25*, 907–913. [CrossRef]
63. Strecher, V.J.; DeVellis, B.M.; Becker, M.H.; Rosenstock, I.M. The Role of Self-Efficacy in Achieving Health Behavior Change. *Health Educ. Q.* **1986**, *13*, 73–92. [CrossRef] [PubMed]
64. Jeffery, R.W. How can Health Behavior Theory be made more useful for intervention research? *Int. J. Behav. Nutr. Phys. Act.* **2004**, *1*, 10. [CrossRef]
65. Clark, M.M.; Abrams, D.B.; Niaura, R.S.; Eaton, C.A.; Rossi, J.S. Self-efficacy in weight management. *J. Consult. Clin. Psychol.* **1991**, *59*, 739–744. [CrossRef]
66. Hillman, C.H.; Snook, E.M.; Jerome, G.J. Acute cardiovascular exercise and executive control function. *Int. J. Psychophysiol.* **2003**, *48*, 307–314. [CrossRef]
67. Pontifex, M.B.; Scudder, M.R.; Drollette, E.S.; Hillman, C.H. Fit and vigilant: The relationship between sedentary behavior and failures in sustained attention during preadolescence. *Neuropsychology* **2012**, *26*, 407–413. [CrossRef]
68. Hillman, C.; Pontifex, M.; Raine, L.; Castelli, D.; Hall, E.; Kramer, A. The effect of acute treadmill walking on cognitive control and academic achievement in preadolescent children. *Neuroscience* **2009**, *159*, 1044–1054. [CrossRef] [PubMed]
69. Bartholomew, J.B.; Jowers, E.M. Physically active academic lessons in elementary children. *Prev. Med.* **2011**, *52*, S51–S54. [CrossRef] [PubMed]
70. Mahar, M.T.; Murphy, S.K.; Rowe, D.A.; Golden, J.; Shields, A.T.; Raedeke, T.D. Effects of a Classroom-Based Program on Physical Activity and On-Task Behavior. *Med. Sci. Sports Exerc.* **2006**, *38*, 2086–2094. [CrossRef]
71. Budde, H.; Voelcker-Rehage, C.; Pietrabyk-Kendziorra, S.; Ribeiro, P.; Tidow, G. Acute coordinative exerciseimprovesattentional performance in adolescents. *Neurosci. Lett.* **2008**, *441*, 219–223. [CrossRef]
72. Ellemberg, D.; St-Louis-Deschênes, M. The effect of acute physical exercise on cognitive function during development. *Psychol. Sport Exerc.* **2010**, *11*, 122–126. [CrossRef]
73. Hillman, C.H.; Buck, S.M.; Themanson, J.R.; Pontifex, M.B.; Castelli, D.M. Aerobic fitness and cognitive development: Event-related brain potential and task performance indices of executive control in preadolescent children. *Dev. Psychol.* **2009**, *45*, 114–129. [CrossRef] [PubMed]
74. Pesce, C.; Crova, C.; Cereatti, L.; Casella, R.; Bellucci, M. Physical activity and mental performance in preadolescents: Effects of acute exercise on free-recall memory. *Ment. Health Phys. Act.* **2009**, *2*, 16–22. [CrossRef]
75. Tomporowski, P.D. Effects of acute bouts of exercise on cognition. *Acta Psychol.* **2003**, *112*, 297–324. [CrossRef]
76. Fredericks, C.R.; Kokot, S.J.; Krog, S. Using a developmental movement programme to enhance academic skills in grade 1 learners. *South Afr. J. Res. Sport, Phys. Educ. Recreat.* **2006**, *28*, 29–42. [CrossRef]
77. Kamijo, K.; Pontifex, M.B.; O'leary, K.C.; Scudder, M.R.; Wu, C.-T.; Castelli, D.M.; Hillman, C.H. The effects of an afterschool physical activity program on working memory in preadolescent children. *Dev. Sci.* **2011**, *14*, 1046–1058. [CrossRef]
78. Masini, A.; Marini, S.; Gori, D.; Leoni, E.; Rochira, A.; Dallolio, L. Evaluation of school-based interventions of active breaks in primary schools: A systematic review and meta-analysis. *J. Sci. Med. Sport* **2019**, *23*, 377–384. [CrossRef]
79. Watson, A.; Timperio, A.; Brown, H.; Best, K.; Hesketh, K.D. Effect of classroom-based physical activity interventions on academic and physical activity outcomes: A systematic review and meta-analysis. *Int. J. Behav. Nutr. Phys. Act.* **2017**, *14*, 114. [CrossRef] [PubMed]
80. Gallotta, M.C.; Bonavolontà, V.; Zimatore, G.; Iazzoni, S.; Guidetti, L.; Baldari, C. Effects of Open (Racket) and Closed (Running) Skill Sports Practice on Children's Attentional Performance. *Open Sports Sci. J.* **2020**, *13*, 105–113. [CrossRef]
81. Shephard, R.J. Habitual Physical Activity and Academic Performance. *Nutr. Rev.* **1996**, *54*, S32–S36. [CrossRef] [PubMed]

82. Dishman, R.K.; McIver, K.L.; Dowda, M.; Saunders, R.P.; Pate, R.R. Self-efficacy, beliefs, and goals: Moderation of declining physical activity during adolescence. *Health Psychol. Off. J. Div. Health Psychol. Am. Psychol. Assoc.* **2019**, *38*, 483–493. [CrossRef]
83. Dishman, R.K.; Saunders, R.P.; Motl, R.W.; Dowda, D.M.; Pate, R.R. Self-Efficacy Moderates the Relation Between Declines in Physical Activity and Perceived Social Support in High School Girls. *J. Pediatr. Psychol.* **2008**, *34*, 441–451. [CrossRef] [PubMed]
84. Verloigne, M.; Cardon, G.; De Craemer, M.; D'haese, S.; De Bourdeaudhuij, I. Mediating Effects of Self-Efficacy, Benefits and Barriers on the Association between Peer and Parental Factors and Physical Activity among Adolescent Girls with a Lower Educational Level. *PLoS ONE* **2016**, *11*, e0157216. [CrossRef]
85. Strauss, R.S. Childhood Obesity and Self-Esteem. *Pediatrics* **2000**, *105*, e15. [CrossRef]
86. Ahn, S.; Fedewa, A.L. A Meta-Analysis of the Relationship Between Children's Physical Activity and Mental Health. *J. Pediatr. Psychol.* **2011**, *36*, 385–397. [CrossRef]
87. Castelli, D.M.; Hillman, C.H.; Buck, S.M.; Erwin, H.E. Physical Fitness and Academic Achievement in Third- and Fifth-Grade Students. *J. Sport Exerc. Psychol.* **2007**, *29*, 239–252. [CrossRef]
88. Chaddock, L.; Erickson, K.I.; Prakash, R.S.; Kim, J.S.; Voss, M.W.; Vanpatter, M.; Pontifex, M.B.; Raine, L.B.; Konkel, A.; Hillman, C.H.; et al. A neuroimaging investigation of the association between aerobic fitness, hippocampal volume, and memory performance in preadolescent children. *Brain Res.* **2010**, *1358*, 172–183. [CrossRef] [PubMed]
89. Singh, A.; Uijtdewilligen, L.; Twisk, J.W.R.; Van Mechelen, W.; Chinapaw, M.J.M. Physical Activity and Performance at School. A Systematic Review of the Literature Including a Methodological Quality Assessment. *Arch. Pediatr. Adolesc. Med.* **2012**, *166*, 49–55. [CrossRef] [PubMed]
90. Esteban-Cornejo, I.; Reilly, J.; Ortega, F.B.; Matusik, P.; Mazur, A.; Erhardt, E.; Forslund, A.; Vlachopapadopoulou, E.A.; Caroli, M.; Boyland, E.; et al. Paediatric obesity and brain functioning: The role of physical activity—A novel and important expert opinion of the European Childhood Obesity Group. *Pediatr. Obes.* **2020**, *15*, e12649. [CrossRef] [PubMed]
91. Kimm, S.Y.; A Barton, B.; Berhane, K.; Ross, J.W.; Payne, G.H.; Schreiber, G.B. Self-esteem and adiposity in black and white girls: The NHLBI growth and health study. *Ann. Epidemiol.* **1997**, *7*, 550–560. [CrossRef] [PubMed]
92. Crosnoe, R.; Muller, C. Body Mass Index, Academic Achievement, and School Context: Examining the Educational Experiences of Adolescents at Risk of Obesity. *J. Health Soc. Behav.* **2004**, *45*, 393–407. [CrossRef]
93. Swallen, K.C.; Reither, E.N.; Haas, S.A.; Meier, A.M. Overweight, Obesity, and Health-Related Quality of Life Among Adolescents: The National Longitudinal Study of Adolescent Health. *Pediatrics* **2005**, *115*, 340–347. [CrossRef]
94. NCD Risk Factor Collaboration (NCD-RisC); Iurilli, M.L.; Zhou, B.; E Bennett, J.; Carrillo-Larco, R.M.; Sophiea, M.K.; Rodriguez-Martinez, A.; Bixby, H.; Solomon, B.D.; Taddei, C.; et al. Heterogeneous contributions of change in population distribution of body mass index to change in obesity and underweight. *eLife* **2021**, *10*, e60060. [CrossRef]
95. Wassenaar, T.M.; Wheatley, C.M.; Beale, N.; Salvan, P.; Meaney, A.; Possee, J.B.; Atherton, K.E.; Duda, J.L.; Dawes, H.; Johansen-Berg, H. Effects of a programme of vigorous physical activity during secondary school physical education on academic performance, fitness, cognition, mental health and the brain of adolescents (Fit to Study): Study protocol for a cluster-randomised trial. *Trials* **2019**, *20*, 189. [CrossRef]
96. Gortmaker, S.L.; Must, A.; Perrin, J.M.; Sobol, A.M.; Dietz, W.H. Social and Economic Consequences of Overweight in Adolescence and Young Adulthood. *New Engl. J. Med.* **1993**, *329*, 1008–1012. [CrossRef]
97. Masini, A.; Ceciliani, A.; Dallolio, L.; Gori, D.; Marini, S. Evaluation of feasibility, effectiveness, and sustainability of school-based physical activity "active break" interventions in pre-adolescent and adolescent students: A systematic review. *Can. J. Public Health* **2022**, *113*, 713–725. [CrossRef]

nutrients

MDPI

Article

Phytanic Acid Intake and Lifestyle Modifications on Quality of Life in Individuals with Adult Refsum Disease: A Retrospective Survey Analysis

Jeffrey J. Li *, Jane J. Kim and Fauzia Nausheen

Department of Education, California University of Science and Medicine, Colton, CA 92324, USA
* Correspondence: jeffreyli923@gmail.com

Abstract: Adult Refsum disease (ARD) is a rare peroxisomal biogenesis disorder inherited in an autosomal recessive fashion and is often characterized by retinitis pigmentosa, cerebellar ataxia, and polyneuropathy. Many patients with ARD require diet modification, psychosocial support, and various specialist visits to manage their symptoms. In this study, we explored the quality of life in individuals with ARD by analyzing retrospective survey data collected by the Coordination of Rare Diseases at Sanford (CoRDS) Registry and Global Defeat Adult Refsum Everywhere (DARE) Foundation. Statistical tests used were frequencies, mean, and median. There were 32 respondents, ranging between 11 and 32 responses for each question. The mean age at diagnosis was 35.5 ± 14.5 years (range 6–64) with 36.4% male and 63.6% female respondents. The average age for retinitis pigmentosa diagnosis was 22.8 ± 15.7 years (range 2–61). Dieticians were the most frequently seen (41.7%) for management of low-phytanic-acid diets. Most participants exercise at least once per week (92.5%). Depression symptoms were reported in 86.2% of the participants. Early diagnosis of ARD is important for managing symptoms and preventing progression of visual impairment due to phytanic acid buildup. Interdisciplinary approach should be used for patients to address physical and psychosocial impairments of ARD.

Keywords: adult Refsum disease; rare disease; phytanic acid; quality of life; retinitis pigmentosa; PHYH; patient registry; chronic disease; physical function; dietary modification

Citation: Li, J.J.; Kim, J.J.; Nausheen, F. Phytanic Acid Intake and Lifestyle Modifications on Quality of Life in Individuals with Adult Refsum Disease: A Retrospective Survey Analysis. *Nutrients* **2023**, *15*, 2551. https://doi.org/10.3390/nu15112551

Academic Editors: Domenico Tafuri and Francesca Latino

Received: 10 May 2023
Revised: 26 May 2023
Accepted: 27 May 2023
Published: 30 May 2023

1. Introduction

Refsum disease is divided into two subclasses: adult Refsum disease (ARD) and infantile Refsum disease (IRD). Adult Refsum disease, also called heredopathia atactica polyneuritiformis, is a very rare peroxisomal biogenesis disorder inherited in an autosomal recessive fashion and often characterized by retinitis pigmentosa, cerebellar ataxia, polyneuropathy, and increased protein in cerebrospinal fluid (CSF) without an increase in cell count [1]. In ARD, patients have an increase in phytanic acid (3,7,11,15-tetramethylhexadecanoic acid), a saturated branched-chain fatty acid, in their bodies due to a defect in the alpha-oxidation of phytanic acid [2,3]. While the majority of patients with ARD have a deficiency in phytanoyl-CoA hydroxylase (PAHX), which is encoded by the phytanoyl-CoA 2-hydroxylase (PHYH) gene, some patients instead have a deficiency of the type 2 peroxisomal targeting signal (PTS2) receptor encoded by peroxin (PEX)7. The alpha oxidation pathway of phytanic acid is presented in Figure 1.

In IRD, mutations in 12 different genetic loci have been identified as contributing to IRD's presentation, including peroxisome ATPases PEX1, PEX2, and PEX26. The primary mutation of PAHX implicated in ARD results in increases of primarily phytanic acid, whereas the multiple mutations in IRD lead to increases in very long-chain fatty acids (VLCFA), di- and tri-hydroxycholestanoic acid, pipecolic acid, as well as phytanic acid [4]. While IRD patients typically present in the first year of life, ARD often presents from late childhood to adulthood. However, there have been cases of both IRD patients presenting in adulthood and ARD patients presenting in childhood [5,6]. Both subclasses have similar presentations of visual changes, hearing loss, and cerebellar ataxia. However, patients with IRD can have additional signs of craniofacial dysmorphism, intellectual disability, and gastrointestinal disorders, whereas patients with ARD may present with ichthyosis, kidney malfunction, and bilateral shortening of the metacarpals or metatarsals [7]. There are no official estimates of ARD prevalence. Most cases of ARD in literature reside in the United Kingdom or Norway, where ARD is better known than in other countries. ARD prevalence in the United Kingdom is estimated to be one in 1,000,000 with even lower prevalence in other Caucasian countries, such as the United States [8].

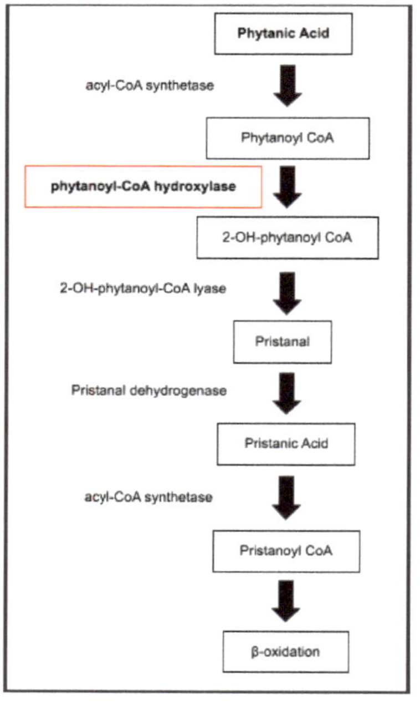

Figure 1. Phytanic acid oxidation pathway. The red box indicates the enzyme deficient with PHYH mutations in adult Refsum disease.

ARD has a wide range of phenotypic presentations, but traditionally presents in late childhood with night vision deterioration, retinitis pigmentosa, and anosmia. The diagnosis for ARD is often made through clinical correlation of symptoms and biallelic mutations in either PHYH or PEX7 [2]. If genetic testing is not diagnostic, then a decrease in activity of phytanoyl-CoA hydroxylase enzyme can also be diagnostic. From 10 to 15 years after diagnosis, symptoms include ataxia, deafness, polyneuropathy, ichthyosis, and cardiac arrhythmias. Bony changes such as shortened metacarpals or metatarsals can also indicate possible ARD. After clinical or molecular diagnosis is made, ARD is normally confirmed with elevated phytanic acid levels (>200 μmol/L) [9]. Phytanic acid causes damage by excessively depositing in various tissues. Food products rich in phytanic acid include meat from ruminant animals, such as beef, lamb, mutton, and goat [2]. Additionally, almonds, coconuts, and peanuts are high in phytanic acid. In terms of treatment, ARD is treated with reduction of phytanic acid via reducing phytanic acid in the diet or eliminating phytanic acid via plasmapheresis or lipid apheresis [10]. Although decreasing plasma phytanic acid may symptomatically treat ARD patients in terms of ichthyosis, sensory neuropathy, and ataxia, decreasing serum phytanic acid may not resolve symptoms of retinitis pigmentosa, deafness, and anosmia [11,12].

Many patients with ARD modify their diet, undergo routine plasma exchange or lipid apheresis, and require accommodations in their daily lives. Some patients prefer regular plasma exchange alongside a low-phytanic-acid diet, whereas others prefer dietary modification only. Although plasmapheresis effectively removes phytanic acid in the short-term, the procedure is expensive and increases the risk of morbidity because of its invasive mechanism. However, diet modifications can be difficult for patients due to strict food restriction, although dietary recommendations for patients with ARD have grown more comprehensive since its inception [13]. Since treatment is not efficacious in reducing retinitis pigmentosa, many patients with ARD suffer from vision loss due to the early onset of visual symptoms in the disease course. The quality of life for patients with ARD is often decreased even with regular treatment and dietary compliance. However, recent advances in stem-cell therapy for retinitis pigmentosa have shown increased retinal neurogenesis and decreased retinal inflammation in mice, indicating a possible clinical treatment for retinitis pigmentosa in the future [14]. Another possible treatment for ARD overall would be chemical co-substrate rescue of PAHX deficiencies due to PHYH mutations, as illustrated in vitro by previous studies [15].

Increasing awareness of ARD can increase early diagnosis of ARD in more patients, as well as support new methods for treatment and symptomatic accommodation. Our study explored the quality of life in individuals with ARD by analyzing retrospective survey data. As one of the first studies to explore global survey data from individuals with ARD, we sought to increase understanding of the clinical and lifestyle trends in individuals with ARD.

2. Materials and Methods

We would like to thank the Coordination of Rare Diseases at Sanford (CoRDS) Registry and the Global Defeat Adult Refsum Everywhere (DARE) Foundation for where this data originated. Participants with ARD answered both the "CoRDS Questionnaire" and "Global DARE Foundation Patient Survey Questionnaire". The "CoRDS Questionnaire" was a 79-question form surveying patient demographics. The "Global DARE Foundation Patient Survey Questionnaire" was a 63-question survey tailored to asking questions specifically about ARD.

The Refsum Patient Registry is the first ever ARD registry and was established by the Global DARE Foundation in August 2020 in partnership with Sanford Research. Submissions were accepted starting July 2020. This study took place from July 2020 to February 2023. The collected data was de-identified by Sanford Research. The data obtained were approved for publication by the CoRDS registry and Global DARE Foundation; Institutional Review Board approval HS-2023-02 was obtained from the California University of Science and Medicine (CUSM). Access to this data can be acquired through approval from the CoRDS Registry website.

The aim of this retrospective survey analysis was to discuss the effects of ARD on quality of life. The requirement was that the participant have a diagnosis of ARD. Participant ages ranged from 7 to 79 years. Legal guardians filled out the forms if the participant was under 18 years. The number of responses to the different questions in these surveys varied because not every question was answered by each individual. Quality of life factors extracted and reported from this survey included demographics, symptoms, details surrounding ARD diagnosis, specialists seen, and medical tests conducted. Also described were the impacts of ARD on physical health, mental health, physical activity, and diet. Statistical tests used were frequencies, mean, and median.

3. Results

3.1. Demographics

Thirty-two individual participants responded to these self-reported surveys. The number of responses for each question varied between 11 and 32. The mean age was 47.4 ± 17.2 years (range 7–79) at time of survey submission with 36.4% male and 63.6%

female respondents (Table 1). The mean age at diagnosis was 35.5 ± 14.5 years (range 6–64) and the mean age of first symptom onset was 18.5 ± 16.8 years (range 1–64). There were nine different countries that participants were born in. The participants' country of birth with the most responses came from the United States of America (23.8%) and the United Kingdom (23.8%). Responses also came from Austria (14.3%), Australia (9.5%), South Africa (9.5%), Canada (4.8%), Netherlands (4.8%), Ireland (4.8%), and Argentina (4.8%). For employment status, 29.6% were employed full-time, 22.2% were disabled and unable to work, 14.8% were students, 14.8% were retired, 11.1% were employed part-time, 3.7% were stay-at-home parents, and 3.7% were not employed and not looking for work. Of the individuals who were not employed, most (61.1%) were not able to work due to ARD.

Table 1. Demographics of survey responses.

Age at survey submission, years			
n = 32			
Mean	Standard deviation (SD)	47.4	17.2
Median	min, max	48	7, 79
Age at diagnosis, years			
n = 24			
Mean	SD	35.5	14.5
Median	min, max	37	6, 64
Age at first symptoms, years			
n = 15			
Mean	SD	18.5	16.8
Median	min, max	15	1, 61
Race			
n = 28		n	%
White		26	92.9
Asian		1	3.6
Other/Refuse to answer		1	3.6
Sex at birth			
n = 22		n	%
Male		8	36.4
Female		14	63.6
Country of birth			
n = 21		n	%
United States of America		5	23.8
Australia		2	9.5
Canada		1	4.8
UK		5	23.8
Netherlands		1	4.8
South Africa		2	9.5
Ireland		1	4.8
Argentina		1	4.8
Austria		3	14.3
Which of the following categories best describes the participant's employment status?			
27 responses		n	%
Student		4	14.8
Employed, part-time		3	11.1
Employed, full-time		8	29.6
Stay at home parent		1	3.7
Not employed, looking for work		0	0
Not employed, not looking for work		1	3.7
Retired		4	14.8
Disabled, not able to work		6	22.2

Table 1. *Cont.*

If not working or working part-time, is this due to Refsum disease?		
n = 18	n	%
Yes	11	61.1
No	7	38.9

Parameters reported were age at survey submission, diagnosis, and first symptoms. Other collected data were race, sex at birth, country of birth, and employment status.

3.2. Common Symptoms

The most common symptom at diagnosis of ARD was vision impairment (85.7%), followed by shortened metacarpals or metatarsals (67.9%), dry skin (64.3%), anosmia (64.3%), peripheral neuropathy (53.6%), ataxia (50.0%), joint issues (42.9%), hearing impairment (39.3%), cardiac issues (17.9%), and other (7.1%) (Figure 2). The most common symptom of ARD at survey submission was also vision impairment (92.6%), followed by anosmia (74.1%), shortened metacarpals or metatarsals (70.4%), peripheral neuropathy (63.0%), hearing impairment (55.6%), ataxia (51.9%), joint issues (51.95%), dry skin (48.1%), cardiac issues (18.5%), and other (3.7%).

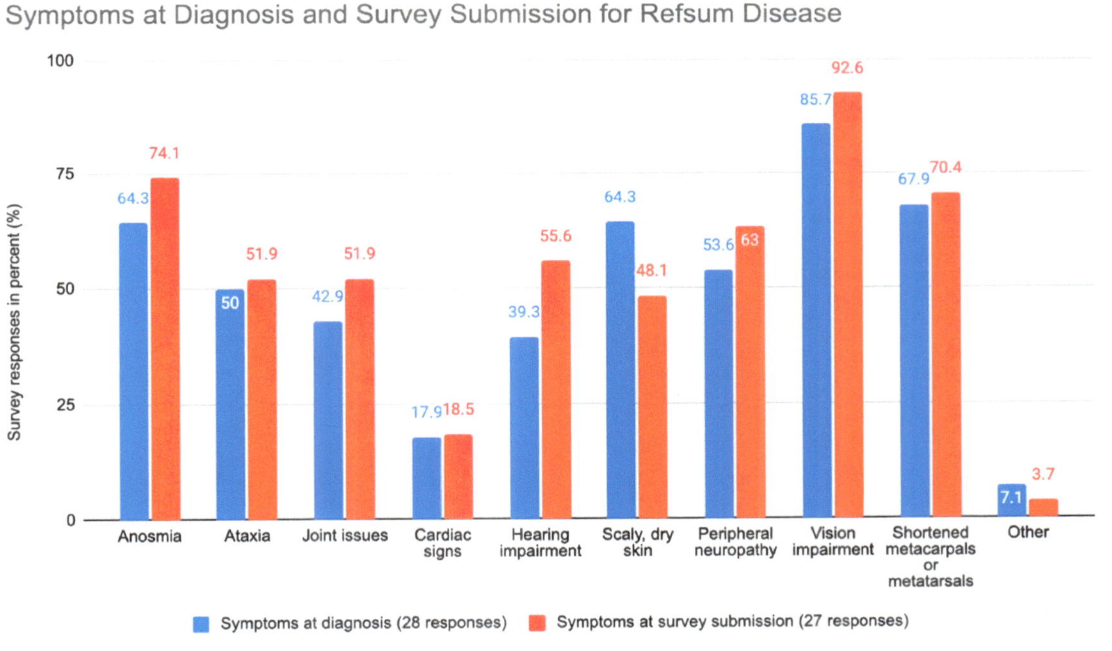

Figure 2. Graph illustrating the frequency of ARD symptoms at diagnosis and survey submission.

3.3. Retinitis Pigmentosa, Gene Mutations, and Hospitalizations

The mean age of retinitis pigmentosa diagnosis was 22.8 ± 15.7 years (range 2–61) (Table 2). Average phytanic acid levels at ARD diagnosis were 689.1 ± 1188.2 µmol/L (range 16–4243). The most common gene mutation was PHYH (69.6%) and the least common was PEX7 (4.3%). A total of 26.1% of respondents did not know their type of gene mutation. Some participants required visual accommodations, such as signal canes (20.8%), long canes (12.5%), and guide dogs (8.3%). A total of 64.6% received cataract surgery. For ARD-related hospitalizations, 12.0% were hospitalized more than five times, 20.0% were hospitalized fewer than five times, 20.0% were hospitalized only at diagnosis, and 48.0% never had any ARD-related hospitalizations.

Table 2. Details about adult Refsum disease diagnosis.

Age at retinitis pigmentosa diagnosis, years			
n = 23			
Mean	Standard deviation (SD)	22.8	15.7
Median	min, max	18	2, 61
Phytanic acid levels at diagnosis, µmol/L			
n = 11			
Mean	Standard deviation (SD)	689.1	1188.2
Median	min, max	375	16, 4243
Gene mutation			
n = 23		n	%
PHYH		16	69.6
PEX7		1	4.3
Unknown		6	26.1
Participant received cataract surgery			
n = 25		n	%
Yes		16	64.0
No		9	36.0
Participant's vision required use of the following:			
n = 24		n	%
Guide dog		2	8.3
Long cane		3	12.5
Signal cane		5	20.8
None		14	58.3
Refsum-related hospitalizations			
n = 25		n	%
None		12	48.0
Only at diagnosis		5	20.0
<5 since diagnosis		5	20.0
>5 since diagnosis		3	12.0

Additional information surrounding the participants' ARD diagnosis. Reported parameters included age of retinitis pigmentosa diagnosis, phytanic acid levels at diagnosis, gene mutation involved, any cataract surgery, accommodations required vision loss, and hospitalizations related to ARD.

3.4. Specialists and Medical Tests

As shown in Figure 3, the most common specialist seen "more frequently than yearly" was the dietician (41.7%) and the most common specialist seen "yearly" was the audiologist (38.1%). At "every 2 years", the neurologist was the most frequently seen (8.3%), and the most frequent for "every three years" was the Ear, Nose, and Throat (ENT) at 8.7%. The most common specialist that participants "never" saw was the audiologist (33.3%).

As shown in Figure 4, the medical tests performed most at "more frequently than yearly" was the visual field test (8.7%). The hearing test was the most common test that participants received annually (50.0%). At every two years, the optical coherence tomography was the most frequent test (15.0%), and the most frequent every three years was the hearing test again (9.1%). The least common test was the darkness adaptation test with 68.4% of participants having never received it.

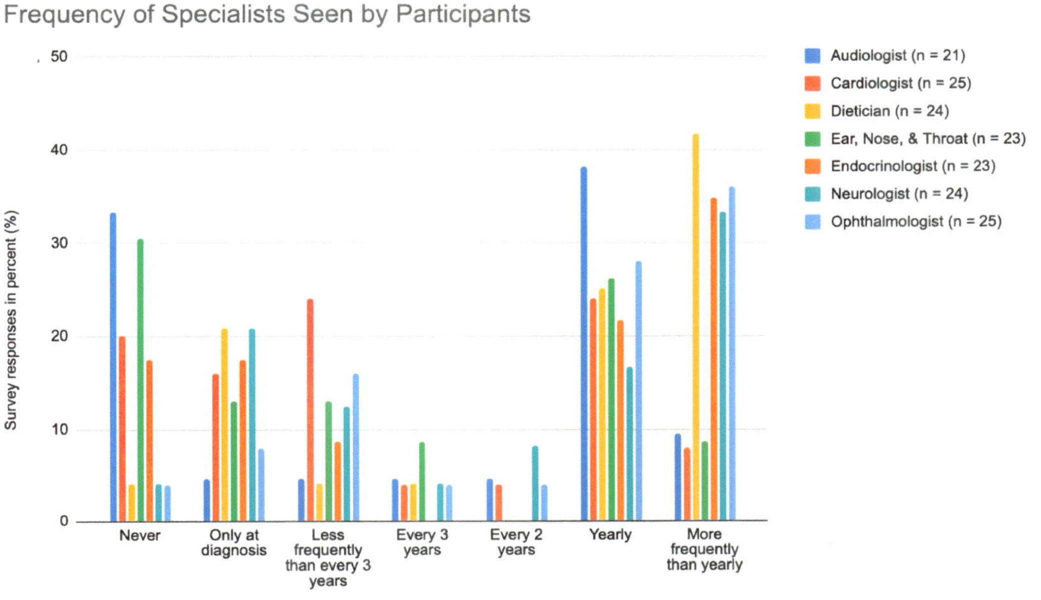

Figure 3. Graph illustrating the frequency participants visit the audiologist, cardiologist, dietician, ENT, endocrinologist, neurologist, and ophthalmologist, where "n" is the number of responses.

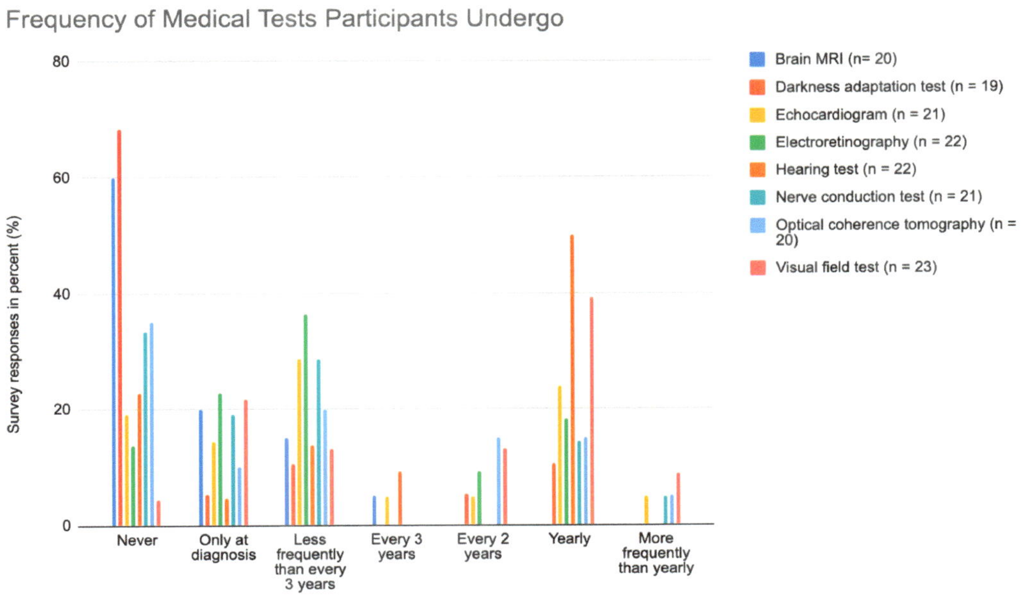

Figure 4. Graph illustrating the frequency participants undergo brain MRI, darkness adaptation testing, echocardiograms, electroretinographies, hearing tests, nerve conduction tests, optical coherence tomographies, and visual field tests, where "n" is the number of responses.

3.5. Limitations with Health, Activity, and Pain

A total of 68.9% of the participants rated their general health as "good" or better while 31.0% rated their health as only "fair" or "poor" (Table 3). A total of 34.5% responded that their health "always" limited them from doing vigorous activities. A total of 44.8% stated that their health "often" or "sometimes" limited their ability to partake in vigorous activities. Only 20.7% "rarely" or "never" had their activities limited by their health. The majority also had daily pain levels of moderate (11.5%) and low (42.3%). A total of 46.2% had no daily pain. Pain interfered with the participants' enjoyment of life always (3.4%), often (20.7%), sometimes (20.7%), rarely (34.5%), and never (20.7%). Most of the participants were discouraged with having ARD with 88.4% being discouraged "a little bit of the time" up to "all of the time". Only 11.5% did not feel discouraged at all with their diagnosis. Similarly, 88.5% of participants were worried about having ARD "a little bit of the time" up to "all of the time" with 11.5% worried "none of the time". When asked how often the participants felt depressed, 6.9% responded "always", 6.9% answered "often", 20.7% reported "sometimes", 51.7% answered "rarely", and 13.8% replied "never".

Table 3. Impact of Refsum disease on daily physical and mental health.

In general, would the participant say his/her health is . . .		
n = 29	n	%
Poor	4	13.8
Fair	5	17.2
Good	13	44.8
Very good	6	20.7
Excellent	1	3.4
Does the participant's health currently limit him/her in doing vigorous activities?		
n = 29	n	%
Always	10	34.5
Often	5	17.2
Sometimes	8	27.6
Rarely	4	13.8
Never	2	6.9
What is the participant's typical daily pain level?		
n = 26	n	%
None	12	46.2
Low (1 to 3)	11	42.3
Moderate (4 to 6)	3	11.5
High (7 to 10)	0	0
How much does pain interfere with the participant's enjoyment of life?		
n = 29	n	%
Always	1	3.4
Often	6	20.7
Sometimes	6	20.7
Rarely	10	34.5
Never	6	20.7
Is the participant discouraged about having Refsum disease?		
n = 26	n	%
None of the time	3	11.5
A little bit of the time	7	26.9
Some of the time	10	38.5
A good bit of the time	4	15.4
Most of the time	1	3.8
All of the time	1	3.8

Table 3. *Cont.*

Is the participant worried about having Refsum disease?		
n = 26	n	%
None of the time	3	11.5
A little bit of the time	6	23.1
Some of the time	12	46.2
A good bit of the time	2	7.7
Most of the time	0	0
All of the time	3	11.5
How often does the participant feel depressed?		
n = 29	n	%
Always	2	6.9
Often	2	6.9
Sometimes	6	20.7
Rarely	15	51.7
Never	4	13.8

Perceptions about the participants' own health, ability to participate in activities, pain level, effects on enjoyment of daily life, and overall mental well-being.

3.6. Mobility and Exercise

According to Table 4, 59.3% of participants had high mobility and 40.7% had impaired mobility categorized as "moderate" mobility, "limited mobility", or "no mobility". When asked about their level of balance, 40.7% had "good" balance, 44.4% had "moderate" balance, and 14.8% had "poor" balance. The majority of participants also exercise at least once per week (92.5%). A total of 3.7% exercise once per week, 33.3% exercise two to three days per week, 33.3% exercise four to six days per week, and 22.2% exercise the full seven days per week. Participants were also asked how many minutes per week they exercise. The majority (42.3%) exercised more than 150 min per week. When asked about the intensity of the exercises, the responses were "no exercise" (3.8%), "light" exercise (57.7%), "moderate" exercise (26.9%), and "hard" exercise (11.5%). A total of 52% reported that ARD affected their ability to perform certain exercises.

Table 4. Impact of Refsum disease on physical activity.

What is the participant's level of mobility?		
n = 27	n	%
No mobility	1	3.7
Limited (requires full assistance)	2	7.4
Moderate (walk semi-assisted, need a railing to walk up and down stairs)	8	29.6
High (can walk unassisted, very steady, able to walk on uneven surfaces)	16	59.3
How would the participant describe the current level of balance?		
n = 27	n	%
Good	11	40.7
Moderate	12	44.4
Poor	4	14.8
How many days in a week does the participant usually exercise?		
n = 27	n	%
0 days	2	7.4
1 day	1	3.7
2–3 days	9	33.3
4–6 days	9	33.3
7 days	6	22.2

Table 4. *Cont.*

Over the course of one week, what is the total number of minutes the participant exercise or does an activity?		
n = 26	n	%
Not active/does not exercise	1	3.8
<30 min	0	0
31–60 min	4	15.4
61–90 min	7	26.9
91–120 min	2	7.7
121–150 min	1	3.8
>150 min	11	42.3
What is the participant's maximum exercise intensity level?		
n = 26	n	%
No exercise	1	3.8
Light (yoga, walking)	15	57.7
Moderate (cycling, swimming)	7	26.9
Hard (running, interval training)	3	11.5
Has Refsum disease prevented the participant from doing certain type(s) of exercise?		
n = 25	n	%
Yes	13	52.0
No	12	48.0

The parameters measured were the participants' mobility, level of balance, amount of exercise, level of exercise, and limitations on exercise.

3.7. Modifications in Diet

The frequency of phytanic acid level measurement and effects of ARD on diet were discussed in Table 5. Phytanic acid levels were measured "more often than every month" (3.8%), "every month" (7.7%), "every other month" (3.8%), "every 3 months" (11.5%), "every 6 months" (23.1%), "yearly" (26.9%), "never" (3.8%), and "unknown" (19.2%). A total of 96.2% were currently on a low-phytanic-acid diet; one participant (3.8%) was currently not on a low-phytanic-acid diet. A total of 60.0% were "very strict" with their compliance of their low-phytanic-acid diet, 36.0% were "good" with their compliance, and 4.0% (one response) were "intermittent" in terms of following the diet. A total of 21.7% snack fewer than three times a day, 73.9% snack three to six times a day, and 4.3% snack more than six times per day. The longest periods where individuals did not eat anything were less than two hours (8.7%), two to three hours (26.1%), four to six hours (56.5%), and greater than six hours (8.7%). The impacts of low-phytanic-acid diets on participants' quality of life have been high (13.0%), moderate (26.1%), low (47.8%), and none (13.0%).

Table 5. Effects of Refsum disease on diet.

How often does the participant have a Phytanic Acid Level done?		
n = 26	n	%
More often than every month	1	3.8
Every month	2	7.7
Every other month	1	3.8
Every 3 months	3	11.5
Every 6 months	6	23.1
Yearly	7	26.9
Never	1	3.8
Unknown	5	19.2
Is the participant on a low-phytanic-acid diet?		
n = 26	n	%
Yes	25	96.2
No	1	3.8

Table 5. *Cont.*

What is the participant's compliance with the low-phytanic-acid diet within the last year?		
n = 25	n	%
Very strict	15	60
Good	9	36
Intermittent	1	4
Poor	0	0
Not following the diet	0	0
How many times a day does the participant eat, including snacks?		
n = 23	n	%
<3	5	21.7
3–6	17	73.9
>6	1	4.3
On average, what is the longest period during the day the participant would go without eating between meals/snacks?		
n = 23	n	%
<2 h	2	8.7
2–3 h	6	26.1
4–6 h	13	56.5
>6 h	2	8.7
What impact does the low-phytanic-acid diet have on the participant's quality of life? (e.g., eating out, social activities)		
n = 23	n	%
No impact	3	13
Low impact	11	47.8
Moderate impact	6	26.1
High impact	3	13

Table shows the frequency of phytanic acid levels and the effects of having adult Refsum disease on diet.

4. Discussion

This study provides an analysis of the general lifestyle of individuals with ARD, including the psychological, physical, and social toll ARD played in their daily lives. We used the data from the Global Patient Registry for Refsum Disease, a database associated with the Global DARE Foundation.

Although ARD has heavy clinical implications for patients' daily lives, it is rarely diagnosed and treated at onset. The first onset of symptoms until medical diagnosis of ARD in the survey group had an average gap of 17 years, demonstrating the need for increased awareness of ARD and its early diagnosis. The youngest survey respondent was 6 years old, supporting the notion that ARDS can be diagnosed in childhood and therefore treated earlier due to the disease's mechanism through inherited gene mutations [6]. In addition, symptoms such as retinitis pigmentosa were diagnosed at an early age (22.8 years). However, despite diagnosing retinitis pigmentosa early, the underlying cause of ARD was rarely identified, and visual symptoms worsened with 41.7% of participants requiring a cane or guide dog because of their deteriorated vision. As one of the most prevalent symptoms with severe consequences towards daily life, patients with retinitis pigmentosa may benefit from early testing for ARD to expedite treatment and prevent progression of vision loss.

Within the questionnaire answers, participants responded to the survey at an average of 12 years after ARD diagnosis. At the time of ARD diagnosis, 85.7% of participants had vision impairment and 64.3% had dry skin. By the time of survey submission, vision impairment increased to 92.6% of participants, and dry skin decreased to 48.1% of participants. The percentage of individuals who acquired hearing loss increased by over 15% from time of diagnosis to time of survey submission. These data support previous research that stated a low-phytanic-acid diet is effective towards improving dry skin but relatively ineffective in improving vision impairment, deafness, or anosmia [11]. Therefore, early diagnosis is vital to prevent symptoms that are irreversible with the current available treatment regimens.

In total, 69.6% of participants reported their known gene mutation was PHYH, despite previous literature stating PHYH gene mutations were the cause of ARD in greater than 90% of ARD patients [16]. The likely explanation for this discrepancy was that 26.1% of participants did not know their ARD-related gene mutation. Patients who did not understand the details of their own condition may feel a lack of control and motivation over their condition and outcomes; promoting understanding in patients with chronic conditions can help improve feelings of autonomy as well as treatment adherence [17].

Due to the wide range of severe clinical manifestations of ARD, a fifth (22.2%) of the participants were disabled and not able to work, and many participants required visiting more than one specialist regularly to manage symptoms and screen for new possible physical deterioration. The most frequently visited specialist was the dietician, with 41.7% patients visiting regularly. Since one of the hallmarks of ARD treatment is dietary modifications to lower phytanic acid intake, regular dietician visits likely increase compliance and provide necessary dietary adjustments and phytanic acid level monitoring. Following dietician visits, ophthalmologist visits were the next most frequent specialist visits (36.0%) for ARD patients. As of now, vision impairment is not effectively treated with current therapy and was the most common symptom (92.6%) among survey respondents [18]. In addition, visual impairment affected the daily lives of ARD patients in mobility and work more than symptoms such as anosmia or bony malformations, increasing the likelihood of compliance with vision-related specialist visits. Along a similar vein, the most common screening among ARD patients were yearly hearing tests, likely due to the progression of deafness caused by ARD and its burden on daily life.

The majority of participants dealt with daily pain and suffered from depressive symptoms. In this study, 53.8% of respondents had everyday "low" or "moderate" amounts of pain, and 40.7% had their mobility affected by the disease. In addition, depression symptoms were reported in 86.2% of the participants, aligning with the high rates of depression in individuals with other chronic conditions, such as COPD, heart disease, and multiple sclerosis [19–21]. As a chronic condition with no definitive cure, ARD patients would likely benefit from psychosocial support from regular therapy sessions with specialists, as well as an interdisciplinary approach to help alleviate the various stresses of ARD.

Many participants shifted their daily habits in terms of diet and exercise to manage their physical changes in pain, mobility, and mood. Dietary modifications for ARD patients included not only low phytanic acid intake, but also preventing breakdown of adipose tissue by avoiding a fasting state with frequent meals or snacks [2]. Breakdown of adipose tissue releases stored phytanic acid into the plasma. Most respondents (91.3%) avoided fasting and did not go without meals or snacking for longer than six hours. This most likely helped decrease the chance of releasing more phytanic acid into the blood, reducing exacerbation of ARD-related symptoms. In addition, 42.3% of individuals with ARD exercised more than 150 min a week. Exercise was likely beneficial in maintaining the mobility, muscle strength, and cardiovascular health compromised by ARD. Previous research had shown that exercise improved depressive symptoms in other chronic conditions and was observed to be one of the most helpful adjuvant treatments for depression, with yoga and mindfulness meditation providing lower but comparable benefits as well [21,22].

This study has some limitations. The first limitation of this study was its reliance on self-reported answers. Individual perceptions of symptoms, diet compliance, specialist visits, exercise routines, and physical and mental well-being guided our study and its findings. A study built on observable, measurable variables would likely aid efforts towards creating more objective depictions of ARD clinical trends. For example, monitoring changes in phytanic acid levels after exercising could reveal a possible threshold between the benefits of improving cardiovascular health versus worsening the stress and energy needs of individuals with ARD. The second limitation was the low number of responses. Since 2021, there have been over 200 confirmed cases of ARD [2], but there were only

32 respondents to the ARD questionnaire. This could be explained by a multitude of reasons. Firstly, the Global Patient Registry for Refsum Disease was only created within the past three years. Secondly, the ARD questionnaire is currently only offered in English. Thirdly, many clinicians and patients may not be actively thinking about ARD as a differential due to its rarity in diagnosis. By continuing the existence of this registry for the long-term future, offering the survey in different languages, and raising awareness of ARD, we may be able to increase the number of respondents to further scientific research into ARD. The third limitation was the time-point that these responses were collected. Instead of surveying patients at seemingly random points after their ARD diagnosis, it would be more insightful to be able to follow at symptom- or diagnosis-onset and observe trends at more discrete levels. However, we acknowledge this was most likely not possible because of how recently the Global Patient Registry for Refsum Disease was created and the overall difficulty of following patients long-term. Future research could follow ARD patients over the period of their lives to document the needs of those who have life-altering ARD symptoms.

5. Conclusions

The CoRDS Registry collects useful information on the quality-of-life factors of individuals with ARD. This manuscript will hopefully serve to increase awareness of ARD and the general clinical manifestations and lifestyles of patients with ARD. If readers of this manuscript have or know someone who has a medical diagnosis of ARD, then we encourage them to register with the Global Patient Registry for Refsum Disease to help further research into the rare disease that is Adult Refsum Disease [23].

Author Contributions: Conceptualization, J.J.L. and J.J.K.; methodology, J.J.L. and F.N.; formal analysis, J.J.L. and J.J.K.; investigation, J.J.L. and F.N.; data curation, J.J.L.; writing—original draft preparation, J.J.L. and J.J.K.; writing—review and editing, J.J.L. and F.N.; visualization, J.J.L. and F.N.; supervision, F.N.; project administration, F.N. All authors have read and agreed to the published version of the manuscript.

Funding: This research received no external funding.

Institutional Review Board Statement: The study was conducted in accordance with the Declaration of Helsinki and approved by the Institutional Review Board (or Ethics Committee) of California University of Science and Medicine (protocol code HS-2023-02; 20 January 2023).

Informed Consent Statement: Informed consent was obtained from all subjects involved in the study.

Data Availability Statement: The data that support the findings of this study are available from Coordination of Rare Diseases at Sanford (CoRDS), but restrictions apply to the availability of these data, which were used under license for the current study, and so are not publicly available. Data are, however, available from CoRDS upon reasonable request and approval.

Acknowledgments: The authors would like to thank the participants who registered with the Global Patient Registry for Refsum Disease.

Conflicts of Interest: The authors declare no conflict of interest.

References

1. Jansen, G.A.; Oftnan, R.; Ferdinandusse, S.; Ijlst, L.; Muijsers, A.O.; Skjeldal, O.H.; Stokke, O.; Jakobs, C.; Besley, G.T.; Wraith, J.E.; et al. Refsum disease is caused by mutations in the phytanoyl-CoA hydroxylase gene. *Nat. Genet.* **1997**, *17*, 190–193. [CrossRef] [PubMed]
2. Waterham, H.R.; Wanders, R.J.; Leroy, B.P. Refsum Disease. In *GeneReviews*; Adam, M.P., Ardinger, H.H., Pagon, R.A., Wallace, S.E., Bean, L.J.H., Gripp, K.W., Amemiya, A., Eds.; University of Washington: Seattle, WA, USA, 2021. Available online: https://www.ncbi.nlm.nih.gov/books/NBK1353/ (accessed on 14 March 2023).
3. Wanders, R.J.; Jansen, G.A.; Skjeldal, O.H. Refsum Disease, Peroxisomes and Phytanic Acid Oxidation: A Review. *J. Neuropathol. Exp. Neurol.* **2001**, *60*, 1021–1031. [CrossRef] [PubMed]
4. Kumar, R.; De Jesus, O. Refsum Disease. In *StatPearls*; StatPearls Publishing: Tampa, FL, USA, 2023. Available online: https://www.ncbi.nlm.nih.gov/books/NBK560618/ (accessed on 14 March 2023).
5. Pakzad-Vaezi, K.L.; Maberley, D.A.L. Infantile Refsum disease in a young adult: Case presentation and brief review. *Retin. Cases Brief Rep.* **2014**, *8*, 56–59. [CrossRef] [PubMed]

6. Tranchant, C.; Aubourg, P.; Mohr, M.; Rocchiccioli, F.; Zaenker, C.; Warter, J.M. A new peroxisomal disease with impaired phytanic and pipecolic acid oxidation. *Neurology* **1993**, *43*, 2044–2048. [CrossRef] [PubMed]
7. Mandel, H.; Meiron, D.; Schutgens, R.B.; Wanders, R.J.; Berant, M. Infantile refsum disease: Gastrointestinal presentation of a peroxisomal disorder. *J. Pediatr. Gastroenterol. Nutr.* **1992**, *14*, 83–85. Available online: https://pubmed.ncbi.nlm.nih.gov/1374125/ (accessed on 14 March 2023). [CrossRef] [PubMed]
8. Gibberd, F.B.; Feher, M.D.; Sidey, M.C.; Wierzbicki, A.S. Smell testing: An additional tool for identification of adult Refsum's disease. *J. Neurol. Neurosurg. Psychiatr.* **2004**, *75*, 1334–1336. [CrossRef] [PubMed]
9. Wierzbicki, A.S.; Lloyd, M.D.; Schofield, C.J.; Feher, M.D.; Gibberd, F.B. Refsum's disease: A peroxisomal disorder affecting phytanic acid alpha-oxidation. *J. Neurochem.* **2002**, *80*, 727–735. [CrossRef] [PubMed]
10. Gutsche, H.U.; Siegmund, J.B.; Hoppmann, I. Lipapheresis: An immunoglobulin-sparing treatment for Refsum's disease. *Acta Neurol. Scand.* **1996**, *94*, 190–193. [CrossRef] [PubMed]
11. Gibberd, F.B.; Billimoria, J.D.; Page, N.G.; Retsas, S. Heredopathia atactica polyneuritiformis (refsum's disease) treated by diet and plasma-exchange. *Lancet* **1979**, *1*, 575–578. [CrossRef] [PubMed]
12. Kohlschütter, A.; Santer, R.; Lukacs, Z.; Altenburg, C.; Kemper, M.J.; Rüther, K. A child with night blindness: Preventing serious symptoms of Refsum disease. *J. Child Neurol.* **2012**, *27*, 654–656. [CrossRef] [PubMed]
13. Baldwin, E.J.; Gibberd, F.B.; Harley, C.; Sidey, M.C.; Feher, M.D.; Wierzbicki, A.S. The effectiveness of long-term dietary therapy in the treatment of adult Refsum disease. *J. Neurol. Neurosurg. Psychiatr.* **2010**, *81*, 954–957. [CrossRef] [PubMed]
14. Brown, C.; Agosta, P.; McKee, C.; Walker, K.; Mazzella, M.; Alamri, A.; Svinarich, D.; Chaudhry, G.R. Human primitive mesenchymal stem cell-derived retinal progenitor cells improved neuroprotection, neurogenesis, and vision in rd12 mouse model of retinitis pigmentosa. *Stem Cell Res. Ther.* **2022**, *13*, 148. [CrossRef] [PubMed]
15. Mukherji, M.; Kershaw, N.J.; MacKinnon, C.H.; Clifton, I.J.; Wierzbicki, A.S.; Schofield, C.J.; Lloyd, M.D. "Chemical co-substrate rescue" of phytanoyl-CoA 2-hydroxylase mutants causing Refsum's Disease. *Chem. Commun.* **2001**, 972–973. [CrossRef]
16. Heydari, A.; Ziaee, E.S.; Gazrani, A. Relationship between Awareness of Disease and Adherence to Therapeutic Regimen among Cardiac Patients. *Int. J. Community Based Nurs. Midwifery* **2015**, *3*, 23–30. Available online: https://pubmed.ncbi.nlm.nih.gov/25553331/ (accessed on 14 March 2023). [PubMed]
17. O'Neal, T.B.; Luther, E.E. Retinitis Pigmentosa. In *StatPearls*; StatPearls Publishing: Tampa, FL, USA, 2023.
18. Pozuelo, L.; Tesar, G.; Zhang, J.; Penn, M.; Franco, K.; Jiang, W. Depression and heart disease: What do we know, and where are we headed? *Cleve. Clin. J. Med.* **2009**, *76*, 59–70. [CrossRef] [PubMed]
19. Wang, H. Micrornas, multiple sclerosis, and depression. *Int. J. Mol. Sci.* **2021**, *22*, 7802. [CrossRef] [PubMed]
20. Yohannes, A.M.; Alexopoulos, G.S. Depression and anxiety in patients with COPD. *Eur. Respir. Rev.* **2014**, *23*, 345–349. [CrossRef] [PubMed]
21. Tian, J.; Kang, Y.; Liu, P.; Yu, H. Effect of Physical Activity on Depression in Patients with Parkinson's Disease: A Systematic Review and Meta-Analysis. *Int. J. Environ. Res. Public Health* **2022**, *19*, 6849. [CrossRef] [PubMed]
22. Saeed, S.A.; Cunningham, K.; Bloch, R.M. Depression and Anxiety Disorders: Benefits of Exercise, Yoga, and Meditation | AAFP. *American Family Physician*. 15 May 2019. Available online: https://www.aafp.org/pubs/afp/issues/2019/0515/p620.html (accessed on 14 March 2023).
23. Global Patient Registry for Refsum Disease. Available online: https://www.defeatadultrefsumeverywhere.org/refsum-patient-registry (accessed on 12 March 2023).

nutrients

MDPI

Article

Assessment of Lifestyle, Eating Habits and the Effect of Nutritional Education among Undergraduate Students in Southern Italy

Fiorenzo Moscatelli [1], Antonella De Maria [2], Luigi Antonio Marinaccio [1], Vincenzo Monda [3], Antonietta Messina [2], Domenico Monacis [4], Giusi Toto [4], Pierpaolo Limone [5], Marcellino Monda [2], Giovanni Messina [1], Antonietta Monda [2,*] and Rita Polito [1,*]

1 Department of Clinical and Experimental Medicine, University of Foggia, 71013 Foggia, Italy; fiorenzo400@gmail.com (F.M.); luigimarinaccio@me.com (L.A.M.); giovanni.messina@unifg.it (G.M.)
2 Department of Experimental Medicine, Section of Human Physiology, Unit of Dietetics and Sports Medicine, Università degli Studi della Campania "Luigi Vanvitelli", 80131 Naples, Italy; antodemar82@gmail.com (A.D.M.); antonietta.messina@unicampania.it (A.M.); marcellino.monda@unicampania.it (M.M.)
3 Department of Movement Sciences and Wellbeing, University of Naples "Parthenope", 80133 Naples, Italy; vincenzo.monda@uniparthenope.it
4 Department of Humanities, Letters, Cultural Heritage, Educational Sciences, University of Foggia, 71100 Foggia, Italy; domenico.monacis@unifg.it (D.M.); giusi.toto@unifg.it (G.T.)
5 Department of Humanities, Telematic University "Pegaso", 80143 Naples, Italy; pierpaolo.limone@unipegaso.it
* Correspondence: antoniettamonda99@gmail.com (A.M.); rita.polito@unifg.it (R.P.)

Citation: Moscatelli, F.; De Maria, A.; Marinaccio, L.A.; Monda, V.; Messina, A.; Monacis, D.; Toto, G.; Limone, P.; Monda, M.; Messina, G.; et al. Assessment of Lifestyle, Eating Habits and the Effect of Nutritional Education among Undergraduate Students in Southern Italy. *Nutrients* **2023**, *15*, 2894. https://doi.org/10.3390/nu15132894

Academic Editor: Josep A. Tur

Received: 3 June 2023
Revised: 22 June 2023
Accepted: 25 June 2023
Published: 26 June 2023

Abstract: Background: The years spent at university represent a critical period that can influence both the quality of lifestyle and the eating habits of subsequent adulthood, and also, in the long term, the health of the individual. The aim of this study was to investigate the lifestyle of university students living away from home. Methods: Each subject recruited for the study was given a questionnaire to obtain general information, eating habits and physical activity levels before (T0) and after six month of training seminars (T1). Blood pressure, body composition and questionnaire responses were investigated. Results: The main findings of this study are a significant decrement in blood pressure; an increment in physical activity practice; an increased number of subjects who pay attention to the calorific value of food and also an improvement in BIA parameters. Conclusions: In conclusion, this study demonstrated the challenges that university students face in leading a healthy lifestyle and caring for their nutritional needs, particularly when they are away from their families. No intervention specifically targets young adults, even though much emphasis is placed on the promotion of a healthy lifestyle based on a varied and balanced diet and sufficient exercise. Our study showed that it is possible to improve lifestyle through educational events aimed at making students aware of the health risks deriving from unhealthy lifestyles.

Keywords: lifestyle; health; university students; nutritional education; correct nutrition; physical activity

1. Introduction

The "healthy lifestyle" (HL) is a way of living, characterized by a methodical approach to behavior management across various domains [1]. Although there have been significant advancements in the prevention and management of chronic diseases, it is widely acknowledged that lifestyle factors, such as smoking, eating habits, stress management, and inactivity, play a significant role in the onset and progression of many chronic diseases [2]. Although clinicians may advise behavioral modifications such weight loss, quitting smoking, and exercising or engaging in physical activity (PA), convincing patients to follow this advice can be difficult [3]. However, even when efficient pharmaceutical, behavioral, and psychological therapies are available, it can be challenging to inspire people to start and continue lifestyle modifications. Adherence, which is also known as compliance [4], is described as "the degree to which a person's behavior (in terms of taking drugs, following a diet, exercising, or adopting other lifestyle changes) agrees with medical or health advice" [5].The effectiveness of medical interventions can be negatively affected, and even completely eliminated, by non-adherence or by conflict between health practices and health advice. According to some estimates in general populations, up to 40% of patients do not follow medical advice to the letter, and when considerable lifestyle changes or difficult behavioral changes are necessary, this prevalence rate can reach 70% or higher [6]. Non-compliance can make treatment ineffective and erode trust in the healthcare practitioner. According to estimates, the total yearly costs of non-adherence to healthy lifestyles in the United States are in the hundreds of billions of dollars [7], with insufficient PA accounting for 10% of total health care spending [8].

A segment of the population particularly exposed to changes in lifestyle is university students who live outside their own families [9]. University students may be a group at biological and psychosocial risk, frequently engaging in unhealthy eating, insufficient resting habits, and substance addiction [10]. A shift in eating habits, such as a decrease in the intake of fresh or minimally processed meals and an increase in the consumption of ultra-processed goods, was brought on by changes in lifestyle during the previous several decades. A bad diet and other health issues, such as being overweight, are signs of new eating habits. The high frequency of stress in this demographic, emotional dysregulation with a lack of self-control, and identity concerns can all be linked to all these maladaptive behaviors among students. Modifying these unhealthy behaviors has the potential to be the focus of intervention while also enhancing academic success. Regarding this aspect, a recent multicenter study with 6222 college students reveals an intriguing trend of health-related behaviors across nations [11]. As observed in Romania, Lebanon, Turkey, and Croatia, students in some nations exhibited a cluster of unhealthier behaviors, such as a higher prevalence of smoking, skipping breakfast, and sleeping for shorter amounts of time on working days. Students from Italy and Spain, on the other hand, showed greater adherence to the Mediterranean diet, consumed more breakfast each day, smoked less, and slept for longer periods of time on working days. Furthermore, even in these young, generally healthy university students, we discovered a relationship between lifestyle and health outcomes, such as body mass index (BMI) and self-perceived health.

It is interesting to note that BMI was also positively correlated with total self-reported METs per week, which represents total physical activity. Since BMI is only a crude weight indicator for height and does not account for body composition, it is possible that the positive correlation between self-reported physical activity and BMI is caused by greater muscle mass [12]. Furthermore, the median METs per week for all countries suggested that the group overall had a moderate to high level of physical activity [13]. The practice of sports, even by overweight individuals, has been documented to contribute to less sedentary behavior in the past, but it has not been linked to healthy eating habits [14]. However, given that research typically shows a negative correlation between BMI and levels of physical activity [15], a more logical explanation would be that students with higher BMIs, even those within the normal range, were more involved in practicing their sports [15]. Moreover, only a marginal correlation between physical activity and BMI in non-obese patients has been shown previously [16], lending support to this latter hypothesis. Mobile phone use and perceived stress level, both of which were strongly correlated, were also positively associated with BMI. The reports from a recent systematic review and meta-analysis, showing that college students with mobile phone addiction were more likely to experience increased levels of anxiety, depression, and impulsivity as well as being more likely to have poor sleep quality [17], lend support to this association. The increasing popularity of smartphones has led to many problems due to excessive use. Excessive smartphone use can interfere with concentration at school or work and can cause physical difficulties, such as neck stiffness, blurred vision, wrist or back pain, and sleep disturbances. It can also reduce in-person social interaction and academic achievement and lead to relationship problems. Studies on smartphone use have shown that among smartphone users, 45.8% feel anxiety when they are not holding their smartphone, 27.1% spend too much time using their smartphone, and 22.6% have repeatedly attempted to reduce their smartphone use but have failed. Moreover, 21.0% of smartphone users reported difficulties with school or work due to excessive smartphone use. Further, these percentages were higher for individuals in their teens and twenties. Considering that addiction is a phenomenon characterized by tolerance, withdrawal symptoms, dependence, and social problems, the research described above suggests the concept of "smartphone addiction" [10]. In fact, data have revealed that the relationship between mobile phone addiction and numerous psychological and behavioral difficulties, including stress, anxiety and depression, might be mediated by interpersonal problems [18]. According to the interpersonal theory, people who are addicted to their mobile phones typically neglect real-world social networking, which leads to a lack of social resources and lower levels of physical activity [19]. In order to understand smartphone addiction, knowledge of risk and protective factors for such addiction is essential [20,21]. A person's risk of obesity and cardiovascular problems may increase if they engage in little physical activity [22]. On the other hand, it has been hypothesized that physical activity can be used to reduce and regulate body fat [23]. More generally, regular exercise was shown to be an efficient way to lower a variety of health risk factors, particularly those connected to metabolic syndrome and cardiovascular diseases [24]. In order to maintain a suitable level of cardio-respiratory fitness, the American Academy of Sports Medicine specifically advises people to engage in at least 150 min per week of moderate intensity cardiovascular exercise and at least 75 min per week of strenuous intensity training. Also advised is resistance exercise 2–3 days per week [25]. Human health has long been a major concern, and research in the fields of medicine, biology, psychology, education, and social and philosophical inquiry have all focused on this issue. The ability to maintain health, prevent sickness, and lead a healthy lifestyle has always been correlated with the level of social development. It takes some instructional effort during sports and health and fitness activities to help children develop the understanding and belief that health is the highest value in a person and must be consistently kept. Given that one of the objectives set by the World Health Organization by 2020 is to promote a healthy lifestyle [26] and one of the key strategies for promoting health is to adopt a healthy lifestyle, it is important to design interventions to change unhealthy lifestyles and promote the dimensions and behaviors associated with health-promoting lifestyles. One of the approaches available for educational interventions in health promotion is the intervention mapping approach. This approach evaluates and intervenes on health-related problems from a problem solving and ecological perspective. Students are a large part of their community and the social capital of that community. On the one hand, studying the promotion of students' lifestyles is effective in designing promotional interventions to promote healthy behaviors of these individuals. On the other hand, students with healthy lifestyles can become role models for other people in society [26].

Therefore, considering the situation of students who live outside their own households, and the importance that lifestyle has on health, the aim of this study was to evaluate the lifestyle and eating habits and the effect of nutritional education among undergraduate students in southern Italy.

2. Materials and Methods

2.1. Participant

For this study 80 subjects (age range: 18 to 28) were enrolled (in Table 1 are reported the anthropometric parameters and blood pressure values). However, the second evaluation was carried out on only 70 subjects, as 10 dropped out. All subjects were students at the University of Foggia and were recruited at ADISU Puglia (Agency for the Right to University Study) in cooperation with the Department of Clinical Experimental. All recruited subjects joined the "Lifestyle and Planet life: improving your life while saving the planet" project. To achieve this goal, a pilot study was conducted, which aimed to detect the physical activity level and lifestyle at the ADISU Residences. No exclusion criteria were applied for this investigation. Participants were provided with comprehensive information regarding the project and were assured that they were free to withdraw from the study at any time.

Table 1. Anthropometric parameters and blood pressure values of subjects.

Parameters	Female	Male
Number (*n*)	49	21
Systolic pressure	124.25 mmHg ± 20.9	129.03 mmHg ± 19.8
Diastolic pressure	74.53 mmHg ± 13.1	76.92 mmHg ± 10.6
Height (cm)	162.2 ± 6.3	176.3 ± 8.2
Weight (kg)	61.9 ± 15.2	73.6 ± 13.5

All students provided written informed consent before the beginning of the investigation and after having had a detailed explanation of the tests. The study was approved by the Institutional Ethics Committee of the University of Foggia on 22 May 2018, no.440/DS, and conducted according to the ethical principles of the Declaration of Helsinki.

2.2. Study Design

Each subject recruited for the study was given a questionnaire (Appendix A) to obtain general information, eating habits and physical activity levels before (T0) and after (T1) six months of training seminars. The training seminars were carried out via an electronic platform. The seminars focused on the following topics: nutrition and reduction of risk factors, physical activity and correct lifestyles. However, after the six-month training period, the number of subjects who underwent T1 measurements dropped to 70. Using Google forms (with the possibility of multiple answers), a questionnaire was administered under the supervision of researchers involved in the investigation. The questionnaire consisted of two multiple choice questions on physical activity levels, in which students could choose whether they practiced sports and, in the event of a positive answer, they could choose weekly attendance. Moreover, blood pressure was measured in each student at T0 and T1 by a team of doctors using a manual sphygmomanometer (Sfigmomanometer Aneroide Erka Perfect). Furthermore, before (T0) and after (T1) training seminars, each student was subjected to body composition analysis using a Quantum V Segmental Bioelectrical Impedence Analyzer (A-Wave, Santeramo in Colle, Bari, Italy). The topics covered during the training seminars concerned nutritional education, the composition of foods in terms of macronutrients and micronutrients, and the beneficial effects of physical activity, especially in the prevention and treatment of non-communicable diseases, such as obesity and cardiovascular diseases. Statistical analysis was performed using GraphPad 6 Software, Inc., Boston, MA, USA, for Windows, version 6.01, Dotmatics, R&D scientific. The data are presented as mean (M) ± standard deviation (SD), and statistical significance was set at $p < 0.05$. The Shapiro–Wilk test was used to check the normal distribution of variables. The paired samples t-test was performed to investigate the differences between T0 and T1 if data were normally distributed and the Wilcoxon test was performed if data were not normally distributed. The chi-square test was performed to investigate differences in questionnaire answers.

3. Results

The first parameter analyzed was blood pressure. The results showed significant difference between T0 and T1. Systolic blood pressure decreased from T0 to T1 (130.40 ± 17.81 vs. 120.0 ± 9.58; $p < 0.01$), while diastolic blood pressure decreased from T0 to T1 (78.69 ± 12.60 vs. 72.89 ± 8.50, $p < 0.01$) (Figure 1).

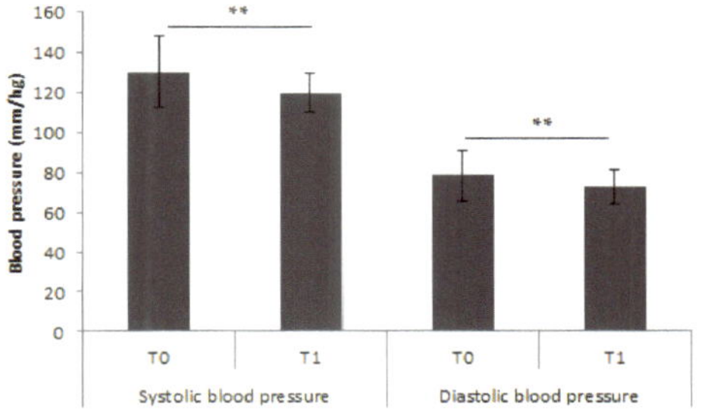

Figure 1. Differences in blood pressure between To and T1. ** indicates p-value < 0.01.

Body composition, measured via bioelectrical impedance, showed significant differences between T0 and T1 (Table 2).

Table 2. BIA analysis.

BIA Parameters	T0		T1		*p*-Value	
Weight (Kg)	66.18	±14.34	58.49	±9.54	0.0012	**
FFM%	72.31	±9.06	76.68	±6.59	0.1354	ns
FM%	27.69	±14.03	23.32	±6.34	0.0251	*
TBW%	54.26	±6.46	55.81	±4.63	0.0926	ns
ECW%	44.81	±3.43	45.68	±3.34	0.0654	ns
ICW%	54.53	±6.93	51.44	±12.08	0.0381	*
MM%	37.72	±6.89	38.45	±5.11	0.3144	ns
BCM%	27.04	±6.97	24.36	±6.32	0.0127	*
BM (kcal)	1533.94	±202.48	1456.64	±183.48	0.0130	*
BMI (kg/m^2)	23.81	±4.39	21.85	±2.20	0.0022	**

* Indicates p-value < 0.05; ** Indicates p-value < 0.01.

Regarding physical activity practice (Question 7), the results showed significant differences between T0 and T1; in fact, the number of the students that declared they practiced physical activity changed from yes 35, no 45 (T0) to yes 50, no 20 (T1) (Chi-square, df = 5.208, 1; z = 2.282; $p < 0.05$).

The percentage of subjects who tried to lose weight increased (Question 9). In fact, it went from 51% at T0 to 60% at T1; however, no significant differences emerged (Chi-square, df = 1.640, 1; z = 1.281; $p > 0.05$). Regarding question number 10, the percentage of subjects who declared that they paid attention to the caloric value of food increased from 63.75% to 78.0% (Chi-square, df = 4.104, 1; z = 2.026; $p < 0.05$).

The portions of sweets consumed per week (Question 11) changed from T0 (<1 per week = 33.75%; 2 per week = 33.75%; >2 per week = 32.5%) to T1 (<1 per week = 49.1%; 2 per week = 38.9%; >2 per week 12%) (Figure 2).

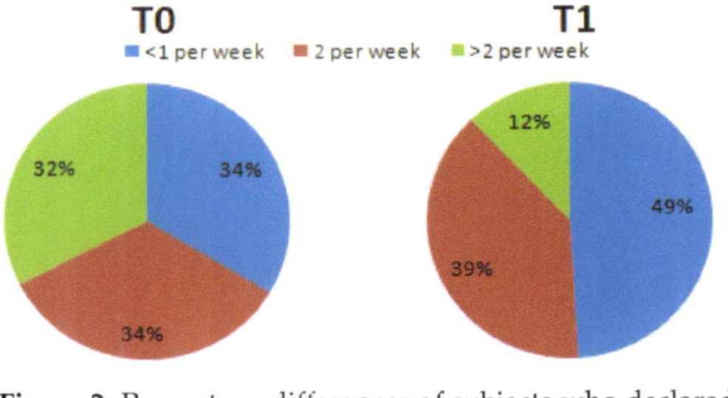

Figure 2. Percentage differences of subjects who declared they consumed sweets during the week (Chi-square, df = 12.69, 2; $p < 0.01$).

Regarding question number 13, servings of red meat consumed per week, the percentages changed from T0 (<1 per week = 28.1%; 2 per week = 40.9%; >2 per week = 31.0%) to T1 (<1 per week = 45.0%; 2 per week = 35.2%; >2 per week 19.8%) (Figure 3).

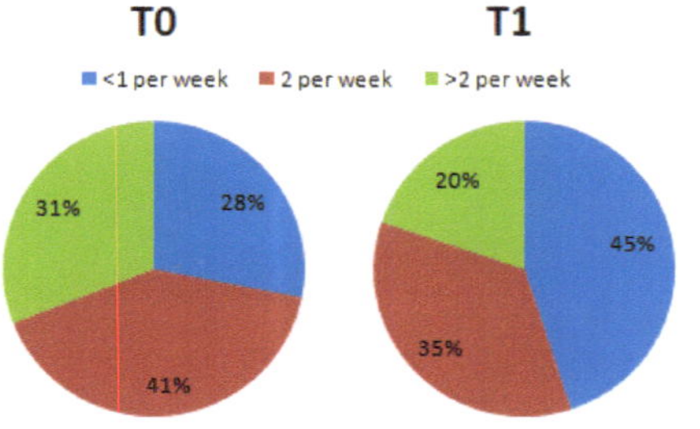

Figure 3. Percentage differences of subjects who declared they consumed red meat during the week (Chi-square, df = 6.805, 2; $p < 0.05$).

Regarding question number 13, the percentage of processed meat consumed per week changed from T0 (<1 per week = 27.5%; 2 per week = 45.6%; >2 per week = 26.9%) to T1 (<1 per week = 45.5%; 2 per week = 39.5%; >2 per week = 15.0%) (Figure 4).

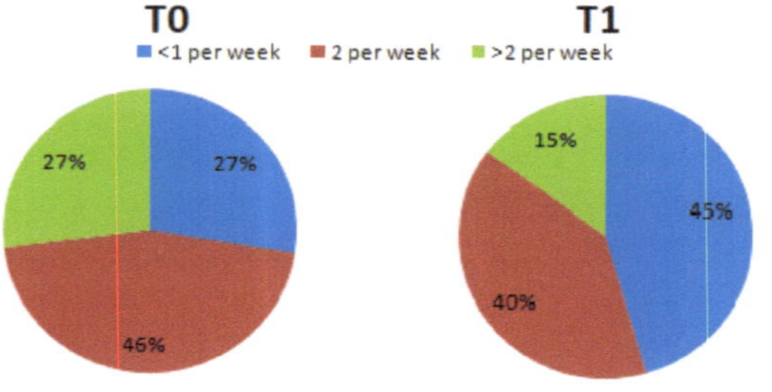

Figure 4. Percentage differences of subjects who declared they consumed processed meat during the week (Chi-square, df = 7.875, 2; $p < 0.05$).

Regarding question number 14, the number of eggs consumed per week, the results showed no significant differences between T0 (<1 per week = 67.5%; 2 per week = 25.0%; >2 per week = 7.5%) and T1 (<1 per week = 67.5%; 2 per week = 25.5%; >2 per week = 7.0%). Regarding question number 15, the percentages of legumes consumed per week, the results showed significant differences between T0 (<1 per week = 20.0%; 1 per week = 21.25%; 2 per week = 36.25%; >2 per week = 22.5%) and T1 (<1 per week = 10.5%; 1 per week = 32.5%; 2 per week = 44.25%; >2 per week = 12.75%) (Figure 5).

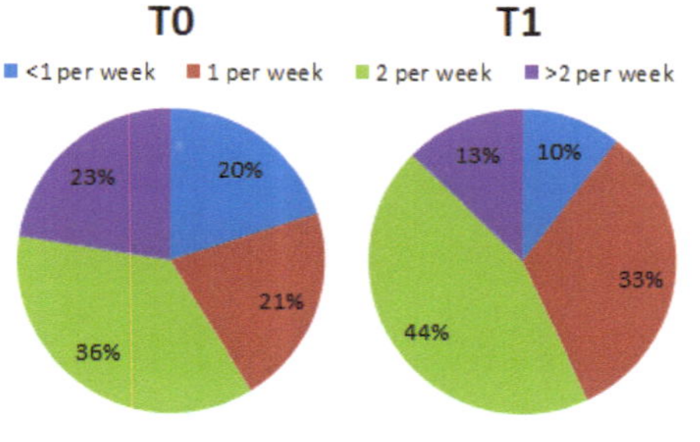

Figure 5. Percentage differences of subjects who declared they consumed legumes during the week (Chi-square, df = 9.353, 3; $p < 0.05$).

Regarding questions 16 (portions of white meat), 17 (fish or seafood) and 18 (dairy products consumed per day), no significant differences emerged between T0 and T1, while regarding question number 19, the percentage of fruit consumed per day, significant differences emerged between T0 (<1 per day = 50.0%; 1–2 per day = 48.75%; 3–4 per day = 1.25%; >4 per day = 0%) and T1 (<1 per day = 20.0%; 1–2 per day = 67.50%; 3–4 per day = 11.25%; >4 per day = 1,25%) (Figure 6).

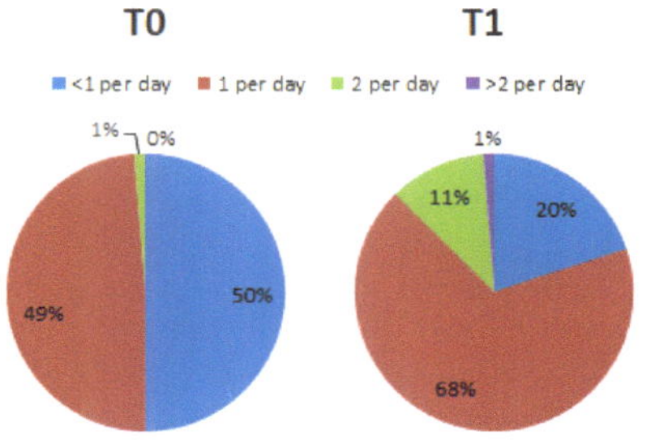

Figure 6. Percentage differences of subjects who declared they consumed fruits during the week (Chi-square, df = 25.33, 3; $p < 0.001$).

Finally, regarding questions 20 (vegetables consumed per day), 21 (olive oil consumed per day), 22 (cereals consumed per day), 23 (glasses of water drunk every day), 24 (glasses of wine drunk every day), 25 (salt consumed every day), 26 (snacks consumed during meals), 27 (sugar intake) and 28 (hours spent watching TV a day), the results showed no significant differences between T0 and T1.

4. Discussion

The main purpose of this study was to evaluate the effect and impact of a training period focused on improving lifestyles. The main findings of this study are: (1) a significant decrement in blood pressure; (2) an increment in physical activity practice; (3) an increased number of subjects who paid attention to the calorific value of food; (4) a percentage decrease in subjects who declared they consumed sweets during the week; (5) a percentage decrease in subjects who declared they consumed red meat during the week; (6) a percentage decrease in processed meat consumed per week; (7) a percentage increase in subjects who declared they consumed legumes during the week; (8) a percentage increase in fruit consumed per day; (9) an improvement in BIA parameters. Our study, in addition to providing an overview of the lifestyle of students who live away from home, seems to show the effectiveness of training events as a tool to achieve improvement in lifestyles.

The results of our study are in line with previously published results by WHO [27]. In fact, given that the 2010 Global Recommendations on Lifestyle for Health state that one in four (27.5%) adults and more than three quarters (81%) of adolescents do not meet the recommendations for physical exercise and diet [28], there is an urgent need to increase the priority and investment in services that promote a healthy lifestyle [3]. These statistics also show that gender differences are significant and that participation levels have not increased globally over the past 20 years. Inequalities in involvement by age, gender, socioeconomic level and geography are also frequently seen in national data, which highlights the urgency of increasing financial support for health [4]. For current and future generations to remain healthy throughout their lifespans, proper diet is crucial. A balanced diet lowers the risk of chronic diseases while promoting healthy growth and development. Adults who follow a healthy diet have a lower risk of obesity, heart disease, type 2 diabetes, and several malignancies. They also live longer. People with chronic diseases can control their conditions and prevent complications by eating healthfully.

When there are no healthy options, consumers could select for items that are higher in calories and less nutritious. People from low-income areas and some racial and ethnic groups frequently do not have easy access to establishments that provide inexpensive healthier foods. Consequently, it is important to encourage people of all ages to eat healthily [29]. A healthy diet, according to the WHO (2016), should, for example, include a high intake of fruits, vegetables, and whole grains, as well as a low intake of saturated fats, salt, and processed carbohydrates [30]. Given that many healthy behaviors are formed and established during the transition from adolescence to young adulthood, this time may be a crucial moment for health promotion measures, including the promotion of good eating. Furthermore, excessive weight gain has been observed among young adults, particularly university students [31–33]. The transition from high school to university also coincides with new living conditions, which could cause eating habits to change [34]. However, few researchers have examined potential shifts in eating habits post matriculation.

Sport was the recreational activity students engaged in the most, despite doing so less frequently than the weekly average recommended for the maintenance of good health [14], which is consistent with other studies describing a lack of regular sport activities [14] and a decline in all forms of physical activity in correlation with the start of university [15]. According to an important investigation conducted in Southern Italy [11], students who live alone spend less time engaging in sports and other leisure activities overall [35].

From a nutritional standpoint, most respondents, particularly students living away from home, acknowledged that their eating habits had altered while they were attending college. It has been emphasized in other contexts how challenging it is for students to maintain appropriate eating routines [36]. The various factors that affect people's eating choices include changes in lifestyle, the comfort and convenience of fast food, taste, their immediate physical and social environment, gender, attention to weight, and beliefs [37]. Other studies indicate that college students have adopted unhealthy eating habits, particularly with regard to low consumption of fruits and vegetables [38,39], milk and dairy products [40], fish [41], eggs, pulses [42], meat [43], sausages, and sweets [37–40].

College students from various backgrounds and with various dietary habits are affected by the difficulties of adopting a diet that complies with the Guidelines [32,37,40–42].

Although many students are influenced by the shift in eating patterns, those who lived away from their families are primarily impacted [43–45]. University students who lived with their parents consumed significantly more fruits, vegetables, legumes, and fish than those who did not live with their parents, which may be explained by the fact that they were not directly involved in meal planning and preparation while the family offered ongoing encouragement for making healthy food choices [46]. According to other studies [36,47], students who leave their families have a diet that deviates significantly from the ideal model of the Mediterranean diet in terms of their consumption of fruits, vegetables, pulses, and fish, while also consuming more ready-made foods and fries [48,49]. This is particularly true of students who are living on their own. These young people's eating habits may be related to a variety of factors, including their first-ever complete independence [48], lack of expertise in meal planning and preparation, lack of time [49], or financial constraints that compel them to spend less on food [46].

The results of our study seem to suggest that attending university, especially away from the family, could play a role in the onset of unhealthy lifestyles, and that these could, at least in part, be modified through training events aimed at sensitizing this population during this moment of their life.

5. Conclusions

In conclusion, this study demonstrated the challenges that university students face in leading a healthy lifestyle and caring for their nutritional needs, particularly when they are away from their families. No intervention specifically targets young adults, even though much emphasis is placed on the promotion of a healthy lifestyle based on a varied and balanced diet and sufficient exercise.

Our study showed that it is possible to improve lifestyle through educational events aimed at making students aware of the health risks deriving from unhealthy lifestyles.

However, although our study provided important results, there are some criticisms that will need to be investigated in the future. In particular, the sample should be increased to seek gender homogeneity; moreover, students who live in families alongside those who live in university residences should be encouraged to accurately evaluate the effect of the latter. In conclusion, it can be said that it would be appropriate to provide training and support programs in universities for students to improve their lifestyle. In fact, good nutrition is essential to keeping current and future generations healthy across the lifespan. Adults who eat a healthy diet live longer and have a lower risk of obesity, heart disease, type 2 diabetes, and certain cancers. New discoveries in health research happen every day, but one clear message remains consistent: optimal nutrition is imperative to human health. Research shows that an unhealthy diet is one of the major risk factors for a range of chronic diseases, including diabetes, cancer, cardiovascular diseases, and other conditions linked to obesity.

This is a study that comes from a pilot project that allowed us to obtain a statistical report on the university population of Foggia. Being a pilot project, this could be extended to other Apulian and national universities. Furthermore, because of the educational initiatives implemented during the project, improvements were highlighted both in the anthropometric parameters evaluated in the population and in the levels of physical activity, as well as through the effectiveness of the programmed training events, demonstrating that education on a healthy lifestyle and monitoring of body composition have a fundamental role in the conduct of a healthy lifestyle for the prevention of non-communicable diseases.

Author Contributions: Conceptualization, F.M. and R.P.; methodology, A.D.M.; software, L.A.M.; validation, M.M. and A.M. (Antonietta Monda); formal analysis, D.M. and V.M.; investigation, A.M. (Antonietta Messina) and G.T.; resources, G.M.; data curation, R.P.; writing—original draft preparation, F.M.; writing—review and editing, R.P.; visualization, G.M.; supervision, M.M. and P.L.; project administration, G.M.; funding acquisition, G.M. All authors have read and agreed to the published version of the manuscript.

Funding: This research received no external funding.

Institutional Review Board Statement: The study was conducted in accordance with the Declaration of Helsinki, and approved by the Institutional Ethics Committee of the University of Foggia on 22 May 2018, no.440/DS.

Informed Consent Statement: Informed consent was obtained from all subjects involved in the study.

Data Availability Statement: The data presented in this study are available on request from the corresponding author.

Acknowledgments: We thank all ADISU Puglia staff for having contributed to the realization of the project.

Conflicts of Interest: The authors declare no conflict of interest.

Appendix A

	n (%)	*n* (%)	*n* (%)	*n* (%)
Question 1: Sex	Male	Female		
Question 2: Age				
Question 3: Smoke	Yes	No		
Question 4: COVID Vaccination	Yes	No		
Question 5: COVID Positivity	Yes	No		
Question 6: Gastrointestinal disorders/Other disease	Yes	No		
Question 7: Do you practice physical activity?	Yes	No		

	n (%)	*n* (%)	*n* (%)	*n* (%)
Question 8: If yes, how many times a week?	<1 time per week	1–2 times per week	2–3 times per week	>3 times per week
Question 9: Have you been trying to lose weight in the past three months? If not: Have you tried to avoid gaining weight?	No attempt to either lose or avoid gaining weight in the past three months	Attempts to both lose weight and avoid gaining weight in the past three months for reasons related to body shape or weight		
Question 10: Do you pay close attention to the calorific value of food?	Yes	No		
Question 11: How many portions of sweets do you consume per week? (Chocolates (1 portion = 3 g) biscuits and similar)	<1 serving per week	2 servings per week	>2 servings per week	
Question 12: How many servings of red meat do you eat per week? (Beef, pork, lamb, 1 portion 100–150 g)	<1 serving per week	2 servings per week	>2 servings per week	
Question 13: How many servings of processed meat do you consume per week? (sausages and similar; 1 portion = 50 g)	<1 serving per week	2 servings per week	>2 servings per week	
Question 14: How many eggs do you eat per week? (1 serving = 1 egg)	<1 serving per week	2 servings per week	>2 servings per week	
Question 15: How many servings of legumes do you consume per week? (Lentils, beans, peas, chickpeas (1 serving = 1 plate or 150 g)	<1 serving per week	1 serving per week	2 servings per week	>2 servings per week
Question 16: How many portions of white meat do you eat per week? (Poultry, rabbit—1 serving = 150 g)	<1 serving per week	2 servings per week	>2 servings per week	
Question 17: How many servings of fish or seafood do you eat per week? (White/fatty fish (1 serving = 100–150 g), canned fish (1 serving = 1 can or 50 g), seafood (1 serving = 200 g))	<1 serving per week	2 servings per week	>2 servings per week	
Question 18: How many dairy products do you consume per day? (Milk—1 portion = 200 mL milk, two yogurts, 1 portion of spreadable cheese)	<1 serving per day	2 servings per day	>2 servings per day	
Question 19: How much fruit do you consume per day? (All fruit and fresh fruit juices—1 serving = 150–200 g)	<1 serving per day	1–2 servings per day	3–4 servings per day	>4 servings per day
Question 20: How many servings of vegetables do you consume per day? (1 portion = 150–200 g)	<1 serving per day	1–2 servings per day	>2 servings per day	
Question 21: How many tablespoons of olive oil do you consume per day (cooking or salad dressing)? (Olive oil, extra virgin olive oil—1 serving = 1 tablespoon)	<1 serving per day	1–2 servings per day	>2 servings per day	

	n (%)	*n* (%)	*n* (%)	*n* (%)
Question 22: How many servings of cereals do you consume per day? (White and wholemeal bread—1 serving = 40 g, half sandwich, cereals—1 serving = 1 plate of rice, pasta or 40 g of breakfast cereal)	<1 serving per day	1–2 servings per day	>2 servings per day	
Question 23: Do you drink more than 6 glasses of water or at least one cup of tea a day? (Water or tea—1 glass serving)	Yes	No		
Question 24: Do you drink wine every day with meals? (White/red wine (1 serving = 1 glass of wine))	<1 serving per day	1–2 servings per day	>2 servings per day	
Question 25: Do you limit adding salt to meals?	Yes	No		
Question 26: Do you usually use nibbling between meals? Do you consume snacks between meals?	Yes	No		
Question 27: Do you limit your sugar intake in drinks? (including sugary drinks)	Yes	No		
Question 28: How many hours do you spend watching TV a day?	<1 h per day	1 h a day	2 h a day	>2 h a day

References

1. Stonerock, G.L.; Blumenthal, J.A. Role of Counseling to Promote Adherence in Healthy Lifestyle Medicine: Strategies to Improve Exercise Adherence and Enhance Physical Activity. *Prog. Cardiovasc. Dis.* **2017**, *59*, 455–462. [CrossRef]
2. Pratt, M.; Perez, L.G.; Goenka, S.; Brownson, R.C.; Bauman, A.; Sarmiento, O.L.; Hallal, P.C. Can Population Levels of Physical Activity Be Increased? Global Evidence and Experience. *Prog. Cardiovasc. Dis.* **2015**, *57*, 356–367. [CrossRef] [PubMed]
3. Sallis, R.; Franklin, B.; Joy, L.; Ross, R.; Sabgir, D.; Stone, J. Strategies for Promoting Physical Activity in Clinical Practice. *Prog. Cardiovasc. Dis.* **2015**, *57*, 375–386. [CrossRef] [PubMed]
4. Mifsud, J.L.; Galea, J.; Garside, J.; Stephenson, J.; Astin, F. Motivational interviewing to support modifiable risk factor change in individuals at increased risk of cardiovascular disease: A systematic review and meta-analysis. *PloS ONE* **2020**, *15*, e0241193. [CrossRef] [PubMed]
5. McDonald, H.P.; Garg, A.X.; Haynes, R.B. Interventions to Enhance Patient Adherence to Medication Prescriptions: Scientific Review. *JAMA* **2002**, *288*, 2868–2879. [CrossRef]
6. Martin, L.R.; Williams, S.L.; Haskard, K.B.; Dimatteo, M.R. The Challenge of Patient Adherence. *Ther. Clin. Risk Manag.* **2005**, *1*, 189–199.
7. Krienke, R. Adherence to Medication. *N. Engl. J. Med.* **2005**, *353*, 1972–1974.
8. Carlson, S.A.; Fulton, J.E.; Pratt, M.; Yang, Z.; Adams, E.K. Inadequate Physical Activity and Health Care Expenditures in the United States. *Prog. Cardiovasc. Dis.* **2015**, *57*, 315–323. [CrossRef]
9. Cena, H.; Porri, D.; De Giuseppe, R.; Kalmpourtzidou, A.; Salvatore, F.P.; El Ghoch, M.; Itani, L.; Kreidieh, D.; Brytek-Matera, A.; Pocol, C.B.; et al. How Healthy Are Health-Related Behaviors in University Students: The HOLISTic Study. *Nutrients* **2021**, *13*, 675. [CrossRef]
10. Lyzwinski, L.N.; Caffery, L.; Bambling, M.; Edirippulige, S. The Mindfulness App Trial for Weight, Weight-Related Behaviors, and Stress in University Students: Randomized Controlled Trial. *JMIR mHealth uHealth* **2019**, *7*, e12210. [CrossRef]
11. Cena, H.; Tesone, A.; Niniano, R.; Cerveri, I.; Roggi, C.; Turconi, G. Prevalence Rate of Metabolic Syndrome in a Group of Light and Heavy Smokers. *Diabetol. Metab. Syndr.* **2013**, *5*, 28. [CrossRef] [PubMed]
12. Mertens, E.; Deriemaeker, P.; Van Beneden, K. Adjustments in Food Choices and Physical Activity during Lockdown by Flemish Adults. *Nutrients* **2021**, *13*, 3794. [CrossRef]
13. Haddad, C.; Hallit, R.; Akel, M.; Honein, K.; Akiki, M.; Kheir, N.; Obeid, S.; Hallit, S. Validation of the Arabic Version of the ORTO-15 Questionnaire in a Sample of the Lebanese Population. *Eat. Weight Disord.-Stud. Anorex. Bulim. Obes.* **2020**, *25*, 951–960. [CrossRef] [PubMed]
14. Hemmingsson, E.; Ekelund, U. Is the Association between Physical Activity and Body Mass Index Obesity Dependent? *Int. J. Obes.* **2007**, *31*, 663–668. [CrossRef] [PubMed]

15. Li, Y.; Li, G.; Liu, L.; Wu, H. Correlations between Mobile Phone Addiction and Anxiety, Depression, Impulsivity, and Poor Sleep Quality among College Students: A Systematic Review and Meta-Analysis. *J. Behav. Addict.* **2020**, *9*, 551–571. [CrossRef]
16. Taheri, S.; Lin, L.; Austin, D.; Young, T.; Mignot, E. Short Sleep Duration Is Associated with Reduced Leptin, Elevated Ghrelin, and Increased Body Mass Index. *PLoS Med.* **2004**, *1*, e62. [CrossRef]
17. Remskar, M.; Western, M.J.; Maynard, O.M.; Ainsworth, B. Exercising body but not mind: A qualitative exploration of attitudes to combining physical activity and mindfulness practice for mental health promotion. *Front. Psychol.* **2022**, *13*, 984232. [CrossRef]
18. Zaccagni, L.; Barbieri, D.; Gualdi-Russo, E. Body Composition and Physical Activity in Italian University Students. *J. Transl. Med.* **2014**, *12*, 120. [CrossRef]
19. Sacheck, J.M.; Kuder, J.F.; Economos, C.D. Physical Fitness, Adiposity, and Metabolic Risk Factors in Young College Students. *Med. Sci. Sports Exerc.* **2010**, *42*, 1039–1044. [CrossRef]
20. Wagner, A.; Dallongeville, J.; Haas, B.; Ruidavets, J.B.; Amouyel, P.; Ferrières, J.; Simon, C.; Arveiler, D. Sedentary Behaviour, Physical Activity and Dietary Patterns Are Independently Associated with the Metabolic Syndrome. *Diabetes Metab.* **2012**, *38*, 428–435. [CrossRef]
21. Garber, C.E.; Blissmer, B.; Deschenes, M.R.; Franklin, B.A.; Lamonte, M.J.; Lee, I.M.; Nieman, D.C.; Swain, D.P. Quantity and Quality of Exercise for Developing and Maintaining Cardiorespiratory, Musculoskeletal, and Neuromotor Fitness in Apparently Healthy Adults: Guidance for Prescribing Exercise. *Med. Sci. Sports Exerc.* **2011**, *43*, 1334–1359. [CrossRef] [PubMed]
22. WHO. *Global Recommendations on Physical Activity for Health*; WHO: Geneva, Switzerland, 2010; Volume 60. [CrossRef]
23. Guthold, R.; Stevens, G.A.; Riley, L.M.; Bull, F.C. Worldwide Trends in Insufficient Physical Activity from 2001 to 2016: A Pooled Analysis of 358 Population-Based Surveys with 1·9 Million Participants. *Lancet. Glob. Health* **2018**, *6*, e1077–e1086. [CrossRef]
24. Hilger, J.; Loerbroks, A.; Diehl, K. Eating Behaviour of University Students in Germany: Dietary Intake, Barriers to Healthy Eating and Changes in Eating Behaviour since the Time of Matriculation. *Appetite* **2017**, *109*, 100–107. [CrossRef] [PubMed]
25. Beard, J.R.; Officer, A.; de Carvalho, I.A.; Sadana, R.; Pot, A.M.; Michel, J.-P.; Lloyd-Sherlock, P.; Epping-Jordan, J.E.; Peeters, G.M.E.E.G.; Mahanani, W.R.; et al. The World Report on Ageing and Health: A Policy Framework for Healthy Ageing. *Lancet* **2016**, *387*, 2145–2154. [CrossRef] [PubMed]
26. Mensink, G.B.M.; Schienkiewitz, A.; Haftenberger, M.; Lampert, T.; Ziese, T.; Scheidt-Nave, C. Overweight and Obesity in Germany: Results of the German Health Interview and Examination Survey for Adults (DEGS1). *Bundesgesundheitsblatt. Gesundheitsforschung. Gesundh.* **2013**, *56*, 786–794. [CrossRef]
27. Nelson Laska, M.; Larson, N.I.; Neumark-Sztainer, D.; Story, M. Dietary Patterns and Home Food Availability during Emerging Adulthood: Do They Differ by Living Situation? *Public Health Nutr.* **2010**, *13*, 222–228. [CrossRef]
28. Lupi, S.; Bagordo, F.; Stefanati, A.; Grassi, T.; Piccinni, L.; Bergamini, M.; De Donno, A. Assessment of Lifestyle and Eating Habits among Undergraduate Students in Northern Italy. *Ann. Ist. Super. Sanita* **2015**, *51*, 154–161.
29. Cluskey, M.; Grobe, D. College Weight Gain and Behavior Transitions: Male and Female Differences. *J. Am. Diet. Assoc.* **2009**, *109*, 325–329. [CrossRef]
30. Driskell, J.A.; Kim, Y.-N.; Goebel, K.J. Few Differences Found in the Typical Eating and Physical Activity Habits of Lower-Level and Upper-Level University Students. *J. Am. Diet. Assoc.* **2005**, *105*, 798–801. [CrossRef]
31. Nicklas, T.A.; Baranowski, T.; Cullen, K.W.; Berenson, G. Eating Patterns, Dietary Quality and Obesity. *J. Am. Coll. Nutr.* **2001**, *20*, 599–608. [CrossRef]
32. Debate, R.D.; Topping, M.; Sargent, R.G. Racial and Gender Differences in Weight Status and Dietary Practices among College Students. *Adolescence* **2001**, *36*, 819–833. [PubMed]
33. Stefanikova, Z.; Sevcikova, L.; Jurkovicova, J.; Sobotova, L.; Aghova, L. Positive and Negative Trends in University Students' Food Intake. *Bratisl. Lek. Listy* **2006**, *107*, 217–220. [PubMed]
34. Colić Barić, I.; Satalić, Z.; Lukesić, Z. Nutritive Value of Meals, Dietary Habits and Nutritive Status in Croatian University Students According to Gender. *Int. J. Food Sci. Nutr.* **2003**, *54*, 473–484. [CrossRef] [PubMed]
35. Hercberg, S.; Preziosi, P.; Galan, P.; Deheeger, M.; Papoz, L.; Dupin, H. Dietary Intake of a Representative Sample of the Population of Val-de-Marne; III. Mineral and Vitamin Intake. *Rev. Epidemiol. Sante Publique* **1991**, *39*, 245–261.
36. Skemiene, L.; Ustinaviciene, R.; Piesine, L.; Radisauskas, R. Peculiarities of Medical Students' Nutrition. *Medicina* **2007**, *43*, 145–152.
37. Huang, T.T.K.; Harris, K.J.; Lee, R.E.; Nazir, N.; Born, W.; Kaur, H. Assessing Overweight, Obesity, Diet, and Physical Activity in College Students. *J. Am. Coll. Health* **2003**, *52*, 83–86. [CrossRef]
38. Anding, J.D.; Suminski, R.R.; Boss, L. Dietary Intake, Body Mass Index, Exercise, and Alcohol: Are College Women Following the Dietary Guidelines for Americans? *J Am. Coll. Health* **2001**, *49*, 167–171. [CrossRef]
39. Mammas, I.; Bertsias, G.; Linardakis, M.; Moschandreas, J.; Kafatos, A. Nutrient Intake and Food Consumption among Medical Students in Greece Assessed during a Clinical Nutrition Course. *Int. J. Food Sci. Nutr.* **2004**, *55*, 17–26. [CrossRef]
40. Racette, S.B.; Deusinger, S.S.; Strube, M.J.; Highstein, G.R.; Deusinger, R.H. Weight Changes, Exercise, and Dietary Patterns during Freshman and Sophomore Years of College. *J. Am. Coll. Health* **2005**, *53*, 245–251. [CrossRef]
41. Di Palma, D.; Tafuri, D. Special needs and inclusion in sport management: A specific literature review | Posebne potrebe i uključenje u sportskom menadžmentu: Specifičan pregled literature. *Sport Sci.* **2016**, *9*, 24–31.
42. Mazzeo, F.; Santamaria, S.; Monda, V.; Messina, G.; Monda, M. Dietary supplements use in competitive and non-competitive boxer: An exploratory study. *Biol. Med.* **2016**, *8*, 1000294.

43. Gaetano, R.; Domenico, T.; Gaetano, A. Physical activity and its relation to body and ludic expression in childhood Mediterranean. *J. Soc. Sci.* **2015**, *6*, 293–296.
44. Monda, M.; Viggiano, A.; Vicidomini, C.; Tafuri, D.; De Luca, B. Expresso coffee increases parasympathetic activity in young, healthy people. *Nutr. Neurosci.* **2009**, *12*, 43–48. [CrossRef] [PubMed]
45. Sirangelo, I.; Borriello, M.; Liccardo, M.; Scafuro, M.; Russo, P.; Iannuzzi, C. Hydroxytyrosol Selectively Affects Non-Enzymatic Glycation in Human Insulin and Protects by AGEs Cytotoxicity. *Antioxidants* **2021**, *10*, 1127. [CrossRef] [PubMed]
46. Kresić, G.; Kendel Jovanović, G.; Pavicić Zezel, S.; Cvijanović, O.; Ivezić, G. The Effect of Nutrition Knowledge on Dietary Intake among Croatian University Students. *Coll. Antropol.* **2009**, *33*, 1047–1056.
47. Arroyo Izaga, M.; Rocandio Pablo, A.M.; Ansotegui Alday, L.; Pascual Apalauza, E.; Salces Beti, I.; Rebato Ochoa, E. Diet Quality, Overweight and Obesity in University Students. *Nutr. Hosp.* **2006**, *21*, 673–679.
48. Zykova, S.N.; Storhaug, H.M.; Toft, I.; Chadban, S.J.; Jenssen, T.G.; White, S.L. Cross-Sectional Analysis of Nutrition and Serum Uric Acid in Two Caucasian Cohorts: The AusDiab Study and the Tromsø Study. *Nutr. J.* **2015**, *14*, 49. [CrossRef]
49. Pan, Y.L.; Dixon, Z.; Himburg, S.; Huffman, F. Asian Students Change Their Eating Patterns after Living in the United States. *J. Am. Diet. Assoc.* **1999**, *99*, 54–57. [CrossRef]

nutrients

MDPI

Article

Causal Associations between Dietary Habits and Chronic Pain: A Two-Sample Mendelian Randomization Study

Ren Zhou, Lei Zhang, Yu Sun, Jia Yan * and Hong Jiang *

Department of Anesthesiology, The Ninth People's Hospital of Shanghai, Jiao Tong University School of Medicine, Shanghai 200011, China; zhouren77@126.com (R.Z.); weiymzhl@126.com (L.Z.); dr_sunyu@163.com (Y.S.)
* Correspondence: mzkyanj@163.com (J.Y.); jianghongjiuyuan@163.com (H.J.)

Abstract: Chronic pain is a prevalent and debilitating condition with significant impacts on individuals and society. While the role of diet in chronic pain is well-known, the relationship between special dietary choices and chronic pain remains unclear. This study investigates the causal associations between 20 dietary habits and chronic pain using a Mendelian randomization (MR) approach. Publicly available genome-wide association study data from the UK Biobank dataset were utilized for secondary analysis, and genetic instrumental variables strongly correlated with 20 different dietary habits were selected. Multisite chronic pain (MCP) scores were used as the primary outcome, with site-specific chronic pain (SSCP) including back pain, headache, knee pain, neck pain, and hip pain as secondary outcomes. The inverse-variance-weighted (IVW) method was the primary method used in the MR. The weighted median (WM) and Mendelian randomization pleiotropy residual sum and outlier test (MR-PRESSO) methods were used as sensitivity analyses. This study identified causal associations between specific dietary habits and chronic pain. A high intake of cheese, cereal, dried fruits, and fresh fruits was associated with lower MCP scores. Conversely, high alcohol, salt, pork, and poultry intakes were associated with higher MCP scores. Similar associations between special dietary habits and some types of SSCP, such as back and neck pain, were also observed. The findings were consistent across different statistical methods, and sensitivity analyses confirmed the reliability of the results. In conclusion, our study provides evidence of a causal relationship between various dietary habits and different types of chronic pain based on secondary analysis of the UK Biobank dataset. Adhering to an anti-inflammatory diet, including increased consumption of fruits and cereal while reducing salt and pork intake, may potentially alleviate chronic pain symptoms.

Keywords: chronic pain; dietary habits; Mendelian randomization; multisite chronic pain

Citation: Zhou, R.; Zhang, L.; Sun, Y.; Yan, J.; Jiang, H. Causal Associations between Dietary Habits and Chronic Pain: A Two-Sample Mendelian Randomization Study. *Nutrients* **2023**, *15*, 3709. https://doi.org/10.3390/nu15173709

Academic Editors: Domenico Tafuri, Francesca Latino and Carmen Mannucci

Received: 16 June 2023
Revised: 11 August 2023
Accepted: 17 August 2023
Published: 24 August 2023

1. Introduction

Chronic pain (CP) is a complex condition that can arise from various causes and is characterized by persistent or recurring pain lasting for over three months [1,2]. It significantly impacts an individual's physical and emotional well-being. Globally, approximately 30% of the population experiences CP, with an average of three reported pain locations [3]. In the United States, more than 50 million adults (20.5%) suffer from pain on most days [4]. The burden of CP on society and health systems continues to increase, especially with an aging population. While the exact mechanisms underlying CP are not fully understood, certain risk factors such as smoking, being overweight, and lifestyle choices have been identified [5,6].

The role of dietary habits in overall health and well-being is crucial. A balanced and nutritious diet has been associated with numerous health benefits, including weight management, disease prevention, and improved immune function [7,8]. Conversely, poor dietary choices characterized by excessive consumption of processed foods, added salts, and unhealthy fats have been linked to an increased risk of chronic conditions such as obesity, diabetes, cardiovascular diseases, and certain types of cancer [9–12]. Moreover, studies have suggested that an unhealthy diet may contribute to systemic inflammation, oxidative stress [13], and impaired gut health [14], all of which are implicated in the development and maintenance of chronic pain [15,16]. Therefore, investigating the relationship between dietary habits and CP is essential for a comprehensive understanding of how food choices impact overall health outcomes.

Dietary behavior and intake have the potential to influence the occurrence, maintenance, and perception of chronic pain. Unhealthy dietary behavior, often associated with overweight and obesity, is a primary contributing factor to adverse nutritional status in chronic pain patients [17,18]. Furthermore, studies have explored the connection between specific nutrients and CP, such as the correlation between chronic lower back pain or rheumatoid arthritis and diminished intake of vitamin D [19], alkaline minerals [20], and omega-3 polyunsaturated fatty acids [21]. Certain dietary habits, such as the consumption of dried fruits, have also been causally associated with chronic pain [22]. However, systematic studies focusing specifically on the examination of dietary habits in the context of CP are lacking.

In this study, we used Mendelian randomization to assess the effects of 20 dietary habits on chronic pain, including multi-site chronic pain (MCP), chronic back pain (BP), chronic headache (Headache), chronic knee pain (KP), chronic neck/shoulder pain (NP), and chronic hip pain (HP). Our aim is to provide insights into the role of dietary habits in CP prevention and guide recommendations regarding food source quality.

2. Method

2.1. Study Design

This study is a secondary analysis of the UK biobank dataset. We used a two-sample Mendelian randomization (MR) approach to investigate the causal relationships between dietary habits and CP. Figure 1 provides a schematic overview of the study design and data sources. Figure 2 provides a summary experimental protocol of our study. We exclusively used publicly available genome-wide association study (GWAS) summary statistics, eliminating the need for additional ethical approval or informed consent.

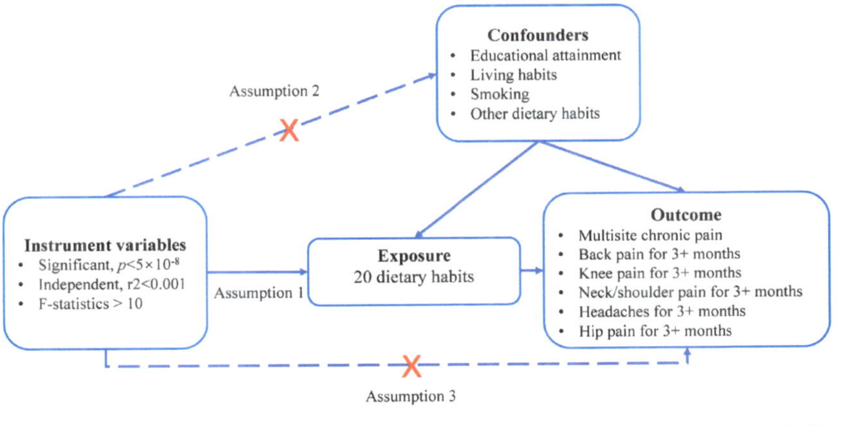

Figure 1. A directed acyclic graph is used to illustrate the hypothetical impact of dietary habits on chronic pain. The presence of a dotted line indicates a potential direct causal relationship or pleiotropic effect between the exposure and the outcome.

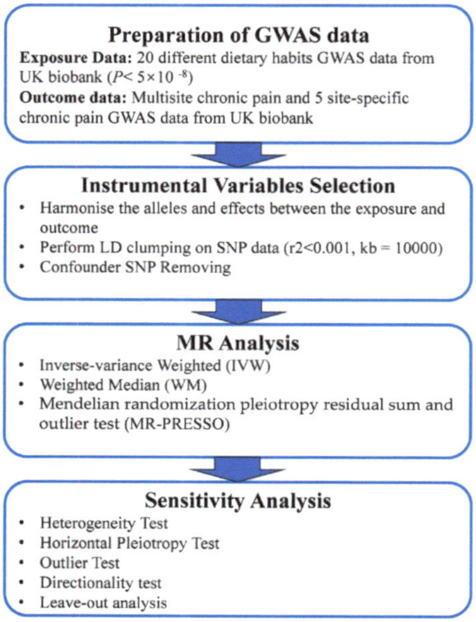

Figure 2. Flowchart of the MR analysis.

2.2. Exposure Data

In this study, we considered 20 different dietary habits as exposures based on previous studies [23,24], including beef intake, bread intake, cereal intake, cheese intake, coffee intake, cooked vegetable intake, dried fruit intake, fresh fruit intake, lamb/mutton intake, non-oily fish intake, oily fish intake, pork intake, poultry intake, processed meat intake, salad/raw vegetable intake, salt added to food, water intake, hot drink temperature, tea intake, and alcohol intake frequency. We obtained summary data from the GWAS conducted by the UK Biobank, a large prospective study comprising approximately 500,000 participants aged between 38 years and 73 years who provided their genetic and phenotypic information and were recruited between the years 2006 and 2010 from across the UK [25]. A full description of the study design, participants, and quality control (QC) methods has previously been described in detail [26]. The data for each dietary pattern consisted of integer variables, such as, on average, the number of cups of coffee per day, or categorical variables, such as how often one eats poultry. Unreasonable responses were excluded during data submission. The dietary habit assessment questions and specific definition of units (e.g., one tablespoon or one cup) used in our study can be found in the publicly available information on the UK Biobank website (http://biobank.ctsu.ox.ac.uk/crystal/label.cgi?id=100052, accessed on 10 August 2023, https://biobank.ctsu.ox.ac.uk/crystal/field.cgi?id=1558, accessed on 10 August 2023). The content of the specific questionnaires can be seen in Table S1. Summary GWAS data for dietary habits were collected from OpenGWAS, conducted by Elsworth et al., at MRC-IEU, and are maintained by MRC-IEU. Information on the collection, cleaning, and analysis of these raw GWAS data can be found in the original documentation at MRC-IEU.

2.3. Outcome Data

Our primary outcome was MCP, for which we obtained GWAS data from a large study conducted by Johnston et al. [27] involving 387,649 European individuals. MCP is defined as self-reported pain lasting at least three months in seven distinct bodily regions (head, face, neck/shoulder, back, stomach/abdomen, hip, and knee).

Secondary outcomes included specific site-specific chronic pain (SSCP) lasting more than three months. Summary GWAS data for SSCP were collected from OpenGWAS, by

Elsworth et al., at MRC-IEU, and maintained by MRC-IEU. Information on the collection, cleaning, and analysis of this raw GWAS data can be found in the original documentation at MRC-IEU. QC files are also included in the documentation. The GWAS data related to headache, chronic hip pain, chronic neck/shoulder pain, chronic back pain, and chronic knee pain were included in our study. The CP assessment questions used in our study can be found in publicly available information from the UK Biobank website (https://biobank.ctsu.ox.ac.uk/crystal/label.cgi?id=154, accessed on 10 August 2023).

2.4. Selection of Instrumental Variables

To explore the causal impact of dietary habits on chronic pain (CP), we employed a two-sample MR design (see Figure 1). We established five criteria to screen single-nucleotide polymorphisms (SNPs) used as instrumental variables:

(a) We selected SNPs significantly correlated with dietary habits ($p < 5 \times 10^{-8}$), ensuring their independence from each other (r2 < 0.001) within a clumping distance of 10,000 kb;
(b) SNPs associated with the outcomes of interest ($p < 5 \times 10^{-8}$) were excluded from the instrumental variables;
(c) SNPs associated with potential confounding factors, such as educational attainment, living habits (e.g., usual walking pace, time spent watching television), smoking, and other dietary habits, were excluded from all analyses. The relevant information was obtained from the PhenoScanner database V2 (http://www.phenoscanner.medschl.cam.ac.uk/, accessed on 10 June 2023);
(d) To ensure a strong correlation between instrumental variables and exposure factors, we verified that the F-statistic of each SNP was larger than 10 using the formula $F = \beta^2/SE_\beta{}^2$;
(e) Palindromic SNPs with intermediate allele frequencies were excluded from the analysis to maintain consistency between the effects of SNPs on exposures and outcomes.

Supplementary Tables S2–S5 provide detailed information on the selected SNPs.

2.5. Statistical Analysis

All statistical analyses were performed using R version 4.0.5 (R Foundation for Statistical Computing, Vienna, Austria) and using the TwoSampleMR and MRPRESSO R packages. When SNPs were unavailable in the outcome summary data, we used proxy SNPs with an r2 $\geq$ 0.8 in the highest linkage disequilibrium. We harmonized the exposure and outcome data and aligned the effect allele in the exposure and outcome GWAS using the TwoSampleMR package.

The primary Mendelian randomization (MR) analysis relied on the inverse-variance-weighted (IVW) method. This approach employs a meta-analysis technique to combine the Wald ratios of individual SNPs. It assumes that instrumental variables (IVs) solely affect outcomes through specific exposures. In the absence of horizontal pleiotropy, the IVW method provides reliable results. The IVW method is commonly employed as a major method in MR studies. To ensure the robustness of our findings, we conducted several complementary analyses using the weighted median (WM) method and Mendelian randomization pleiotropy residual sum and outlier test (MR-PRESSO) methods [28] in subsequent sensitivity analyses. The IVW method only gives consistent estimates if all genetic variants are included. Additionally, the weighted median requires 50% of the weight to come from valid IVs. The MR-PRESSO analysis detects and attempts to reduce horizontal pleiotropy by removing significant outliers. Data are presented as β, standard errors (SE) of β, and p-values. A positive β means that an increased intake of a certain food is positively associated with pain, while a negative β means that increased intake of certain food is negatively associated with pain. In our study, we only considered a dietary habit to be associated with a specific CP when significant results were obtained for IVW and the results for WM and MR-PRESSO showed the same trend.

Heterogeneity in MR analysis refers to inconsistencies in the estimates derived from different instrumental variables, reflecting the compatibility of the instrumental variables with the causal inference being made. We calculated the Cochran Q statistic to quantify the heterogeneity of causal estimates generated from both the IVW and MR-Egger regression methods. A p-level > 0.05 means there is no significant heterogeneity detected in this analysis. The Mendelian randomization pleiotropy residual sum and outlier test (MR-PRESSO) [28] was also used to detect horizontal pleiotropic outliers in the multi-instrument MR study and to correct for horizontal pleiotropy via outlier removal.

We used the Steiger directionality test to confirm the causal direction between dietary habits and CP; a p-level <0.05 meant the causal direction was correct in this analysis. To validate the reliability of the estimates, we performed a leave-one-out sensitivity analysis by recalculating the overall effect size and removing each SNP one at a time until reaching significant results for the primary outcome.

For our primary outcome, we adjusted the p-values using the false detection rate (FDR) [29]. We considered associations with an FDR of less than 0.05 as significant. However, for the secondary outcome, we considered both the raw p-value and the FDR in our analysis.

3. Results

3.1. Genetic Instruments for 20 Dietary Habits

Table 1 provides comprehensive information about each participating GWAS study. The analyses included a total of 20 different dietary habits as exposures. The number of single nucleotide polymorphisms (SNPs) considered for each dietary habit varied between 3 and 87. Supplementary Tables S2–S5 contain detailed information regarding the instrumental variables (IVs) employed for these 20 dietary habits. We assessed the impact of various dietary exposures on the outcome, and, in all cases, the F statistics of the identified SNPs surpassed the empirical threshold of 10, ranging from 29.74 to 811.86. This finding suggests that the obtained results are less susceptible to deviations caused by weak IVs.

Table 1. Detailed information on the GWAS datasets used in this MR study.

ID or PMID	Food Types	Trait	Sample Size	Case	Control	SNPs (N)	Consortium (Author)	Population	Category	Year
ukb-b-2862	Meat and Poultry	Beef intake	461,053	NA	NA	9,851,867	MRC-IEU (Elsworth et al.)	European	Categorical (single)	2018
ukb-b-11348	Bread	Bread intake	452,236	NA	NA	9,851,867	MRC-IEU (Elsworth et al.)	European	Integer, slices/week	2018
ukb-b-15926	Grains, Nuts, and Seeds	Cereal intake	441,640	NA	NA	9,851,867	MRC-IEU (Elsworth et al.)	European	Integer, bowls/week	2018
ukb-b-1489	Dairy Products	Cheese intake	451,486	NA	NA	9,851,867	MRC-IEU (Elsworth et al.)	European	Categorical (single)	2018
ukb-b-5237	Drinks	Coffee intake	428,860	NA	NA	9,851,867	MRC-IEU (Elsworth et al.)	European	Integer, cups/day	2018
ukb-b-8089	Vegetable	Cooked vegetable intake	448,651	NA	NA	9,851,867	MRC-IEU (Elsworth et al.)	European	Integer, tablespoons/day	2018
ukb-b-16576	Fruit	Dried fruit intake	421,764	NA	NA	9,851,867	MRC-IEU (Elsworth et al.)	European	Integer, pieces/day	2018
ukb-b-3881	Fruit	Fresh fruit intake	446,462	NA	NA	9,851,867	MRC-IEU (Elsworth et al.)	European	Integer, pieces/day	2018
ukb-b-14179	Meat and Poultry	Lamb/mutton intake	460,006	NA	NA	9,851,867	MRC-IEU (Elsworth et al.)	European	Categorical (single)	2018
ukb-b-17627	Seafood	Non-oily fish intake	460,880	NA	NA	9,851,867	MRC-IEU (Elsworth et al.)	European	Categorical (single)	2018
ukb-b-2209	Seafood	Oily fish intake	460,443	NA	NA	9,851,867	MRC-IEU (Elsworth et al.)	European	Categorical (single)	2018
ukb-b-5640	Meat and Poultry	Pork intake	460,162	NA	NA	9,851,867	MRC-IEU (Elsworth et al.)	European	Categorical (single)	2018
ukb-b-8006	Meat and Poultry	Poultry intake	461,900	NA	NA	9,851,867	MRC-IEU (Elsworth et al.)	European	Categorical (single)	2018
ukb-b-6324	Meat and Poultry	Processed meat intake	461,981	NA	NA	9,851,867	MRC-IEU (Elsworth et al.)	European	Categorical (single)	2018
ukb-b-1996	Vegetable	Salad/raw vegetable intake	435,435	NA	NA	9,851,867	MRC-IEU (Elsworth et al.)	European	Integer, tablespoons/day	2018
ukb-b-8121	Food additive	Salt added to food	462,630	NA	NA	9,851,867	MRC-IEU (Elsworth et al.)	European	Categorical (single)	2018

Table 1. *Cont.*

ID or PMID	Food Types	Trait	Sample Size	Case	Control	SNPs (N)	Consortium (Author)	Population	Category	Year
ukb-b-14898	Drinks	Water intake	427,588	NA	NA	9,851,867	MRC-IEU (Elsworth et al.)	European	Integer, glasses/day	2018
ukb-b-14203	-	Hot drink temperature	457,873	NA	NA	9,851,867	MRC-IEU (Elsworth et al.)	European	Categorical (single)	2018
ukb-b-6066	Drinks	Tea intake	447,485	NA	NA	9,851,867	MRC-IEU (Elsworth et al.)	European	Integer, cups/day	2018
ukb-b-5779	Drinks	Alcohol intake frequency.	462,346	NA	NA	9,851,867	MRC-IEU (Elsworth et al.)	European	Categorical (single)	2018
31194737	-	Multisite chronic pain	387,649	NA	NA	9,926,106	Johnston et al. [27]	European	Categorical Ordered	2019
ukb-b-8463	-	Back pain for 3+ months	117,404	80,588	36,816	9,851,867	MRC-IEU (Elsworth et al.)	European	Binary	2018
ukb-b-8906	-	Knee pain for 3+ months	97,889	76,910	20,979	9,851,867	MRC-IEU (Elsworth et al.)	European	Binary	2018
ukb-b-16118	-	Neck/shoulder pain for 3+ months	105,396	72,887	32,509	9,851,867	MRC-IEU (Elsworth et al.)	European	Binary	2018
ukb-b-13092	-	Headache for 3+ months	91,269	41,719	49,550	9,851,867	MRC-IEU (Elsworth et al.)	European	Binary	2018
ukb-b-133	-	Hip pain for 3+ months	51,516	40,152	11,364	9,851,867	MRC-IEU (Elsworth et al.)	European	Binary	2018

3.2. Causal Effects of Dietary Habits on CP

The summary of causal effects of dietary habits on different types of CP are shown in Figure 3. Therein, 8 causal associations between 20 dietary habits and MCP were identified based on the IVW method with adjusted p-values (Table 2). Among them, cheese intake ($\beta = -0.12$, $SE_\beta = 0.04$, $p_{Raw} = 0.002$, $p_{Adjusted} = 0.008$), cereal intake ($\beta = -0.17$, $SE_\beta = 0.06$, $p_{Raw} = 0.004$, $p_{Adjusted} = 0.015$), and high intake of dried fruit ($\beta = -0.22$, $SE_\beta = 0.07$, $p_{Raw} = 0.001$, $p_{Adjusted} = 0.008$) and fresh fruit ($\beta = -0.17$, $SE_\beta = 0.07$, $p_{Raw} = 0.017$, $p_{Adjusted} = 0.041$) reduced MCP scores. In contrast, high alcohol intake ($\beta = 0.22$, $SE_\beta = 0.07$, $P_{Raw} = 0.001$, $P_{Adjusted} = 0.008$), salt intake ($\beta = 0.10$, $SE_\beta = 0.03$, $p_{Raw} = 0.006$, $p_{Adjusted} = 0.016$), pork intake ($\beta = 0.36$, $SE_\beta = 0.12$, $p_{Raw} = 0.003$, $p_{Adjusted} = 0.013$), and poultry intake ($\beta = 0.73$, $SE_\beta = 0.14$, $p_{Raw} < 0.001$, $p_{Adjusted} < 0.001$) increased MCP scores. Further, our study found an association between dietary habits with different types of CP. Six dietary habits were associated with NP and four of them were still significant after adjusting the p-value (Table 3) with the IVW method. Five dietary habits were associated with BP and one was still significant after adjusting the p-value (Table 4) with the IVW method. Based on the IVW method, HP, KP, and headaches were associated with three, one, and one dietary habits, respectively. However, none of these was still significant after the adjusted p-value was applied (Supplementary Tables S6–S8). The results of the WM method and PRESSO method mostly showed a similar trend with those of the IVW method, though some p-values were slightly above 0.05 (Supplementary Tables S9–S20). We also constructed scatter plots to visualize the main results (Figure 4).

There was no evidence for potential pleiotropy in our study. However, a widespread heterogeneity effect was shown in our study. Thus, we used the random effect IVW as the major analysis method. For those dietary habits that did not show a significant heterogeneity effect in MR analysis, the fixed effect IVW was used. The MR-PRESSO global test showed no significant outliers in some results; however, all those results remained the same after excluding the outlier SNPs (Supplementary Tables S21–S26). The Steiger directionality test confirmed all the causal directions of dietary habits and CP (Supplementary Tables S21–S26).

In our sensitivity analysis of the primary outcome, we conducted a leave-one-out analysis, excluding one significant SNP each time. For all the significant associations with MCP, the leave-one out estimates were also statistically significant (Figure 5). Funnel plots also proved the credibility of the results of the IVW method (Figure S1).

Exposure	MCP	Knee	Back	Head	Neck	Hip
Beef intake	-	-	-	-	-	-
Bread intake	-	-	-	-	-	-
Cereal intake	↓ ↓	-	↓	-	↓	-
Cheese intake	↓ ↓	-	↓	-	-	↓
Coffee intake	-	-	-	-	-	-
Cooked vegetable intake	-	-	-	-	-	-
Dried fruit intake	↓ ↓	-	-	-	-	-
Fresh fruit intake	↓ ↓	-	-	-		-
Lamb/mutton intake	-	↓	-	-	↓ ↓	↓
Non-oily fish intake	-	-	-	-	-	-
Oily fish intake	-	-	-	-	↓ ↓	↓
Pork intake	↑ ↑	-	↑ ↑	-	-	-
Poultry intake	↑ ↑	-	↑	-	↑ ↑	-
Processed meat intake	-	-	-	-	-	-
Salad / raw vegetable intake	-	-	-	-	-	-
Salt added to food	↑ ↑	-	-	-	↑	-
Water intake	-	-	-	-	-	-
Hot drink temperature	-	-	-	-	-	-
Tea intake	-	-	-	↓	-	-
Alcohol intake frequency	↑ ↑	-	↑	-	↑ ↑	-

↑ means a significant positive association with dietary habits and Pain only in raw-*p*.
↑ ↑ means a significant positive association with dietary habits and Pain in adjusted-*p*.
↓ means a significant negative association with dietary habits and Pain only in raw-*p*.
↓ ↓ means a significant negative association with dietary habits and Pain in adjusted-*p*.
- means no significant association with dietary habits and Pain

Figure 3. Summary results of casual association between dietary habits and chronic pain according to the inverse-variance-weighted method.

Table 2. MR results of the IVW method for the association of dietary habits with multi-site chronic pain.

Exposure	Number of SNPs	β	$SE_β$	*p*-Value	Adjusted *p*-Value
Beef intake	13	0.10	0.11	0.389	0.648
Bread intake	26	−0.01	0.07	0.882	0.936
Cereal intake	30	−0.17	0.06	**0.004**	**0.015**
Cheese intake	46	−0.12	0.04	**0.002**	**0.008**
Coffee intake	29	−0.04	0.06	0.554	0.791
Cooked vegetable intake	11	0.11	0.11	0.294	0.534
Dried fruit intake	29	−0.22	0.07	**0.001**	**0.008**
Fresh fruit intake	39	−0.17	0.07	**0.017**	**0.041**
Lamb/mutton intake	29	−0.06	0.08	0.479	0.737
Non-oily fish intake	6	−0.02	0.22	0.924	0.936
Oily fish intake	50	−0.06	0.05	0.214	0.428
Pork intake	11	0.36	0.12	**0.003**	**0.013**
Poultry intake	3	0.73	0.14	**<0.001**	**<0.001**
Processed meat intake	18	0.01	0.08	0.936	0.936
Salad/raw vegetable intake	14	−0.16	0.11	0.158	0.351
Salt added to food	87	0.10	0.03	**0.006**	**0.016**
Water intake	33	−0.01	0.06	0.807	0.936
Hot drink temperature	57	0.02	0.07	0.777	0.936
Tea intake	31	−0.03	0.07	0.685	0.914
Alcohol intake frequency	14	0.22	0.07	**0.001**	**0.008**

MR, Mendelian randomization; IVW, Inverse-variance-weighted. Bold indicated a *p* value less than 0.05.

Table 3. MR results of IVW method for association of dietary habits with chronic neck/shoulder pain.

Exposure	Number of SNPs	β	$SE_β$	*p*-Value	Adjusted *p*-Value
Beef intake	13	0.08	0.05	0.09	0.301
Bread intake	27	−0.01	0.03	0.795	0.837
Cereal intake	30	−0.08	0.04	**0.037**	0.148
Cheese intake	46	−0.06	0.02	**0.007**	0.069
Coffee intake	29	−0.02	0.03	0.607	0.828
Cooked vegetable intake	11	0.06	0.06	0.34	0.617
Dried fruit intake	29	−0.05	0.03	0.14	0.354
Fresh fruit intake	39	−0.04	0.04	0.274	0.61
Lamb/mutton intake	29	0.01	0.04	0.749	0.837
Non-oily fish intake	6	0.11	0.08	0.142	0.354
Oily fish intake	50	−0.02	0.02	0.329	0.617
Pork intake	11	0.23	0.06	**<0.001**	**0.01**
Poultry intake	3	0.23	0.11	**0.037**	0.148
Processed meat intake	18	−0.02	0.04	0.662	0.828
Salad/raw vegetable intake	14	0.02	0.06	0.779	0.837
Salt added to food	87	0.01	0.02	0.647	0.828
Water intake	34	−0.02	0.03	0.55	0.828
Hot drink temperature	57	0	0.04	0.986	0.986
Tea intake	32	−0.02	0.02	0.449	0.748
Alcohol intake frequency	15	0.06	0.03	**0.023**	0.148

MR, Mendelian randomization; IVW, Inverse-variance-weighted. Bold indicated a *p* value less than 0.05.

Table 4. MR results of IVW method for association of dietary habits with chronic back pain.

Exposure	Number of SNPs	β	$SE_β$	*p*-Value	Adjusted *p*-Value
Beef intake	13	0.08	0.05	0.09	0.301
Bread intake	27	−0.01	0.03	0.795	0.837
Cereal intake	30	−0.08	0.04	**0.037**	0.148
Cheese intake	46	−0.06	0.02	**0.007**	0.069
Coffee intake	29	−0.02	0.03	0.607	0.828
Cooked vegetable intake	11	0.06	0.06	0.34	0.617

Table 4. *Cont.*

Exposure	Number of SNPs	β	$SE_β$	*p*-Value	Adjusted *p*-Value
Dried fruit intake	29	−0.05	0.03	0.14	0.354
Fresh fruit intake	39	−0.04	0.04	0.274	0.61
Lamb/mutton intake	29	0.01	0.04	0.749	0.837
Non-oily fish intake	6	0.11	0.08	0.142	0.354
Oily fish intake	50	−0.02	0.02	0.329	0.617
Pork intake	11	0.23	0.06	**<0.001**	**0.01**
Poultry intake	3	0.23	0.11	**0.037**	0.148
Processed meat intake	18	−0.02	0.04	0.662	0.828
Salad/raw vegetable intake	14	0.02	0.06	0.779	0.837
Salt added to food	87	0.01	0.02	0.647	0.828
Water intake	34	−0.02	0.03	0.55	0.828
Hot drink temperature	57	0	0.04	0.986	0.986
Tea intake	32	−0.02	0.02	0.449	0.748
Alcohol intake frequency	15	0.06	0.03	**0.023**	0.148

MR, Mendelian randomization; IVW, Inverse-variance-weighted. Bold indicated a *p* value less than 0.05.

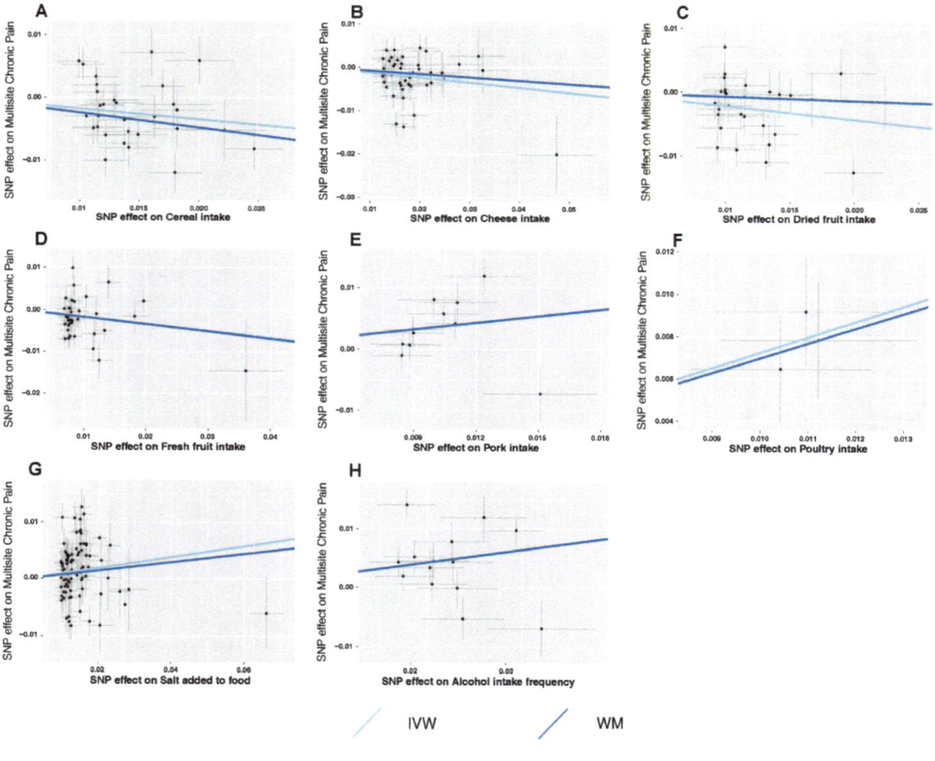

Figure 4. Scatter plots depicting the results of Mendelian randomization (MR) analyses investigating the association between dietary habits and multisite chronic pain. Each line in the plot represents a different MR method, and the slope of each line represents the estimated association between the two variables. (**A**) Scatter plot between cereal intake and multisite chronic pain; (**B**) Scatter plot between cheese intake and multisite chronic pain; (**C**) Scatter plot between dried fruit intake and multisite chronic pain; (**D**) Scatter plot between fresh fruit intake and multisite chronic pain; (**E**) Scatter plot between pork intake and multisite chronic pain; (**F**) Scatter plot between poultry intake and multisite chronic pain; (**G**) Scatter plot between salt added in food and multisite chronic pain; (**H**) Scatter plot between alcohol intake frequency and multisite chronic pain. IVW, inverse-variance-weighted; WM, weight median.

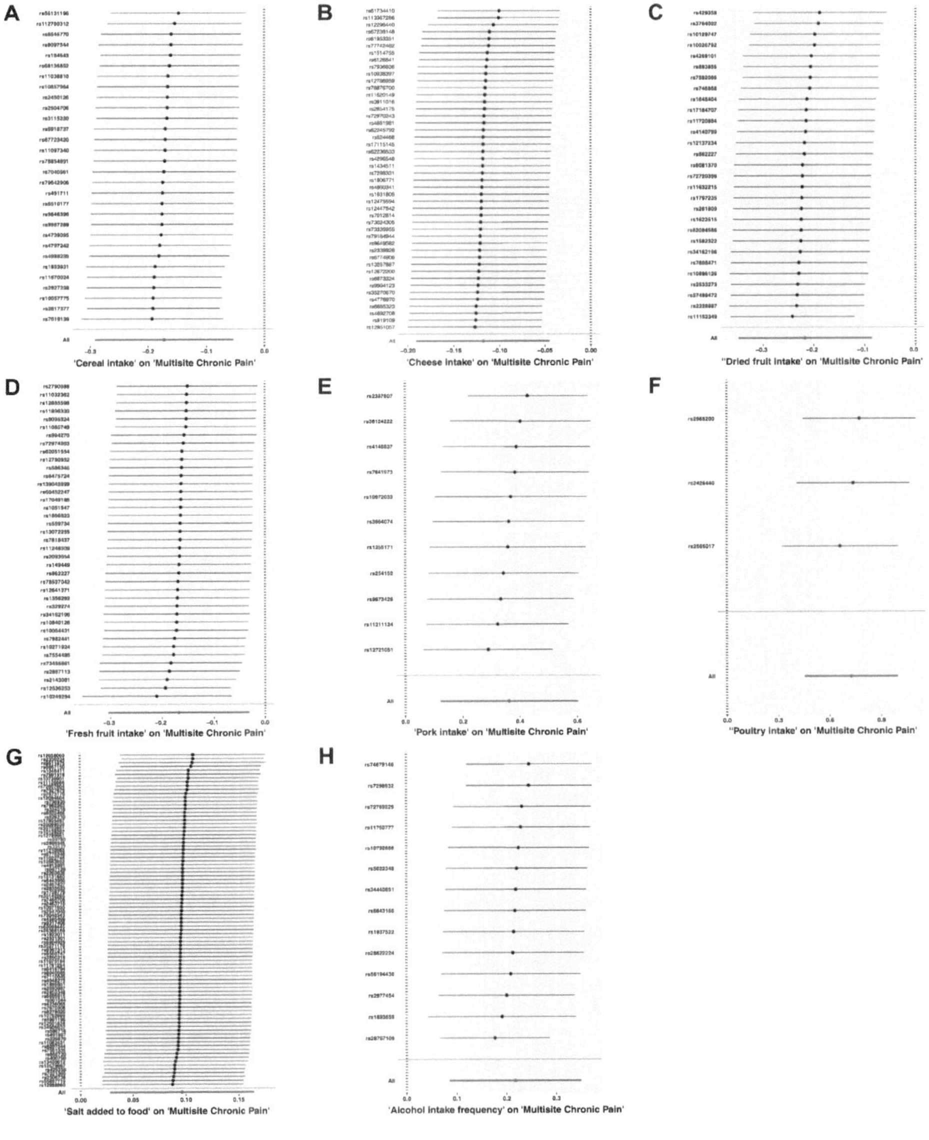

Figure 5. The results of a leave-one-out analysis on Mendelian randomization (MR). Each black line in the figure corresponds to the outcome of the MR analysis when one single nucleotide polymorphism (SNP) is removed from the analysis, while the remaining SNPs are used on the left. (**A**) Leave-one-out analysis between cereal intake and multisite chronic pain; (**B**) Leave-one-out analysis between cheese intake and multisite chronic pain; (**C**) Leave-one-out analysis between dried fruit intake and multisite chronic pain; (**D**) Leave-one-out analysis between fresh fruit intake and multisite chronic pain; (**E**) Leave-one-out analysis between pork intake and multisite chronic pain; (**F**) Leave-one-out analysis between poultry intake and multisite chronic pain; (**G**) Leave-one-out analysis between salt added in food and multisite chronic pain; (**H**) Leave-one-out analysis between alcohol intake frequency and multisite chronic pain. Red line reference the overall effect.

4. Discussion

In this study, we conducted a secondary analysis of the UK biobank dataset to investigate the impact of 20 dietary habits on chronic pain, including multi-site pain (MCP), back pain (BP), neck pain (NP), headache, hip pain (HP), and knee pain (KP). The results of our study revealed causal associations between various dietary habits and different types of chronic pain. For the primary outcome of MCP, we identified that cheese intake, cereal intake, and high consumption of dried and fresh fruit were associated with lower MCP scores. Conversely, high alcohol intake, salt intake, pork intake, and poultry intake were associated with higher MCP scores. Additionally, we found some correlations between dietary habits and other types of chronic pain other than headache.

Our main findings suggest that an inflammatory diet plays a role in chronic pain. Previous studies have extensively demonstrated the relationship between dietary intake and chronic inflammation [30,31]. Certain foods, such as vegetables and fruits, are considered anti-inflammatory, while others, including processed meats and trans-fatty acids, are known to promote inflammation and are referred to as pro-inflammatory foods. In our study, we observed a causal association between lower cereal and fruit intake and higher MCP scores. Both cereal and fruit are recognized as anti-inflammatory foods based on previous research [32–34]. On the other hand, high intakes of alcohol and salt, which showed a significant positive association with MCP scores in our study, are widely known to induce inflammatory responses in the body [35–37]. Some studies have also linked pork and poultry intake in the context of a high-fat diet to organismal inflammation [38]. While the association between chronic pain and many chronic inflammatory conditions is well-established [15,39], it is important to note that certain factors associated with dietary inflammation, such as processed meat intake, had a lesser impact on chronic pain [39]. Further research is needed to understand whether chronic inflammation is the sole mediator of chronic pain related to dietary habits.

Interestingly, our study revealed a negative association between cheese intake and MCP. Dairy fats, which are rich in saturated fatty acids, have been traditionally believed to have adverse effects on the body by raising LDL cholesterol levels. However, recent studies have suggested that certain types of dairy fat intake, such as cheese, may have a protective effect against the development of certain diseases such as heart disease [40]. In our study, we found an inverse causal association between cheese intake and MCP scores, indicating a potential protective effect of cheese consumption on MCP. In addition to saturated fatty acids, cheese is a good source of protein, calcium, and vitamin D [41], which are essential for maintaining bone and muscle health [42]. This is particularly relevant for individuals with chronic pain, as musculoskeletal issues often contribute to their discomfort.

Our study also demonstrated a positive causal association between pork and poultry intake and MCP scores. Pork and poultry may contain higher levels of pro-inflammatory compounds, which can contribute to increased inflammation in the body. Previous studies have suggested that red meat intake can contribute to the development of chronic inflammation in the body [43], and, although there is less research on poultry intake and inflammation, one study also found that high intake of chicken and pork proteins aggravated high-fat-diet-induced inflammation [38]. Chronic pain conditions often involve an underlying inflammatory response, and consuming foods that promote inflammation can worsen pain symptoms and increase discomfort. Interestingly, our study did not find a strong association between beef and lamb intake and CP, and lamb intake even showed a protective effect on certain types of chronic pain. The differential effects of different types of meat intake on health outcomes have also been observed in previous studies. For instance, one study found a beneficial effect of beef intake on depression, while pork and poultry did not show a similar effect [6]. Another study highlighted the potential negative impact of high chicken or pork intake in a high-fat diet on neuronal integrity and neurodegenerative disorders [38]. It is essential for future studies to further differentiate between various types of meat intake when assessing their health effects.

It is important to acknowledge that our study shares common limitations with Mendelian randomization studies. Assumptions may be violated due to horizontal pleiotropy, where a genetic variant affecting the outcome through a different pathway from the investigated exposure could introduce biased estimates. However, we conducted several sensitivity analyses, including MR-PRESSO, an intercept test in MR-Egger regression, a heterogeneity test, and leave-one-out analysis, which consistently supported our results. Another limitation is that the analyzed GWASs primarily focused on individuals of European ancestry, which may limit the generalizability of our findings to other ethnicities. Furthermore, due to the lack of detailed data in the original studies, additional non-linear regression and subgroup analyses were not feasible.

5. Conclusions

In conclusion, our study provides evidence of a causal relationship between various dietary habits and different types of chronic pain. Adhering to an anti-inflammatory diet, including increased consumption of fruit and cereal while reducing salt and pork intake, may potentially alleviate chronic pain symptoms.

Supplementary Materials: The following supporting information can be downloaded at: https://www.mdpi.com/article/10.3390/nu15173709/s1, Figure S1: The figure displays the results of funnel plots in Mendelian randomization (MR). Table S1 shown summary of 20 dietary habits questionnaire; Table S2–S5 shown the details of SNPs select; Table S6–S8 shown the association between dietary habits and headache, chronic knee pain and chronic hip pain measured by IVW method; Table S9–S14 shown the association between dietary habits and chronic pain measured by weight median method; Table S15–S20 shown the association between dietary habits and chronic pain measured by weight median method; Table S21–S26 shown the results of Heterogeneity, Pleiotropy, and directionality test.

Author Contributions: Project administration: Y.S. Validation: J.Y. Software: R.Z. Supervision: H.J. Visualization: L.Z. Original Draft: R.Z. All authors have read and agreed to the published version of the manuscript.

Funding: This work was supported by the Science and Technology Commission of Shanghai Municipality (STCSM) (23YF1422700).

Institutional Review Board Statement: We used the publicly available GWAS catalog in this study. No additional ethical approval was required for re-analysis of the data.

Informed Consent Statement: Not applicable.

Data Availability Statement: The data supporting the findings of this study are available in IEU open GWAS project websites (https://gwas.mrcieu.ac.uk/) and can also obtained from the corresponding author upon reasonable request.

Acknowledgments: This study utilized data from the publicly available GWAS data, and the authors would like to thank all the contributors to and participants in the data.

Conflicts of Interest: The authors declare no actual or potential competing financial interest.

References

1. Cohen, S.P.; Vase, L.; Hooten, W.M. Chronic pain: An update on burden, best practices, and new advances. *Lancet* **2021**, *397*, 2082–2097. [CrossRef]
2. Jackson, T.; Thomas, S.; Stabile, V.; Shotwell, M.; Han, X.; McQueen, K. A Systematic Review and Meta-Analysis of the Global Burden of Chronic Pain Without Clear Etiology in Low- and Middle-Income Countries: Trends in Heterogeneous Data and a Proposal for New Assessment Methods. *Anesth. Analg.* **2016**, *123*, 739–748. [CrossRef]
3. Kozak-Szkopek, E.; Broczek, K.; Slusarczyk, P.; Wieczorowska-Tobis, K.; Klich-Raczka, A.; Szybalska, A.; Mossakowska, M. Prevalence of chronic pain in the elderly Polish population—Results of the PolSenior study. *Arch. Med. Sci.* **2017**, *13*, 1197–1206. [CrossRef]
4. Yong, R.J.; Mullins, P.M.; Bhattacharyya, N. Prevalence of chronic pain among adults in the United States. *Pain* **2022**, *163*, e328–e332. [CrossRef] [PubMed]

5. Wong, C.K.; Mak, R.Y.; Kwok, T.S.; Tsang, J.S.; Leung, M.Y.; Funabashi, M.; Macedo, L.G.; Dennett, L.; Wong, A.Y. Prevalence, Incidence, and Factors Associated with Non-Specific Chronic Low Back Pain in Community-Dwelling Older Adults Aged 60 Years and Older: A Systematic Review and Meta-Analysis. *J. Pain* **2022**, *23*, 509–534. [CrossRef]
6. Robinson, C.L.; Kim, R.S.; Li, M.; Ruan, Q.Z.; Surapaneni, S.; Jones, M.; Pak, D.J.; Southerland, W. The Impact of Smoking on the Development and Severity of Chronic Pain. *Curr. Pain Headache Rep.* **2022**, *26*, 575–581. [CrossRef]
7. Castro-Barquero, S.; Lamuela-Raventós, R.M.; Doménech, M.; Estruch, R. Relationship between Mediterranean Dietary Polyphenol Intake and Obesity. *Nutrients* **2018**, *10*, 1523. [CrossRef] [PubMed]
8. Venter, C.; Eyerich, S.; Sarin, T.; Klatt, K.C. Nutrition and the Immune System: A Complicated Tango. *Nutrients* **2020**, *12*, 818. [CrossRef]
9. Lane, M.M.; Gamage, E.; Travica, N.; Dissanayaka, T.; Ashtree, D.N.; Gauci, S.; Lotfaliany, M.; O'Neil, A.; Jacka, F.N.; Marx, W. Ultra-Processed Food Consumption and Mental Health: A Systematic Review and Meta-Analysis of Observational Studies. *Nutrients* **2022**, *14*, 2568. [CrossRef] [PubMed]
10. Juul, F.; Vaidean, G.; Parekh, N. Ultra-processed Foods and Cardiovascular Diseases: Potential Mechanisms of Action. *Adv. Nutr.* **2021**, *12*, 1673–1680. [CrossRef]
11. Zhou, L.; Wen, X.; Peng, Y.; Zhao, L.; Yu, Y. Salt added to food and body mass index: A bidirectional Mendelian randomisation study. *Nutr. Diet.* **2021**, *78*, 315–323. [CrossRef]
12. Elizabeth, L.; Machado, P.; Zinöcker, M.; Baker, P.; Lawrence, M. Ultra-Processed Foods and Health Outcomes: A Narrative Review. *Nutrients* **2020**, *12*, 1955. [CrossRef]
13. Liu, D.; Zhou, L.; Yang, M.; McIntyre, R.S.; Cao, B. Oxidative Stress Mediates the Association Between Dietary Fat Intake and Cognition in US Older Adults. *Am. J. Geriatr. Psychiatry* **2022**, *30*, 761–773. [CrossRef] [PubMed]
14. Malesza, I.J.; Malesza, M.; Walkowiak, J.; Mussin, N.; Walkowiak, D.; Aringazina, R.; Bartkowiak-Wieczorek, J.; Mądry, E. High-Fat, Western-Style Diet, Systemic Inflammation, and Gut Microbiota: A Narrative Review. *Cells* **2021**, *10*, 3164. [CrossRef]
15. Ji, R.R.; Nackley, A.; Huh, Y.; Terrando, N.; Maixner, W. Neuroinflammation and Central Sensitization in Chronic and Widespread Pain. *Anesthesiology* **2018**, *129*, 343–366. [CrossRef]
16. Kaushik, A.S.; Strath, L.J.; Sorge, R.E. Dietary Interventions for Treatment of Chronic Pain: Oxidative Stress and Inflammation. *Pain. Ther.* **2020**, *9*, 487–498. [CrossRef]
17. Wasser, J.G.; Vasilopoulos, T.; Zdziarski, L.A.; Vincent, H.K. Exercise Benefits for Chronic Low Back Pain in Overweight and Obese Individuals. *PM&R* **2017**, *9*, 181–192. [CrossRef]
18. Elma, Ö.; Yilmaz, S.T.; Deliens, T.; Clarys, P.; Nijs, J.; Coppieters, I.; Polli, A.; Malfliet, A. Chronic Musculoskeletal Pain and Nutrition: Where Are We and Where Are We Heading? *PM&R* **2020**, *12*, 1268–1278. [CrossRef]
19. Lombardo, M.; Feraco, A.; Ottaviani, M.; Rizzo, G.; Camajani, E.; Caprio, M.; Armani, A. The Efficacy of Vitamin D Supplementation in the Treatment of Fibromyalgia Syndrome and Chronic Musculoskeletal Pain. *Nutrients* **2022**, *14*, 3010. [CrossRef] [PubMed]
20. Vormann, J.; Worlitschek, M.; Goedecke, T.; Silver, B. Supplementation with alkaline minerals reduces symptoms in patients with chronic low back pain. *J. Trace Elem. Med. Biol.* **2001**, *15*, 179–183. [CrossRef] [PubMed]
21. MacFarlane, L.A.; Cook, N.R.; Kim, E.; Lee, I.M.; Iversen, M.D.; Gordon, D.; Buring, J.E.; Katz, J.N.; Manson, J.E.; Costenbader, K.H. The Effects of Vitamin D and Marine Omega-3 Fatty Acid Supplementation on Chronic Knee Pain in Older US Adults: Results from a Randomized Trial. *Arthritis Rheumatol.* **2020**, *72*, 1836–1844. [CrossRef] [PubMed]
22. Huang, J.; Xie, Z.F. Dried fruit intake causally protects against low back pain: A Mendelian randomization study. *Front. Nutr.* **2023**, *10*, 1027481. [CrossRef] [PubMed]
23. Chen, T.T.; Chen, C.Y.; Fang, C.P.; Cheng, Y.C.; Lin, Y.F. Causal influence of dietary habits on the risk of major depressive disorder: A diet-wide Mendelian randomization analysis. *J. Affect. Disord.* **2022**, *319*, 482–489. [CrossRef]
24. Mao, X.; Huang, C.; Wang, Y.; Mao, S.; Li, Z.; Zou, W.; Liao, Z. Association between Dietary Habits and Pancreatitis among Individuals of European Ancestry: A Two-Sample Mendelian Randomization Study. *Nutrients* **2023**, *15*, 1153. [CrossRef] [PubMed]
25. Sudlow, C.; Gallacher, J.; Allen, N.; Beral, V.; Burton, P.; Danesh, J.; Downey, P.; Elliott, P.; Green, J.; Landray, M.; et al. UK biobank: An open access resource for identifying the causes of a wide range of complex diseases of middle and old age. *PLoS Med.* **2015**, *12*, e1001779. [CrossRef] [PubMed]
26. Collins, R. What makes UK Biobank special? *Lancet* **2012**, *379*, 1173–1174. [CrossRef]
27. Johnston, K.J.A.; Adams, M.J.; Nicholl, B.I.; Ward, J.; Strawbridge, R.J.; Ferguson, A.; McIntosh, A.M.; Bailey, M.E.S.; Smith, D.J. Genome-wide association study of multisite chronic pain in UK Biobank. *PLoS Genet.* **2019**, *15*, e1008164. [CrossRef]
28. Verbanck, M.; Chen, C.Y.; Neale, B.; Do, R. Detection of widespread horizontal pleiotropy in causal relationships inferred from Mendelian randomization between complex traits and diseases. *Nat. Genet.* **2018**, *50*, 693–698. [CrossRef]
29. Storey, J.D. The positive false discovery rate: A Bayesian interpretation and the *q*-value. *Ann. Stat.* **2003**, *31*, 2013–2035. [CrossRef]
30. Christ, A.; Lauterbach, M.; Latz, E. Western Diet and the Immune System: An Inflammatory Connection. *Immunity* **2019**, *51*, 794–811. [CrossRef]
31. Galland, L. Diet and inflammation. *Nutr. Clin. Pract.* **2010**, *25*, 634–640. [CrossRef] [PubMed]
32. Lee, Y.M.; Han, S.I.; Song, B.C.; Yeum, K.J. Bioactives in Commonly Consumed Cereal Grains: Implications for Oxidative Stress and Inflammation. *J. Med. Food* **2015**, *18*, 1179–1186. [CrossRef] [PubMed]

33. Soujanya, K.V.; Jayadeep, A.P. Obesity-associated biochemical markers of inflammation and the role of grain phytochemicals. *J. Food Biochem.* **2022**, *46*, e14257. [CrossRef]
34. Wallace, T.C.; Bailey, R.L.; Blumberg, J.B.; Burton-Freeman, B.; Chen, C.O.; Crowe-White, K.M.; Drewnowski, A.; Hooshmand, S.; Johnson, E.; Lewis, R.; et al. Fruits, vegetables, and health: A comprehensive narrative, umbrella review of the science and recommendations for enhanced public policy to improve intake. *Crit. Rev. Food Sci. Nutr.* **2020**, *60*, 2174–2211. [CrossRef] [PubMed]
35. Ramos-Lopez, O.; Milagro, F.I.; Riezu-Boj, J.I.; Martinez, J.A. Epigenetic signatures underlying inflammation: An interplay of nutrition, physical activity, metabolic diseases, and environmental factors for personalized nutrition. *Inflamm. Res.* **2021**, *70*, 29–49. [CrossRef]
36. Azak, A.; Huddam, B.; Gonen, N.; Yilmaz, S.R.; Kocak, G.; Duranay, M. Salt intake is associated with inflammation in chronic heart failure. *Int. Cardiovasc. Res. J.* **2014**, *8*, 89–93.
37. Kuang, R.; O'Keefe, S.J.D.; Ramos Del Aguila de Rivers, C.; Koutroumpakis, F.; Binion, D.G. Is Salt at Fault? Dietary Salt Consumption and Inflammatory Bowel Disease. *Inflamm. Bowel Dis.* **2023**, *29*, 140–150. [CrossRef]
38. Zhang, M.; Song, S.; Zhao, D.; Shi, J.; Xu, X.; Zhou, G.; Li, C. High intake of chicken and pork proteins aggravates high-fat-diet-induced inflammation and disorder of hippocampal glutamatergic system. *J. Nutr. Biochem.* **2020**, *85*, 108487. [CrossRef]
39. Ji, R.R.; Chamessian, A.; Zhang, Y.Q. Pain regulation by non-neuronal cells and inflammation. *Science* **2016**, *354*, 572–577. [CrossRef]
40. Hu, M.J.; Tan, J.S.; Gao, X.J.; Yang, J.G.; Yang, Y.J. Effect of Cheese Intake on Cardiovascular Diseases and Cardiovascular Biomarkers. *Nutrients* **2022**, *14*, 2936. [CrossRef]
41. Benedik, E. Sources of vitamin D for humans. *Int. J. Vitam. Nutr. Res.* **2022**, *92*, 118–125. [CrossRef] [PubMed]
42. Agostini, D.; Zeppa Donati, S.; Lucertini, F.; Annibalini, G.; Gervasi, M.; Ferri Marini, C.; Piccoli, G.; Stocchi, V.; Barbieri, E.; Sestili, P. Muscle and Bone Health in Postmenopausal Women: Role of Protein and Vitamin D Supplementation Combined with Exercise Training. *Nutrients* **2018**, *10*, 1103. [CrossRef] [PubMed]
43. Samraj, A.N.; Pearce, O.M.; Läubli, H.; Crittenden, A.N.; Bergfeld, A.K.; Banda, K.; Gregg, C.J.; Bingman, A.E.; Secrest, P.; Diaz, S.L.; et al. A red meat-derived glycan promotes inflammation and cancer progression. *Proc. Natl. Acad. Sci. USA* **2015**, *112*, 542–547. [CrossRef] [PubMed]

nutrients

MDPI

Article

Examining the Nutrition of Cardiological Patients in Hospitals: Evaluating the Discrepancy between Received Diets and Reference Diet Based on ESC 2021 Guidelines—Hospital Diet Medical Investigation) (HDMI) Study

Daniel Śliż [1,2], Alicja Monika Jodczyk [1,2,*], Klaudia Łakoma [1,2], Alicja Kucharska [3], Mariusz Panczyk [4], Olga Maria Rostkowska [5], Karolina Turlej [3], Agnieszka Młynarska [6], Jarosław Drożdż [7], Milena Jarzębska-Wódka [7], Piotr Wierzbiński [2], Marcin Grabowski [8], Anna Ukleja [8], Natalia Adamczyk [2], Alicja Baska [2,9], Szczepan Wiecha [10], Marcin Barylski [11], Adam Rafał Poliwczak [11] and Artur Mamcarz [1]

1 3rd Department of Internal Medicine and Cardiology, Medical University of Warsaw, 04-749 Warsaw, Poland
2 Polish Society of Lifestyle Medicine, 00-388 Warsaw, Poland
3 Department of Human Nutrition, Medical University of Warsaw, 01-445 Warsaw, Poland
4 Department of Education and Research in Health Sciences, Medical University of Warsaw, 00-518 Warsaw, Poland
5 Department of Transplantation Medicine, Nephrology and Internal Medicine, Medical University of Warsaw, 02-006 Warsaw, Poland
6 Department of Gerontology and Geriatric Nursing, Medical University of Silesia, 40-635 Katowice, Poland
7 Department of Cardiology, Medical University of Lodz, 92-213 Lodz, Poland
8 1st Chair and Department and of Cardiology Medical University of Warsaw, 02-097 Warsaw, Poland
9 School of Public Health, Centre of Postgraduate Medical Education, 01-813 Warsaw, Poland
10 Department of Physical Education and Health in Biala Podlaska, Faculty in Biala Podlaska, Jozef Pilusdski University of Physical Education, 00-968 Warsaw, Poland
11 Department of Internal Medicine and Cardiac Rehabilitation, Medical University of Lodz, 90-647 Lodz, Poland
* Correspondence: alajodczyk4@gmail.com

Citation: Śliż, D.; Jodczyk, A.M.; Łakoma, K.; Kucharska, A.; Panczyk, M.; Rostkowska, O.M.; Turlej, K.; Młynarska, A.; Drożdż, J.; Jarzębska-Wódka, M.; et al. Examining the Nutrition of Cardiological Patients in Hospitals: Evaluating the Discrepancy between Received Diets and Reference Diet Based on ESC 2021 Guidelines—Hospital Diet Medical Investigation) (HDMI) Study. *Nutrients* **2023**, *15*, 4606. https://doi.org/10.3390/nu15214606

Academic Editor: Domenico Tafuri

Received: 9 October 2023
Revised: 25 October 2023
Accepted: 27 October 2023
Published: 30 October 2023

Abstract: Cardiovascular diseases (CVDs) are the leading causes of death worldwide. CVDs have become the dominant cause of death and have been a significant health challenge since the second half of the 20th century in the Polish population. The aim of our HDMI (hospital diet medical investigation) study was to examine the quality of the hospital diets given to cardiac patients and assess how much they adhere to the European Society of Cardiology (ESC) 2021 guidelines. By comparing the diets received by patients with the recommended dietary patterns outlined in the ESC 2021 guidelines, we sought to identify discrepancies. The study was conducted in two steps: creating a 7-day model menu and comparing it with the received diets and then making comparisons with ESC 2021 guidelines. Additionally, we designed a survey to obtain the characteristics of the hospitals. The results show that the nutrition in hospitals remains substandard. None of the diets had an appropriate salt supply or predominance of plant-based food patterns. Only 1/7 diets avoided sweetened beverages, and 2/7 diets had an appropriate amount of fiber. This underscores a gap in the healthcare system to improve patients' health by implementing dietary interventions that foster the development of healthy eating habits.

Keywords: cardiology; nutrition; hospital; diet

1. Introduction

Cardiovascular diseases (CVDs) are a leading cause of mortality worldwide, requiring specialized care for cardiological patients, leading significantly to excessive costs for the healthcare system [1–3]. In Poland, CVDs have been a major health problem since the second half of the 20th century, and they remain the main cause of death among Polish residents [4,5]. In 2019, CVDs accounted for 39.4% of deaths, but due to the COVID-19 pandemic in 2021, the data show that they decreased to 34.8% [4].

Proper nutrition plays a pivotal role in managing and treating these patients, influencing their recovery and long-term outcomes [1]. Hospitalized cardiac patients often rely on dietary interventions provided by healthcare facilities to aid their recovery and overall cardiovascular health. It should be pointed out that the patient's poor nutritional status in hospitals reduces the effectiveness of their treatment and increases the risk of complications [6]. It has already been proven that targeted nutritional intervention can prevent unplanned readmissions [7]. Moreover, it has already been shown that the patient's nutritional education starts in the hospital and is crucial in improving lifestyle and well-being [8]. For this reason, the importance of evidence-based healthcare including nutrition care in a hospital setting is emphasized [1]. Promoting healthy dietary habits is a crucial aspect of patient care that extends beyond the hospital. It is essential to underscore the pivotal role that a patient's daily dietary choices and lifestyle play in their overall well-being. Notably, the prevailing diet among the typical Polish population leans more toward a Western dietary pattern than the healthier Mediterranean diet.

It is noteworthy that a mere 24% of the Polish population can be classified as adhering to a wholesome diet, while the majority of individuals encounter various hurdles. Alarmingly, 29.1% openly express their reluctance to embrace healthier eating habits. These statistics strongly underscore the imperative need for healthcare professionals to step in and provide guidance and intervention [4]. Regrettably, nutritional and dietary interventions still tend to be undervalued in the routine clinical management of chronic diseases [9].

In the case of cardiac diseases, proper nutrition constitutes not only an element of prevention but also an essential part of treatment, as confirmed by the latest 2021 European Society of Cardiology (ESC 2021) recommendations [10]. However, the extent to which these hospital diets align with the latest guidelines remains unclear. In addition, in 2018, the Supreme Audit Office alerted that diets in Polish hospitals were poorly balanced [11]. Only 17% of hospitals examined in this report had an appropriate caloric value. In addition, the diets lacked important nutrients such as calcium, iron, magnesium and potassium. Deficiencies of the above ingredients were accompanied by an increased supply of salt (142–374% of the norm) and vitamin A. The meals provided in the hospitals did not fulfill their essential function—supporting the treatment and recovery process—and sometimes could constitute a harmful factor. It was highlighted that not only a well-planned diet but also the support of a dietician was indicated as the key factor determining the improvement in the patient's health. The existing gap within the healthcare system was identified in the form of inadequate hospital nutrition that failed to meet the patient's specific needs. It was also emphasized that, ultimately, hospital nutrition frequently fell short in contributing to the improvement in the patient's health [11]. Given the lapse of five years since the issuance of the 2018 report, our pilot study tries to meticulously assess the quality of hospital diets, thus making a comprehensive comparison with the current ESC guidelines in the context of a cardiac patient.

This publication is part of an HDMI (hospital diet medical investigation) study and aims to investigate the nutrition provided to cardiological patients in hospitals and assess how much it adheres to the ESC 2021 guidelines. By comparing the actual diets received by patients with the recommended dietary patterns outlined in the ESC 2021 guidelines, we seek to identify any gaps and discrepancies. Understanding the existing disparities between received diets and evidence-based recommendations is crucial for optimizing the care and outcomes of cardiological patients.

A healthy diet is recognized as a fundamental pillar of CVD prevention, with an emphasis on several key principles that include a focus on plant-based foods, the limitation of saturated fatty acids to less than 10% of energy intake, aiming for a daily fiber intake of 30–45 g, incorporating fish into the diet 1–2 times per week, restricting salt intake to less than 5 g per day and limiting the consumption of alcohol, red meat and sugary beverages [10]. A randomized, single-blind secondary prevention trial, the Lyon Diet Heart Study, showed that the Mediterranean diet reduced the recurrence rates after the first myocardial infarction compared to a prudent Western-type diet. Patients randomized to the Mediterranean diet experienced a 72% reduction in recurrent non-fatal myocardial infarction and a 56% decline in mortality compared to the control diet [12]. One of the biggest challenges in the dietary prevention of CVDs lies in devising more effective strategies to motivate individuals to initiate dietary changes and, equally importantly, sustain them over time [13].

We hypothesized that the diet served to cardiological patients in Polish hospitals did not fulfill the ESC 2021 recommendations and was not properly balanced.

2. Materials and Methods

2.1. Design and Setting

Our comparative cross-sectional study was conducted between January and August 2022. The study was carried out in four stages: creating a survey, preparing the reference diet, evaluating diets in "Dieta 6" computer program and comparing the results with ESC recommendations.

2.2. Sample

Hospital workers (heads of the departments, specialists and resident doctors) from various districts in Poland were invited to participate in a study through email, which contained information about the purpose of the study, terms of the participation and required information. To qualify for participation, hospitals had to be public, have a general medicine ward and complete the questionnaire. Private hospitals or hospitals specializing in certain illnesses or groups of patients, such as military personnel, were excluded. Filling in the questionnaire and clicking the "send" button were tantamount to informed consent to participate in the study (proper information was mentioned in the background of the questionnaire).

2.3. Bioethical Consideration

The study adheres to the regulations set by the Institutional Ethics Committee of the Medical University of Warsaw and aligns with the principles outlined in the Helsinki Declaration (1964). It was not necessary to acquire approval from the Institutional Review Board, but a thorough assessment was obtained. Respondents did not receive any monetary or material rewards for their participation. Anonymity was emphasized at the beginning of the questionnaire, along with a comprehensive explanation of the study's objectives.

2.4. Instrument and Data Collection

The study involved (1) a survey with questions about general information and the organization of hospital food service within the hospital, as well as (2) submitting menus of patient meals for the following 10 days. The self-made questionnaire, created using Google Forms, consisted of 12 questions. It included information about the hospital's degree of reference (1st/2nd/3rd degree of reference—the division is based on the specific regulations in Poland and classifies hospitals by the number of beds, the competence of medical personnel, access to equipment, diagnostic and therapeutic procedures; 1st is the lowest degree with limited and basic medical procedures, and 3rd is the highest with the most extended and specialized procedures), number of beds in total, internal and cardiology unit, presence of a person providing nutritional education (nurse/physician/dietician), location of meal preparation (hospital kitchen/catering) and availability of special diet dedicated for patients with cardiological problems. The data collection period was from 26th January 2022 to 3rd August 2022. Data sent via Google Forms and email were exported as an Excel file for further analysis. The questionnaire is attached in the Supplementary S1.

The received menus were evaluated using "Dieta 6", which is a unique computer program standardized for the Polish population. It was designed by the National Institute of Public Health—National Institute of Hygiene in Warsaw. It was based on national composition tables and the nutritional value of food products and meals. It enables researchers to calculate the average nutritional value and composition of the consumed diet.

We created a model seven-day menu (standardized menu) to compare different hospital menus with the European Society of Cardiology (ESC) recommendations [10]. We aimed to create a genuine and achievable diet with the inclusion of universal, widely available and relatively low-cost ingredients. This menu is attached in the Supplementary S2. The guidelines were published in 2021 [10] to assist healthcare professionals in reducing the burden of CVD on both individual patients and the population. They emphasize the importance of educating patients (class IIa according to ESC 2021 recommendations), intervening in CVD risk factors and promoting a healthy diet. Table 1, based on the guidelines, outlines the main features of a healthy diet, which is recommended for patients with cardiological problems.

Table 1. Features of a healthy diet according to ESC guidelines [10].

The Predominance of Plant-Based Food Patterns and the Reduction in Animal-Based Ones
Limitation of saturated fatty acids to <10% of energy intake and their replacement by PUFAs, MUFAs and carbohydrates from whole grains
Minimalization of the amount of unsaturated trans fatty acids in the diet and exclusion of highly processed products
<5g total salt intake per day
30–45 g of fiber per day, optimally from whole grains
≥200 g of fruit per day (≥2–3 servings) ≥200 g of vegetables per day (≥2–3 servings)
Reduction in the consumption of red meat to the maximum 350–500 g per week, especially processed meat
Eating fish 1–2 times a week, especially fatty fish
Consuming 30 g of unsalted nuts per day
Limitation of alcohol consumption to a maximum of 100 g/week
Avoidance of sweetened beverages, including fruit juices and sweet carbonated and non-carbonated drinks

Abbreviations: ESC—European Society of Cardiology.

2.5. Statistical Analysis

Statistical analyses were conducted utilizing STATISTICA™ 13.3 software (TIBCO Software, Palo Alto, CA, USA). Nutritional values were computed from menu data using DIETA Software version 6.0 (National Institute of Public Health—National Institute of Hygiene in Warsaw). Descriptive statistics, including mean (M), standard deviation (SD) and range (Min–Max), were calculated for each nutritional value. One-way ANOVA with post hoc Dunnett's two-tailed test was employed to assess the differences in means for nutrients and products between hospitals and the standardized menu. Dunnett's test is a multiple comparison procedure that compares each of several means with a single control in many-to-one comparisons. The null hypothesis was rejected for p-values less than 0.05, indicating that the tested difference in means was statistically significant.

3. Results

3.1. The Outcomes from the Survey

The study involved seven hospitals, out of which five were classified as third degree of reference, one as second degree and one as first degree. While six hospitals had a separate cardiology ward, only Hospital 5 provided a specific diet for cardiological patients. Nutrition education was provided in six hospitals, with only four having a dietician present. In terms of meal preparation, four hospitals prepared their own meals, while three ordered catering services. Table 2 displays the general characteristics of the hospitals, such as hospital type, bed capacity (total, internal medicine and cardiology unit), availability of nutrition education and meal preparation location. For further analysis, due to the lack of a specific cardiological diet, the general diet from Hospitals 1, 2, 3, 4, 6 and 7 was taken into consideration. From Hospital 5, a cardiological diet was examined.

Table 2. The characteristics of the hospitals.

Variables	Hospital						
	1	2	3	4	5	6	7
Type of the hospital (degree of reference)	3rd	2nd	3rd	3rd	3rd	1st	3rd
Total number of beds	1035	452	683	677	363	176	321
Number of beds in internal medicine unit	386	96	21	87	45	52	50
Number of beds in cardiology unit	63	32	20	250	76	0	171
Presence of dietician	No	No	No	Yes	Yes	Yes	Yes
Nutrition education	Yes	Yes	No	Yes	Yes	Yes	Yes
The person responsible for nutrition education	Specialists outside the hospital (occasionally)	Doctor	-	Dietician, doctor, nurse	Dietician, doctor, nurse	Dietician, doctor, nurse	Dietician, doctor, nurse
Place of meal preparation	Catering	Catering	Kitchen in the hospital	Kitchen in the hospital	Catering	Kitchen in the hospital	Kitchen in the hospital

3.2. *Differences in the Content of Nutrients and product Groups*

Table 3 demonstrates energy and carbohydrate content. A one-way ANOVA revealed that there was a statistically significant difference in the mean of energy (kcal) ($F = 12307.61$; $p < 0.001$), the mean of total carbohydrates ($F = 14789.56$; $p < 0.001$), the mean percentage of energy from carbohydrates ($F = 16905.28$; $p < 0.001$), the mean percentage of digestible carbohydrates ($F = 13424.08$; $p < 0.001$) and the mean percentage of fiber ($F = 5870.27$; $p < 0.001$) between hospitals. Compared to the standardized menu, there was a statistically greater mean of energy in Hospital 2 ($p = 0.001$), a greater mean of total carbohydrates in Hospitals 2 ($p = 0.041$), 6 ($p < 0.001$) and 7 ($p = 0.001$), a greater mean percentage of energy from carbohydrates in Hospitals 1 ($p = 0.012$), 3 ($p = 0.005$), 5 ($p = 0.001$), 6 ($p < 0.001$) and 7 ($p < 0.001$), a greater mean of digestive carbohydrates in Hospitals 1 ($p = 0.001$), 2 ($p = 0.001$), 6 ($p < 0.001$) and 7 ($p < 0.001$) and a lower mean of fiber in all hospitals ($p < 0.001$).

Table 3. Energy and carbohydrate content.

Hospital	M	SD	Mini	Max	*p*-Value **
	Energy (kcal) ($F = 12307.61$; $p < 0.001$ *)				
Hospital 1	2102.07	225.04	1816.34	2558.38	0.764
Hospital 2	2358.78	69.31	2226.81	2448.24	0.001
Hospital 3	1966.56	96.46	1811.74	2087.37	0.997
Hospital 4	1783.75	57.43	1701.71	1881.67	0.078
Hospital 5	1930.80	94.88	1812.60	2155.87	0.914
Hospital 6	2214.67	200.95	1936.56	2610.27	0.073

Table 3. *Cont.*

Hospital	M	SD	Mini	Max	p-Value **
Hospital 7	2147.87	193.29	1810.47	2448.25	0.376
Reference (standardized menu)	2005.36	20.64	1984.31	2039.82	-
Total amount of carbohydrates (F = 14789.56; p < 0.001 *)					
Hospital 1	305.31	23.42	266.56	335.13	0.152
Hospital 2	313.61	17.84	297.40	342.39	0.041
Hospital 3	296.36	19.22	262.96	320.93	0.606
Hospital 4	256.89	11.35	236.96	273.61	0.204
Hospital 5	295.79	15.72	267.12	319.26	0.644
Hospital 6	330.95	23.53	308.43	378.07	<0.001
Hospital 7	323.93	24.75	295.07	378.07	0.001
Reference (standardized menu)	281.09	11.37	266.26	298.17	-
Percentage of energy from carbohydrates (F = 16905.28; p < 0.001 *)					
Hospital 1	53.72	4.20	47.39	61.55	0.012
Hospital 2	48.44	2.18	45.86	51.66	1.000
Hospital 3	54.29	2.20	52.08	58.99	0.005
Hospital 4	51.73	2.24	47.29	53.92	0.243
Hospital 5	55.11	2.41	52.06	58.66	0.001
Hospital 6	56.08	4.86	46.41	63.33	<0.001
Hospital 7	56.55	3.61	52.18	62.12	<0.001
Reference (standardized menu)	47.91	1.89	45.21	50.17	-
Digestible carbohydrates (F = 13424.08; p < 0.001 *)					
Hospital 1	279.68	21.63	243.90	302.89	0.001
Hospital 2	284.35	17.34	268.19	314.86	0.001
Hospital 3	265.74	17.78	236.36	283.61	0.058
Hospital 4	229.35	10.92	210.00	240.88	0.946
Hospital 5	264.23	14.21	237.12	283.48	0.083
Hospital 6	307.76	23.65	286.53	355.29	<0.001
Hospital 7	301.46	23.18	280.08	355.29	<0.001
Reference	238.42	10.78	224.78	253.90	-
Fiber (F = 5870.27; p < 0.001 *)					
Hospital 1	25.64	3.49	19.60	32.23	<0.001
Hospital 2	29.26	3.11	25.78	34.70	<0.001
Hospital 3	30.61	3.73	26.59	38.50	<0.001
Hospital 4	27.54	3.08	24.04	32.72	<0.001
Hospital 5	31.56	2.75	25.95	35.78	<0.001
Hospital 6	23.20	2.15	19.40	27.16	<0.001
Hospital 7	22.47	3.11	14.99	27.16	<0.001
Reference (standardized menu)	42.68	2.55	39.39	45.99	-

M—mean, SD—standard deviation, * one-way ANOVA, ** post hoc Dunnett's two-tailed test (multiple comparisons to a control are also referred to as many-to-one comparisons).

Table 4 illustrates the protein content of the hospitals. A statistically significant difference was observed between hospitals in the mean of total protein (F = 7647.82; $p < 0.001$), the mean percentage of energy from protein (F = 6927.64; $p < 0.001$), the mean of plant-based protein (F = 5740.80; $p < 0.001$) and the mean of animal-based protein

($F = 2150.84$; $p < 0.001$). Compared to the standardized menu, Hospitals 1, 3, 4, 5, 6 and 7 had a lower mean of total protein ($p < 0.001$), a lower mean percentage of energy from protein ($p < 0.001$ for Hospitals 1, 2, 3, 5, 6 and 7, and $p = 0.003$ for Hospital 4) and a lower mean of plant-based protein ($p < 0.001$). Hospital 2 had a significantly higher mean of animal-based protein ($p = 0.007$).

Table 4. Protein content.

Hospital	M	SD	Mini	Max	p-Value **
Total amount of protein (F = 7647.82; p < 0.001 *)					
Hospital 1	73.95	9.93	59.16	92.37	<0.001
Hospital 2	99.05	7.06	88.81	110.67	0.737
Hospital 3	76.19	8.58	60.14	89.47	<0.001
Hospital 4	79.58	9.24	65.39	94.59	<0.001
Hospital 5	80.93	6.33	72.44	89.17	<0.001
Hospital 6	83.14	5.97	74.43	92.17	<0.001
Hospital 7	82.07	8.00	64.63	92.17	<0.001
Reference (standardized menu)	104.63	5.42	99.66	110.50	-
Percentage of energy from protein (F = 6927.64; p < 0.001 *)					
Hospital 1	14.14	1.65	11.04	16.39	<0.001
Hospital 2	16.74	0.84	15.02	17.72	<0.001
Hospital 3	15.64	1.69	12.88	18.35	<0.001
Hospital 4	17.62	1.55	15.21	19.44	0.003
Hospital 5	16.95	1.21	15.51	18.90	<0.001
Hospital 6	15.20	2.10	11.83	19.19	<0.001
Hospital 7	15.40	1.86	12.34	19.19	<0.001
Reference (standardized menu)	21.06	1.05	20.04	22.46	-
Plant-based protein (F = 5740.80; p < 0.001 *)					
Hospital 1	31.69	5.06	25.88	40.58	<0.001
Hospital 2	34.04	4.38	29.80	42.46	<0.001
Hospital 3	32.83	5.31	25.94	42.07	<0.001
Hospital 4	35.50	3.95	31.85	43.69	<0.001
Hospital 5	36.46	1.96	33.84	40.67	<0.001
Hospital 6	34.04	2.51	30.74	40.06	<0.001
Hospital 7	33.86	2.58	30.74	40.06	<0.001
Reference (standardized menu)	55.79	4.69	48.32	60.81	-
Animal-based protein (F = 2150.84; p < 0.001 *)					
Hospital 1	42.00	10.52	30.00	59.57	0.659
Hospital 2	64.75	8.67	52.26	79.11	0.007
Hospital 3	43.16	8.19	28.66	54.04	0.831
Hospital 4	43.84	10.18	31.72	61.10	0.932
Hospital 5	44.22	6.43	35.25	50.96	0.941
Hospital 6	48.93	5.71	41.53	58.15	1.000
Hospital 7	48.03	7.39	32.78	58.15	1.000
Reference (standardized menu)	48.04	9.86	38.28	62.00	-

M—mean, SD—standard deviation, * one-way ANOVA, ** post hoc Dunnett's two-tailed test (multiple comparisons to a control are also referred to as many-to-one comparisons).

Table 5 displays the fatty acid content, and there were significant variations among hospitals in the mean of total fat (F = 1697.67; $p < 0.001$ *), the mean percentage of energy from fat (F = 3635.99; $p < 0.001$ *), the mean percentage of saturated fatty acids (F = 2027.01; $p < 0.001$ *), the mean percentage of polyunsaturated fatty acids (F = 1000.07; $p < 0.001$ *), the mean percentage of n-3 acids (F = 328.40; $p < 0.001$ *) and the mean percentage of n-6 acids (F = 1251.11; $p < 0.001$ *). Compared to the standardized menu, Hospital 2 had a significantly greater mean of total fat content ($p = 0.011$). The variation in the mean percentage of energy from fat was not statistically significant ($p > 0.05$). All hospitals except Hospital 4 had a greater mean of saturated fatty acids ($p < 0.001$), and each hospital had a lower mean of polyunsaturated fatty acids ($p < 0.001$) and n-6 acids ($p < 0.001$). Hospital 5 had a lower mean of n-3 acids ($p = 0.022$).

Table 5. Fatty acid content.

Hospital	M	SD	Mini	Max	p-Value **
	Total amount of fat (F = 1697.67; $p < 0.001$ *)				
Hospital 1	70.61	17.99	52.48	113.35	0.601
Hospital 2	85.13	6.03	76.65	91.61	0.011
Hospital 3	59.77	5.40	53.47	70.09	1.000
Hospital 4	55.01	4.66	47.42	60.16	0.953
Hospital 5	53.97	5.69	47.26	65.92	0.869
Hospital 6	67.20	20.62	39.60	114.07	0.913
Hospital 7	63.20	14.57	39.60	81.07	1.000
Reference (standardized menu)	60.90	6.08	54.62	69.48	-
	Percentage of energy from fat (F = 3635.99; $p < 0.001$ *)				
Hospital 1	29.80	4.30	24.03	39.67	0.608
Hospital 2	32.47	1.94	29.53	34.83	0.076
Hospital 3	27.12	2.21	24.44	31.17	1.000
Hospital 4	27.73	2.39	24.70	31.26	1.000
Hospital 5	24.83	1.76	22.49	27.19	0.840
Hospital 6	26.74	6.09	18.16	40.03	1.000
Hospital 7	26.07	4.44	18.16	32.37	0.998
Reference (standardized menu)	26.97	2.77	23.81	30.82	-
	Saturated fatty acids (F = 2027.01; $p < 0.001$ *)				
Hospital 1	29.24	4.21	23.59	38.38	<0.001
Hospital 2	32.83	3.76	27.55	36.96	<0.001
Hospital 3	27.85	2.92	23.41	33.35	<0.001
Hospital 4	18.66	2.09	15.44	20.88	0.542
Hospital 5	24.20	2.63	20.98	29.12	0.003
Hospital 6	32.84	7.66	21.16	46.71	<0.001
Hospital 7	30.26	6.51	21.16	42.16	<0.001
Reference (standardized menu)	14.56	2.86	11.51	18.65	-
	Polyunsaturated fatty acids (F = 1000.07; $p < 0.001$ *)				
Hospital 1	9.91	4.27	4.09	20.09	<0.001
Hospital 2	12.10	2.04	9.97	15.94	<0.001
Hospital 3	7.87	2.06	4.79	10.77	<0.001
Hospital 4	12.71	1.27	10.51	14.32	<0.001
Hospital 5	7.47	1.67	5.41	10.15	<0.001

Table 5. *Cont.*

Hospital	M	SD	Mini	Max	*p*-Value **
Hospital 6	7.34	3.12	4.34	14.81	<0.001
Hospital 7	7.25	3.02	4.08	14.35	<0.001
Reference (standardized menu)	20.92	2.33	18.22	24.56	-
n-3 acids (F = 328.40; *p* < 0.001 *)					
Hospital 1	2.31	1.39	0.65	5.64	0.979
Hospital 2	2.95	0.84	2.25	4.61	0.987
Hospital 3	1.60	0.75	0.59	2.74	0.169
Hospital 4	1.90	0.41	1.26	2.37	0.575
Hospital 5	1.20	0.48	0.67	1.87	0.022
Hospital 6	1.51	0.95	0.65	3.83	0.111
Hospital 7	1.48	0.94	0.53	3.71	0.098
Reference (standardized menu)	2.63	0.22	2.42	2.94	-
n-6 acids (F = 1251.11; *p* < 0.001 *)					
Hospital 1	7.59	3.02	3.45	14.41	<0.001
Hospital 2	9.14	1.35	7.56	11.31	<0.001
Hospital 3	6.26	1.68	3.85	8.41	<0.001
Hospital 4	10.81	0.90	9.24	11.94	<0.001
Hospital 5	6.27	1.26	4.65	8.27	<0.001
Hospital 6	5.82	2.18	3.68	10.96	<0.001
Hospital 7	5.76	2.10	3.55	10.61	<0.001
Reference (standardized menu)	18.28	2.35	15.28	21.77	-

M—mean, SD—standard deviation, * one-way ANOVA, ** post hoc Dunnett's two-tailed test (multiple comparisons to a control are also referred to as many-to-one comparisons).

The mean of sodium and salt intake varied among the hospitals (F = 3411.58, $p < 0.001$, and F = 3416.12, $p < 0.001$), but the difference was not statistically significant ($p > 0.05$). Precise values are presented in Table 6.

The hospitals showed significant differences in the mean mass of fruits (F = 723.54; $p < 0.001$ *), the mean mass of vegetables (F = 976.51; $p < 0.001$ *), the mean mass of whole-grain products (F = 65.67; $p < 0.001$ *), the mean mass of processed meat (F = 717.13; $p < 0.001$ *) and the mean mass of sugar (F = 1327.33; $p < 0.001$ *). A lower mean mass of fruits was provided in Hospital 1 ($p = 0.022$) and Hospitals 3, 4, 5, 6 and 7 ($p < 0.001$). A lower mean mass of vegetables was provided in Hospitals 1, 3, 4, 5, 6 and 7 ($p < 0.001$). A lower mean mass of whole-grain products was provided in Hospitals 1 ($p = 0.007$), 3 ($p = 0.012$), 6 ($p = 0.002$) and 7 ($p = 0.002$). Due to shortages in particular menus, an analysis of legumes, nuts and beans, fish and nuts was not conducted. Every hospital provided red meat in their diet and the reference diet did not. Fish was provided only in one meal once a week in most of the hospitals. Nuts and beans were provided only in Hospitals 1, 3 and 4. Precise values are presented in Table 7.

Table 6. Salt and sodium content.

Hospital	M	SD	Mini	Max	*p*-Value **
		Sodium (F = 3411.58; *p* < 0.001 *)			
Hospital 1	4775.18	959.62	2915.91	5998.96	0.768
Hospital 2	4607.82	637.18	3644.99	5297.30	0.984
Hospital 3	4431.87	730.49	3112.99	5729.69	1.000
Hospital 4	4938.30	542.37	3941.12	5492.84	0.508
Hospital 5	4479.40	698.99	3009.38	5470.33	1.000
Hospital 6	4619.53	377.62	3797.37	5069.05	0.971
Hospital 7	4601.40	404.12	3678.83	5065.45	0.981
Reference (standardized menu)	4367.64	194.53	4218.51	4591.51	-
		Salt (F = 3416.12; *p* < 0.001 *)			
Hospital 1	11.94	2.40	7.30	15.00	0.768
Hospital 2	11.53	1.59	9.12	13.25	0.984
Hospital 3	11.09	1.83	7.79	14.33	1.000
Hospital 4	12.35	1.36	9.86	13.74	0.509
Hospital 5	11.20	1.75	7.53	13.68	1.000
Hospital 6	11.56	0.94	9.50	12.68	0.971
Hospital 7	11.51	1.01	9.20	12.67	0.981
Reference (standardized menu)	10.93	0.49	10.55	11.49	-

M—mean, SD—standard deviation, * one-way ANOVA, ** post hoc Dunnett's two-tailed test (multiple comparisons to a control are also referred to as many-to-one comparisons).

Table 7. Products divided into subgroups.

Hospital	M	SD	Mini	Max	*p*-Value **
		Fruits (F = 723.54; *p* < 0.001 *)			
Hospital 1	222.13	82.60	138.75	334.95	0.022
Hospital 2	302.08	38.68	228.75	341.25	0.998
Hospital 3	181.06	58.06	138.75	301.23	<0.001
Hospital 4	123.16	68.55	36.43	178.10	<0.001
Hospital 5	159.66	79.48	111.00	336.00	<0.001
Hospital 6	159.40	39.92	138.75	261.25	<0.001
Hospital 7	140.73	13.00	122.50	174.75	<0.001
Reference (standardized menu)	316.93	19.07	300.00	346.20	-
		Vegetables (F = 976.51; *p* < 0.001 *)			
Hospital 1	405.29	122.86	188.75	578.40	<0.001
Hospital 2	618.17	103.37	492.24	767.15	0.198
Hospital 3	505.80	174.03	290.73	897.74	0.001
Hospital 4	390.16	95.71	265.26	490.29	<0.001
Hospital 5	344.32	82.91	209.17	438.37	<0.001
Hospital 6	288.80	92.30	73.18	393.36	<0.001
Hospital 7	280.55	95.49	73.18	393.36	<0.001
Reference (standardized menu)	759.97	132.29	574.61	861.45	-
		Whole-grain products (F = 65.67; *p* < 0.001 *)			
Hospital 1	24.00	40.61	0.00	100.00	0.005
Hospital 2	42.00	45.50	0.00	110.00	0.095

Table 7. *Cont.*

Hospital	M	SD	Mini	Max	*p*-Value **
Hospital 3	47.50	34.92	0.00	80.00	0.012
Hospital 4	101.00	48.64	0.00	130.00	0.804
Hospital 5	135.00	129.07	0.00	260.00	1.000
Hospital 6	15.00	24.15	0.00	50.00	0.002
Hospital 7	15.00	24.15	0.00	50.00	0.002
Reference (standardized menu)	135.00	69.76	50.00	200.00	-
		Legumes			
Hospital 1	28.99	12.04	20.48	37.50	
Hospital 2	41.25	-	41.25	41.25	
Hospital 3	25.42	4.69	20.00	28.13	
Hospital 4	62.50	-	62.50	62.50	
Hospital 5	-	-	-	-	
Hospital 6	43.75	-	43.75	43.75	
Hospital 7	43.75	-	43.75	43.75	
Reference (standardized menu)	43.25	11.17	26.25	52.50	
		Nuts and beans			
Hospital 1	10.27	-	10.27	10.27	
Hospital 2	-	-	-	-	
Hospital 3	7.50	3.54	5.00	10.00	
Hospital 4	11.90	-	11.90	11.90	
Hospital 5	-	-	-	-	
Hospital 6	-	-	-	-	
Hospital 7	-	-	-	-	
Reference (standardized menu)	30.00	0.00	30.00	30.00	
		Fish			
Hospital 1	61.08	23.81	40.00	83.33	-
Hospital 2	233.33	-	-	-	-
Hospital 3	58.81	42.48	9.76	83.33	-
Hospital 4	153.33	-	-	-	-
Hospital 5	120.00	-	-	-	-
Hospital 6	88.24	-	-	-	-
Hospital 7	88.24	-	-	-	-
Reference (standardized menu)	216.00	-	-	-	-
		Red meat			
Hospital 1	53.68	31.03	25.67	102.27	
Hospital 2	89.11	49.25	50.53	159.90	
Hospital 3	84.34	41.99	32.08	131.20	
Hospital 4	175.36	23.20	158.95	191.76	
Hospital 5	72.78	40.32	32.08	127.92	
Hospital 6	79.90	23.40	36.90	98.18	
Hospital 7	79.90	23.40	36.90	98.18	
Reference (standardized menu)	-	-	-	-	

Table 7. *Cont.*

Hospital	M	SD	Mini	Max	*p*-Value **
	Processed meat (F = 717.13; $p < 0.001$ *)				
Hospital 1	36.63	11.43	17.70	46.08	0.001
Hospital 2	47.23	12.14	29.50	57.82	0.053
Hospital 3	39.39	5.78	33.98	46.08	0.002
Hospital 4	37.29	10.24	22.66	46.26	0.003
Hospital 5	44.63	17.39	28.32	76.80	0.015
Hospital 6	33.48	10.45	22.66	47.20	<0.001
Hospital 7	33.48	10.45	22.66	47.20	<0.001
Reference (standardized menu)	67.20	19.20	38.40	76.80	-
	Sugar (F = 1327.33; $p < 0.001$ *)				
Hospital 1	54.46	12.79	41.25	83.95	<0.001
Hospital 2	52.46	3.61	48.13	56.57	<0.001
Hospital 3	49.02	5.66	44.89	60.09	<0.001
Hospital 4	2.07	1.45	0.40	3.75	1.000
Hospital 5	42.98	6.12	35.20	56.10	<0.001
Hospital 6	49.43	3.41	46.25	56.25	<0.001
Hospital 7	45.80	9.38	20.00	54.25	<0.001
Reference (standardized menu)	3.09	1.28	1.88	4.43	-

M—mean, SD—standard deviation, * one-way ANOVA, ** post hoc Dunnett's two-tailed test (multiple comparisons to a control are also referred to as many-to-one comparisons).

3.3. Fulfilling the ESC Guidelines

The hospitals were evaluated based on the ESC recommendations [10] mentioned earlier, and the findings are reported in Table 8. All hospitals successfully met four of the criteria, which included intake of vegetables and fish, reduced consumption of red meat and avoidance of alcohol. However, none of the hospitals provided a diet that predominantly consisted of plant-based food and limited consumption of animal-based food, ensured intake of unsalted nuts or limited salt intake to less than 5 g. Hospital 4 met six out of eleven criteria, while Hospitals 2, 3 and 5 met five out of eleven criteria. Hospitals 1, 6 and 7 could only fulfill four out of eleven criteria.

Table 8. The summary of the ESC criteria met by hospitals [10].

Recommendation	Hospital						
	1	2	3	4	5	6	7
The predominance of plant-based and reduction in animal-based food pattern	−	−	−	−	−	−	−
Saturated fatty acids should account for <10% of energy intake	−	−	−	+	−	−	−
<5 g total salt intake per day	−	−	−	−	−	−	−
30–45 g of fiber per day	−	−	+	−	+	−	−
≥200 g of fruit per day (≥2–3 servings)	−	+	−	−	−	−	−
≥200 g of vegetables per day (≥2–3 servings)	+	+	+	+	+	+	+
Reduction In the consumption of red meat to the maximum 350–500 g per week, especially processed meat	+	+	+	+	+	+	+
Eat fish 1–2 times a week	+	+	+	+	+	+	+
30 g of unsalted nuts per day	−	−	−	−	−	−	−
Alcohol consumption should be limited to a maximum of 100 g/week	+	+	+	+	+	+	+
Sweetened beverages, including fruit juices and sweet carbonated drinks and non-carbonated, should be avoided	−	−	−	+	−	−	−

Red colour—a recommendation not fulfiled; Green colour—a recommendation fulfiled.

3.4. Estimated Number of Patients

Each hospital has a specific number of beds, taking into account the province in which it is located, and there is an annual occupancy rate. Based on these factors, it is possible to estimate the total number of patient beds and person days of hospitalization. For example, Hospital 1 has 1035 beds, Hospital 2 has 452 beds, Hospital 3 has 683 beds, Hospital 4 has 677 beds, Hospital 5 has 363 beds, Hospital 6 has 176 beds, and Hospital 7 has 321 beds. By summing up these data, we can determine the number of cardiological patients who represent a missed opportunity for nutrition education, ultimately having a negative impact on the healthcare system. Hospitals 1, 2 and 3 are in the Mazowieckie Voivodeship, where there is an average of 40.8 beds per year, which gives us (2170 × 40.8) 88536patients. Hospitals 4, 6 and 7 are in the Śląskie Voivodeship, where there is an average of 36.2 beds per year, i.e., (1174 × 36.2) 42498,8patients, and Hospital 5 is in the Łódzkie Voivodeship, where the index is 42.4, which means approx. 15391 patients per year [14]. Adding up the number of patients, we get an estimated result: 146425,8patients per year.

4. Discussion

4.1. A Summary of Main Results

Based on the data mentioned above, it is crucial to acknowledge that despite the ESC recommendations with the latest introduced in 2021 [10], there remains a significant oversight in incorporating them into dietary planning in hospitals, consequently having an adverse impact on the health of cardiac patients. The collected data highlight a substantial disparity within the Polish healthcare system. Firstly, certain hospitals do not provide specialized cardiological diets, instead offering general diets that fail to meet the specific nutritional requirements and deviate significantly from the proposed standards. Secondly, hospitals should serve as an environment for initiating nutrition education, exposing patients to well-balanced diets that acquaint them with heart-healthy eating habits. However, the absence of a standardized dietary model poses a risk of patients making nutritional errors that may have a detrimental effect on their health. Thirdly, it is important to highlight that numerous hospitals lack the presence of a qualified dietitian. This absence not only hinders access to essential nutritional guidance but also diminishes the patient's ability to comprehend the dietary requirements they should adhere to. This underscores the critical need for a dietitian to be an integral part of the cardiac patient's healthcare team, given the profound significance of proper nutrition in their treatment and recovery. The aforementioned analysis underscores the critical need for improvement in the domain of nutrition, as the current state of affairs impedes the comprehensive care of cardiac patients in hospitals. Recognizing that these diseases rely heavily on primary and secondary prevention measures is crucial. Compared to the recommendations, the primary issue regarding hospital nutrition lies in the reliance on meat-based dietary patterns rather than plant-based alternatives. Hospitals provide patients with products that are detrimental to their health, such as containing excessive amounts of salt, saturated fatty acids and sugar. The recommendations explicitly state that sweetened beverages should be avoided, including fruit juices as well as sugary carbonated and non-carbonated drinks. However, hospitals continue to offer these to patients. Furthermore, there is suboptimal inclusion of fiber and healthy fatty acids, and none of the hospitals have implemented the recommendation of consuming 30 g of unsalted nuts daily. In the majority of hospitals, fish was only included in one meal per week, which also fails to meet the recommended dietary guidelines. The reference diet developed during this study, based on the recommendations mentioned above, demonstrates that creating such meals does not require hard-to-find ingredients and is even more cost-effective than the dishes currently proposed in existing diets.

4.2. Strengths and Limitations

The positivity of the study lies in its pilot nature, paving the way for future research, a path that a dedicated research team is already ready to follow. The limited number of hospitals included in this study is a potential limitation, which may indicate a low level of representativeness. However, even with this small sample size, it highlights the magnitude of the problem when viewed from a certain perspective. Our calculations assume that approximately 146,425 patients per year did not receive exposure to an appropriate diet. As indicated by the data, inadequately nourished cardiac patients have a higher rate of hospital readmissions, leading to increased healthcare costs. Improper nutrition is often associated with prolonged hospital stays. Moreover, malnutrition can affect patients with both low and excessive body weight. One of the notable strengths of this study lies in its acknowledgment of the genuine challenges faced by cardiac patients within the Polish healthcare system. It sheds light on various aspects, starting from the lack of specialized wards catering to their specific needs to the provision of diets that are detrimental to their health and the absence of crucial nutrition education, which plays a vital role in the recovery process and could potentially alleviate the burden on the healthcare system. The study effectively demonstrates the reality of the meals that patients receive as part of their hospital diet therapy.

4.3. Agreement and Disagreement with Other Studies

The findings from various studies, including our own, provide evidence that the current nutrition provided to cardiac patients in hospitals is insufficient [15,16]. Lifestyle factors undoubtedly play a critical role in determining the occurrence and recurrence of cardiovascular events, with diet being the most extensively studied and substantiated component. It is noteworthy that diet holds equal significance in both primary and secondary prevention of CVD. The CORDIOPREV research has demonstrated that patients can effectively modify their dietary patterns when they receive adequate support from healthcare professionals and comprehensive nutritional education tailored to their cardiological condition [17]. One potential solution to address the nutrition problem among cardiac patients in the hospital is establishing an interdisciplinary team that focuses on the complexity of the issue [18]. Taking a comprehensive approach to nutrition is crucial, as interventions targeting individual nutrients alone, such as vitamin D or omega-3 fatty acids, have not shown sufficient effectiveness in reducing the incidence of CVDs [19]. Accumulating evidence suggests that incorporating targeted nutrition education for patients, emphasizing dietary modifications, has immense potential as a preventive strategy. This approach can decrease reliance on antihypertensive medications, optimize drug dosage and administration and enhance the overall effectiveness of pharmacological interventions. Moreover, this multifaceted approach can have far-reaching implications by positively impacting individuals' cardiovascular risk profiles in both primary and secondary prevention of CVD [20]. Research confirms that patients are most receptive to lifestyle changes after a life-changing event, such as a myocardial infarction, and exposure to a well-planned diet is crucial [21,22]. However, one of the major challenges in nutritional CVD prevention is developing more effective strategies to inspire patients to change their diet and maintain these changes [13,22].

Multiple studies have shown that many hospitals serve fried foods, processed meats and sugar-sweetened beverages on a daily basis [16], which is also supported by our own data. A recent survey conducted in New York City hospitals revealed that none of the eight hospitals surveyed met all the required health standards. The study found that the diets provided to patients contained excessive calories from fat and saturated fat, inadequate fiber intake and sodium levels exceeding the recommended limits [23]. It is well established that a high sodium intake significantly increases the susceptibility to CVD [24]. For cardiac patients, gradually transitioning to less salty foods is advisable to encourage salt reduction in their ongoing management [20]. The hospital diet should serve as a model and an encouragement to change habits; unfortunately, the currently proposed diet in the surveyed Polish hospitals contains an excessively high amount of salt. Moreover, patients consuming food in healthcare settings expect the available options to be nutritious and in line with recommended guidelines [16].

The 2021 guidelines from the European Society of Cardiology emphasize the importance of proper nutrition in the treatment of cardiac patients. A healthy diet based on the Mediterranean diet has been shown to reduce the risk of CVDs and other chronic diseases [10]. The health benefits of this dietary approach have also been demonstrated in primary and secondary cardiovascular prevention [25]. Furthermore, research has shown that adherence to the Mediterranean diet not only provides significant health benefits but also leads to reduced hospitalization durations [26]. Therefore, implementing these dietary recommendations in hospital settings is of paramount importance. It should be emphasized that poor dietary quality has surpassed all other risk factors, contributing to approximately 11 million deaths and accounting for nearly 50% of global CVD mortality [27]. Consequently, addressing dietary quality becomes a critical step in reducing morbidity and mortality associated with CVDs.

In everyday clinical practice for chronic diseases, nutritional issues are often overlooked [9]. Compelling evidence has demonstrated the utmost significance of nutrition for patients perceiving that their meals have been meticulously crafted to meet their specific nutritional requirements. Dissatisfaction with hospital meals has been correlated with decreased consumption, which subsequently increases vulnerability to malnutrition and rehospitalization [15]. The role of diet and nutrition directly impacts cardiovascular health, which, in turn, has the potential to influence readmission rates for heart disease. This provides hospitals with an economic incentive to invest in conscientious dietary options [28]. There is also a new trend of incorporating an all-vegetarian menu into hospital patient populations, which has been shown in Florida hospitals to have positive health benefits for both patients and the hospital budget. These data demonstrate that offering patients a healthy and adequate diet does not increase healthcare costs but has rather positive consequences [28,29].

It is worth emphasizing that as early as 2018 [11], concerns were raised regarding the poor quality of nutrition in hospitals, and as our data show, no significant improvements have been made, with diets in many hospitals still being poorly planned. This indicates that the topic is not sufficiently publicized, and the importance of nutrition is not fully understood. It is crucial that the diet is properly tailored to the patient's nutritional status and health, as being either underweight or obese has been associated with a longer stay [6]. Giving due importance to the issue of hospital nutrition and adhering to recommended guidelines are imperative.

A study has shown that implementing individualized nutritional support, as opposed to standard hospital food, significantly reduces the risk of mortality and major cardiovascular events. Therefore, emphasizing and implementing these recommendations can substantially impact patient outcomes [30].

5. Conclusions

This study shows a clear gap in the Polish healthcare system, wherein a substantial educational potential remains squandered given the considerable influx of patients requiring cardiological care admitted to Polish hospitals. Moreover, it highlights significant deficiencies in the nutritional care provided to cardiac patients in the Polish healthcare system. Hospitals continue to neglect recommended dietary plans, adversely affecting patient health. Many hospitals lack specialized cardiological diets, offering instead general diets that fail to meet patient needs. Nutrition education during hospital stays is inadequate, leading to potential errors in post-discharge diets. Improvements are needed in adopting plant-based diets, reducing processed meat, saturated fatty acids, sugar and salt and including fiber and healthy fatty acids. Implementing recommended guidelines, such as regular fish consumption, avoiding sweetened beverages and providing unsalted nuts, is crucial for better care and patient well-being. It is currently well established that a healthy diet is a cornerstone for CVD primary and secondary prevention. To effectively tackle these challenges, it is imperative to enact policy reforms, launch comprehensive educational campaigns and enhance the synergy among healthcare professionals. By acknowledging and proactively resolving the disparities highlighted in this research, we can enhance the management of cardiac patients and ultimately enhance their overall well-being. This pilot study not only sheds light on current issues but also lays the foundation for promising future research avenues.

Supplementary Materials: The following supporting information can be downloaded at https://www.mdpi.com/article/10.3390/nu15214606/s1, Supplementary S1: The questionnaire, Supplementary S2: Reference diet.

Author Contributions: Conceptualization, D.Ś., A.M.J., A.K., K.Ł. and N.A.; methodology, M.P. and A.M.J.; formal analysis, M.P.; investigation, A.M.J., K.Ł., D.Ś., A.K., K.T. and N.A.; resources, D.Ś., A.K., K.T., A.M.(Artur Mamcarz), A.M. (Agnieszka Młynarska), M.J.-W., A.U., O.M.R., A.M., J.D., M.J.-W., A.U. and P.W.; data curation, M.P.; writing—original draft preparation, A.M.J. and K.Ł.; writing—review and editing, A.M.J., K.Ł., D.Ś., A.K., A.B., O.M.R., P.W., M.G., S.W., M.B. and A.R.P.; visualization, A.M.J., K.Ł., D.Ś. and M.P.; supervision, D.Ś. and A.M.; project administration, A.M.J., K.Ł., D.Ś. and A.K. All authors have read and agreed to the published version of the manuscript.

Funding: This research received no external funding.

Institutional Review Board Statement: This study was conducted according to the guidelines of the Declaration of Helsinki and approved by the Institutional Review of the Bioethical Committee at the Medical University of Warsaw (date: 11 September 2023, number: AKBE/258/2023).

Informed Consent Statement: Informed consent was obtained from all subjects involved in the study.

Data Availability Statement: The raw data supporting the conclusions of this article will be made available by the authors, without undue reservation.

Conflicts of Interest: The authors declare that the research was conducted in the absence of any commercial or financial relationships that could be construed as a potential conflict of interest.

References

1. Vaduganathan, M.; Mensah, G.A.; Turco, J.V.; Fuster, V.; Roth, G.A. The Global Burden of Cardiovascular Diseases and Risk. *J. Am. Coll. Cardiol.* **2022**, *80*, 2361–2371. [CrossRef] [PubMed]
2. Roth, G.A.; Mensah, G.A.; Fuster, V. The Global Burden of Cardiovascular Diseases and Risks. *J. Am. Coll. Cardiol.* **2020**, *76*, 2980–2981. [CrossRef] [PubMed]
3. WHO. Cardiovascular Diseases (CVDs). Available online: https://www.who.int/news-room/fact-sheets/detail/cardiovascular-diseases-(cvds) (accessed on 1 July 2023).
4. Wojtyniak, B.; Goryński, P. Health Status of Polish Population and It's Determinants 2022. Available online: https://www.pzh.gov.pl/raport-sytuacja-zdrowotna-ludnosci-polski-i-jej-uwarunkowania/ (accessed on 1 July 2023).
5. The Polish Central Statistical Office. Mortality in the First Half of 2021. Available online: https://stat.gov.pl/obszary-tematyczne/ludnosc/statystyka-przyczyn-zgonow/umieralnosc-w-2021-roku-zgony-wedlug-przyczyn-dane-wstepne,10,3.html (accessed on 1 July 2023).
6. Czapla, M.; Juárez-Vela, R.; Łokieć, K.; Wleklik, M.; Karniej, P.; Smereka, J. The Association between Nutritional Status and Length of Hospital Stay among Patients with Hypertension. *Int. J. Environ. Res. Public Health* **2022**, *19*, 5827. [CrossRef] [PubMed]
7. Sharma, Y.; Miller, M.; Kaambwa, B.; Shahi, R.; Hakendorf, P.; Horwood, C.; Thompson, C. Factors Influencing Early and Late Readmissions in Australian Hospitalised Patients and Investigating Role of Admission Nutrition Status as a Predictor of Hospital Readmissions: A Cohort Study. *BMJ Open* **2018**, *8*, e022246. [CrossRef] [PubMed]
8. Paterick, T.E.; Patel, N.; Tajik, A.J.; Chandrasekaran, K. Improving Health Outcomes Through Patient Education and Partnerships with Patients. *Bayl. Univ. Med. Cent. Proc.* **2017**, *30*, 112–113. [CrossRef] [PubMed]
9. Boban, M.; Bulj, N.; Kolačević Zeljković, M.; Radeljić, V.; Krcmar, T.; Trbusic, M.; Delić-Brkljačić, D.; Alebic, T.; Vcev, A. Nutritional Considerations of Cardiovascular Diseases and Treatments. *Nutr. Metab. Insights* **2019**, *12*, 117863881983370. [CrossRef]
10. McDonagh, T.A.; Metra, M.; Adamo, M.; Gardner, R.S.; Baumbach, A.; Böhm, M.; Burri, H.; Butler, J.; Čelutkienė, J.; Chioncel, O.; et al. 2021 ESC Guidelines for the Diagnosis and Treatment of Acute and Chronic Heart Failure. *Eur. Heart J.* **2021**, *42*, 3599–3726. [CrossRef] [PubMed]
11. NIK—The Supreme Audit Office. Nutrition of Patients in Hospitals. Available online: https://www.nik.gov.pl/aktualnosci/nik-o-zywieniu-w-szpitalach.html (accessed on 1 July 2023).
12. de Lorgeril, M.; Salen, P.; Martin, J.L.; Monjaud, I.; Delaye, J.; Mamelle, N. Mediterranean diet, traditional risk factors, and the rate of cardiovascular complications after myocardial infarction: Final report of the Lyon Diet Heart Study. *Circulation* **1999**, *99*, 779–785. [CrossRef] [PubMed]

13. Kjeldsen, E.W.; Thomassen, J.Q.; Rasmussen, K.L.; Nordestgaard, B.G.; Tybjærg-Hansen, A.; Frikke-Schmidt, R. Impact of Diet on Ten-Year Absolute Cardiovascular Risk in a Prospective Cohort of 94 321 Individuals: A Tool for Implementation of Healthy Diets. *Lancet Reg. Health-Eur.* **2022**, *19*, 100419. [CrossRef]
14. Ministry of Health. Statistical Bulletin 2022. Available online: https://ezdrowie.gov.pl/portal/home/badania-i-dane/biuletyn-statystyczny (accessed on 1 July 2023).
15. Osman, N.S.; Md Nor, N.; Md Sharif, M.S.; Hamid, S.B.A.; Rahamat, S. Hospital Food Service Strategies to Improve Food Intakes among Inpatients: A Systematic Review. *Nutrients* **2021**, *13*, 3649. [CrossRef]
16. Petersen, M.R.; Freeman, A.M.; Madrid, M.; Aggarwal, M. Strategies for Incorporating Lifestyle Medicine in Everyday Hospital Practice. *Am. J. Lifestyle Med.* **2021**, *15*, 531–537. [CrossRef] [PubMed]
17. Delgado-Lista, J.; Alcala-Diaz, J.F.; Torres-Peña, J.D.; Quintana-Navarro, G.M.; Fuentes, F.; Garcia-Rios, A.; Ortiz-Morales, A.M.; Gonzalez-Requero, A.I.; Perez-Caballero, A.I.; Yubero-Serrano, E.M.; et al. Long-Term Secondary Prevention of Cardiovascular Disease with a Mediterranean Diet and a Low-Fat Diet (CORDIOPREV): A Randomised Controlled Trial. *Lancet* **2022**, *399*, 1876–1885. [CrossRef] [PubMed]
18. Driggin, E.; Cohen, L.P.; Gallagher, D.; Karmally, W.; Maddox, T.; Hummel, S.L.; Carbone, S.; Maurer, M.S. Nutrition Assessment and Dietary Interventions in Heart Failure. *J. Am. Coll. Cardiol.* **2022**, *79*, 1623–1635. [CrossRef] [PubMed]
19. Vadiveloo, M.; Lichtenstein, A.H.; Anderson, C.; Aspry, K.; Foraker, R.; Griggs, S.; Hayman, L.L.; Johnston, E.; Stone, N.J.; Thorndike, A.N.; et al. Rapid Diet Assessment Screening Tools for Cardiovascular Disease Risk Reduction Across Healthcare Settings: A Scientific Statement From the American Heart Association. *Circ. Cardiovasc. Qual. Outcomes* **2020**, *13*, e000094. [CrossRef] [PubMed]
20. Cicero, A.F.G.; Veronesi, M.; Fogacci, F. Dietary Intervention to Improve Blood Pressure Control: Beyond Salt Restriction. *High Blood Press. Cardiovasc. Prev.* **2021**, *28*, 547–553. [CrossRef] [PubMed]
21. Steca, P.; Monzani, D.; Greco, A.; Franzelli, C.; Magrin, M.E.; Miglioretti, M.; Sarini, M.; Scrignaro, M.; Vecchio, L.; Fattirolli, F.; et al. Stability and Change of Lifestyle Profiles in Cardiovascular Patients after Their First Acute Coronary Event. *PLoS ONE* **2017**, *12*, e0183905. [CrossRef]
22. Diab, A.; Dastmalchi, L.N.; Gulati, M.; Michos, E.D. A Heart-Healthy Diet for Cardiovascular Disease Prevention: Where Are We Now? *Vasc. Health Risk Manag.* **2023**, *19*, 237–253. [CrossRef]
23. Moran, A.; Lederer, A.; Johnson Curtis, C. Use of Nutrition Standards to Improve Nutritional Quality of Hospital Patient Meals: Findings from New York City's Healthy Hospital Food Initiative. *J. Acad. Nutr. Diet.* **2015**, *115*, 1847–1854. [CrossRef]
24. Wang, Y.-J.; Yeh, T.-L.; Shih, M.-C.; Tu, Y.-K.; Chien, K.-L. Dietary Sodium Intake and Risk of Cardiovascular Disease: A Systematic Review and Dose-Response Meta-Analysis. *Nutrients* **2020**, *12*, 2934. [CrossRef]
25. Minelli, P.; Montinari, M.R. The Mediterranean Diet and Cardioprotection: Historical Overview And Current Research. *J. Multidiscip. Healthc.* **2019**, *12*, 805–815. [CrossRef]
26. Lampropoulos, C.E.; Konsta, M.; Dradaki, V.; Roumpou, A.; Dri, I.; Papaioannou, I. Effects of Mediterranean Diet on Hospital Length of Stay, Medical Expenses, and Mortality in Elderly, Hospitalized Patients: A 2-Year Observational Study. *Nutrition* **2020**, *79–80*, 110868. [CrossRef] [PubMed]
27. Afshin, A.; Sur, P.J.; Fay, K.A.; Cornaby, L.; Ferrara, G.; Salama, J.S.; Mullany, E.C.; Abate, K.H.; Abbafati, C.; Abebe, Z.; et al. Health Effects of Dietary Risks in 195 Countries, 1990–2017: A Systematic Analysis for the Global Burden of Disease Study 2017. *Lancet* **2019**, *393*, 1958–1972. [CrossRef] [PubMed]
28. Saldivar, B.; Al-Turk, B.; Brown, M.; Aggarwal, M. Successful Incorporation of a Plant-Based Menu Into a Large Academic Hospital. *Am. J. Lifestyle Med.* **2022**, *16*, 311–317. [CrossRef]
29. Aggarwal, M.; Grady, A.; Desai, D.; Hartog, K.; Correa, L.; Ostfeld, R.J.; Freeman, A.M.; McMacken, M.; Gianos, E.; Reddy, K.; et al. Successful Implementation of Healthful Nutrition Initiatives into Hospitals. *Am. J. Med.* **2020**, *133*, 19–25. [CrossRef] [PubMed]
30. Hersberger, L.; Dietz, A.; Bürgler, H.; Bargetzi, A.; Bargetzi, L.; Kägi-Braun, N.; Tribolet, P.; Gomes, F.; Hoess, C.; Pavlicek, V.; et al. Individualized Nutritional Support for Hospitalized Patients With Chronic Heart Failure. *J. Am. Coll. Cardiol.* **2021**, *77*, 2307–2319. [CrossRef]

nutrients

MDPI

Article

Evaluation of Dietary and Alcohol Drinking Patterns in Patients with Excess Body Weight in a Spanish Cohort: Impact on Cardiometabolic Risk Factors

Maite Aguas-Ayesa [1,2], Patricia Yárnoz-Esquiroz [1,2,3], Laura Olazarán [1,2,3], Carolina M. Perdomo [1,2], Marta García-Goñi [1], Patricia Andrada [1], Javier Escalada [1], Camilo Silva [1,2,3], Ascensión Marcos [4] and Gema Frühbeck [1,2,3,*]

1. Department of Endocrinology & Nutrition, Clínica Universidad de Navarra, 31008 Pamplona, Spain
2. CIBER Fisiopatología de la Obesidad y Nutrición (CIBEROBN), Instituto de Salud Carlos III, 31008 Pamplona, Spain
3. Obesity and Adipobiology Group, Instituto de Investigación Sanitaria de Navarra (IdiSNA), 31008 Pamplona, Spain
4. Immunonutrition Research Group, Department of Metabolism and Nutrition, Institute of Food Science and Technology and Nutrition (ICTAN)—CSIC, 28040 Madrid, Spain

* Correspondence: gfruhbeck@unav.es; Tel.: +34-948-255-400

Abstract: Unhealthy dietary habits and sedentarism coexist with a rising incidence of excess weight and associated comorbidities. We aimed to analyze the dietary and drinking patterns of patients with excess weight, their main characteristics, plausible gender differences and impact on cardiometabolic risk factors, with a particular focus on the potential contribution of beer consumption. Data from 200 consecutive volunteers (38 ± 12 years; 72% females) living with overweight or class I obesity attending the obesity unit to lose weight were studied. Food frequency questionnaires and 24 h recalls were used. Reduced-rank regression (RRR) analysis was applied to identify dietary patterns (DPs). Anthropometry, total and visceral fat, indirect calorimetry, physical activity level, comorbidities and circulating cardiometabolic risk factors were assessed. Study participants showed high waist circumference, adiposity, insulin resistance, dyslipidemia, pro-inflammatory adipokines and low anti-inflammatory factors like adiponectin and interleukin-4. A low-fiber, high-fat, energy-dense DP was observed. BMI showed a statistically significant ($p < 0.05$) correlation with energy density (r = 0.80) as well as percentage of energy derived from fat (r = 0.61). Excess weight was associated with a DP low in vegetables, legumes and whole grains at the same time as being high in sweets, sugar-sweetened beverages, fat spreads, and processed meats. RRR analysis identified a DP characterized by high energy density and saturated fat exhibiting negative loadings (>−0.30) for green leafy vegetables, legumes, and fruits at the same time as showing positive factor loadings (>0.30) for processed foods, fat spreads, sugar-sweetened beverages, and sweets. Interestingly, for both women and men, wine represented globally the main source of total alcohol intake ($p < 0.05$) as compared to beer and distillates. Beer consumption cannot be blamed as the main culprit of excess weight. Capturing the DP provides more clinically relevant and useful information. The focus on consumption of single nutrients does not resemble real-world intake behaviors.

Keywords: overweight; obesity; dietary pattern; alcoholic beverages; beer; adiposity; gender differences; cardiometabolic risk factors

Citation: Aguas-Ayesa, M.; Yárnoz-Esquiroz, P.; Olazarán, L.; Perdomo, C.M.; García-Goñi, M.; Andrada, P.; Escalada, J.; Silva, C.; Marcos, A.; Frühbeck, G. Evaluation of Dietary and Alcohol Drinking Patterns in Patients with Excess Body Weight in a Spanish Cohort: Impact on Cardiometabolic Risk Factors. *Nutrients* **2023**, *15*, 4824. https://doi.org/10.3390/nu15224824

Academic Editors: Domenico Tafuri and Francesca Latino

Received: 6 October 2023
Revised: 14 November 2023
Accepted: 16 November 2023
Published: 17 November 2023

1. Introduction

A dramatic increase in obesity prevalence has been witnessed in the last decades [1–5]. The World Health Organization (WHO) recognizes that currently 39% of the world's population presents overweight, while 13% is living with obesity [6]. Similar demographic trends are observed in Spain, with an estimated prevalence of overweight of 39.3% and obesity of 21.6% in Spanish adults aged 25–64 years [7,8]. Noteworthily, excess weight is an independent risk factor for type 2 diabetes (T2D), cardiovascular diseases (CVD), hypertension, non-alcoholic fatty liver, dyslipidemia, and certain cancers, among others [9,10]. Thus, excess weight and its accompanying comorbidities embody one of the main health problems worldwide [11,12].

Unhealthy dietary habits and sedentarism coexist with the rising incidence of overweight, obesity, the associated comorbidities and chronic disease [13–15]. After decades of proposed policies, obesity prevalence has not been successfully decreased. Despite the knowledge of obesity as a chronic, treatable disease, its management still faces challenges regarding the most adequate treatment approach [16]. In this context, nutrition, among others, is a key element in health and disease with a maintained caloric imbalance between energy expenditure and intake over time promoting the development of obesity [17,18]. The diverse components that influence total energy expenditure (TEE) include thermic effect of food (TEF), resting energy expenditure (REE) and physical activity [19,20]. While sophisticated techniques like doubly labelled water or direct calorimetry can be applied to measure TEE, they are mainly being carried out for research purposes [21,22]. In the clinical setting, however, REE can be determined by indirect calorimetry, which combined with physical activity level (PAL) assessment can inform about TEE [23].

Dietary interventions systematically offered during decades to lose weight have lacked patient experience, thereby being vastly ineffective in the long term. Lifestyle choices and in particular dietary changes are often seen as something that people living with obesity (PlwO) should improve by themselves, representing one of the lacunae in management. Traditionally, emphasis has been placed on avoiding particularly detrimental foods. Overconsumption of energy based on nutrient-poor, energy-dense foods and drinks reportedly elevates the risk of excess body weight and also increases the risk of obesity comorbidities [24]. Noteworthily, these dietary and drinking patterns rely on regional and cultural beliefs, as do some unsupervised weight-loss attempts. In Spain, people attempting to lose weight usually eliminate bread and beer from their diets, since these food items are popularly considered as the main culprits of excess weight. However, this belief has not been supported by scientific evidence specifically addressing this question.

In this context, we hypothesized that the excess energy intake cannot be blamed on a single food item like beer. Therefore, the rationale of the present study relied on the fact that capturing the multiple dietary options comprising a dietary pattern (DP) is key for better understanding energy imbalance in excess weight. To that end, we aimed to analyze the dietary and drinking patterns of patients with excess weight to decipher their main characteristics and the potential existence of gender differences and their impact on cardiometabolic risk factors, with a particular focus on the potential contribution of beer consumption. This information will enable the possibility of establishing more evidence-based dietary recommendations based on actual DPs to achieve a more holistic body weight regulation rather than condemning single foods.

2. Materials and Methods

2.1. Patient Selection and Study Design

This one-arm, open-label, exploratory, cross-sectional study collected momentary data from 200 consecutive male and female White patients with overweight or class I obesity attending the Endocrinology and Nutrition Department of the University of Navarra Clinic to lose weight. Criteria for inclusion consisted of being of either gender, a non-smoker, aged between 18–65 years, having a body mass index (BMI) between 25.0–35.0 kg/m^2, not taking medication, and having no psychiatric pathology. All subjects involved in the study provided informed consent. Criteria for exclusion included other severe systemic diseases unrelated to excess weight, infections or inflammatory processes, cancer, or marked nephro- or hepatopathy, other than metabolic-associated fatty liver disease, pregnancy or lactation, being vegetarian, vegan, or persons with special dietary requirements. The sample population thus represents a homogeneous study group comprised of the habitual patients attending our department and clinic who share a common ethnic, cultural and socioeconomic background. In this line, neither physical activity nor lifestyle factors accounted as potential confounders due to the shared sedentarism and daily living style habits. This is a substudy of the trial registered with ClinicalTrials.gov, number NCT01055626. The study design was accepted from a scientific and ethical point of view by the Ethical Committee of the Hospital, receiving approval number 30/2006 further updated in number 2020.041. All procedures conformed to the guidelines of the Declaration of Helsinki as well as its subsequent revisions; the research protocol was approved by the Universidad de Navarra's Ethical Committee (protocol 2020.041).

2.2. *Anthropometry, Body Composition and Resting Energy Expenditure Determination*

To obtain the BMI, weight was measured in kg with a 0.1 kg precision and height was determined in meters with a Harpenden stadiometer (Holtain Ltd., Crymych, UK). To measure waist circumference (WC), a non-elastic tape was placed on the midaxillary line at the midpoint between the rib cage and the iliac crest. In order to ensure uniformity in anthropometric measurements and to limit interobserver variability, anthropometric measurements were obtained by the same researcher with the ISAK (International Society for the Advancement of Kinanthropometry) certification. Total percent body fat was determined using the Bod-Pod®, based on air-displacement plethysmography (Life Measurements, Concord, CA, USA) applying the equation of Siri, as described before [25]. Air-displacement plethysmography reportedly closely agrees with hydrodensitometry, or underwater weighing, traditionally considered the gold standard [26]. Trunk fat and visceral adiposity were quantified by abdominal bioimpedance using a Tanita AB-140 device (Tanita Corp., Tokyo, Japan), as previously reported [27]. Trunk fat includes both subcutaneous and intra-abdominal adipose tissue, while visceral fat is expressed in arbitrary units, which, as indicated by the manufacturer, multiplied by 10 equates to the visceral adiposity area assessed by computed tomography expressed in cm^2. Blood pressure was determined in a semi-sitting position with a sphygmomanometer at the non-dominant arm at at least 3 different time intervals to obtain the mean following a 5 min rest. After a 12 h fasting period, production of carbon dioxide and consumed oxygen were measured at thermoneutrality by indirect calorimetry (Vmax29, SensorMedics Corporation, Yorba Linda, CA, USA) to determine REE [23]. A validated questionnaire [28] was applied to estimate the physical activity level (PAL) to obtain the TEE of all study participants.

2.3. *Dietetic History and Nutritional Information*

Information on dietary intake was collected by trained dietitians using 24 h recall records together with a standardized dietary intake toolkit to help with quantification. Interviews covered both workdays and weekends to ensure a proportional representation. All types of drinks (including alcoholic beverages) and foods were registered, calculating the size of portions with habitual brands, household estimates of package weights, and where appropriate, also collecting ways of cooking [29]. Foods consumed in quantities over 15 g or one tablespoon were included. Mean intakes of the single macronutrients, specific food groups and individual DPs were calculated across the 24 h recalls. While 24 h recalls are more appropriate to gauge the eating patterns of a specific demography, food frequency questionnaires (FFQs) are more appropriate for tracking individual dietary patterns. Foods not consumed on a daily basis but still relevant to a person's diet were captured via FFQs that rely on a longer recall period. Bias in FFQs was minimized by conducting dietitian-led interviews instead of self-reported questionnaires.

In order to develop the cluster analysis, the food group information needs to be obtained. The 24 h recall data identified 16 food groups, which included eggs, dairy products and milk, seafood and fish, meat, cold meats, legumes, cereals, starchy staples, leafy dark green vegetables, roots and tubers, other vegetables, fruits, nuts and seeds, oils, fats and spreads, sweets, and beverages. Mean dietary intake was obtained by the multiple-pass approach applied, including the three-steps method, namely 'quick listing', 'detailed food and beverage describing' and 'final reviewing', to enhance full and precise recalling to decrease respondent burden. Dietary adequacy was analyzed via individual diet diversity scores by means of summing the number of nine aggregated food groups (eggs, meats, fish, milk and dairy products, starchy staples, vegetables, fruits, legumes, nuts and seeds) [30]. The actual complexities of the general population's eating habits can be better captured by DP analysis, as it accounts for the interactive and synergistic effects of different nutrients and food groups [31,32]. Reduced-rank regression (RRR) uses data-driven analysis of real-world eating behaviors aimed at identifying DPs in food intake reports via the combination of current knowledge and plausible mechanisms of diet-related disease [33]. Thereby, establishment of the relative relevance of a number of nutrient-related mechanisms plausibly linking diet to weight or cardiometabolic risk can be achieved by RRR [34]. The combination of nutritional information is important, since to reliably assess diet–disease relations the within-person variability in dietary consumption needs to be contemplated [35].

2.4. Blood Analyses

Anthropometric as well as body composition determinations were carried out on the same day together with the blood extraction. To avoid plausible confounders based on hormone rhythmicity, blood samples were taken in the morning after an overnight fast of at least 10 h. An automated analyzer (Roche, Basel, Switzerland/Hitachi, Tokyo, Japan Modular P800) was used to quantify plasma glucose [36]. An enzyme-amplified chemiluminescence assay (Immulite®, Diagnostic Products Corp., Los Angeles, CA, USA) was used to measure insulin concentrations, as in prior studies [37,38]. Insulin sensitivity and resistance were estimated via the quantitative insulin-sensitivity check index (QUICKI) [39] and the homeostatic model assessment (HOMA) [40], respectively [41,42]. The lipid profile was assessed via enzymatic spectrophotometric methods (Roche, Basel, Switzerland) for triglyceride and total cholesterol determination [43]. Colorimetry (Beckman Synchron® CX analyzer, Beckman Instruments, Ltd., Bucks, UK) was applied to determine high-density lipoprotein (HDL-cholesterol), while low-density lipoprotein (LDL-cholesterol) was obtained from the Friedewald formula. The ultrasensitive assay (Tina-quant® CRP, Roche, Basel, Switzerland) was used to measure high-sensitivity C-reactive protein (CRP). The Clauss method was applied to analyze circulating fibrinogen levels (Hemoliance®, Instrumentation Laboratory, Barcelona, Spain). A fluorescence polarization immunoassay (Axis Biochemicals ASA, Oslo, Norway) was used for homocysteine determination with an IMX analyzer (Abbott, Abbott Park, IL, USA). An automated analyzer (Roche/Hitachi Modular P800) analyzed aspartate aminotransferase (AST), alanine aminotransferase (ALT), γ-glutamyltransferase (γ-GT), uric acid, and creatinine [44,45].

Targeted adipokines and cytokines were selected to reflect the impact on inflammation of the DP identified. Furthermore, well-known cardiometabolic risk factors were also included to evaluate the effects of excess weight. Determination of the circulating concentrations of pro-inflammatory adipo-cytokines and interleukines (IL) included leptin, which works also as a lipostat, by quantification via a double-antibody radioimmunoassay method (Linco Research, Inc., St. Charles, MO, USA) with 5.0% and 4.5% intra-and inter-assay coefficients of variation (CV), respectively [46]. An enzyme-linked immunosorbent assay (ELISA) was applied for the determination of adiponectin, which operates as an insulin sensitizer and cardioprotective factor (BioVendor, Brno, Czech Republic) with intra-and inter-assay CVs of 6.7% and 7.8%, respectively. Chitinase-3-like protein 1 (YKL-40), osteopontin (OPN), lipocalin-2 (LCN-2) and tenascin C (TNC) are well-known proinflammatory factors and were assessed with commercially available ELISA kits of R&D systems (Minneapolis, MN, USA) and IBL International GMBH (Hamburg, Germany), which exhibited the following intra- and interassay CVs: 4.6% and 6.0% for YKL-40; 3.2 and 5.9% for OPN; 3.7% and 6.5% for LCN-2; 4.9% and 5.4% for TNC, respectively. A high-sensitivity ELISA kit was used to measure TNF-α (no. HSTAA00D, R&D Systems, Minneapolis, MN, USA), which also has relevant autocrine modulation, with intra- and inter-assay CVs of 5.0% and 5.7%, respectively. Since some interleukins have proinflammatory effects while others exert anti-inflammatory properties, representative cytokines of both of them were assessed. IL-1β

concentrations were determined by ELISA (RAF052R, BioVendor), while IL-6 was analyzed by ELISA (DY206-05, R&D Systems) with an intra-assay CV of 5.4% and an inter-assay CV of 11.4%. Monocyte chemoattractant protein-1 (MCP-1) and IL-10 serum levels were analyzed via the Human Quantikine® Immunoassay kits (R&D Systems), while IL-4 and IL-8 were determined with ELISA commercial kits (Genzyme, Cambridge, MA, USA) and all within-assay CV being less than 8%. Similarly, circulating concentrations of IL-32 were assessed with an ELISA commercial kit (CUSABIO, College Park, MD, USA) yielding intra- and interassay CVs < 8.0 and 10.0%, respectively.

2.5. Statistical Analyses

Mean ± standard errors of the mean (SEM) were used to present data. The Kolmogorov–Smirnov test assessed the normal distribution while Levene's test analyzed homogeneity of variance. Student's *t* tests were applied for comparison of two groups. ANOVA and χ^2 analysis, as pertinent, were carried out to analyze gender distribution differences, type of alcoholic beverage and frequency distribution with categorical variables. Pearson's correlation coefficients (*r*) interrogated the correlations between anthropometric measures, energy density and source. SPSS version 23 (SPSS, Chicago, IL, USA) was used for all calculations. Because of the homogeneity of the study sample, the possibility of potential confounders due to differences in race, ethnicity, lifestyle factors, educational level, and socioeconomic background, among others, was not observed. Statistical significance was established at 0.05.

The mean intakes of specific nutrients, individual food groups and DPs were analyzed across the 24 h dietary assessment. Additionally to comparing reproducibility estimates for consumption of food groups determined by the 24 h assessment as well as the FFQ, the comparison with the rankings of usual food group intake was established. DPs were derived from RRR in order to establish the linear function of food groups (as predictor variables) via maximization of the explained response variation [34]. Different DPs were derived from each analysis, encompassing the whole spectrum from the 'more healthy' to the 'less healthy' profile. No notable differences between the patterns derived from percentage energy, adjusting weight for intake of total energy or directly using weight values, were observed [47]. DP scores were considered in categorical and continuous forms applying the Statistical Analysis System software version 9.4 (SAS Institute, Cary, NC, USA). Analyses of weighted Pearson correlation evaluated the correlations between DP scores and their subcomponents. DP z-scores were calculated for each participant by the RRR model as a linear, weighted combination of all of their standardized food group intakes with unique weights for each DP. Individual z-scores for each participant were obtained for each DP derived, whereby a higher z-score corresponds to a higher adherence to the DP identified. Thus, intake of foods with a negative factor loading decreased the DP z-score, while consumption of foods with a positive factor loading increased the DP z-score. Factor loadings ≤ -0.20 and ≥ 0.20 for food groups were significant and taken as the more prominent negative and positive contributors to the DP z-scores, respectively.

3. Results

3.1. Study Cohort Characteristics

The first 200 participants attending the Endocrinology and Nutrition Department for weight loss exhibited a mean age of 38 ± 12 years with a majority of females (72%). This finding is in agreement with males and females exhibiting different behaviors and attitudes as regards weight perception, with males being less likely to perceive themselves as overweight than females [48]. Moreover, this misperception translates into males being less likely than females to seek advice for weight loss [49]. The anthropometric and clinical characteristics of study participants are summarized in Table 1. Participants belonged to both the overweight and obesity categories, exhibiting a mean BMI within class I obesity accompanied by a high WC together with an elevated total and visceral adiposity. Although the mean baseline glycemia was within the normal range, the high HOMA index evidenced that it was achieved at the expense of a high insulin resistance accompanied by a low insulin sensitivity, as indicated by the QUICKI index. Participants had a normal hepatic and renal function. Mean blood pressure, both systolic and diastolic, were within the normal range. As expected with the high adiposity of the participants, metabolic alterations consistent in dyslipidemia, elevated proinflammatory adipokines like leptin, osteopontin, lipocalin 2, tenascin C, YKL-40, TNF-α, MCP-1, IL-1β, and IL-32, as well as high circulating concentrations of cardiometabolic risk factors like CRP, von Willebrandt factor, homocysteine and fibrinogen together with low anti-inflammatory adipokines like adiponectin and cytokines such as interleukin-4, were observed.

Table 1. Anthropometric and clinical characteristics of the study participants.

Variables	Mean ± SEM (n = 200)
Sex (males/females)	56/144
Age (years)	38 ± 12
BMI (kg/m^2)	32.2 ± 3.5
Body fat (%)	40.0 ± 4.3
Visceral adiposity (a.u.)	13.9 ± 3.7
WC (cm)	97 ± 15
Systolic BP (mm Hg)	131 ± 16
Diastolic BP (mm Hg)	85 ± 10
Glucose (mg/dL)	98 ± 17
Insulin (μU/mL)	11.9 ± 3.5
HOMA-IR	2.87 ± 0.91
QUICKI	0.330 ±0.044
Triglycerides (mg/dL)	103 ± 10
Total cholesterol (mg/dL)	208 ± 22
HDL-cholesterol (mg/dL)	53 ± 8
LDL-cholesterol (mg/dL)	130 ± 9
Free fatty acids (mEq/L)	0.5 ± 0.1
Creatinine (mg/dL)	0.80 ± 0.11
Uric acid (mg/dL)	4.5 ± 0.2
Urea (mg/dL)	12.4 ± 1.3
ALT (UI/L)	19 ± 7
AST (UI/L)	25 ± 8
AP (UI/L)	66 ± 9
GGT (UI/L)	28 ± 9
Leptin (ng/mL)	36.5 ± 6.1
Adiponectin (μg/mL)	6.7 ± 2.1
Osteopontin (ng/mL)	68.8 ± 9.4
Lipocalin 2 (ng/mL)	70.2 ± 3.6
Tenascin C (ng/mL)	63.1 ± 3.7
YKL-40 (ng/mL)	36.0 ± 4.4
TNF-α (pg/mL)	6.37 ± 3.52
MCP-1 (pg/mL)	486 ± 200
IL-1β (pg/mL)	5.15 ± 1.02
IL-4 (ng/mL)	3.41 ± 1.18
IL-6 (pg/mL)	4.42 ± 0.79
IL-8 (pg/mL)	4.31 ± 1.25
IL-10 (pg/mL)	6.52 ± 2.38
IL-32 (pg/L)	10.5 ± 2.3
CRP (mg/L)	5.3 ± 2.7
vW Factor Ag (%)	133 ± 28
Homocysteine (μmol/L)	8.72 ± 0.79
Fibrinogen (mg/dL)	286 ± 25

BMI, body mass index; a.u., arbitrary units; WC, waist circumference; BP, blood pressure; HOMA-IR, homeostasis model assessment—insulin resistance; QUICKI, quantitative insulin sensitivity check index; ALT, alanine aminotransferase; AST, aspartate aminotransferase; AP, alkaline phosphatase; Gamma-GT, gamma-glutamil transferase; YKL-40, chitinase 3-like protein-1; TNF-α, tumor necrosis factor alpha; MCP-1, monocyte chemotaxis protein 1; IL, interleukin; CRP, C reactive protein; vW factor Ag, von Willebrand factor antigen.

3.2. Energy Imbalance, Dietary and Drinking Patterns

Globally, participants exhibited an excess energy intake over expenditure. Furthermore, gender-specific differences were observed, with men showing a significantly higher ($p < 0.05$) intake of calories derived from alcohol than women. However, energy intake derived from alcohol was not high (2.9 and 4.5% of total energy intake for females and males, respectively) showing that this was not the main factor explaining the energy imbalance between intake and expenditure (Table 2). This was further corroborated, as 5% of the participants attending the Department for weight loss were nondrinkers. The minimal and maximal values show the wide range in intake of both females and males.

Table 2. Total caloric intake derived from alcohol consumption.

Energy	Females			Males			Stat. Sign.
	Mean	Minimum	Maximum	Mean	Minimum	Maximum	
Total energy intake (TEI) (kcal/d)	2252	1219	3562	2494	1915	4158	*
EI from alcohol (kcal/d)	73	0	430	96	0	482	*
EI from alc. (% of total)	2.9	0	12.1	4.5	0	15.4	*

EI from alc, energy intake from alcohol. Data are presented as mean followed by minimum and maximum values in brackets. Stat. sign., statistical significance; * $p < 0.05$ assessed by Chi-square (χ^2) analysis.

Interestingly, for both women and men, wine represented globally the main source of total alcohol intake ($p < 0.05$), as compared to beer and distillates (Table 3). This pattern was maintained in female participants that had either less than one drink per week or 5–7 drinks per week, whereas in males having less than 1 drink per week beer was the main type of alcoholic beverage. Data in low- and high-frequency consumption provide clinically interesting information. As expected, the proportion of males consuming beer was significantly higher ($p < 0.05$) than that of women in global terms, while females consumed significantly more ($p < 0.05$) wine than men without significant differences between sexes regarding liquors and distilled spirits.

Table 3. Proportion of total alcohol intake according to type of beverage and frequency of consumption.

Type of Beverage	Females Frequency (Drinks/Week)			Males Frequency (Drinks/Week)			Stat. Sign.
	(%)	<1	5–7	(%)	<1	5–7	
Beer	18.8	23.7	13.8 #	41.0	48.0 §	38.6	*
Wine	69.3 §	52.6 §	79.4 §	46.3	31.6	51.0 §	*
Distillates	11.9	23.7	6.9 #	12.7 §	20.3 §	10.5 #	ns

Distillates, liquors and distilled spirits. Data are presented as mean percentage values. Stat. sign., statistical significance; Chi-square (χ^2) analysis was applied to determine differences in gender and categorical variables; ns, non-significant for gender; $p < 0.05$ assessed by χ^2 for * gender, § type of beverage and # frequency of consumption.

Figure 1 shows the macronutrient distribution derived from the total energy intake for both males and females obtained during the dietetic history. Both genders exhibited an inadequate intake resulting from a high intake of fats and a low intake of complex carbohydrates as compared to healthy dietary recommendations.

The intake of sugar-sweetened beverages was above the optimal intake for both males and females. In addition, the energy consumption of healthy foods was suboptimal in the cohort studied (Tables 4 and 5). Noteworthily, frequency of intake of specific foods in relation to the frequency of consumption of alcoholic drinks per week yields valuable information as regards the impact of healthy and unhealthy food groups. Men exhibited generally a higher consumption of both unhealthy and healthy foods than women. Overweight and obesity were both associated with a DP low in vegetables, legumes and whole grains at the same time as being high in sweets, sugar-sweetened beverages, spreads, and processed meats (Figure 2). RRR analysis identified a DP characterized by high energy density and saturated fat exhibiting negative loadings (>−0.30) for green leafy vegetables, legumes, and fruits at the same time as showing positive factor loadings (>0.30) for processed foods, fat spreads, sugar-sweetened beverages, and sweets (Figure 3). A low-fiber, high-fat, energy-dense DP was observed in the study participants. BMI showed a statistically significant ($p < 0.05$) correlation with energy density (r = 0.80) as well as percent of energy derived from fat (r = 0.61), while positive high factor loadings significantly ($p < 0.05$) related to adverse biomarkers profiles.

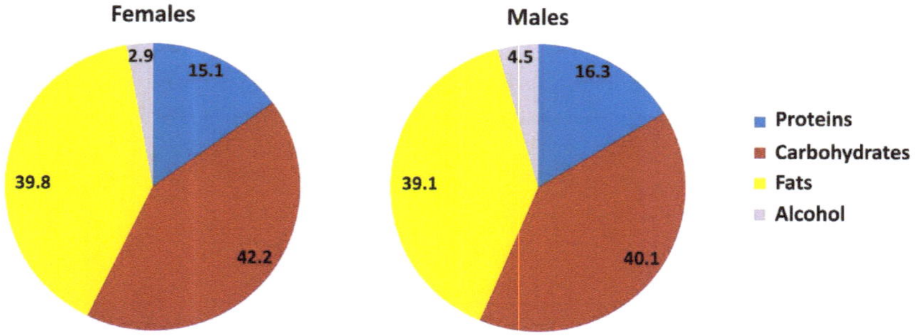

Figure 1. Percentage distribution of macronutrients of the total energy intake. Mean values expressed as percentage (%) of total energy intake.

Table 4. Global proportion of participants with a low intake of foods in relation to frequency of alcoholic drinks per week.

Diet Poor in	Females Freq. (Drinks/Week)			Males Freq. (Drinks/Week)			Stat. Sign.
	(%)	<1	5–7	(%)	<1	5–7	
Fib. and cer.	71.4 §	70.2	79.5	74.8 §	76.3	65.8	ns
Vegetables	66.8 §	61.3	72.0	71.0 §	74.8	53.7 #	ns
Salads	55.2 §	57.5	53.1	69.4 §	68.6	63.1	*
Legumes	87.3 §	85.4	89.6	85.3 §	88.7	83.6	ns
Fruits	20.0	15.2	24.1 #	39.2	39.0	38.5	*
Fish	60.2	63.6	57.8	68.1	67.8	60.0	ns

Fib. and cer., Fiber and cereals; Freq., Frequency. Data are presented as mean percentage values. Stat. sign., statistical significance; Chi-square (χ^2) analysis was applied to determine differences in gender and categorical variables; ns, non-significant for gender; $p < 0.05$ assessed by χ^2 for * gender, § type of beverage and # frequency of consumption.

Table 5. Global proportion of participants with elevated intake of foods in relation to frequency of alcoholic drinks per week.

Diet High in	Females Fre. (Drinks/Week)			Males Freq. (Drinks/Week)			Stat. Sign.
	(%)	<1	5–7	(%)	<1	5–7	
Sweets	93.1 §	92.4	91.7	79.6 §	80.3	82.1	*
Solid fats	76.5	75.3	77.2	77.8	79.0	75.4	ns
Ch./cold mts	49.9 §	50.6 §	48.8 §	83.4 §	81.2	84.7	*
Sug.-sw. bev.	80.7 §	80.2	81.0	81.2 §	80.3	83.6	ns

Solid fats, butter and margarine; Ch./cold mts, cheese/cold meats; Sug.-sw. bev., sugar-sweetened beverages. Data are presented as mean percentage values. Stat. sign., statistical significance; Chi-square (χ^2) analysis was applied to determine differences in gender and categorical variables; ns, non-significant for gender; $p < 0.05$ assessed by χ^2 for * gender, § type of beverage.

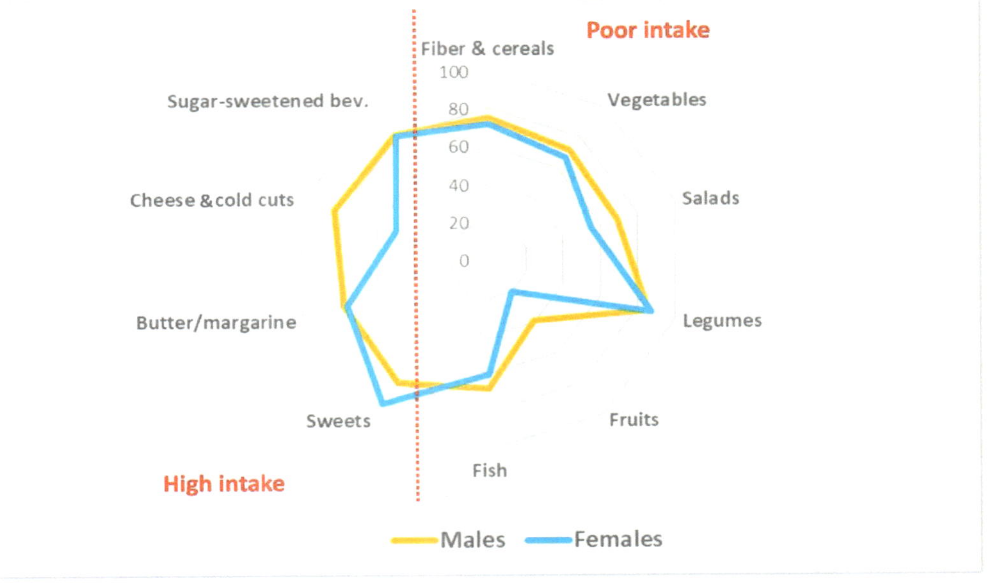

Figure 2. Radial plot illustrating the differences between males and females in the intake of some of the main elements characterizing dietary pattern. The different radial lines (0–100) indicate the percentage of study participants exhibiting on the right side of the graph a poor intake of fibers and cereals, vegetables, salads, legumes, fruits and fish, whereas on the left side the graph reflects the percentage of study participants showing a high intake of sugar-sweetened beverages, cheese, butter and fat spreads, and sweets.

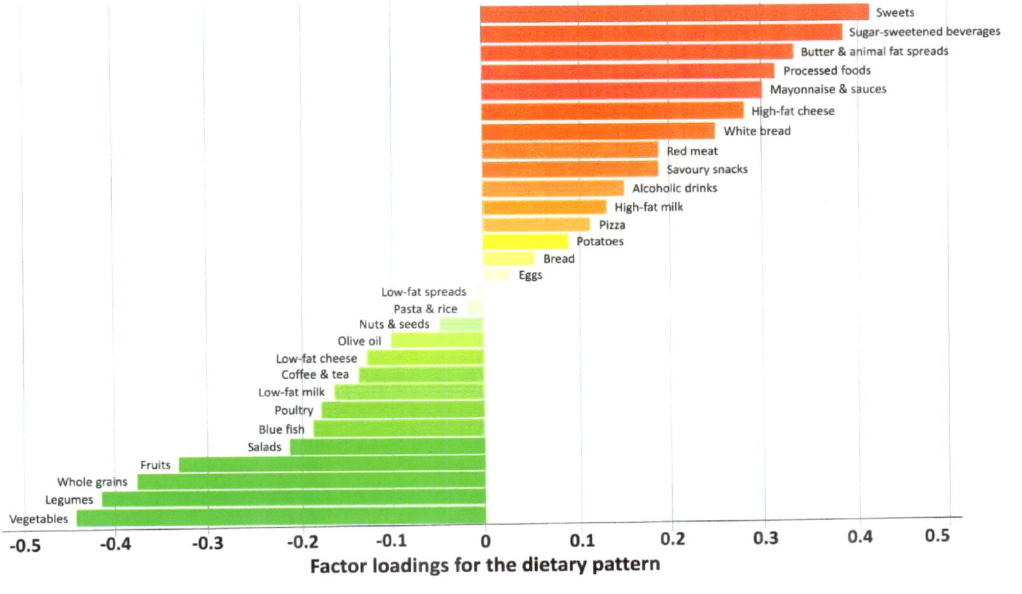

Figure 3. Dietary pattern factor loadings derived from reduced-rank regression.

4. Discussion

The present study provides evidence about the anthropometric and cardiometabolic profile of patients with excess weight of this cohort. The detailed nutritional information gathered within the experimental design combining 24 h recalls, FFQs and RRR analysis allows the detailed description of their quantitative and qualitative characteristics. Macronutrient analysis highlighted the high energy intake coming from fats at the expense of a low consumption of complex carbohydrates. Noteworthily, excess weight was associated with a DP low in vegetables, legumes, whole grains and fruits at the same time as being high in sweets, sugar-sweetened beverages, fat spreads, and processed foods. RRR analysis identified a DP characterized by high energy density and saturated fat exhibiting negative loadings (>-0.30) for green leafy vegetables, legumes, and fruits at the same time as showing positive factor loadings (>0.30) for processed foods, fat spreads, sugar-sweetened beverages, and sweets. A high intake of sugar-sweetened beverages was shown in both males and females. Interestingly, high intakes of low-fiber bread, confectionery, chocolate, butter and added sugars, together with low intakes of vegetables and fresh fruits, conform to DPs that have been associated with a higher T2D incidence, in particular in PlwO and among younger people [24]. Similarly, an obesity-related DP has been prospectively associated with an elevated risk of CVD, an increased incidence of site-specific and overall cancers as well as all-cause mortality [50–52]. Likewise, higher consumption of ultra-processed food (UPF) in prospective cohort studies reportedly increases the risk of obesity and obesity-related outcomes [53]. Higher UPF consumption is inversely associated with vegetable, fruit, legume and seafood intake. Taken together, these findings show that the DP, more strongly than individual foods, impacts health outcomes.

As regards alcohol consumption, men exhibited a higher intake of calories derived from alcohol than women, which was in line with previously reported findings [54]. Whilst gender-specific differences were observed, in both cases, the energy intake derived from alcohol was not high, thereby demonstrating that it was not the main factor driving the energy imbalance between intake and expenditure. In fact, some studies have observed that PlwO exhibited lower rates of alcohol consumption as compared to the general population [55,56] showing that the higher the BMI, the lower the alcohol intake, probably in relation to overeating competing at brain-reward sites with alcohol [56,57]. This contrasts, however, with rodent studies in which some neuropeptides increased consumption of both high-fat foods and alcohol [58,59]. Noteworthily, frequency and quantity of consumption define hazardous drinking as well as its association with morbidity [60–62]. In this context, attention has been drawn to the relevance of drinking patterns and type of alcoholic beverages consumed, since some profiles of frequency and amount are associated with higher BMIs [63–67]. In our study, irrespective of gender, wine represented globally the main source of alcoholic beverages, with sugar-sweetened intake being above the optimal level in both genders. It was observed that beer intake was more frequent among people with a lower alcohol consumption frequency (<1 d/week) as compared to participants with a higher alcohol consumption frequency (5–7 d/week). In this respect, some authors suggest that in terms of contribution to body weight regulation, no relationship between alcohol intake and body weight exists [68,69]. However, via analyses of 5-year changes in alcohol consumption in relation to 5-year changes in WC and BMI, other authors have identified that the association of alcohol intake with measures of obesity is complex [70]. Thus, while in females intake of wine and distillates/mixed drinks exhibited contrasting associations with changes in WC and BMI, in males weekly alcohol consumption reduction with a special emphasis on stopping excessive intake could have beneficial effects on WC and BMI management. Of note, mean alcohol volume intake may mask the relationship between amount and frequency with health outcomes [63,71]. In the last years, an increase in alcohol consumption in adult females has been observed, thereby narrowing the gender gap [54]. Additionally, in spite of drinking less than men, women can reach higher circulating alcohol concentrations, thus experiencing more hazardous effects. However, except for hypertension risk, no gender-related effects of alcohol intake on coronary heart disease and stroke risk have been identified [62]. While the relationship is relatively linear between alcohol intake and hypertension for males, for women it does have a J shape, further evidencing the complexity of the gender-specific differences in alcohol intake. In this context, a higher adherence to a healthy beverage index reportedly minimizes the risk of metabolic alterations [72].

Suboptimal diets reportedly account for chronic disease development and more mortality globally than any other risks [3,18,73–77]. In the case of overweight and obesity, the focus of people pursuing weight loss has traditionally been on restricting single food items, which are blamed for leading to excess weight. In this respect, cultural beliefs frequently only blame beer [78], perpetuating conjectures and presumptions of unsubstantiated beliefs in the obesity domain [79]. Our study provides evidence that focusing on dietary and drinking patterns better captures the individual's reality. Undoubtedly, in a complex activity like eating in which multiple homeostatic and hedonic factors are involved, focusing on a single item that can be blamed as the main culprit of excess weight represents a simplistic approach [80,81]. Data show that quantifying food consumption by both grams or binary variables yielded meaningful DPs, with each method providing different advantages; while weight accounts for food amount consumed, binary intake preferentially describes general food choices, which are potentially easier to be modified and therefore useful in public health settings [47]. In this context, governments should learn from their earlier policy failures [82] to prioritize policies that make minimal demands on individuals and have the potential for a population-wide reach so as to maximize their potential for equitable impacts. Moreover, policies should be proposed in ways that readily lead to implementation and evaluation.

Strengths of the study include the detailed patient phenotyping, the combination of tools aimed at achieving realistic nutritional information, and the experience of the research team in the field, which ensures reproducible study conditions. Our results provide clinically useful information relevant for therapeutic purposes. Given that foods are not consumed in isolation, identification of DPs based on actual eating behaviors yields more valuable interpretation and understanding of the diet-related causes of excess weight to help in the management of patients. Capturing the real DPs will help to develop a better understanding of the complexity of intake and to build trust in order to deliver effective evidence-based health interventions. Nonetheless, the present work also has some limitations. One potential limitation of the present study pertains to the generalizability to other populations, since the research was carried out in the homogeneous sample population attending our Department for weight loss, thereby overturning the potential effect of confounding factors such as race, ethnicity, educational level, cultural beliefs, lifestyle factors and socioeconomic status, among others. However, this aspect may be also viewed as a strength, since the sample homogeneity ensures the lack of potential confounders. The findings would need to be confirmed in other populations in order to determine potential ethnic or race-specific differences as regards DPs, body composition and associations with the cardiometabolic risk factors.

5. Conclusions

The focus on consumption of single nutrients does not resemble real-world intake behaviors. As hypothesized, capturing the DP provides more clinically relevant and useful information. The obtained data can provide useful information for strategic approaches to improve the effectiveness of obesity policies. Healthcare practitioners should be encouraged to offer dietary policies that focus on the promotion of diet components for which non-optimal intakes are evident, rather than only targeting individual or specific food items. The evidence provided in the present study can help to broaden and implement policy decision making for obesity management with DP modification, yielding more long-term benefits in terms of sustainability and feasibility of interventions.

Author Contributions: Conceptualization, G.F. and A.M.; methodology, M.A.-A., P.Y.-E., L.O., C.M.P., M.G.-G., P.A., J.E., C.S., A.M. and G.F.; investigation, M.A.-A., P.Y.-E., L.O., C.M.P., M.G.-G., P.A., J.E., C.S., A.M. and G.F.; formal analysis, M.A.-A., P.Y.-E., L.O., C.S. and G.F.; writing—original draft preparation, G.F.; writing—review and editing, M.A.-A., P.Y.-E., L.O., C.M.P., M.G.-G., P.A., J.E., C.S., A.M. and G.F.; supervision, G.F. and A.M.; project administration, G.F.; funding acquisition, A.M. and G.F. All authors have read and agreed to the published version of the manuscript.

Funding: This study was funded by the "Centro de Información Cerveza y Salud", Spanish Institute of Health ISCIII (Subdirección General de Evaluación and Fondos FEDER project PI22/00745) and CIBEROBN. The funding organization had no role in data collection, analysis and interpretation, manuscript writing or the decision to submit it for publication.

Institutional Review Board Statement: All procedures carried out followed the Declaration of Helsinki as well as its subsequent revisions. The study design was accepted from a scientific and ethical point of view by the Ethical Committee of the Hospital receiving approval number 30/2006 further updated in number 2020.041, approval date: 22 July 2020.

Informed Consent Statement: All participants in the study provided written informed consent.

Data Availability Statement: The data presented in this study are available upon reasonable request from the corresponding author. The data are not publicly available due to privacy restrictions.

Acknowledgments: The technical assistance of Sonia Romero is gratefully acknowledged. We also thank the valuable collaboration of all the staff of the Department of Endocrinology and Nutrition of the University of Navarra Clinic.

Conflicts of Interest: The authors declare no conflict of interest. The funders had no role in the design of the study; in the collection, analyses, or interpretation of data; in the writing of the manuscript; or in the decision to publish the results.

Abbreviations

ALT: alanine aminotransferase; AST: aspartate aminotransferase; AP: alkaline phosphatase; BMI: body mass index; BP: blood pressure; CRP: C reactive protein; CV: coefficient of variation CVD: cardiovascular diseases; DP: dietary pattern; ELISA: enzyme-linked immunosorbent assay; FFQs: food frequency questionnaires; Gamma-GT: gamma-glutamil transferase; HOMA-IR: homeostasis model assessment—insulin resistance; IL: interleukin; ISAK: International Society for the Advancement of Kinanthropometry; LCN-2: lipocalin-2; MCP-1: monocyte chemotaxis protein 1; OPN: osteopontin; PAL: physical activity level; PlwO: people living with obesity; QUICKI: quantitative insulin sensitivity check index; REE: resting energy expenditure; RRR: reduced-rank regression; TEE: total energy expenditure; TEF: thermic effect of food; TNC: tenascin C; T2D: type 2 diabetes; TNF-α: tumor necrosis factor alpha; YKL-40: chitinase 3-like protein-1; vW factor Ag, von Willebrand factor antigen; WC: waist circumference; WHO: World Health Organization.

References

1. Frühbeck, G.; Gómez-Ambrosi, J. Control of body weight: A physiologic and transgenic perspective. *Diabetologia* **2003**, *46*, 143–172. [CrossRef]
2. NCD Risk Factor Collaboration (NCD-RisC). Worldwide trends in body-mass index, underweight, overweight, and obesity from 1975 to 2016: A pooled analysis of 2416 population-based measurement studies in 128.9 million children, adolescents, and adults. *Lancet* **2017**, *390*, 2627–2642. [CrossRef] [PubMed]
3. GBD 2017 Diet Collaborators. Health effects of dietary risks in 195 countries, 1990–2017: A systematic analysis for the Global Burden of Disease Study 2017. *Lancet* **2019**, *393*, 1958–1972. [CrossRef]
4. GBD 2019 Diseases & Injuries-Collaborators. Global burden of 369 diseases and injuries in 204 countries and territories, 1990–2019: A systematic analysis for the Global Burden of Disease Study 2019. *Lancet* **2020**, *396*, 1204–1222. [CrossRef] [PubMed]
5. Catalán, V.; Avilés-Olmos, I.; Rodríguez, A.; Becerril, S.; Fernández-Formoso, J.A.; Kiortsis, D.; Portincasa, P.; Gómez-Ambrosi, J.; Frühbeck, G. Time to consider the "exposome hypothesis" in the development of the obesity pandemic. *Nutrients* **2022**, *14*, 1597. [CrossRef]
6. World Health Organization. Fact Sheet: Obesity and Overweight. Available online: http://www.who.int/mediacentre/factsheets/fs311/en/ (accessed on 9 June 2021).

7. Gutiérrez-Fisac, J.L.; Guallar-Castillón, P.; León-Munoz, L.M.; Graciani, A.; Banegas, J.R.; Rodríguez-Artalejo, F. Prevalence of general and abdominal obesity in the adult population of Spain, 2008–2010: The ENRICA study. *Obes. Rev.* **2012**, *13*, 388–392. [CrossRef]
8. Talukdar, D.; Seenivasan, S.; Cameron, A.J.; Sacks, G. The association between national income and adult obesity prevalence: Empirical insights into temporal patterns and moderators of the association using 40 years of data across 147 countries. *PLoS ONE* **2020**, *15*, e0232236. [CrossRef]
9. Heymsfield, S.B.; Wadden, T.A. Mechanisms, pathophysiology, and management of obesity. *N. Engl. J. Med.* **2017**, *376*, 254–266. [CrossRef] [PubMed]
10. Frühbeck, G.; Busetto, L.; Dicker, D.; Yumuk, V.; Goossens, G.H.; Hebebrand, J.; Halford, J.G.C.; Farpour-Lambert, N.J.; Blaak, E.E.; Woodward, E.; et al. The ABCD of obesity: An EASO Position Statement on a diagnostic term with clinical and scientific implications. *Obes. Facts* **2019**, *12*, 131–136. [CrossRef] [PubMed]
11. de Heredia, F.P.; Gómez-Martínez, S.; Marcos, A. Obesity, inflammation and the immune system. *Proc. Nutr. Soc.* **2012**, *71*, 332–338. [CrossRef]
12. Yárnoz-Esquiroz, P.; Olazarán, L.; Aguas-Ayesa, M.; Perdomo, C.M.; García-Goñi, M.; Silva, C.; Fernández-Formoso, J.A.; Escalada, J.; Montecucco, F.; Portincasa, P.; et al. 'Obesities': Position statement on a complex disease entity with multifaceted drivers. *Eur. J. Clin. Investig.* **2022**, *52*, e13811. [CrossRef] [PubMed]
13. Cooremans, K.; Geuens, M.; Pandelaere, M. Cross-national investigation of the drivers of obesity: Re-assessment of past findings and avenues for the future. *Appetite* **2017**, *114*, 360–367. [CrossRef] [PubMed]
14. Mozaffarian, D. The White House Conference on hunger, nutrition, and health—A new national strategy. *N. Engl. J. Med.* **2022**, *387*, 2014–2017. [CrossRef] [PubMed]
15. Perdomo, C.M.; Frühbeck, G.; Escalada, J. Impact of nutritional changes on nonalcoholic fatty liver disease. *Nutrients* **2019**, *11*, 677. [CrossRef]
16. Perdomo, C.M.; Cohen, R.V.; Sumithran, P.; Clément, K.; Frühbeck, G. Contemporary medical, device, and surgical therapies for obesity in adults. *Lancet* **2023**, *401*, 1116–1130. [CrossRef] [PubMed]
17. Hill, J.O.; Wyatt, H.R.; Peters, J.C. Energy balance and obesity. *Circulation* **2012**, *126*, 126–132. [CrossRef]
18. Shan, Z.; Wang, F.; Li, Y.; Baden, M.Y.; Bhupathiraju, S.N.; Wang, D.D.; Sun, Q.; Rexrode, K.M.; Rimm, E.B.; Qi, L.; et al. Healthy eating patterns and risk of total and cause-specific mortality. *JAMA Intern. Med.* **2023**, *183*, 142–153. [CrossRef]
19. Hall, K.D.; Heymsfield, S.B.; Kemnitz, J.W.; Klein, S.; Schoeller, D.A.; Speakman, J.R. Energy balance and its components: Implications for body weight regulation. *Am. J. Clin. Nutr.* **2012**, *95*, 989–994. [CrossRef]
20. Soares, M.J.; Müller, M.J. Resting energy expenditure and body composition: Critical aspects for clinical nutrition. *Eur. J. Clin. Nutr.* **2018**, *72*, 1208–1214. [CrossRef]
21. Kenny, G.P.; Notley, S.R.; Gagnon, D. Direct calorimetry: A brief historical review of its use in the study of human metabolism and thermoregulation. *Eur. J. Appl. Physiol.* **2017**, *117*, 1765–1785. [CrossRef]
22. Westerterp, K.R. Exercise, energy expenditure and energy balance, as measured with doubly labelled water. *Proc. Nutr. Soc.* **2018**, *77*, 4–10. [CrossRef] [PubMed]
23. Lam, Y.Y.; Ravussin, E. Indirect calorimetry: An indispensable tool to understand and predict obesity. *Eur. J. Clin. Nutr.* **2017**, *71*, 318–322. [CrossRef] [PubMed]
24. Biltoft-Jensen, A.; Matthiessen, J.; Hess Ygil, K.; Christensen, T. Defining energy-dense, nutrient-poor food and drinks and estimating the amount of discretionary energy. *Nutrients* **2022**, *14*, 1477. [CrossRef]
25. Gomez-Ambrosi, J.; Gallego-Escuredo, J.M.; Catalan, V.; Rodriguez, A.; Domingo, P.; Moncada, R.; Valenti, V.; Salvador, J.; Giralt, M.; Villarroya, F.; et al. FGF19 and FGF21 serum concentrations in human obesity and type 2 diabetes behave differently after diet- or surgically-induced weight loss. *Clin. Nutr.* **2017**, *36*, 861–868. [CrossRef] [PubMed]
26. Fields, D.A.; Higgins, P.B.; Radley, D. Air-displacement plethysmography: Here to stay. *Curr. Opin. Clin. Nutr. Metab. Care* **2005**, *8*, 624–629. [CrossRef] [PubMed]
27. Gomez-Ambrosi, J.; Gonzalez-Crespo, I.; Catalan, V.; Rodriguez, A.; Moncada, R.; Valenti, V.; Romero, S.; Ramirez, B.; Silva, C.; Gil, M.J.; et al. Clinical usefulness of abdominal bioimpedance (ViScan) in the determination of visceral fat and its application in the diagnosis and management of obesity and its comorbidities. *Clin. Nutr.* **2018**, *37*, 580–589. [CrossRef] [PubMed]
28. Johansson, G.; Westerterp, K.R. Assessment of the physical activity level with two questions: Validation with doubly labeled water. *Int. J. Obes.* **2008**, *32*, 1031–1033. [CrossRef] [PubMed]
29. Wrieden, W.L.; Barton, K.L.; Cochrane, L.; Adamson, A.J. *Calculation and Collation of Typical Food Portion Sizes for Adults Aged 19–64 and Older People Aged 65 and Over*; Final Technical Report to the Food Standards Agency; Food Standards Agency: London, UK, 2006; Volume 5.
30. Kennedy, G.; Ballard, T.; Dop, M.E. *Guidelines for Measuring Household and Individual Dietary Diversity*; Food and Agriculture Organization of the United Nations: Rome, Italy, 2013.
31. Hu, F.B. Dietary pattern analysis: A new direction in nutritional epidemiology. *Curr. Opin. Lipidol.* **2002**, *13*, 3–9. [CrossRef] [PubMed]
32. Newby, P.K.; Tucker, K.L. Empirically derived eating patterns using factor or cluster analysis: A review. *Nutr. Rev.* **2004**, *62*, 177–203. [CrossRef] [PubMed]

33. Hoffmann, K.; Schulze, M.B.; Schienkiewitz, A.; Nöthlings, U.; Boeing, H. Application of a new statistical method to derive dietary patterns in nutritional epidemiology. *Am. J. Epidemiol.* **2004**, *159*, 935–944. [CrossRef]
34. Lazarova, S.V.; Jessri, M. Associations between dietary patterns and cardiovascular disease risk in Canadian adults: A comparison of partial least squares, reduced rank regression, and the simplified dietary pattern technique. *Am. J. Clin. Nutr.* **2022**, *116*, 362–377. [CrossRef] [PubMed]
35. Carter, J.L.; Lewington, S.; Piernas, C.; Bradbury, K.; Key, T.J.; Jebb, S.A.; Arnold, M.; Bennett, D.; Clarke, R. Reproducibility of dietary intakes of macronutrients, specific food groups, and dietary patterns in 211 050 adults in the UK Biobank study. *J. Nutr. Sci.* **2019**, *8*, e34. [CrossRef] [PubMed]
36. Frühbeck, G.; Catalán, V.; Rodríguez, A.; Ramírez, B.; Becerril, S.; Salvador, J.; Colina, I.; Gómez-Ambrosi, J. Adiponectin-leptin ratio is a functional biomarker of adipose tissue inflammation. *Nutrients* **2019**, *11*, 454. [CrossRef]
37. Frühbeck, G.; Alonso, R.; Marzo, F.; Santidrián, S. A modified method for the indirect quantitative analysis of phytate in foodstuffs. *Anal. Biochem.* **1995**, *225*, 206–212. [CrossRef] [PubMed]
38. Moreno-Navarrete, J.M.; Escote, X.; Ortega, F.; Serino, M.; Campbell, M.; Michalski, M.C.; Laville, M.; Xifra, G.; Luche, E.; Domingo, P.; et al. A role for adipocyte-derived lipopolysaccharide-binding protein in inflammation- and obesity-associated adipose tissue dysfunction. *Diabetologia* **2013**, *56*, 2524–2537. [CrossRef] [PubMed]
39. Katz, A.; Nambi, S.S.; Mather, K.; Baron, A.D.; Follmann, D.A.; Sullivan, G.; Quon, M.J. Quantitative insulin sensitivity check index: A simple, accurate method for assessing insulin sensitivity in humans. *J. Clin. Endocrinol. Metab.* **2000**, *85*, 2402–2410. [CrossRef]
40. Matthews, D.R.; Hosker, J.P.; Rudenski, A.S.; Naylor, B.A.; Treacher, D.F.; Turner, R.C. Homeostasis model assessment: Insulin resistance and β-cell function from fasting plasma glucose and insulin concentrations in man. *Diabetologia* **1985**, *28*, 412–419. [CrossRef] [PubMed]
41. Gayoso-Diz, P.; Otero-González, A.; Rodriguez-Alvarez, M.X.; Gude, F.; García, F.; De Francisco, A.; Quintela, A.G. Insulin resistance (HOMA-IR) cut-off values and the metabolic syndrome in a general adult population: Effect of gender and age: EPIRCE cross-sectional study. *BMC Endocr. Disord.* **2013**, *13*, 47. [CrossRef]
42. Tahapary, D.L.; Pratisthita, L.B.; Fitri, N.A.; Marcella, C.; Wafa, S.; Kurniawan, F.; Rizka, A.; Tarigan, T.J.E.; Harbuwono, D.S.; Purnamasari, D.; et al. Challenges in the diagnosis of insulin resistance: Focusing on the role of HOMA-IR and Tryglyceride/glucose index. *Diabetes Metab. Syndr. Obes.* **2022**, *16*, 102581. [CrossRef] [PubMed]
43. Sabater, M.; Moreno-Navarrete, J.M.; Ortega, F.J.; Pardo, G.; Salvador, J.; Ricart, W.; Frühbeck, G.; Fernández-Real, J.M. Circulating pigment epithelium-derived factor levels are associated with insulin resistance and decrease after weight loss. *J. Clin. Endocrinol. Metab.* **2010**, *95*, 4720–4728. [CrossRef]
44. Catalán, V.; Gómez-Ambrosi, J.; Rodríguez, A.; Ramírez, B.; Rotellar, F.; Valentí, V.; Silva, C.; Gil, M.J.; Fernández-Real, J.M.; Salvador, J.; et al. Increased levels of calprotectin in obesity are related to macrophage content: Impact on inflammation and effect of weight loss. *Mol. Med.* **2011**, *17*, 1157–1167. [CrossRef] [PubMed]
45. Rodríguez, A.; Gómez-Ambrosi, J.; Catalán, V.; Rotellar, F.; Valentí, V.; Silva, C.; Mugueta, C.; Pulido, M.R.; Vázquez, R.; Salvador, J.; et al. The ghrelin O-acyltransferase-ghrelin system reduces TNF-α-induced apoptosis and autophagy in human visceral adipocytes. *Diabetologia* **2012**, *55*, 3038–3050. [CrossRef] [PubMed]
46. Muruzábal, F.J.; Frühbeck, G.; Gómez-Ambrosi, J.; Archanco, M.; Burrell, M.A. Immunocytochemical detection of leptin in non-mammalian vertebrate stomach. *Gen. Comp. Endocrinol.* **2002**, *128*, 149–152. [CrossRef] [PubMed]
47. Smith, A.D.; Emmett, P.M.; Newby, P.K.; Northstone, K. Dietary patterns obtained through principal components analysis: The effect of input variable quantification. *Br. J. Nutr.* **2013**, *109*, 1881–1891. [CrossRef] [PubMed]
48. Lemon, S.C.; Rosal, M.C.; Zapka, J.; Borg, A.; Andersen, V. Contributions of weight perceptions to weight loss attempts: Differences by body mass index and gender. *Body Image* **2009**, *6*, 90–96. [CrossRef] [PubMed]
49. Pagoto, S.L.; Schneider, K.L.; Oleski, J.L.; Luciani, J.M.; Bodenlos, J.S.; Whited, M.C. Male inclusion in randomized controlled trials of lifestyle weight loss interventions. *Obesity* **2012**, *20*, 1234–1239. [CrossRef] [PubMed]
50. Calder, P.C.; Ahluwalia, N.; Brouns, F.; Buetler, T.; Clement, K.; Cunningham, K.; Esposito, K.; Jönsson, L.S.; Kolb, H.; Lansink, M.; et al. Dietary factors and low-grade inflammation in relation to overweight and obesity. *Br. J. Nutr.* **2011**, *106*, S5–S78. [CrossRef]
51. Maimaitiyiming, M.; Yang, H.; Zhou, L.; Zhang, X.; Cai, Q.; Wang, Y. Associations between an obesity-related dietary pattern and incidence of overall and site-specific cancers: A prospective cohort study. *BMC Med.* **2023**, *21*, 251. [CrossRef]
52. Maimaitiyiming, M.; Yang, H.; Li, H.; Xu, C.; Li, S.; Zhou, L.; Zhang, X.; Wang, Y. The association of obesity-related dietary patterns and main food groups derived by reduced-rank regression with cardiovascular diseases incidence and all-cause mortality: Findings from 116,711 adults. *Eur. J. Nutr.* **2023**, *62*, 2605–2619. [CrossRef]
53. Dicken, S.J.; Batterham, R.L. The role of diet quality in mediating the association between ultra-processed food intake, obesity and health-related outcomes: A review of prospective cohort studies. *Nutrients* **2021**, *14*, 23. [CrossRef]
54. White, A.M. Gender differences in the epidemiology of alcohol use and related harms in the United States. *Alcohol Res.* **2020**, *40*, 1–3. [CrossRef]
55. Mannisto, S.; Uusitalo, K.; Roos, E.; Fogelholm, M.; Pietinen, P. Alcohol beverage drinking, diet and body mass index in a cross-sectional survey. *Eur. J. Clin. Nutr.* **1997**, *51*, 326–332. [CrossRef] [PubMed]
56. Kleiner, K.D.; Gold, M.S.; Frost-Pineda, K.; Lenz-Brunaman, B.; Perri, M.G.; Jacobs, W.S. Body mass index and alcohol use. *J. Addict. Dis.* **2004**, *23*, 105–118. [CrossRef]

57. Volkow, N.D.; Michaelides, M.; Baler, R. The neuroscience of drug reward and addiction. *Physiol. Rev.* **2019**, *99*, 2115–2140. [CrossRef] [PubMed]
58. Leibowitz, S.F. Regulation and effects of hypothalamic galanin: Relation to dietary fat, alcohol ingestion, circulating lipids and energy homeostasis. *Neuropeptides* **2005**, *39*, 327–332. [CrossRef]
59. Karatayev, O.; Barson, J.R.; Carr, A.J.; Baylan, J.; Chen, Y.W.; Leibowitz, S.F. Predictors of ethanol consumption in adult Sprague-Dawley rats: Relation to hypothalamic peptides that stimulate ethanol intake. *Alcohol* **2010**, *44*, 323–334. [CrossRef] [PubMed]
60. MacKillop, J.; Agabio, R.; Feldstein Ewing, S.W.; Heilig, M.; Kelly, J.F.; Leggio, L.; Lingford-Hughes, A.; Palmer, A.A.; Parry, C.D.; Ray, L.; et al. Hazardous drinking and alcohol use disorders. *Nat. Rev. Dis. Primers* **2022**, *8*, 80. [CrossRef] [PubMed]
61. Marcos, A.; Serra-Majem, L.; Pérez-Jiménez, F.; Pascual, V.; Tinahones, F.J.; Estruch, R. Moderate consumption of beer and its effects on cardiovascular and metabolic health: An updated review of recent scientific evidence. *Nutrients* **2021**, *13*, 879. [CrossRef]
62. Piano, M.R.; Thur, L.A.; Hwang, C.L.; Phillips, S.A. Effects of alcohol on the cardiovascular system in women. *Alcohol Res.* **2020**, *40*, 12. [CrossRef] [PubMed]
63. Breslow, R.A.; Smothers, B.A. Drinking patterns and body mass index in never smokers: National Health Interview Survey, 1997–2001. *Am. J. Epidemiol.* **2005**, *161*, 368–376. [CrossRef]
64. Kerr, W.C.; Greenfield, T.K.; Tujague, J.; Brown, S.E. A drink is a drink? Variation in the amount of alcohol contained in beer, wine and spirits drinks in a US methodological sample. *Alcohol. Clin. Exp. Res.* **2005**, *29*, 2015–2021. [CrossRef] [PubMed]
65. Kerr, W.C.; Greenfield, T.K.; Tujague, J. Estimates of the mean alcohol concentration of the spirits, wine, and beer sold in the United States and per capita consumption: 1950 to 2002. *Alcohol. Clin. Exp. Res.* **2006**, *30*, 1583–1591. [CrossRef] [PubMed]
66. Tujague, J.; Kerr, W.C. Energy intake estimates of respondent-measured alcoholic beverages. *Alcohol Alcohol.* **2009**, *44*, 34–41. [CrossRef] [PubMed]
67. Lukasiewicz, E.; Mennen, L.I.; Bertrais, S.; Arnault, N.; Preziosi, P.; Galan, P.; Hercberg, S. Alcohol intake in relation to body mass index and waist-to-hip ratio: The importance of type of alcoholic beverage. *Public Health Nutr.* **2005**, *8*, 315–320. [CrossRef] [PubMed]
68. Pirola, R.C.; Lieber, C.S. The energy cost of the metabolism of drugs, including ethanol. *Pharmacology* **1972**, *7*, 185–196. [CrossRef]
69. Kromhout, D. Energy and macronutrient intake in lean and obese middle-aged men (the Zutphen study). *Am. J. Clin. Nutr.* **1983**, *37*, 295–299. [CrossRef] [PubMed]
70. Butler, J.L.; Gordon-Larsen, P.; Steffen, L.M.; Shikany, J.M.; Jacobs, D.R.J.; Popkin, B.M.; Poti, J.M. Associations of 5-year changes in alcoholic beverage intake with 5-year changes in waist circumference and BMI in the Coronary Artery Risk Development in Young Adults (CARDIA) study. *PLoS ONE* **2023**, *18*, e0281722. [CrossRef] [PubMed]
71. Room, R. Measuring drinking patterns: The experience of the last half century. *J. Subst. Abus.* **2000**, *12*, 23–31. [CrossRef]
72. Rasaei, N.; Hosseininasab, D.; Shiraseb, F.; Gholami, F.; Noori, S.; Ghaffarian-Ensaf, R.; Daneshzad, E.; Clark, C.C.T.; Mirzaei, K. The association between healthy beverage index and healthy and unhealthy obesity phenotypes among obese women: A cross-sectional study. *Int. J. Clin. Pract.* **2022**, *2022*, 7753259. [CrossRef]
73. Newby, P.K.; Muller, D.; Hallfrisch, J.; Qiao, N.; Andres, R.; Tucker, K.L. Dietary patterns and changes in body mass index and waist circumference in adults. *Am. J. Clin. Nutr.* **2003**, *77*, 1417–1425. [CrossRef] [PubMed]
74. Villegas, R.; Yang, G.; Gao, Y.T.; Cai, H.; Li, H.; Zheng, W.; Shu, X.O. Dietary patterns are associated with lower incidence of type 2 diabetes in middle-aged women: The Shanghai Women's Health Study. *Int. J. Epidemiol.* **2010**, *39*, 889–899. [CrossRef] [PubMed]
75. Ng, M.; Fleming, T.; Robinson, M.; Thomson, B.; Graetz, N.; Margono, C.; Mullany, E.C.; Biryukov, S.; Abbafati, C.; Abera, S.F.; et al. Global, regional, and national prevalence of overweight and obesity in children and adults during 1980–2013: A systematic analysis for the Global Burden of Disease Study 2013. *Lancet* **2014**, *384*, 766–781. [CrossRef] [PubMed]
76. Kimokoti, R.W.; Millen, B.E. Nutrition for the prevention of chronic diseases. *Med. Clin. N. Am.* **2016**, *100*, 1185–1198. [CrossRef]
77. Sweetman, A.K.; Carter, J.; Perez-Cornago, A.; Gao, M.; Jebb, S.A.; Piernas, C. Dietary pattern adherence in association with changes in body composition and adiposity measurements in the UK Biobank study. *Obes. Res. Clin. Pract.* **2023**, *17*, 233–241. [CrossRef] [PubMed]
78. Bergmann, M.M.; Schutze, M.; Steffen, A.; Boeing, H.; Halkjaer, J.; Tjonneland, A.; Travier, N.; Agudo, A.; Slimani, N.; Rinaldi, S.; et al. The association of lifetime alcohol use with measures of abdominal and general adiposity in a large-scale European cohort. *Eur. J. Clin. Nutr.* **2011**, *65*, 1079–1087. [CrossRef] [PubMed]
79. Casazza, K.; Brown, A.; Astrup, A.; Bertz, F.; Baum, C.; Brown, M.B.; Dawson, J.; Durant, N.; Dutton, G.; Fields, D.A.; et al. Weighing the evidence of common beliefs in obesity research. *Crit. Rev. Food Sci. Nutr.* **2015**, *55*, 2014–2053. [CrossRef] [PubMed]
80. Bellisle, F.; McDevitt, R.; Prentice, A.M. Meal frequency and energy balance. *Br. J. Nutr.* **1997**, *77* (Suppl. S1), S57–S70. [CrossRef] [PubMed]
81. Bellisle, F. Edograms: Recording the microstructure of meal intake in humans-a window on appetite mechanisms. *Int. J. Obes.* **2020**, *44*, 2347–2357. [CrossRef] [PubMed]
82. Theis, D.R.Z.; White, M. Is Obesity Policy in England Fit for Purpose? Analysis of Government Strategies and Policies, 1992–2020. *Milbank Q.* **2021**, *99*, 126–170. [CrossRef]

nutrients

MDPI

Review

The Potential Role of Nutrition in Overtraining Syndrome: A Narrative Review

Maria Ester la Torre [1,†], Antonietta Monda [2,†], Antonietta Messina [3], Maria Ida de Stefano [1], Vincenzo Monda [4], Fiorenzo Moscatelli [5], Francesco Tafuri [6], Emma Saraiello [4], Francesca Latino [5], Marcellino Monda [2], Giovanni Messina [1], Rita Polito [1,*] and Domenico Tafuri [4]

1 Department of Clinical and Experimental Medicine, University of Foggia, 71122 Foggia, Italy; ester.latorre@unifg.it (M.E.l.T.); maria.destefano@unifg.it (M.I.d.S.); giovanni.messina@unifg.it (G.M.)
2 Department of Experimental Medicine, Section of Human Physiology, Unit of Dietetics and Sports Medicine, University of Campania "Luigi Vanvitelli", 80138 Naples, Italy; antoniettamonda99@gmail.com (A.M.); marcellino.monda@unicampania.it (M.M.)
3 Department of Precision Medicine, University of Campania "Luigi Vanvitelli", 80138 Naples, Italy; antonietta.messina@unicampania.it
4 Department of Economics, Law, Cybersecurity, and Sports Sciences, University of Naples "Parthenope", 80131 Naples, Italy; vincenzo.monda@uniparthenope.it (V.M.); emma.saraiello@uniparthenope.it (E.S.); domenico.tafuri@uniparthenope.it (D.T.)
5 Department of Human Sciences, Telematic University Pegaso, 80100 Naples, Italy; fiorenzo400@gmail.com (F.M.); francesca.latino@unipegaso.it (F.L.)
6 Heracle Lab Research in Educational Neuroscience, Niccolò Cusano University, 00166 Roma, Italy; francesco.tafuri@unicusano.it
* Correspondence: rita.polito@unifg.it
† These authors contributed equally to this work.

Abstract: Competition between athletes and an increase in sporting knowledge have greatly influenced training methods while increasing the number of them more and more. As a result, the number of athletes who have increased the number and intensity of their workouts while decreasing recovery times is rising. Positive overtraining could be considered a natural and fundamental process when the result is adaptation and improved performance; however, in the absence of adequate recovery, negative overtraining could occur, causing fatigue, maladaptation, and inertia. One of the earliest forms of fatigue is overreaching. It is considered to be an accumulation of training that leads to reduced sports performance, requiring days or weeks to recover. Overreaching, if followed by adequate recovery, can lead to an increase in athletic performance. Nonetheless, if overreaching becomes extreme, combined with additional stressors, it could lead to overtraining syndrome (OTS). OTS, caused by systemic inflammation, leads to central nervous system (CNS) effects, including depressed mood, further inflammation, central fatigue, and ultimately neurohormonal changes. There are therefore not only physiological, biochemical, and immunological but also psychological symptoms or markers that must be considered, independently or together, being intrinsically linked with overtraining, to fully understand OTS. However, to date, there are very few published studies that have analyzed how nutrition in its specific food aspects, if compromised during OTS, can be both etiology and consequence of the syndrome. To date, OTS has not yet been fully studied, and the topic needs further research. The purpose of this narrative review is therefore to study how a correct diet and nutrition can influence OTS in all its aspects, from prevention to treatment.

Keywords: overtraining syndrome (OTS); overreaching; physical activity; nutrition; dietary intake

Citation: la Torre, M.E.; Monda, A.; Messina, A.; de Stefano, M.I.; Monda, V.; Moscatelli, F.; Tafuri, F.; Saraiello, E.; Latino, F.; Monda, M.; et al. The Potential Role of Nutrition in Overtraining Syndrome: A Narrative Review. *Nutrients* **2023**, *15*, 4916. https://doi.org/10.3390/nu15234916

Academic Editor: Matthew Barnes

Received: 31 October 2023
Revised: 13 November 2023
Accepted: 23 November 2023
Published: 24 November 2023

1. Introduction

In the competitive sport sector, it is natural for athletes to increase their training loads to improve performance [1]. This increase in muscular work is tolerated only through time intervals consisting of muscle recovery and rest and periodization of training [2].

Overtraining syndrome (OTS) occurs following a prolonged period of overreaching (OR) combined with additional stressors [3], where OR is defined as an accumulation of training load leading to decreased performance and with days to weeks for muscle recovery [3,4]. OR can be subdivided into functional OR (FOR) and non-functional OR (NFOR), where FOR refers to the positive improvement in fitness and performance since recovery has occurred, while NFOR is associated with the impairment of performance and fitness due to a lack of muscle recovery [4–6]. The difficulty, therefore, is in understanding whether an athlete has achieved OTS or NFOR status. The difference between FOR, NFOR, and OTS depends on the time necessary for the individual and the body to recover; in fact, OR generally requires days or weeks, NFOR requires weeks or months, and OTS can require months or even years of recovery [3–7]. Therefore, OTS can be defined, as reported by Sims et al. an accumulation of stress whether or not due to training, which results in a long-term decrease in performance capacity with or without associated physiological or psychological signs and symptoms of overtraining, in which recovery of performance capacity may take several weeks or months [8].

The causes of OTS can be varied, such as systemic inflammation and consequent effects on the central nervous system including neurohormonal changes, depressed mood, and a sense of fatigue [9] (Figure 1). Furthermore, other causes of the onset of OTS could be depletion of glycogen and tissue trauma, and increased levels of cytokines but also depletion of glutamine in the muscles with consequent muscle pain, weight loss, and mood changes or frequent illnesses which can accompany the decrease in performance [10–14].

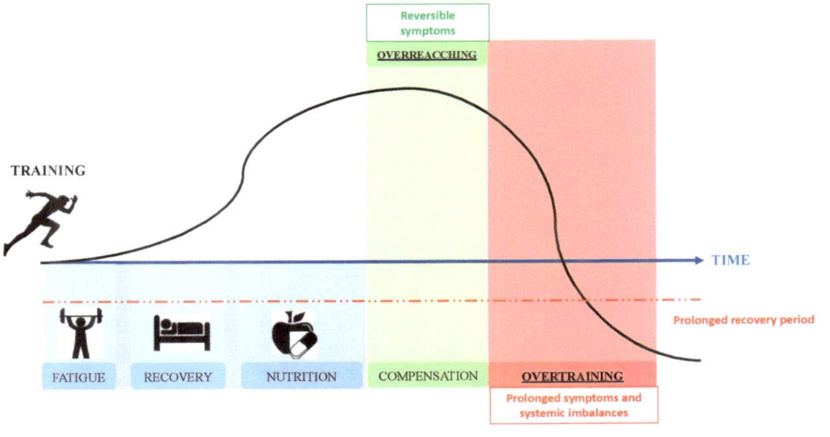

Figure 1. Process of development of overtraining syndrome in athletes.

Epidemiologically, OTS is extremely rare, but given the difficulty in diagnosis, exact data on prevalence and incidence are currently lacking [15,16]. Epidemiological analyses have found that episodes of NFO and OTS can occur during the athlete's career; for example, according to the literature, about 35% of adolescent swimmers had been overtrained at least once in their career [17], 5–30% had experienced staleness [16], and approximately 60% of elite male and female runners experienced NFOR symptoms compared to 33% in non-elite runners [18]. Furthermore, the risk of OTS appears to be significantly increased in individual sports and low-intensity sports [19]. Therefore, overtraining is still an underestimated phenomenon, with consequent lower interest in the scientific literature in this regard; indeed, to date, OTS has not been fully studied and the topic currently requires further scientific research. The purpose of this narrative review, therefore, is to study and explore how correct nutrition and integration can influence and modulate OTS in all its aspects, from prevention to treatment, to reduce the onset of the phenomenon, contributing to a primary prevention perspective, in line with global standards.

2. Pathophysiology of OTS

In general, regular exercise of mild to moderate intensity is used as a non-pharmacological intervention strategy to prevent various pathological medical conditions such as type 2 diabetes, insulin resistance, pulmonary and cardiovascular diseases, colon cancer and breast cancer, dementia, and depression, and to improve the symptoms [20–26]. Despite this, it is necessary to highlight the paradox of the overtrained athlete; as already stated, excessive training can aggravate the decline in sports performance [20,27]. Numerous theories on the causes of this phenomenon have been proposed and studied.

2.1. Cytokine Hypothesis

In addition to the association between overtraining and skeletal muscle injury that was established about two decades ago, the cytokine hypothesis was introduced, which integrates and explains most of the signs and symptoms that appear to be linked to the decline in sports performance induced by excessive training [11,21,28]. Therefore, the hypothesis states that high-load training associated with insufficient recovery could induce musculoskeletal trauma and subsequently determine the production and release of cytokines and interleukins such as IL-1 beta, IL-6, and TNF-alpha with subsequent multi-organ interaction [11,12]. This theory has been extensively studied through various works in multiple fields. For example, elevated levels of IL-6 and TNF-alpha in skeletal muscle have been shown in mouse studies to be related to impairment of the insulin signaling pathway, resulting in insulin resistance, muscle atrophy, and activation of endoplasmic reticulum (ER) stress [20,29]. Furthermore, high levels of cytokines such as IL-1 beta, IL-6, and TNF-alpha, derived from peripheral inflammatory stimuli, at the hypothalamic level are linked to an increased restriction of food intake among the cytokines. TNF-alpha plays a fundamental role in reduced food intake and increased cellular catabolism, causing difficulty in increasing body weight [30–34]. Furthermore, based on the theoretical correlation between inflammation and ER stress [29], a study by Pinto et al. states that the activation of hypothalamic ER stress is present above all following excessive training [35]. These data strengthen the theory of proinflammatory cytokines, as they could act as an initial stimulus in the central nervous system, inducing some maladaptation soon after a period of excessive training, affecting physical performance, food restriction, and cellular catabolism [36]. Finally, the inflammatory role of cytokines may also be linked to the development of pathological cardiac hypertrophy, due to the increase in IL-6 [31]. Indeed, recently, there has been an increase in IL-6 protein levels, a reduction in the activation of AMPK-activated protein kinase, mTOR, and ribosomal protein rpS6, and signs of left ventricular fibrosis in mice [37]. The results suggest that excessive training can induce signs of pathological cardiac hypertrophy, thus contributing to impairment and reduction in physical performance [37]. Thus, it is difficult for athletes to manage and recover by modulating overtraining at a systemic level due to the proinflammatory action of cytokines.

2.2. Glutamine Hypothesis

The production of cytokines is not the only reason for the onset of overtraining syndrome; there is also a relationship between hormones and overtraining syndrome, the latter of which can lead to complex hormonal dysfunctions [3]. Nonetheless, the diagnosis of OTS is complicated and can only be made after the exclusion of other more common medical conditions. Physiological, biochemical, and immunological markers for overtraining have received much attention in recent years and could be of potential use if routinely measured in the laboratory as a medical aid in preventing overtraining [38]. Glutamine is among the various molecules that could be used as biomarkers of overtraining. Glutamine is an essential amino acid for the function of immune cells, in particular [4,12,39]. It also plays a key role in DNA and RNA synthesis, nitrogen transport, and acid–base balance and gluconeogenesis [4,13,39]. The decrease in glutamine after physical exercise could be responsible for an increase in the incidence of infections of the respiratory system and of the upper respiratory tract (URTI) in overtrained athletes [4,11,40]. Prolonged training or repeated periods of high-intensity training could transiently reduce plasma glutamine concentrations, resulting in greater sensitivity to URTIs [11,13,39]. As described in the cytokine theory, plasma glutamine also plays a key role in various processes controlled by cytokines [41]. Glutamine is a molecular precursor of the synthesis of inflammatory proteins which are upregulated with overtraining [12]. Furthermore, systemic inflammation, given by the increase in proinflammatory cytokines, induces a catabolic state with a consequent increase in both glucose and protein metabolism. Therefore, glutamine is a crucial element in gluconeogenesis. Furthermore, alterations in the hypothalamic–pituitary–adrenal (HPA) and gonadal (HPG) axis, following some studies in the literature, appear to be responsible for OTS. In particular, endurance athletes might show subtle changes in HPA axis function and thus show alterations in the levels of cortisol, ACTH, testosterone, and other hormones [42–45]. In a study by Barron et al., an adrenocortical deficit was also highlighted in athletes with overtraining syndrome [44]. That is, the data show that growth hormone, prolactin, and ACTH responses to insulin-induced hypoglycemia were lower in overtrained athletes than in healthy, well-trained controls. Unfortunately, however, current data and research are in some cases inconsistent regarding the patterns of these hormonal changes, as they depend on various factors including individual training capacity, intrinsic vulnerability to stressful factors, and other hormonal levels [45].

2.3. Central Fatigue Theory

OTS is often characterized by mood, sleep–wake cycle, and behavioral disturbances [15,43,46,47]. The neurotransmitter involved in the regulation of these mechanisms is serotonin (5-HT); 5-HT alterations could trigger OTS [3,12,46–48]. Central fatigue theory hypothesizes that OTS is caused by an increase in the synthesis of 5-hydroxytryptomine (5-HT) in the central nervous system (CNS). Physical exercise decreases the levels of branched-chain amino acids (BCAA) due to an increase in their oxidation into glucose, favoring the entry of tryptophan into the brain and the subsequent conversion into 5-HT as both BCAAs and tryptophan use the same transporter to cross the blood–brain barrier [48]. Therefore, a decrease in plasma BCAAs with a consequent increase in plasma tryptophan lead to an increase in CNS tryptophan levels [43,48]. 5-HT derives from tryptophan. Budgett et al. studied how the administration of serotonin reuptake inhibitors to athletes increases the levels of 5-HT with consequent reduction in the athletes' performance [48], while the integration of branched-chain amino acids therefore reduced 5-HT production and resulted in increased physical and mental energy [49].

2.4. Glycogen Hypothesis

Another phenomenon that could interfere with an alteration of the central neurotransmitters involved in fatigue during OTS could be a reduction in muscle glycogen due to overtraining [50]. Glycogen is the key energy source during moderate to vigorous exercise [50]. Low muscle glycogen also results in an increase in oxidative processes and a decrease in total concentrations of BCAA, as reported, leading to central fatigue [50]. Some studies report that, during OTS, there are lower amounts of lactate in the blood in overtrained athletes compared to well-trained athletes [51,52]. Therefore, post-exercise blood lactate measurement could be used as a tool to identify athletes at risk of overtraining, especially if such testing were performed on a regular basis [51,52].

2.5. Autonomic Nervous System Hypothesis

Certainly, an imbalance in the autonomic nervous system, reducing sympathetic activation and parasympathetic dominance, could lead to some symptoms of OTS such as decreased performance, fatigue, depression, and bradycardia, but also to a reduced nocturnal urinary excretion of catecholamines [4,12,43]. Generally, catecholamine excretion decreases with increasing training and fatigue, returning to baseline levels during the recovery period [4,53]. A reduced organ sensitivity to catecholamines could therefore be an indicator of symptoms of reduced sympathetic activation [43]. Furthermore, studies have stated that the effects of intense training on autonomic nervous system control may be reversible, even after a week of rest [54]. Heart rate variability (HRV) monitoring could therefore be an indicator of autonomic function and a predictor of overtraining [45,55].

2.6. Oxidative Stress Hypothesis

Intense physical exercise could generate reactive oxygen species (ROS) and consequently increase oxidative stress [56]. ROS are related to post-exercise inflammatory response mechanisms and to the propagation of muscle damage with promotion of ROS-induced infiltration of neutrophils and macrophages into muscle [57,58]. Neutrophils and macrophages generate a free radical such as superoxide, which in turn can be converted into hydrogen peroxide, which is then able to react with the superoxide in the presence of a transition metal to form hydroxyl radicals [58]. Furthermore, the typical markers of protein oxidation and the total oxidizing capacity were elevated, while the levels of antioxidants such as glutathione and coenzyme Q10, and γ-tocopherol and carotenoids were decreased [59–61]. Alterations in redox homeostasis have also been reported in individuals with OTS; however, information on ROS generation in OTS needs further research. The identification of common biomarkers for overtrained athletes could allow an adequate intervention to prevent the progression of fatigue towards a more severe stage of overtraining syndrome [59–61].

3. Dietary Intake in OTS

Overtraining syndrome, and therefore an imbalance between training and recovery, can be aggravated or implied by improper nutritional intake [62–64]. Nutrition therefore plays a fundamental role in maintaining correct body homeostasis, as an imbalance in the quantities and proportions of macro- and micronutrients, but also an incorrect state of hydration, could affect correct energy intake, the maintenance of normal physiological functions, and good body composition, necessary for recovery [5]. However, there are few studies in the literature on the relationship between nutrition and OTS. Some of the published studies state that inadequate nutrition, pathologies, and stress factors and sleep disorders [65] can lead to dysfunctions of the normal metabolic processes involved in immune, inflammatory, neurological, hormonal, and metabolic responses [65]. For example, it has already been demonstrated that overtraining syndrome can reduce an athlete's appetite, producing a counterproductive effect, since if the athlete presents continuous caloric deficits during training, these could elevate both the stress hormone and the response and production of cytokines during physical exercise, limiting the athlete [11,12,66–70]. Furthermore, fatigue, injury, immunosuppression, and decreased performance can occur not only when energy expenditure regularly exceeds daily caloric intake but also when hydration is insufficient [70–72]. In fact, incorrect hydration could have a negative effect not only on performance but also on the metabolism of some macronutrients, such as glycogen, which can only be stored in the muscle when bound to water in a ratio of 1:3 g [72]. It has been found that even low levels of dehydration can compromise both the athlete's exercise capacity and the cardiovascular and thermoregulatory responses [73]. Optimal performance, therefore, is possible only when dehydration as well as hyperthermia is minimized by ingesting large volumes of fluids during exercise. Therefore, the consumption of liquids in volumes similar to the amount of sweat lost results in the maintenance of important physiological functions and significantly improves exercise performance, even during exercise lasting only 1 h. Carbohydrate ingestion also improves exercise performance, an effect that is independent of and additional to the prevention of dehydration [73]. A study conducted by Candegiani et al. demonstrated how an adequate daily caloric intake, regardless of the composition of macronutrients, can have positive bodily influences, for example, raising salivary cortisol 30 min after waking up, improving the speed and quality of muscle recovery, and increasing the elimination rate of muscle recovery markers, mainly creatine kinase (CK) and lactate [74]. Therefore, an increase in caloric intake could lead to a higher quality of sports performance by positively influencing both muscle recovery and the hormonal environment.

3.1. Dietary Fat

High levels of physical activity and an inadequate or unbalanced nutritional intake are implicated in imbalances, above all hormonal [63,75–77]. Therefore, in addition to a correct daily caloric intake, it is also important to maintain an adequate number of macronutrients. Each macronutrient has its own precise role, for example, the daily lipid component plays important physiological roles, especially in the construction of cell membranes, as fuel for exercise, as antioxidants, and as precursors of some hormones [78]. Recent studies have shown that low-fat diets impair sports performance [79–81]. For example, a high-fat diet has been shown in some studies to potentially reduce plasma levels of pro-inflammatory cytokines such as IL-6, IL-1 and TNF-α, due to the anti-inflammatory action of omega-3 fatty acids [82–86]. Furthermore, increasing the amount of lipids in the diet can increase muscular endurance without the appearance of adverse effects, especially on plasma levels of cortisol, interferon-γ, and PGE2 [78]. Some studies state that reducing the fat content in the diet could reduce circulating testosterone levels, implicated in dysregulation of the hypothalamic–pituitary–thyroid (HPTA) axis negatively affecting health and physical performance [84]. A study by Hamalainen et al. reports that reducing dietary fat from a daily intake of approximately 40% of energy to a daily intake of less than 25% of energy resulted in an approximately 15% decrease in total testosterone levels compared to a diet with a high percentage of fat, about 41% of energy intake, and a reduction in polyunsaturated fatty acids [83,85,86]. Despite this, it should be emphasized that a diet rich in fats is highly discouraged due to negative effects on the cardiovascular system, which is why the total quantity of fats in the daily diet must comply with the guidelines provided by the World Health Organization (WHO), in which the percentage of total lipids must not exceed 30% of total calories, combined with an intake of saturated fatty acids of less than about 10%, and an intake of lower trans fatty acids at 1%, favoring the consumption of unsaturated fatty acids [87–90]. Precisely for this reason, new research has focused on how to improve sports performance without increasing the fat component beyond the recommended percentages, and in this case, ketone ester supplementations (KEs) seem to be a valid alternative. Interestingly, in some studies, KEs significantly inhibited the onset of fatigue symptoms while improving the training load tolerated by the athlete, especially resistance exercise performance, and increasing total energy intake. Furthermore, KEs counteracted training-induced sympathetic overdrive, as also evidenced by stable nocturnal catecholamine production. KEs also presented a marked and positive impact on tachycardia following exercise, thus demonstrating that KEs are a powerful nutritional strategy that prevents the development of physiological symptoms of overtraining and determines their overcoming [91]. Currently, medium-chain triglyceride (MCT) supplements as a source of fat and as an energy source have also gained greater attention in the sports field. MCTs could directly enter the mitochondrial level and be used for energy production through the beta-oxidation process [92]. This process would provide the athlete with an immediate source of fat that can be used for energy purposes, thus saving glycogen consumption [36]. Although there are studies that state that MCTs show improved athletic performance, other studies show side effects such as gastrointestinal disorders when taking MCTs compared to carbohydrates [92]. Further research is therefore necessary in the field of sports nutrition supplementation in this area.

3.2. Dietary Protein

The protein component, on the other hand, is essential for the maintenance and growth of muscle tissue [78]. Following high-intensity workouts, the ratio between protein degradation and synthesis increases, resulting in a greater consumption of dietary protein [78,93]. Therefore, there is a linear relationship between protein intake and loss of lean mass during periods of negative energy balance, so a good protein supplementation proves to be an effective strategy to mitigate the loss of lean body mass [94,95]. An imbalance in the mechanisms involved in the defense of the individual has been observed due to moderate and severe protein imbalance [82,96]. It has been reported that diets low in protein (<20%) or, on the contrary, abundant in protein (>60%) have been shown to impair phagocytosis activity and the production of interleukins such as IL-2, resulting in potential damage to the T cell system [82]. Glutamine plays a fundamental role in protein composition, as previously mentioned. Glutamine is mainly used as fuel for lymphocytes and macrophages for the synthesis of DNA and RNA [13,97–100]. Intense exercise can lead to depletion of skeletal muscle stores; therefore, it has been hypothesized that glutamine depletion may be linked to impaired immunity [99,100]. As stated in one study, a low protein intake can be responsible for protein loss, but in reality, even a high protein intake can be harmful and cause an approximately 25% decrease in the amount of plasma glutamine due to an increase in renal absorption [78]. Furthermore, the dietary protein component is an important determinant of body characteristics; in fact, it is positively correlated with parameters such as metabolism and body composition, and with an increase in the basal metabolic rate (BMR), fat oxidation, muscle mass, and consequent hydration. Protein intake is significantly but inversely related to extracellular water, i.e., it protects against water loss, thus preventing edema [74,101–104]. Finally, growth hormone (GH) and insulin-like growth factor-1 (IGF-I) appear to be related to healthy diet and lifestyle, while they appear to be impaired during OTS [105,106]. IGF-1 mediates many of the actions of GHs, which regulate growth processes, maintenance above all of lean mass but also bone mass, cellular differentiation, and mitochondrial homeostasis, a process that is, in turn, key to the metabolic processing of carbohydrates, fats, and amino acids (AAs) [107–109]. During a fasted condition, GH primarily inhibits and reduces protein degradation by stimulating its synthesis in muscle and other tissues [110]. This defense mechanism, to all intents and purposes, could also be mediated by IGF-I, considering that chronic GH exposure increases hepatic IGF-I production [111]. In order, therefore, to maintain protein homeostasis and to support and/or prevent OTS, protein requirements for healthy endurance athletes have been estimated at approximately 1.2-1.4 g/kg and slightly higher for endurance training athletes up to about 1.7 g/kg [112,113]. The acceptable macronutrient distribution range (AMDR) states that proteins can vary from 0.8 g/kg to 2.5 g/kg based on individual needs, converted in percentage from 10% to 35% of energy, but the increase in protein requirement and intake to date remains controversial [114,115]. As regards the integration of BCAAs, whose imbalance appears to be a cause of the emergence of overtraining, currently, in this case the results are not yet convincing. As stated by Gastmann et al., BCAAs, if taken with an adequate dose of carbohydrates during prolonged physical exercise, would not seem to determine improvements in OTS. Furthermore, high doses of ingested BCAAs could increase the secondary production of ammonia, leading to further premature fatigue, as BCAAs are considered the main donors of amino groups at a muscular level. Overall results in this field are currently inconclusive and require more controlled experimental research [116]. Finally, regarding the use of creatine as a sports protein supplement, currently, despite its widespread use in this field, studies and research are still needed to explore the topic in more depth. From what has emerged in the literature, creatine could have a positive effect on muscle recovery, muscle hypertrophy, and strength. In a study by Wang et al. and Ribeiro et al., the use of a higher dose of approximately 20 g per day (period of approximately 5 days), and a subsequent maintenance dose of approximately 3 g per day (51 days), combined with resistance training, led to hypertrophy of the skeletal muscle mass, less fatigue, and reduction in creatine kinase in the creatine-treated group compared to the control group [117,118].

3.3. Dietary Carbohydrate

In addition to lipids and proteins, carbohydrates, according to numerous studies conducted in recent years, play a role as the main macronutrient to support and improve physical performance [119]. Carbohydrates, therefore, play the role of primary fuel, especially for the first 60–90 min of sports activity in the form of muscle and liver glycogen [120]. Similarly, after physical activity, the intake of carbohydrates supplies the muscle cells with the fuel necessary to begin the muscle reconstruction processes. Using the energy gained from burning carbohydrates, the muscles can absorb amino acids from the bloodstream, initiating protein synthesis [119]. Furthermore, carbohydrates stimulate the production and release of insulin, an anabolic hormone that plays a key role both in protein synthesis and in the reduction of protein breakdown [120]. During exercise, muscle cell injury often occurs with a characteristic increase in plasma creatine kinase activity and a consequent increase in post-exercise cytokine release [121]. Furthermore, a reduction in blood glucose levels may be linked to activation of the hypothalamic–pituitary–adrenal (HPA) axis, increased release of adrenocorticotropic hormone (ACTH) and consequently cortisol, increased GH levels, a decrease in insulin levels, and highly variable levels of adrenaline [122,123]. Several studies in the literature suggest that carbohydrates have effects on the production of IL-6 and the IL-1 receptor antagonist (IL-1ra) involved in the inflammatory cascade response to heavy exertion, reducing their plasma levels and altering the release [98,124,125]. In addition, the consumption of carbohydrates during physical activity appears to attenuate the increase in hormones such as cortisol and GH, limiting the degree of immunosuppression following physical exercise [123]. While increasing the amount of carbohydrates has been shown to have predominantly positive effects on the hormonal profile, however, excessive intake could have a pro-inflammatory role [126,127]. In fact, it could potentially induce a pathological increase in the action of aromatase, i.e., the enzymatic system responsible for the conversion of the androgen hormone, testosterone, into estrogen, estradiol, causing an imbalance in the testosterone–estradiol (T/E) ratio [101,102,104,126]. Furthermore, an excessive intake of carbohydrates could lead to an increase in lactate levels and a slight, but not significant, increase in neutrophils [126–129]. In conclusion, the deprivation of carbohydrates can determine reduced and/or delayed hormonal responses, which indirectly compromise the performance of athletes; therefore, to build excellent glycogen stores and help rebuild muscle tissue after physical activity, a rapid and adequate carbohydrate intake is required. If an intake of carbohydrates of less than about 5.0 g/kg/day has harmful effects on the hormonal level and on physical performance, in order to ensure that good muscle reserves and correct physiological functions are associated with it, carbohydrate intake should equal approximately 8 to 10 g/kg of body mass per day, which could represent 60 to 70% of total calories [74,106,127,130]. Furthermore, it is essential to evaluate the use of carbohydrate-based supplements during training and sports. Among the most used supplements are maltodextrins, which have been used in recent years. The effect of maltodextrins (polymers deriving from the hydrolysis process of starches [131]) during sports practice has been investigated in several studies, which however have presented conflicting results and opinions. From the results, it was noted that the use of the carbohydrate solution containing maltodextrins during training did not alter the blood concentrations of circulating leukocytes at an immune level [132], not resulting in an improvement in the immune system, while integration with a mixture of glutamine in combination with maltodextrins 2 h before training could be more effective in preventing a decrease in sports practice compared to the consumption of maltodextrin-based supplements or glutamine-based supplements alone [133]. Therefore, it is concluded that only carbohydrate supplementation in combination with glutamine peptide could improve the physical performance of athletes during competitions.

4. Micronutrients and OTS

Strenuous and intense physical activity can induce pathologies, injuries, and chronic fatigue, symptoms underlying overtraining syndrome, partly due to the toxicity of free radicals (FR) and to the altered response of the immune system [61,134]. In this context, in addition to macronutrients, micronutrients play a key role. A dietary imbalance may not provide an athlete with essential micronutrients, vitamins, and mineral salts for his or her health and physical performance [134].

For example, vitamins C, E, and A play important antioxidant roles in the neutralization of oxygen free radicals (ROS) and in supporting the immune system [126]. Vitamin C, in addition to having an antioxidant effect, thus reducing the risk of colds and other viruses, improves the differentiation but also the proliferation of B and T lymphocytes by increasing the levels of circulating antibodies [135–138]. Furthermore, it also modulates post-exercise cytokine production [138]. Vitamin E is used against chronic stress associated with exercise, as it possesses antioxidant properties against ROS, protecting cells and tissues; it reduces PGE2 production and inhibits COX2 activity by reducing nitric oxide [138]. It improves the T immune system by modulating the Th1/Th2 balance [139]. Vitamin A, on the other hand, has a supportive action for various body functions, including reaction time, muscle recovery, and post-exercise protein synthesis, essential for muscle growth and recovery [140–142]. Studies state that the supplementation of vitamin C (500 mg), vitamin E (270 mg), and β-carotene (vitamin A—18 mg) significantly reduces the probability of infection after prolonged exercise from 40 to 15%, and that tocopherol, when taken with ascorbic acid, has optimal effects. Furthermore, ascorbic acid supplementation (600 to 1000 mg/day for 3 weeks) reduced symptoms of URTI in marathon runners [98,143]. Although the integration with different sources of antioxidants therefore provides greater protection of the tissues against the oxidative stress induced by physical activity, the data present in the literature are contradictory. Indeed, over-supplementation of some vitamins, when in intolerable doses, especially with a single antioxidant nutrient, is not strongly recommended [144–146]. For example, excessive doses of vitamin A could have toxic effects on the liver, vitamin E in excessive doses can cause increases in mortality risk factors, and high levels of vitamin C can act as a pro-oxidant and non-oxidant [145,146]. More evidence is needed to support high dosages of antioxidant vitamins. Vitamin D, on the other hand, predominantly plays a cooperative role in the synthesis of various hormones; moreover, there appears to be a direct relationship between vitamin D concentrations and the performance of athletes, such as speed, muscle tone, and grip strength, but it also improves the use of carbohydrates during exercise, providing athletes with more energy, which can help increase performance [82,147,148]. The addition of calcium together with vitamin D intake also shows a reduction in the rate of stress fractures [149,150]. However, calcium intake is usually lower than recommended doses in athletes' diets, especially low-fat diets [151,152]. Precisely for this reason, in rare circumstances, especially if the diet is low in nutrients, physical activity could endanger bones [153]. Therefore, an adequate intake of vitamin D and calcium improves the athlete's performance and physical form [153,154]. Vitamins of the B complex have a fundamental importance, as they contribute to maintaining both health and optimal performance in athletes, improving brain functioning, concentration, and quality of sleep, by regulating melatonin, and energy levels [154]. Among these, vitamin B12 and folic acid, in particular, have a key role and are essential above all for the normal production of red and white blood cells [134]. A possible lack of vitamin B12 or folic acid could determine a reduced proliferative response of lymphocytes, a reduction in the phagocytic and bacterial capacity of neutrophils, and therefore an altered immune response, as well as being, especially folic acid, crucial in lowering homocysteine levels, a risk factor for heart disease [155,156]. Finally, in addition to vitamins, minerals also play a fundamental role in sports [139]. For example, a low zinc intake could compromise the immune system, as it plays an important role in the development of T lymphocytes and resistance to infections. A good supply of zinc is supported by diets high in fat [78]. Furthermore, the immune function is sensitive to the availability of iron; its deficiency can reduce lymphocyte responses, resulting in IL-1 production by macrophages, reduced interferon production, and natural killer cell activity (NKCA) [82,157]. Iron is essential for producing red blood cells, which are necessary to transport oxygen to the muscles, and is involved in hormonal and physiological functions, particularly in women [158]. Restrictive or unbalanced diets, coupled with inadequate levels of exercise, could lead to anemia and other problems associated with low iron levels, such as fatigue, poor performance, and a reduced ability to perform physical activities [159,160]. Therefore, it is necessary to consume a high-quality and varied diet and foods that include sources rich in iron [161]. Selenium, on the other hand, plays a key role as an antioxidant and cofactor of glutathione peroxidase/reductase, affecting multiple aspects of the immune system, and may help improve an athlete's performance and reduce inflammation [162]. Selenium deficiency is generally associated with elevated serum levels of C-reactive protein (CRP), a biomarker of inflammation [163]. Finally, potassium and magnesium are long-recognized key minerals in athletic performance [164]. Magnesium helps improve energy levels, increasing muscle performance and reducing exercise-induced fatigue [165]. Potassium, on the other hand, helps reduce the amount of lactic acid that is stored in the muscles, reducing the occurrence of cramps in athletes, and is also involved in the breakdown of carbohydrates, keeping energy levels high during physical activity, especially if intense [166,167]. In order not to provide too high a dose of the amounts of these two minerals, according to WHO guidelines, adults should not exceed 2000 mg sodium/day (Na) and have a minimum dose of 3510 mg potassium/day, as they are directly related [167–169]. The recommended dietary amount of magnesium is between 400 and 420 mg/day for males and between 310 and 320 mg/day for females [164]. Therefore, in summary, a multitude of physiological systems depend on micronutrients, which have a fundamental impact on the general health and performance of an athlete, without the need to exceed the values recommended by the guidelines, especially if the athlete consumes an adequate amount of nutrient-rich foods through a good dietary intake [170,171] (Figure 2). Unfortunately, many athletes do not meet the recommended requirements for most micronutrients, and therefore, more attention and correct supplementation are needed in this regard [171].

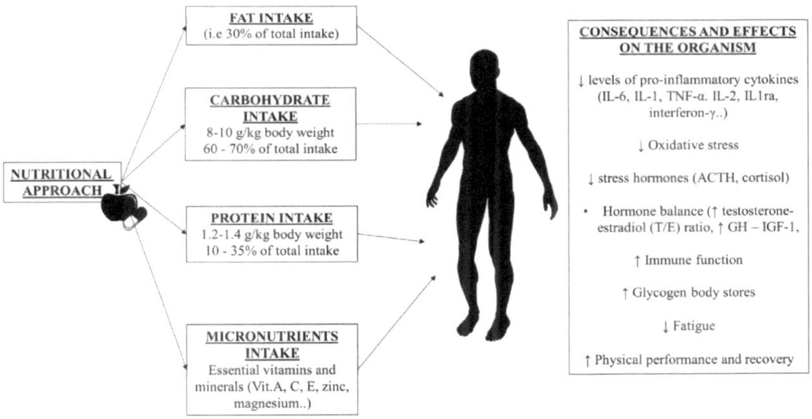

Figure 2. Basic notions and applications of the nutritional approach to preventing overtraining.

5. Discussion

Following a workout, inevitably, the body is urged to interrupt body homeostasis. Through correct training planning, therefore, an individually programmed workload progression associated with sufficient rest, and physical and functional adaptation for the athlete will be achieved, improving physical capacity and consequently performance [172]. However, excessive training and poor rest can undermine this process of physical adaptation and lead to overtraining syndrome, i.e., a condition with a marked decrease in performance despite rigorous training, systemic fatigue, a reduction in the defensive capabilities of the immune system, mood disorders, sleep disorders, biochemical and metabolic alterations, and finally, changes in physiological parameters such as heart rate at rest [6,11,12,41,46,51,88]. The cause of this syndrome is not only overtraining and poor rest but also the sum of additional stressors that could lead to its development. Poor or partially unbalanced nutritional intake could also be a contributing factor to OTS [126]. In general, adequate nutritional intake is involved in the various adaptive processes and above all for the recovery of the organism. Thus, an unbalanced diet could slow down the recovery process and the return to homeostasis [173]. Furthermore, to limit the risk of developing OTS, or to reduce the symptoms related to it, it is important to ensure that from a nutritional point of view there is the right amount of both macro- and micronutrients [173]. For example, an insufficient intake of carbohydrates and a decrease in muscle glycogen stores not compensated by diet could be a determinant of increased fatigue and poor performance with changes in the body's hormonal response (increased release of catecholamines, changes in circulating levels of cortisol and testosterone–estradiol (T/E) ratio etc.) [173,174]. Furthermore, it is now well known that sustaining a workout with a low amount and availability of carbohydrates could accelerate the onset of fatigue [174]. In recent years, proteins have received greater attention, especially in the fields of sports nutrition and nutrition. Proteins and amino acids, especially those such as glutamine, have multiple and fundamental roles [175]. In fact, protein may be especially important for athletes with a negative energy balance, or for athletes with injuries, common endocrine imbalances, fatigue, and immunosuppression [66]. Furthermore, an adequate protein intake also decreases the presence of inflammation at a systemic level by decreasing the amount of circulating pro-inflammatory cytokines, regulating hormone levels such as GH and IGF-1, and strengthening the immune system [176,177]. Adequate protein intake should be considered, first and foremost, by considering individual needs as well as globally recommended guidelines [177]. Furthermore, maintaining an adequate fat intake is essential for correct cellular and tissue functioning; in particular, the intake of the omega-3 family should be increased in athletes [178]. Reducing the dose of dietary fat could compromise the immune system and antioxidant defenses by reducing the expression and production of cytokines [179]. By contrast, an increase in the dose of dietary fat could lead to an increase in the probability of the onset of cardiovascular pathologies [180]. Another aspect linked to the onset of OTS is an uncontrolled increase in oxidative stress with an increase in resting oxidative stress markers and an excess of free radicals. As a result, it is crucial for athletes to ensure adequate antioxidant needs [181]. For example, vitamins A, C, and E as well as minerals such as zinc, selenium, copper, and manganese are essential micronutrients that allow the body to fight against increased oxidative stress, naturally without excess as excessive intake of these micronutrients in the form of food supplements could lead to a negative effect on the body [182]. OTS is an overall body response to an excessive build-up of stressors, training, and lack of rest. While diet and particular dietary regimens are not the only factors determining the development of OTS, it is important to give due attention to nutrition's role in the prevention and management of this condition. Therefore, a well-balanced diet in terms of energy, macronutrients, and micronutrients for an athlete could allow him or her to deal with the various stresses induced by training to favor adequate physiological adaptations and long-term performance. Finally, it is necessary to point out that, because this study is a narrative review, the articles included in it could be subject to selection errors, as a non-systematic approach was used to collect and analyze the articles. This is a possible limitation of this study, making further research and subsequent systematic reviews or meta-analyses necessary, given the importance of the topic covered and the fact that it still receives little attention, especially due to the negative effects of overtraining on human health.

6. Conclusions

Training and performance improvement are key factors in a complex correlation between work overload and recovery time for an athlete's physique. These factors, to improve performance, must be individualized for each athlete, considering their constant evaluation and a suitable treatment plan. If the physical stresses on an athlete are greater than the workload they can tolerate, imbalances could arise and lead to OTS. The importance of a timely diagnosis through the parameters and markers known to date with subsequent care of the athlete is clear in this study. In this context, nutrition can play a fundamental role, both in prevention and as a non-pharmacological treatment, with the aim of helping the athlete in reducing the likelihood of onset of OTS but also in its treatment, improving recovery and physical performance. Therefore, not only consultancy but also continuous updates and scientific research in the field of nutrition and food are essential for increasingly efficient support in the sports field.

Author Contributions: Conceptualization, M.E.l.T. and R.P.; resources, A.M. (Antonietta Messina); data curation, M.I.d.S.; writing—original draft preparation, M.E.l.T. and A.M. (Antonietta Monda); writing—review and editing V.M. and R.P.; visualization, F.M., F.T., E.S. and F.L.; supervision, M.M. and R.P.; project administration, G.M.; funding acquisition, D.T. All authors have read and agreed to the published version of the manuscript.

Funding: This research received no external funding.

Conflicts of Interest: The authors declare no conflict of interest.

References

1. Hamlin, M.J.; Wilkes, D.; Elliot, C.A.; Lizamore, C.A.; Kathiravel, Y. Monitoring Training Loads and Perceived Stress in Young Elite University Athletes. *Front. Physiol.* **2019**, *29*, 10–34. [CrossRef]
2. Kellmann, M. Preventing overtraining in athletes in high-intensity sports and stress/recovery monitoring. *Scand. J. Med. Sci. Sports* **2010**, *20*, 95–102. [CrossRef] [PubMed]
3. Meeusen, R.; Duclos, M.; Foster, C.; Fry, A.; Gleeson, M.; Nieman, D.; Raglin, J.; Rietjens, G.; Steinacker, J.; Urhausen, A. European College of Sport Science, & American College of Sports Medicine. Prevention, diagnosis, and treatment of the overtraining syndrome: Joint consensus statement of the European College of Sport Science and the American College of Sports Medicine. *Med. Sci. Sports Exerc.* **2013**, *45*, 186–205. [PubMed]
4. Halson, S.L.; Jeukendrup, A.E. Does overtraining exist? An analysis of overreaching and overtraining research. *Sports Med.* **2004**, *34*, 967–981. [CrossRef]
5. Carter, J.G.; Potter, A.W.; Brooks, K.A. Overtraining syndrome: Causes, consequences, and methods for prevention. *J. Sport Hum. Perf.* **2014**, *2*, 1–14.
6. Budgett, R.; Newsholme, E.; Lehmann, M.; Sharp, C.; Jones, D.; Jones, T.; Peto, T.; Collins, D.; Nerukar, R.; White, P. Redefining the overtraining syndrome as the unexplained underperformance syndrome. *Br. J. Sports Med.* **2000**, *34*, 67–68. [CrossRef] [PubMed]
7. Nederhof, E.; Lemmink, K.A.P.M.; Visscher, C.; Meeusen, R.; Mulder, T. Psychomotor Speed: Possibly a new marker for overtraining syndrome. *Sports Med.* **2006**, *36*, 817–828. [CrossRef]
8. Sims, S. The overtraining syndrome and endurance athletes. *Strength Cond. J.* **2001**, *23*, 45–46. [CrossRef]
9. Kreider, R. Central Fatigue Hypothesis and Overtraining. In *Overtraining in Sport*; Kreider, R., Fry, A.C., O'Toole, M.L., Eds.; Human Kinetics: Champaign, IL, USA, 1998; pp. 309–334.
10. Snyder, A. Overtraining and glycogen depletion hypothesis. *Med. Sci. Sports Exerc.* **1998**, *30*, 1146–1150. [CrossRef]
11. Smith, L. Tissue trauma: The underlying cause of overtraining syndrome? *J. Strength Cond. Res.* **2004**, *18*, 185–193. [CrossRef]
12. Smith, L. Cytokine hypothesis of overtraining: A physiological adaptation to excessive stress? *Med. Sci. Sports Exerc.* **2000**, *32*, 317–331. [CrossRef] [PubMed]
13. Walsh, N.P.; Blannin, A.K.; Robson, P.J.; Gleeson, M. Glutamine, exercise and immune function. *Sports Med.* **1998**, *26*, 177–191. [CrossRef] [PubMed]
14. Chung, Y.; Hsiao, Y.T.; Huang, W.C. Physiological and Psychological Effects of Treadmill Overtraining Implementation. *Biology* **2021**, *10*, 515. [CrossRef] [PubMed]
15. Hooper, S.; MacKinnon, L.T.; Hanrahan, S. Mood states as an indication of staleness and recovery. *Int. J. Sport Psychol.* **1997**, *28*, 1–12.
16. Koutedakis, Y.; Sharp, N.C. Seasonal variations of injury and overtraining in elite athletes. *Clin. J. Sport Med.* **1998**, *8*, 18–21. [CrossRef] [PubMed]
17. Raglin, J.; Sawamura, S.; Alexiou, S.; Hassmen, P. Training practice and staleness in 13–18-yearold swimmers: A cross-cultural study. *Ped. Exerc. Sci.* **2000**, *12*, 61–70. [CrossRef]
18. Morgan, W.P.; O'Connor, P.; Sparling, P.; Pate, R.R. Psychological characterizations of the elite female distance runner. *Int J Sports Med.* **1987**, *8*, 124–131. [CrossRef]
19. Matos, N.F.; Winsley, R.J.; Williams, C.A. Prevalence of non-functional overreaching/overtraining in young English athletes. *Med. Sci. Sports Exerc.* **2011**, *43*, 1287–1294. [CrossRef]
20. Gleeson, M.; Bishop, N.C.; Stensel, D.J.; Lindley, M.R.; Mastana, S.S.; Nimmo, M.A. The anti-inflammatory effects of exercise: Mechanisms and implications for the prevention and treatment of disease. *Nat. Rev. Immunol.* **2011**, *11*, 607–615. [CrossRef]
21. Hotamisligil, G.S. Inflammation and metabolic disorders. *Nature* **2006**, *444*, 860–867. [CrossRef]
22. Leonard, B.E. Inflammation, depression and dementia: Are they connected? *Neurochem. Res.* **2007**, *32*, 1749–1756. [CrossRef] [PubMed]
23. Rook, G.A.; Dalgleish, A. Infection, immunoregulation, and cancer. *Immunol. Rev.* **2011**, *240*, 141–159. [CrossRef] [PubMed]

24. Shoelson, S.E.; Lee, J.; Goldfine, A.B. Inflammation and insulin resistance. *J. Clin. Investig* **2006**, *116*, 1793–1801. [CrossRef] [PubMed]
25. Vina, J.; Sanchis-Gomar, F.; Martinez-Bello, V.; Gomez-Cabrera, M.C. Exercise acts as a drug; the pharmacological benefits of exercise. *Br. J. Pharmacol.* **2012**, *167*, 1–12. [CrossRef] [PubMed]
26. Jonas, S.; Phillips, E.M. *ACSM's Exercise is Medicine™: A Clinician's Guide to Exercise Prescription*; Lippincott Williams & Wilkins: Philadelphia, PA, USA, 2012.
27. Eichner, E.R. Overtraining: Consequences and prevention. *J. Sports Sci.* **1995**, *13*, 41–48. [CrossRef] [PubMed]
28. Fry, A.C. The role of training intensity in resistance exercise overtraining and overreaching. *Overtrain. Sport* **1998**, *127*, 107–127.
29. Rayavarapu, S.; Coley, W.; Nagaraju, K. Endoplasmic reticulum stress in skeletal muscle homeostasis and disease. *Curr. Rheumatol. Rep.* **2012**, *14*, 238–243. [CrossRef]
30. Gabellec, M.M.; Griffais, R.; Fillion, G.; Haour, F. Expression of interleukin 1 alpha, interleukin 1 beta and interleukin 1 receptor antagonist mRNA in mouse brain: Regulation by bacterial lipopolysaccharide (LPS) treatment. *Brain Res. Mol.* **1995**, *31*, 122–130. [CrossRef]
31. Layé, S.; Parnet, P.; Goujon, E.; Dantzer, R. Peripheral administration of lipopolysaccharide induces the expression of cytokine transcripts in the brain and pituitary of mice. *Brain Res. Mol.* **1994**, *27*, 157–162. [CrossRef]
32. Layé, S.; Gheusi, G.; Cremona, S.; Combe, C.; Kelley, K.; Dantzer, R.; Parnet, P. Endogenous brain IL-1 mediates LPS-induced anorexia and hypothalamic cytokine expression. *Am. J. Physiol. Regul. Integr. Comp. Physiol.* **2000**, *279*, R93–R98. [CrossRef]
33. Romanatto, T.; Cesquini, M.; Amaral, M.E.; Roman, E.A.; Moraes, J.C.; Torsoni, M.A.; Cruz-Neto, A.P.; Velloso, L.A. TNF-alpha acts in the hypothalamus inhibiting food intake and increasing the respiratory quotient--effects on leptin and insulin signaling pathways. *Peptides* **2007**, *28*, 1050–1058. [CrossRef] [PubMed]
34. Pereira, B.C.; da Rocha, A.L.; Pauli, J.R.; Ropelle, E.R.; de Souza, C.T.; Cintra, D.E.; Sant'Ana, M.R.; da Silva, A.S. Excessive eccentric exercise leads to transitory hypothalamic inflammation, which may contribute to the low body weight gain and food intake in overtrained mice. *Neuroscience* **2015**, *311*, 231–242. [CrossRef] [PubMed]
35. Pinto, A.P.; da Rocha, A.L.; Pereira, B.C.; Oliveira, L.D.; Morais, G.P.; Moura, L.P.; Ropelle, E.R.; Pauli, J.R.; da Silva, A.S. Excessive training is associated with endoplasmic reticulum stress but not apoptosis in the hypothalamus of mice. *Appl. Physiol. Nutr. Metab.* **2017**, *42*, 354–360. [CrossRef] [PubMed]
36. da Rocha, A.L.; Pinto, A.P.; Kohama, E.B.; Pauli, J.R.; de Moura, L.P.; Cintra, D.E.; Ropelle, E.R.; da Silva, A.S. The proinflammatory effects of chronic excessive exercise. *Cytokine* **2019**, *119*, 57–61. [CrossRef] [PubMed]
37. Bernardo, B.C.; Weeks, K.L.; Pretorius, L.; McMullen, J.R. Molecular distinction between physiological and pathological cardiac hypertrophy: Experimental findings and therapeutic strategies. *Pharmacol. Ther.* **2010**, *128*, 191–227. [CrossRef]
38. da Rocha, A.L.; Teixeira, G.R.; Pinto, A.P.; de Morais, G.P.; Oliveira, L.D.C.; de Vicente, L.G.; da Silva, L.E.C.M.; Pauli, J.R.; Cintra, D.E.; Ropelle, E.R.; et al. Excessive training induces molecular signs of pathologic cardiac hypertrophy. *J. Cell Physiol.* **2018**, *233*, 8850–8861. [CrossRef] [PubMed]
39. Hiscock, N.; Pedersen, B.K. Exercise-induced immunosuppresion- plasma glutamine is not the link. *J. Appl. Physiol.* **2002**, *93*, 813–822. [CrossRef]
40. Mackinnon, L.T.; Hooper, S.L. Plasma glutamine and upper respiratory tract infection during intensified training in swimmers. *Med. Sci. Sports Exerc.* **1996**, *28*, 285–290.
41. Kreher, J.B.; Schwartz, J.B. Overtraining syndrome: A practical guide. *Sports Health* **2012**, *4*, 128–138. [CrossRef]
42. Angeli, A.; Minetto, M.; Dovio, A.; Paccotti, P. The overtraining syndrome in athletes: A stressrelated disorder. *J. Endocrinol. Investig.* **2004**, *27*, 603–612. [CrossRef]
43. Lehmann, M.; Foster, C.; Keul, J. Overtraining in endurance athletes: A brief review. *Med. Sci. Sports Exerc.* **1993**, *25*, 854–862. [CrossRef] [PubMed]
44. Barron, J.L.; Noakes, T.D.; Levy, W.; Smith, C.; Millar, R. Hypothalamic dysfunction in overtrained athletes. *J. Clin. Endocrinol. Metab.* **1985**, *60*, 803–806. [CrossRef] [PubMed]
45. Urhausen, A.; Kindermann, W. Diagnosis of overtraining—What tools do we have? *Sports Med.* **2002**, *32*, 95–102. [CrossRef] [PubMed]
46. Armstrong, L.E.; VanHeest, J.L. The unknown mechanism of the overtraining syndrome: Clues from depression and psychoneuroimmunology. *Sports Med.* **2002**, *32*, 185–209. [CrossRef] [PubMed]
47. Morgan, W.P.; Brown, D.R.; Raglin, J.S.; O'Connor, P.J.; Ellickson, K.A. Psychological monitoring of overtraining and staleness. *Br. J. Sports Med.* **1987**, *21*, 107–114. [CrossRef]
48. Budgett, R.; Hiscock, N.; Arida, R.M.; Castell, L.M. The effects of the 5-HT2C agonist m-chlorophenylpiperazine on elite athletes with unexplained underperformance syndrome (overtraining). *Br. J. Sports Med.* **2010**, *44*, 280–283. [CrossRef]
49. Castell, L.M.; Poortmans, J.R.; Leclercq, R.; Brasseur, M.; Duchateau, J.; Newsholme, E.A. Some aspects of the acute phase response after a marathon race, and the effects of glutamine supplementation. *Eur. J. Appl. Physiol. Occup. Physiol.* **1997**, *75*, 47–53. [CrossRef]
50. Costill, D.L.; Flynn, M.G.; Kirwan, J.P.; Houmard, J.A.; Mitchell, J.B.; Thomas, R.; Park, S.H. Effects of repeated days of intensified training on muscle glycogen and swimming performance. *Med. Sci. Sports Exerc.* **1988**, *20*, 249–254. [CrossRef]
51. Alghannam, A.F.; Ghaith, M.M.; Alhussain, M.H. Regulation of Energy Substrate Metabolism in Endurance Exercise. *Int. J. Environ. Res. Public Health* **2021**, *7*, 4963. [CrossRef]

52. Gleeson, M. Biochemical and immunological markers of over-training. *J. Sports Sci. Med* **2002**, *1*, 31–41.
53. Kentta, G.; Hassmen, P. Overtraining and recovery: A conceptual model. *Sports Med.* **1998**, *26*, 1–16. [CrossRef]
54. Pichot, V.; Roche, F.; Gaspoz, F.E. Relation between heart rate variability and training load in middle-distance runners. *Med. Sci. Sports Exerc.* **2000**, *32*, 1729–1736. [CrossRef] [PubMed]
55. Fry, R.W.; Grove, J.R.; Morton, A.R.; Zeroni, P.M.; Gaudieri, S.; Keast, D. Psychological and immunological correlates of acute overtraining. *Br. J. Sports Med.* **1994**, *28*, 241–246. [CrossRef] [PubMed]
56. He, F.; Li, J.; Liu, Z.; Chuang, C.C.; Yang, W.; Zuo, L. Redox Mechanism of Reactive Oxygen Species in Exercise. *Front. Physiol.* **2016**, *7*, 486. [CrossRef]
57. Wang, F.; Wang, X.; Liu, Y.; Zhang, Z. Effects of Exercise-Induced ROS on the Pathophysiological Functions of Skeletal Muscle. *Oxid. Med. Cell Longev.* **2021**, *2021*, 3846122. [CrossRef]
58. Mittal, M.; Siddiqui, M.R.; Tran, K.; Reddy, S.P.; Malik, A.B. Reactive oxygen species in inflammation and tissue injury. *Antioxid. Redox Signal.* **2014**, *20*, 1126–1167. [CrossRef] [PubMed]
59. Kajaia, T.; Maskhulia, L.; Chelidze, K.; Akhalkatsi, V.; McHedlidze, T. Implication of relationship between oxidative stress and antioxidant status in blood serum. *Georgian Med. News* **2018**, *284*, 71–76.
60. Lewis, N.A.; Redgrave, A.; Homer, M.; Burden, R.; Martinson, W.; Moore, B.; Pedlar, C.R. Alterations in Redox Homeostasis During Recovery from Unexplained Underperformance Syndrome in an Elite International Rower. *Int. J. Sports Physiol. Perform.* **2018**, *13*, 107–111. [CrossRef]
61. Tanskanen, M.; Atalay, M.; Uusitalo, A. Altered oxidative stress in overtrained athletes. *J. Sports Sci.* **2010**, *28*, 309–317. [CrossRef]
62. Friedl, K.E.; Moore, R.J.; Hoyt, R.W.; Marchitelli, L.J.; Martinez-Lopez, L.E.; Askew, E.W. Endocrine markers of semistarvation in healthy lean men in a multistressor environment. *J. Appl. Physiol.* **2000**, *88*, 1820–1830. [CrossRef]
63. Nindl, B.C.; Barnes, B.R.; Alemany, J.A.; Frykman, P.N.; Shippee, R.L.; Friedl, K.E. Physiological consequences of U.S. Army Ranger training. *Med. Sci. Sports Exerc.* **2007**, *39*, 1380–1387. [CrossRef]
64. Szivak, T.K.; Kraemer, W.J. Physiological Readiness and Resilience: Pillars of Military Preparedness. *J. Strength Cond. Res.* **2015**, *29*, 34–39. [CrossRef] [PubMed]
65. Jensen, A.E.; Arrington, L.J.; Turcotte, L.P.; Kelly, K.R. Hormonal balance and nutritional intake in elite tactical athletes. *Steroids* **2019**, *152*, 108504. [CrossRef]
66. Lowery, L.; Forsythe, C.E. Protein and overtraining: Potential applications for free-living athletes. *J. Int. Soc. Sports Nutr.* **2006**, *3*, 42–50. [CrossRef]
67. Woods, A.L.; Garvican-Lewis, L.A.; Lundy, B.; Rice, A.J.; Thompson, K.G. New approaches to determine fatigue in elite athletes during intensified training: Resting metabolic rate and pacing profile. *PLoS ONE* **2017**, *12*, e0173807. [CrossRef] [PubMed]
68. Melin, A.K.; Heikura, I.A.; Tenforde, A.; Mountjoy, M. Energy Availability in Athletics: Health, Performance, and Physique. *Int. J. Sport Nutr. Exerc. Metab.* **2019**, *29*, 152–164. [CrossRef] [PubMed]
69. Barley, O.R.; Chapman, D.W.; Blazevich, A.J.; Abbiss, C.R. Acute Dehydration Impairs Endurance Without Modulating Neuromuscular Function. *Front. Physiol.* **2018**, *9*, 1562. [CrossRef]
70. Jones, L.C.; Cleary, M.A.; Lopez, R.M.; Zuri, R.E.; Lopez, R. Active dehydration impairs upper and lower body anaerobic muscular power. *J. Strength Cond. Res.* **2008**, *22*, 455–463. [CrossRef]
71. Casa, D.J. Fundamentals of thermal physiology, performance implications, and dehydration. *J. Athl. Train.* **1999**, *34*, 246–252.
72. Murray, B.; Rosenbloom, C. Fundamentals of glycogen metabolism for coaches and athletes. *Nutr. Rev.* **2018**, *76*, 243–259. [CrossRef]
73. Murray, R. Dehydration, hyperthermia, and athletes: Science and practice. *J. Athl. Train.* **1996**, *31*, 248–252. [PubMed]
74. Cadegiani, F.A.; Kater, C.E. Eating, Sleep, and Social Patterns as Independent Predictors of Clinical, Metabolic, and Biochemical Behaviors Among Elite Male Athletes: The EROS-PREDICTORS Study. *Front. Endocrinol.* **2020**, *26*, 414. [CrossRef] [PubMed]
75. Nindl, B.C.; Leone, C.D.; Tharion, W.J.; Johnson, R.F.; Castellani, J.W.; Patton, J.F.; Montain, S.J. Physical performance responses during 72 h of military operational stress. *Med. Sci. Sports Exerc.* **2002**, *34*, 1814–1822. [CrossRef] [PubMed]
76. Kreuz, L.E.; Rose, R.M.; Jennings, J.R. Suppression of plasma testosterone levels and psychological stress. A longitudinal study of young men in Officer Candidate School. *Arch. Gen. Psychiatry* **1972**, *26*, 479–482. [CrossRef] [PubMed]
77. Henning, P.C.; Scofield, D.E.; Spiering, B.A.; Staab, J.S.; Matheny, R.W., Jr.; Smith, M.A.; Bhasin, S.; Nindl, B.C. Recovery of endocrine and inflammatory mediators following an extended energy deficit. *J. Clin. Endocrinol. Metab.* **2014**, *99*, 956–964. [CrossRef] [PubMed]
78. Venkatraman, J.T.; Feng, X.; Pendergast, D.R. Effects of dietary fat and endurance exercise on plasma cortisol, prostaglandin E2, interferon-gamma and lipid peroxides in runners. *J. Am. Coll. Nutr.* **2001**, *20*, 529–536. [CrossRef]
79. Lambert, E.V.; Speechly, D.P.; Dennis, S.C.; Noakes, T.D. Enhanced endurance in trained cyclists during moderate intensity exercise following 2 weeks adaptation to a high fat diet. *Eur. J. Appl. Physiol. Occup. Physiol.* **1994**, *69*, 287–293. [CrossRef]
80. Muoio, D.M.; Leddy, J.J.; Horvath, P.J.; Awad, A.B.; Pendergast, D.R. Effect of dietary fat on metabolic adjustments to maximal VO2 and endurance in runners. *Med. Sci. Sports Exerc.* **1994**, *26*, 81–88. [CrossRef]
81. Horvath, P.J.; Eagen, C.K.; Fisher, N.M.; Leddy, J.J.; Pendergast, D.R. The effects of varying dietary fat on performance and metabolism in trained male and female runners. *J. Am. Coll. Nutr.* **2000**, *19*, 52–60. [CrossRef]
82. Bishop, N.C.; Blannin, A.K.; Walsh, N.P.; Robson, P.J.; Gleeson, M. Nutritional aspects of immunosuppression in athletes. *Sports Med.* **1999**, *28*, 151–176. [CrossRef]

83. Hämäläinen, E.K.; Adlercreutz, H.; Puska, P.; Pietinen, P. Decrease of serum total and free testosterone during a low-fat high-fibre diet. *J. Steroid. Biochem.* **1983**, *18*, 369–370. [CrossRef]
84. Sheng, J.A.; Bales, N.J.; Myers, S.A.; Bautista, A.I.; Roueinfar, M.; Hale, T.M.; Handa, R.J. The Hypothalamic-Pituitary-Adrenal Axis: Development, Programming Actions of Hormones, and Maternal-Fetal Interactions. *Front. Behav. Neurosci.* **2021**, *13*, 601939. [CrossRef]
85. Hämäläinen, E.; Adlercreutz, H.; Puska, P.; Pietinen, P. Diet and serum sex hormones in healthy men. *J. Steroid. Biochem.* **1984**, *20*, 459–464. [CrossRef]
86. Dorgan, J.F.; Judd, J.T.; Longcope, C.; Brown, C.; Schatzkin, A.; Clevidence, B.A.; Campbell, W.S.; Nair, P.P.; Franz, C.; Kahle, L.; et al. Effects of dietary fat and fiber on plasma and urine androgens and estrogens in men: A controlled feeding study. *Am. J. Clin. Nutr.* **1996**, *64*, 850–855. [CrossRef] [PubMed]
87. Visioli, F.; Poli, A. Fatty Acids and Cardiovascular Risk. Evidence, Lack of Evidence, and Diligence. *Nutrients* **2020**, *12*, 3782. [CrossRef] [PubMed]
88. Hooper, L.; Abdelhamid, A.; Bunn, D.; Brown, T.; Summerbell, C.D.; Skeaff, C.M. Effects of total fat intake on body weight. *Cochrane Database Syst. Rev.* **2015**, *2015*, CD011834. [CrossRef]
89. *Diet, Nutrition and the Prevention of Chronic Diseases: Report of a Joint WHO/FAO Expert Consultation*; WHO Technical Report Series, No. 916; World Health Organization: Geneva, Switzerland, 2003.
90. *Nations, Fats and Fatty Acids in Human Nutrition: Report of an Expert Consultation*; FAO Food and Nutrition Paper 91; Food and Agriculture Organization of the United: Rome, Italy, 2010.
91. Poffé, C.; Ramaekers, M.; Van Thienen, R.; Hespel, P. Ketone ester supplementation blunts overreaching symptoms during endurance training overload. *Physiol. J.* **2019**, *597*, 3009–3027. [CrossRef] [PubMed]
92. Kerksick, C.M.; Wilborn, C.D.; Roberts, M.D.; Smith-Ryan, A.; Kleiner, S.M.; Jäger, R.; Collins, R.; Cooke, M.; Davis, J.N.; Galvan, E.; et al. ISSN Exercise&Sports Nutrition Review Update: Research&Recommendations. *J. Int. Soc. Sports Nutr.* **2018**, *15*, 38.
93. Nieman, D.C.; Pedersen, B.K. Exercise and immune function: Recent developments. *Sports Med.* **1999**, *28*, 73–80. [CrossRef]
94. Bopp, M.J.; Houston, D.K.; Lenchik, L.; Easter, L.; Kritchevsky, S.B.; Nicklas, B.J. Lean mass loss is associated with low protein intake during dietary-induced weight loss in postmenopausal women. *J. Am. Diet. Assoc.* **2008**, *108*, 1216–1220. [CrossRef]
95. Cintineo, H.P.; Arent, M.A.; Antonio, J.; Arent, S.M. Effects of Protein Supplementation on Performance and Recovery in Resistance and Endurance Training. *Front. Nutr.* **2018**, *5*, 83. [CrossRef]
96. MacNeil, B.; Hoffman-Goetz, L.; Kendall, A.; Houston, M.; Arumugam, Y. Lymphocyte proliferation responses after exercise in men: Fitness, intensity, and duration effects. *J. Appl. Physiol.* **1991**, *70*, 179–184. [CrossRef]
97. Pedersen, B.K.; Toft, A.D. Effects of exercise on lymphocytes and cytokines. *Br. J. Sports Med.* **2000**, *34*, 246–251. [CrossRef] [PubMed]
98. Nieman, D.C.; Brendle, D.; Henson, D.A.; Suttles, J.; Cook, V.D.; Warren, B.J.; Butterworth, D.E.; Fagoaga, O.R.; Nehlsen-Cannarella, S.L. Immune function in athletes versus non-athletes. *Int. J. Sports Med.* **1995**, *16*, 329–333. [CrossRef] [PubMed]
99. Castell, L.M.; Newsholme, E.A. The effects of oral glutamine supplementation on athletes after prolonged, exhaustive exercise. *Nutrition* **1997**, *13*, 738–742. [CrossRef] [PubMed]
100. Shewchuk, L.D.; Baracos, V.E.; Field, C.J. Dietary L-glutamine does not improve lymphocyte metabolism or function in exercise trained rats. *Med. Sci. Sports Exerc.* **1997**, *29*, 474–481. [CrossRef]
101. Crewther, B.; Keogh, J.; Cronin, J.; Cook, C. Possible stimuli for strength and power adaptation: Acute hormonal responses. *Sports Med.* **2006**, *36*, 215–238. [CrossRef] [PubMed]
102. Durand, R.J.; Castracane, V.D.; Hollander, D.B.; Tryniecki, J.L.; Bamman, M.M.; O'Neal, S.; Hebert, E.P.; Kraemer, R.R. Hormonal responses from concentric and eccentric muscle contractions. *Med. Sci. Sports Exerc.* **2003**, *35*, 937–943. [CrossRef]
103. Hayes, L.D.; Grace, F.M.; Baker, J.S.; Sculthorpe, N. Resting steroid hormone concentrations in lifetime exercisers and lifetime sedentary males. *Aging Male* **2015**, *18*, 22–26. [CrossRef]
104. Shaner, A.A.; Vingren, J.L.; Hatfield, D.L.; Budnar, R.G., Jr.; Duplanty, A.A.; Hill, D.W. The acute hormonal response to free weight and machine weight resistance exercise. *J. Strength Cond. Res.* **2014**, *28*, 1032–1040. [CrossRef]
105. Caputo, M.; Pigni, S.; Agosti, E.; Daffara, T.; Ferrero, A.; Filigheddu, N.; Prodam, F. Regulation of GH and GH Signaling by Nutrients. *Cells* **2021**, *10*, 1376. [CrossRef]
106. Cadegiani, F.A.; Kater, C.E. Hormonal response to a non-exercise stress test in athletes with overtraining syndrome: Results from the Endocrine and metabolic Responses on Overtraining Syndrome (EROS)—EROS-STRESS. *J. Sci. Med. Sport* **2018**, *21*, 648–653. [CrossRef] [PubMed]
107. Bartke, A.; Darcy, J. GH and ageing: Pitfalls and new insights. *Best Pract. Res. Clin. Endocrinol. Metab.* **2017**, *31*, 113–125. [CrossRef] [PubMed]
108. Poudel, S.B.; Dixit, M.; Neginskaya, M.; Nagaraj, K.; Pavlov, E.; Werner, H.; Yakar, S. Effects of GH/IGF on the Aging Mitochondria. *Cells* **2020**, *9*, 1384. [CrossRef] [PubMed]
109. Laron, Z. Insulin-like growth factor 1 (IGF-1): A growth hormone. *Mol. Pathol.* **2001**, *54*, 311–316. [CrossRef]
110. Moøller, N.; Joørgensen, J.O.L. Effects of growth hormone on glucose, lipid, and protein metabolism in human subjects. *Endocr. Rev.* **2009**, *30*, 152–177. [CrossRef]

111. Wurzburger, M.I.; Prelevic, G.M.; Sonksen, P.H.; Balint-Peric, L.A.; Wheeler, M. The effect of recombinant human growth hormone on regulation of growth hormone secretion and blood glucose in insulin-dependent diabetes. *J. Clin. Endocrinol. Metab.* **1993**, *77*, 267–272.
112. Tarnopolsky, M. Protein requirements for endurance athletes. *Nutrition* **2004**, *20*, 662–668. [CrossRef]
113. Fielding, R.A.; Parkington, J. What are the dietary protein requirements of physically active individuals? New evidence on the effects of exercise on protein utilization during post-exercise recovery. *Nutr. Clin. Care* **2002**, *5*, 191–196. [CrossRef]
114. Trumbo, P.; Schlicker, S.; Yates, A.A.; Poos, M. Dietary Reference Intakes for Energy, Carbohydrate, Fiber, Fat, Fatty Acids, Cholesterol, Protein and Amino Acids. *J. Acad. Nutr. Diet.* **2002**, *102*, 1621–1630. [CrossRef]
115. Layman, D.K. Dietary Guidelines should reflect new understandings about adult protein needs. *Nutr. Metab.* **2009**, *13*, 12. [CrossRef] [PubMed]
116. Gastmann, U.A.; Lehmann, M.J. Overtraining and the BCAA hypothesis. *Med. Sci. Sports Exerc.* **1998**, *30*, 1173–1178. [CrossRef] [PubMed]
117. Wang, C.C.; Fang, C.C.; Lee, Y.H.; Yang, M.T.; Chan, K.H. Effects of 4-Week Creatine Supplementation Combined with Complex Training on Muscle Damage and Sport Performance. *Nutrients* **2018**, *10*, 1640. [CrossRef] [PubMed]
118. Ribeiro, A.S.; Avelar, A.; Kassiano, W.; Nunes, J.P.; Schoenfeld, B.J.; Aguiar, A.F.; Trindade, M.C.C.; Silva, A.M.; Sardinha, L.B.; Cyrino, E.S. Creatine supplementation does not influence the ratio between intracellular water and skeletal muscle mass in resistance-trained men. *Int. J. Sport Nutr. Exerc. Metab.* **2020**, *30*, 405–411. [CrossRef] [PubMed]
119. Kanter, M. High-Quality Carbohydrates and Physical Performance: Expert Panel Report. *Nutr. Today* **2018**, *53*, 35–39. [CrossRef] [PubMed]
120. Burke, L.M.; Kiens, B.; Ivy, J.L. Carbohydrates and fat for training and recovery. *J. Sports Sci.* **2004**, *22*, 15–30. [CrossRef]
121. Manfredi, T.G.; Fielding, R.A.; O'Reilly, K.P.; Meredith, C.N.; Lee, H.Y.; Evans, W.J. Plasma creatine kinase activity and exercise-induced muscle damage in older men. *Med. Sci. Sports Exerc.* **1991**, *23*, 1028–1034. [CrossRef]
122. Mitchell, J.B.; Costill, D.L.; Houmard, J.A.; Flynn, M.G.; Fink, W.J.; Beltz, J.D. Influence of carbohydrate ingestion on counterregulatory hormones during prolonged exercise. *Int. J. Sports Med.* **1990**, *11*, 33–36. [CrossRef]
123. Murray, R.; Paul, G.L.; Seifert, J.G.; Eddy, D.E. Responses to varying rates of carbohydrate ingestion during exercise. *Med. Sci. Sports Exerc.* **1991**, *23*, 713–718. [CrossRef]
124. Hadley, A.R.; Tran, L.T.; Fagoaga, O.R.; Nehlsen-Cannarella, S.L.; Yellon, S.M. Sex differences in photoperiod control of antigen-specific primary and secondary humoral immunity in Siberian Hamsters. *J. Neuroimmunol.* **2002**, *128*, 39–48. [CrossRef]
125. Gleeson, M.; Bishop, N.C. Special feature for the Olympics: Effect of exercise on the immune system: Modification of immune responses to exercise by carbohydrate, glutamine and anti-oxidant supplements. *Immunol. Cell Biol.* **2000**, *78*, 554–561.
126. Cadegiani, F.A.; Kater, C.E. Basal hormones and biochemical markers as predictors of OTS: Results from the Endocrine and metabolic Responses on Overtraining Syndrome (EROS) study—EROS-BASAL. *J. Athl. Train.* **2019**, *54*, 906–914. [CrossRef]
127. Ludwig, D.S.; Hu, F.B.; Tappy, L.; Brand-Miller, J. Dietary carbohydrates: Role of quality and quantity in chronic disease. *BMJ* **2018**, *361*, k2340. [CrossRef]
128. Patrice, F.; Céline, K.; Defour, J.P. What is the normal value of the neutrophil-to-lymphocyte ratio? *BMC Res. Notes* **2017**, *10*, 12.
129. Suh, B.; Shin, D.W.; Kwon, H.M.; Yun, J.M.; Yang, H.K.; Ahn, E.; Lee, H.; Park, J.H.; Cho, B. Elevated neutrophil to lymphocyte ratio and ischemic stroke risk in generally healthy adults. *PLoS ONE* **2017**, *12*, e0183706. [CrossRef]
130. Burke, L.M.; Ross, M.L.; Garvican-Lewis, L.A.; Welvaert, M.; Heikura, I.A.; Forbes, S.G.; Mirtschin, J.G.; Cato, L.E.; Strobel, N.; Sharma, A.P.; et al. Low carbohydrate, high fat diet impairs exercise economy and negates the performance benefit from intensified training in elite race walkers. *J. Physiol.* **2017**, *595*, 2785–2807. [CrossRef]
131. Hofman, D.L.; van Buul, V.J.; Brouns, F.J. Nutrition, Health, and Regulatory Aspects of Digestible Maltodextrins. *Crit. Rev. Food Sci. Nutr.* **2016**, *56*, 2091–2100. [CrossRef]
132. Rombaldi, A.J.; Leite, C.F.; Hartleben, C.P.; Medeiros, T.H. Effects of carbohydrate supplementation and different types of exercise training on blood cells concentrations. *Rev. Bras. Med. Esporte* **2013**, *19*, 200–203. [CrossRef]
133. Khorshidi-Hosseini, M.; Nakhostin-Roohi, B. Effect of glutamine and maltodextrin acute supplementation on anaerobic power. *Asian J. Sports Med.* **2013**, *4*, 131–136. [CrossRef] [PubMed]
134. Ghazzawi, H.A.; Hussain, M.A.; Raziq, K.M.; Alsendi, K.K.; Alaamer, R.O.; Jaradat, M.; Alobaidi, S.; Al Aqili, R.; Trabelsi, K.; Jahrami, H. Exploring the Relationship between Micronutrients and Athletic Performance: A Comprehensive Scientific Systematic Review of the Literature in Sports Medicine. *Sports* **2023**, *11*, 109. [CrossRef] [PubMed]
135. Moore, A.; Khanna, D. The Role of Vitamin C in Human Immunity and Its Treatment Potential Against COVID-19: A Review Article. *Cureus* **2023**, *15*, e33740. [CrossRef]
136. Thompson, S.H. Characteristics of the female athlete triad in collegiate cross-country runners. *J. Am. Coll. Health* **2007**, *56*, 129–136. [CrossRef]
137. Li, S.; Fasipe, B.; Laher, I. Potential harms of supplementation with high doses of antioxidants in athletes. *J. Exerc. Sci. Fit.* **2022**, *20*, 269–275. [CrossRef]
138. Holzhauer, P. Micronutrient therapy with selenium, vitamin C, and L carnitine in gynecological oncology: Areas of application and studies. *Gynakologe* **2017**, *50*, 15–21. [CrossRef]

139. De Brito, E.; Teixeira, A.D.O.; Righi, N.C.; Paulitcth, F.D.S.; da Silva, A.M.V.; Signori, L.U. Vitamins C and E Associated with Cryotherapy in the Recovery of the Inflammatory Response After Resistance Exercise: A Randomized Clinical Trial. *J. Strength Cond. Res.* **2022**, *36*, 135–141. [CrossRef] [PubMed]
140. Dobrowolski, H.; Karczemna, A.; Włodarek, D. Nutrition for Female Soccer Players—Recommendations. *Medicina* **2020**, *56*, 28. [CrossRef]
141. Purcell, L.; Society, C.P. Paediatric Sports and Exercise Medicine Section Sport nutrition for young athletes. *Paediatr. Child Health* **2013**, *18*, 200–202. [CrossRef] [PubMed]
142. Mizuma, H.; Tanaka, M.; Nozaki, S.; Mizuno, K.; Tahara, T.; Ataka, S.; Sugino, T.; Shirai, T.; Kajimoto, Y.; Kuratsune, H.; et al. Daily oral administration of crocetin attenuates physical fatigue in human subjects. *Nutr. Res.* **2009**, *29*, 145–150. [CrossRef] [PubMed]
143. Gaeini, A.A.; Rahnama, N.; Hamedinia, M.R. Effects of vitamin E supplementation on oxidative stress at rest and after exercise to exhaustion in athletic students. *J. Sport Med. Phys. Fit* **2006**, *46*, 458.
144. Wierzejska, R.E. Dietary Supplements-For Whom? The Current State of Knowledge about the Health Effects of Selected Supplement Use. *Int. J. Environ. Res. Public Health* **2021**, *18*, 8897. [CrossRef]
145. Erickson, J.M.; Mawson, A.R. Possible role of endogenous retinoid (Vitamin A) toxicity in the pathophysiology of primary biliary cirrhosis. *J. Theor. Biol.* **2000**, *206*, 47–54. [CrossRef]
146. Carlsohn, A.; Braun, H.; Großhauser, M.; König, D.; Lampen, A.; Mosler, S.; Nieß, A.; Oberritter, H.; Schäbethal, K.; Schek, A.; et al. Position of the working group sports nutrition of the German Nutrition Society (DGE): Minerals and vitamins in sports nutrition. *Dtsch. Z. Sportmed.* **2020**, *71*, 208–215. [CrossRef]
147. Brancaccio, M.; Mennitti, C.; Cesaro, A.; Fimiani, F.; Vano, M.; Gargiulo, B.; Caiazza, M.; Amodio, F.; Coto, I.; D'Alicandro, G.; et al. The Biological Role of Vitamins in Athletes' Muscle, Heart and Microbiota. *Int. J. Environ. Res. Public Health* **2022**, *19*, 1249. [CrossRef]
148. Mieszkowski, J.; Kochanowicz, A.; Piskorska, E.; Niespodziński, B.; Siódmiak, J.; Buśko, K.; Stankiewicz, B.; Olszewska-Słonina, D.; Antosiewicz, J. Serum levels of bone formation and resorption markers in relation to vitamin D status in professional gymnastics and physically active men during upper and lower body high-intensity exercise. *J. Int. Soc. Sports Nutr.* **2021**, *18*, 29. [CrossRef]
149. Keen, D.A.; Constantopoulos, E.; Konhilas, J.P. The impact of post-exercise hydration with deep-ocean mineral water on rehydration and exercise performance. *J. Int. Soc. Sports Nutr.* **2016**, *13*, 17. [CrossRef]
150. Peeling, P.; Fulton, S.K.; Binnie, M.; Goodman, C. Training environment and Vitamin D status in athletes. *Int. J. Sports Med.* **2013**, *34*, 248–252. [CrossRef]
151. Vallieres, F.; Tremblay, A.; St-Jean, L. Study of the energy balance and the nutritional status of highly trained female swimmers. *Nutr. Res.* **1989**, *9*, 699–703. [CrossRef]
152. Benson, J.; Gillien, D.M.; Bourdet, K.; Loosli, A.R. Inadequate Nutrition and Chronic Calorie Restriction in Adolescent Ballerinas. *Phys. Sportsmed.* **1985**, *13*, 79–90. [CrossRef] [PubMed]
153. Aminaei, M.; Shamsi, E.H.; Nikoei, R. The impact of eight weeks of calcium intake and vitamin D along with TRX exercise on body composition and lipid profiles of overweight women. *Obes. Med.* **2020**, *19*, 100249. [CrossRef]
154. Hrubša, M.; Siatka, T.; Nejmanová, I.; Vopršalová, M.; Kujovská Krčmová, L.; Matoušová, K.; Javorská, L.; Macáková, K.; Mercolini, L.; Remião, F.; et al. Biological Properties of Vitamins of the B-Complex, Part 1: Vitamins B1, B2, B3, and B5. *Nutrients* **2022**, *14*, 484. [CrossRef] [PubMed]
155. Fernandes, G.; Bysani, C.; Venkatraman, J.T.; Tomar, V.; Zhao, W. Increased TGF-beta and decreased oncogene expression by omega-3 fatty acids in the spleen delays onset of autoimmune disease in B/W mice. *J. Immunol.* **1994**, *152*, 5979–5987. [CrossRef] [PubMed]
156. De Sousa, M.V.; Lundsgaard, A.M.; Christensen, P.M.; Christensen, L.; Randers, M.B.; Mohr, M.; Nybo, L.; Kiens, B.; Fritzen, A.M. Nutritional optimization for female elite football players—Topical review. *Scand. J. Med. Sci. Sport* **2022**, *32*, 81–104. [CrossRef] [PubMed]
157. Endres, S.; Ghorbani, R.; Kelley, V.E.; Georgilis, K.; Lonnemann, G.; van der Meer, J.W.; Cannon, J.G.; Rogers, T.S.; Klempner, M.S.; Weber, P.C. The effect of dietary supplementation with n-3 polyunsaturated fatty acids on the synthesis of interleukin-1 and tumor necrosis factor by mononuclear cells. *N. Engl. J. Med.* **1989**, *320*, 265–271. [CrossRef]
158. Kopp-Woodroffe, S.A.; Manore, M.M.; Dueck, C.A.; Skinner, J.S.; Matt, K.S. Energy and nutrient status of amenorrheic athletes participating in a diet and exercise training intervention program. *Int. J. Sport Nutr.* **1999**, *9*, 70–88. [CrossRef] [PubMed]
159. Pasricha, S.-R.; Tye-Din, J.; Muckenthaler, M.U.; Swinkels, D.W. Iron deficiency. *Lancet* **2021**, *397*, 233–248. [CrossRef]
160. Vincent, J.B.; Neggers, Y.; McClung, J. Roles of Chromium(III), Vanadium, Iron, and Zinc in Sports Nutrition. In *Nutrition and Enhanced Sports Performance*; Elsevier: Amsterdam, The Netherlands, 2018; pp. 653–664.
161. IAlaunyte, I.; Stojceska, V.; Plunkett, A. Iron and the female athlete: A review of dietary treatment methods for improving iron status and exercise performance. *J. Int. Soc. Sport Nutr.* **2015**, *12*, 38. [CrossRef]
162. Rayman, M.P. Selenium and human health. *Lancet* **2012**, *379*, 1256–1268. [CrossRef]
163. Fernández-Lázaro, D.; Fernandez-Lazaro, C.I.; Mielgo-Ayuso, J.; Navascués, L.J.; Martínez, A.C.; Seco-Calvo, J. The Role of Selenium Mineral Trace Element in Exercise: Antioxidant Defense System, Muscle Performance, Hormone Response, and Athletic Performance. A Systematic Review. *Nutrients* **2020**, *12*, 1790. [CrossRef]

164. Volpe, S.L. Magnesium and the Athlete. *Curr. Sports Med. Rep.* **2015**, *14*, 279–283. [CrossRef]
165. Song, J.; She, J.; Chen, D.; Pan, F. Latest research advances on magnesium and magnesium alloys worldwide. *J. Magnes. Alloy.* **2020**, *8*, 1–41. [CrossRef]
166. Barrientos, G.; Alves, J.; Pradas, F.; Robles, M.C.; Muñoz, D.; Maynar, M. Association between Parameters Related to Oxidative Stress and Trace Minerals in Athletes. *Sustainability* **2020**, *12*, 4966. [CrossRef]
167. Mel'Nikov, A.A.; Vikulov, A.D. Relationship between mineral metabolism and blood rheology in athletes. *Fiziol. Cheloveka* **2003**, *29*, 48–56.
168. Sone, R.; Nakazawa, S.; Ohishi, K. Efficacy of mineral-rich antioxidant supplements on oxidative stress markers and exercise performance. *Gazz. Med. Ital. Arch. Sci. Med.* **2022**, *181*, 295–302. [CrossRef]
169. Williams, M.H. Dietary Supplements and Sports Performance: Minerals. *J. Int. Soc. Sports Nutr.* **2005**, *2*, 43. [CrossRef] [PubMed]
170. Jordan, S.L.; Albracht-Schulte, K.; Robert-McComb, J.J. Micronutrient deficiency in athletes and inefficiency of supplementation: Is low energy availability a culprit? *PharmaNutrition* **2020**, *14*, 100229. [CrossRef]
171. Serra, M.C.; Beavers, K.M. Essential and nonessential micronutrients and sport. In *Nutritional Supplements in Sports and Exercise*, 2nd ed.; Springer: Berlin/Heidelberg, Germany, 2015.
172. Burgos-Jara, C.; Cerda-Kohler, H.; Aedo-Muñoz, E.; Miarka, B. Eccentric Resistance Training: A Methodological Proposal of Eccentric Muscle Exercise Classification Based on Exercise Complexity, Training Objectives, Methods, and Intensity. *Appl. Sci.* **2023**, *13*, 7969. [CrossRef]
173. Papadopoulou, S.K. Rehabilitation Nutrition for Injury Recovery of Athletes: The Role of Macronutrient Intake. *Nutrients* **2020**, *12*, 2449. [CrossRef]
174. Noakes, T.D. What Is the Evidence That Dietary Macronutrient Composition Influences Exercise Performance? A Narrative Review. *Nutrients* **2022**, *14*, 862. [CrossRef]
175. Coqueiro, A.Y.; Rogero, M.M.; Tirapegui, J. Glutamine as an Anti-Fatigue Amino Acid in Sports Nutrition. *Nutrients* **2019**, *11*, 863. [CrossRef]
176. Whittaker, J.; Harris, M. Low-carbohydrate diets and men's cortisol and testosterone: Systematic review and meta-analysis. *Nutr. Health* **2022**, *28*, 543–554, Correction in *Nutr. Health* **2022**, *28*, 783. [CrossRef]
177. Food and Nutrition Board (FNB) of the Institute of Medicine. *Dietary Reference Intakes for Energy, Carbohydrate, Fibre, Fat, Fatty Acids, Cholesterol, Protein and Amino Acids (Macronutrients)*; The National Academies Press: Washington, DC, USA, 2005; Available online: https://nap.nationalacademies.org/catalog/10490/dietary-reference-intakes-for-energy-carbohydrate-fiber-fat-fatty-acids-cholesterol-protein-and-amino-acids (accessed on 22 November 2023).
178. Jannas-Vela, S.; Espinosa, A.; Candia, A.A.; Flores-Opazo, M.; Peñailillo, L.; Valenzuela, R. The Role of Omega-3 Polyunsaturated Fatty Acids and Their Lipid Mediators on Skeletal Muscle Regeneration: A Narrative Review. *Nutrients* **2023**, *15*, 871. [CrossRef] [PubMed]
179. Radzikowska, U.; Rinaldi, A.O.; Çelebi Sözener, Z.; Karaguzel, D.; Wojcik, M.; Cypryk, K.; Akdis, M.; Akdis, C.A.; Sokolowska, M. The Influence of Dietary Fatty Acids on Immune Responses. *Nutrients* **2019**, *11*, 2990. [CrossRef]
180. Leong, W.Y.A.; Ngiam, J.N.; Tan, R.S.; Lim, S.L.; Poh, K.K. Controversies and discrepancies in the effect of dietary fat and cholesterol on cardiovascular risk. *Singapore Med. J.* **2021**, *62*, 56–62. [CrossRef]
181. Clemente-Suárez, V.J.; Bustamante-Sanchez, Á.; Mielgo-Ayuso, J.; Martínez-Guardado, I.; Martín-Rodríguez, A.; Tornero-Aguilera, J.F. Antioxidants and Sports Performance. *Nutrients* **2023**, *15*, 2371. [CrossRef] [PubMed]
182. Lobo, V.; Patil, A.; Phatak, A.; Chandra, N. Free radicals, antioxidants and functional foods: Impact on human health. *Pharmacogn. Rev.* **2010**, *4*, 118–126. [CrossRef] [PubMed]

nutrients

MDPI

Article

Microbial Fermentation Enhances the Effect of Black Tea on Hyperlipidemia by Mediating Bile Acid Metabolism and Remodeling Intestinal Microbes

Lingli Sun [1,†], Lianghua Wen [2,†], Qiuhua Li [1], Ruohong Chen [1], Shuai Wen [1], Xingfei Lai [1], Zhaoxiang Lai [1], Junxi Cao [1], Zhenbiao Zhang [1], Mengjiao Hao [1], Fanrong Cao [2] and Shili Sun [1,*]

[1] Tea Research Institute, Guangdong Provincial Key Laboratory of Tea Plant Resources Innovation and Utilization, Guangdong Academy of Agricultural Sciences, Guangzhou 510640, China; sunlingli@tea.gdaas.cn (L.S.); liqiuhua@tea.gdaas.cn (Q.L.); chenruohong@tea.gdaas.cn (R.C.); wenshuai@tea.gdaas.cn (S.W.); laixingfei@tea.gdaas.cn (X.L.); laizhaoxiang@tea.gdaas.cn (Z.L.); caojunxi@tea.gdaas.cn (J.C.); zhangzhenbiao@tea.gdaas.cn (Z.Z.); haomengjiao@gdaas.cn (M.H.)

[2] College of Horticulture, South China Agricultural University, Guangzhou 510000, China; tea15779451015@163.com (L.W.); prcao@scau.edu.cn (F.C.)

* Correspondence: sunshili@zju.edu.cn

† These authors contributed equally to this work.

Abstract: Black tea (BT), the most consumed tea worldwide, can alleviate hyperlipidemia which is a serious threat to human health. However, the quality of summer BT is poor. It was improved by microbial fermentation in a previous study, but whether it affects hypolipidemic activity is unknown. Therefore, we compared the hypolipidemic activity of BT and microbially fermented black tea (EFT). The results demonstrated that BT inhibited weight gain and improved lipid and total bile acid (TBA) levels, and microbial fermentation reinforced this activity. Mechanistically, both BT and EFT mediate bile acid circulation to relieve hyperlipidemia. In addition, BT and EFT improve dyslipidemia by modifying the gut microbiota. Specifically, the increase in *Lactobacillus johnsonii* by BT, and the increase in *Mucispirillum* and *Colidextribacter* by EFT may also be potential causes for alleviation of hyperlipidemia. In summary, we demonstrated that microbial fermentation strengthened the hypolipidemic activity of BT and increased the added value of BT.

Keywords: microbial fermented black tea; hyperlipidemia; bile acid; gut microbes

Citation: Sun, L.; Wen, L.; Li, Q.; Chen, R.; Wen, S.; Lai, X.; Lai, Z.; Cao, J.; Zhang, Z.; Hao, M.; et al. Microbial Fermentation Enhances the Effect of Black Tea on Hyperlipidemia by Mediating Bile Acid Metabolism and Remodeling Intestinal Microbes. *Nutrients* **2024**, *16*, 998. https://doi.org/10.3390/nu16070998

Academic Editors: Franck Gael Carbonero, Domenico Tafuri and Francesca Latino

Received: 1 March 2024
Revised: 22 March 2024
Accepted: 26 March 2024
Published: 28 March 2024

1. Introduction

Economic growth has been accompanied by a dramatic rise in sedentary behavior and high-fat dietary intake, which has caused an increase in metabolic syndromes such as obesity, type 2 diabetes, hyperlipidemia, and non-alcoholic fatty liver disease [1,2]. One sort of dyslipidemia amongst them, hyperlipidemia, is characterized by weight increase and abnormal lipid measures [3,4]. Excessive blood lipid levels have been associated with increased blood viscosity, blood vessel blockage, and insufficient blood flow to some essential organs, and it is additionally connected to coronary heart disease, myocardial infarction, angina pectoris, cerebral thrombosis, stroke, and even a higher risk of COVID-19 and its associated complications [5,6]. Nearly half of all fatalities globally are caused by morbidity and mortality from hyperlipidemia and its consequences, which is a challenge for global public health [7]. Currently, statins and fibrates are the major therapeutic medications employed for managing hyperlipidemia; however, these medications come with a range of side effects, including the development of new-onset type 2 diabetes, the possibility of neurological and neurocognitive consequences, hepatitis, and renal toxicity [8]. As a consequence, developing novel therapies that are safe, effective, and natural has become an increasingly important research area in both academia and industry.

Dietary intervention has been demonstrated to be an efficient remedy for hyperlipidemia. Due to its multiple potential health advantages, including anti-tumor, anti-oxidant, anti-inflammatory, anti-viral, and anti-obesity properties, tea, a medicinal and edible crop, has attracted significant interest on a national and global level [9]. Black tea is the most popular tea in the world, not only for its flavor, but also for its potential health benefits, especially when it comes to fat reduction and weight loss. In fact, various research has reported that black tea alleviates obesity, hyperlipidemia, non-alcoholic fatty liver disease and cardiovascular disease by improving lipid levels [10–12]. Despite the beneficial effects of black tea in alleviating dyslipidemia, the heavy bitter taste, low aroma and poor quality of summer black tea have led to reluctance to consume it, resulting in a waste of resources. In our previous research, we proposed a microbial fermentation to increase the flavor of summer black tea, which reduces the waste of summer black tea to some extent [13], and microbial fermentation, which enhances the output of active chemicals in natural products or creates new active molecules [14,15]. Zhao et al. detected lovastatin, a clinical drug for hyperlipidemia, in microbially fermented pu-erh tea but not in unfermented pu-erh tea [16]. Therefore, we speculate that the employment of microbial fermentation not only improves the quality of black tea, but may also strengthen its hypolipidemic activity. In addition, previous relevant research has focused on the effect of black tea itself on lipid levels [10–12], and there are few reports on the adoption of other methods, especially microbial fermentation, to enhance the lipid-lowering activity of black tea.

The enterohepatic axis is believed to serve a significant role in the management of hyperlipidemia. The communication between the gut and liver depends primarily on the metabolites, with bile acids playing a crucial role [17]. Bile acids are catabolic products of cholesterol in the liver, and their dysregulation is associated with metabolic diseases, intestinal inflammation, and colorectal cancer [18]. Bile acids have even been suggested as treatment options for liver illness and obesity [19]. In addition, bile acids interact with gut microbes, which have been reported to be tightly connected to hyperlipidemia, and may be a new target for the therapeutic treatment of hyperlipidemia [20,21]. Wu et al. indicated that gut microbes are essential for berberine to ameliorate hyperlipidemia [22]. Accordingly, the homeostasis of bile acid metabolism and gut microbial communities is essential for ameliorating hyperlipidemia.

In the current study, we intended to assess whether microbial fermentation, while improving the black tea quality, enhances the efficacy of black tea in alleviating hyperlipidemia induced by a high-fat, high-fructose, and high-cholesterol diet (HFFCD). We evaluated the expression of key genes and proteins for bile acid metabolism using real-time qualitative polymerase chain reaction (RT-qPCR) and Western blot (WB), and assessed the composition of gut microbes using 16S rRNA sequencing. Our study will provide evidence that microbial fermented black tea mediates bile acid metabolism and remodels gut microbes to alleviate hyperlipidemia, and will contribute positively to the search for alternative strategies to ameliorate hyperlipidemia.

2. Materials and Methods

2.1. Preparation of Tea Extracts

The fresh tea was obtained from the Tea Research Institute of Guangdong Academy of Agricultural Sciences (Yingde, Guangdong Province, China). Following traditional processes, fresh tea was processed into BT [13]. The BT is then evenly sprayed with water to preserve its moisture level at 30–35%, stacked for 35 days to allow for natural microbial fermentation, and finally dried, which is processed into EFT [13]. The BT and EFT extracts were referred to the previous methods [23].

2.2. Analysis of the Main Components of Tea

The water extract content of tea was measured according to the Chinese standard (GB/T 8305-2013 [24]); the tea polyphenol content was identified by the Folin-Ciocalteu method (GB/T 8313-2018 [25]); the free amino acid was assessed by the ninhydrin colorimetry method (GB/T 8314-2013 [25]); and total soluble sugars were detected by the anthrone-sulfuric acid colorimetric method [26]; determination of flavonoids by the method of aluminum trichloride [26]; theaflavins, thearubigins, and theabrownins were measured with the People's Republic of China Agricultural Industry Standard (NY/T 3675-2020 [27]).

2.3. High-Performance Liquid Chromatography (HPLC)

The following compounds were created as standard solutions in 50% methanol: gallic acid, gallocatechin, epi-gallocatechin, catechin, caffeine, epi-catechin, epi-gallocatechin-3-gallate, gallocatechin-3-gallate, epi-catechin-3-gallate, and catechin-3-gallate. Then, 95% methanol was used to extract the BT and EFT samples, and the filtrate was run through membranes with pore diameters of 0.45 μm. The Agilent 1260 HPLC system equipped with a Zorbax column (250 mm × 4.6 mm, 5 μm) was used to examine the tea samples. One milliliter per minute was used to elute the samples. Methanol (A) and 0.1% aqueous phosphoric acid (B) were used for elution, with 25% B applied from 0 to 19.5 min. The injection volume was 5 μL, and the sample was detected at 280 nm. At 30 °C, all procedures were completed.

2.4. Animal Experiments

Animal experiments were performed according to the Guidelines for Animal Care and Use of Tea Research Institute of Guangdong Academy of Agricultural Sciences and approved by the Animal Care and Use Committee of the Institute (Serial No. 2019003). Guangdong Zhiyuan Biomedical Technology Co., Ltd. (Guangzhou, China) supplied 50 male C57BL/6J mice that were between 7 and 8 weeks old. The animals were given unrestricted access to food and drink while being kept in a standard environment (25 ± 2 °C temperature, 55 ± 10% humidity, and a 12-h light/dark cycle beginning at 8:00 a.m.). Mice were randomly assigned to one of five treatment groups (n = 10) after 7 days of acclimatization: (1) Con group:normal maintenance chow diet; (2) Mod group: HFFCD; (3) Pos group:HFFCD with a daily intragastric gavage of 20 mg/kg simvastatin; (4) BT group:HFFCD with a daily intragastric gavage of BT extract; and (5) EFT group: HFFCD with a daily intragastric gavage of EFT extract. Diets were obtained from Xietong Bioengineering Co., Ltd. (Nanjing, China), and their compositions and energy densities are listed in Table S1. After 64 days of treatment, all mice were euthanized. Blood and tissue samples were collected, weighed, and rapidly frozen in liquid nitrogen and then stored at −80 °C.

2.5. Biochemical Assays

Blood and liver levels of total cholesterol TC, triglycerides (TG), low-density lipoprotein cholesterol (LDL-C), high-density lipoprotein cholesterol (HDL-C), and (TBA), as well as fecal levels of total bile acids, were measured using commercial kits (Nanjing Jiancheng Bioengineering Institute, Nanjing, China).

2.6. Histopathological Staining

The liver tissues were fixed in 10% formalin solution for 24 h and dehydrated in 75% ethanol for 24 h before being embedded in paraffin. And the blocks were divided into thin slices and stained with hematoxylin and eosin (H&E). The slides were analyzed with an Olympus microscope (Tokyo, Japan, 200× and 400×). Liver sections were evaluated histopathologically using a histologic scoring system for non-alcoholic fatty liver disease in reference to previous research [28].

2.7. Real-Time Quantitative Polymerase Chain Reaction (RT-qPCR)

Total RNA was extracted from liver and ileum using Total RNA Kit I (Omega Bio-Tek, Guangzhou, China), and cDNA was synthesized using a reverse transcription kit (TOYOBO Co., Ltd., Osaka, Japan). RT-qPCR was accomplished on an ABI7500 Real-Time System using SYBR Green PCR Master Mix (YEASEN, Shanghai, China). A three-step amplification program was used: pre-denaturation at 95 °C for 5 min, 40 cycles of 95 °C for 10 s, 60 °C for 20 s, and 72 °C for 20 s. All primer sequences were synthesized by Sangon Biotech Co., Ltd. (Shanghai, China) and are presented in Table S2. The relative gene expression levels were normalized to that of β-actin.

2.8. Western Blot

In total, 990 μL of radio immunoprecipitation assay (RIPA) mixed with 10 μL of phenylmethanesulfonyl fluoride (PMSF) was used to homogenize liver and ileum tissue samples (40 mg). The homogenates were placed on ice for 60 min before being centrifuged at 18,506 g for 20 min at 4 °C to remove debris. Bidicinchoninic acid (BCA, Thermo, Waltham, MA, USA) was utilized for determining the content of protein in the lysates. Equal amounts of each sample were combined with a quarter volume of loading buffer 4× and denatured for 5 min at 98 °C in a water bath. Using polyacrylamide gel electrophoresis (80–120 V), the proteins were separated and then transferred to a polyvinylidene fluoride (PVDF) membrane. After being blocked with 5% defatted milk solution for 2 h, the membranes were incubated overnight with primary antibodies specific for Farnesoid X-Activated Receptor (FXR, proteintech, Wuhan, China, diluted 1:1000), Cytochrome P450 27A1 (CYP27A1, proteintech, Wuhan, China, diluted 1:2000), Cholesterol 7alpha-hydroxylase (CYP7A1, Affinity, Liyang, China, diluted 1:1000), Apical sodium-dependent bile acid transporter (ASBT, Affinity, Liyang, China, diluted 1:500), β-actin (CST, Danvers, MA, USA, diluted 1:1000) and Glyceraldehyde-3-phosphate dehydrogenase (GAPDH, CST, Danvers, MA, USA, diluted 1:500) at 4 °C, followed by the secondary antibody at 37 °C for 50 min. After colorimetric detection and chemiluminescence imaging, the positive bands were quantified by densitometry using Image J software (open source image program for scientific images; http://imagej.net/ImageJ; accessed on 4 May 2023). The results were normalized to the density of GAPDH or β-actin bands.

2.9. Analysis of Gut Microbiota by 16S rRNA Sequencing

DNA from fresh feces was extracted by CTAB and labeled with specific barcodes and amplified with 16S specific primers. TruSeq® DNA PCR-Free Sample Preparation Kit was used to construct the library. The amplified products were sequenced in NovaSeq6000. Sequencing data were filtered, spliced, and compared to obtain Effective Tags, which were generated using QIIME 2 (v2023.2) to generate Amplicon Sequence Variant (ASV) and analyzed for bioinformatics. Taxonomic information was obtained by performing species annotation analysis using the Mothur (v1.48) method with the SSUrRNA database from SILVA138.1. Alpha and beta diversity analyses were performed using R (v4.2.0) to determine species richness and diversity; For species analysis with substantial differences, linear discriminant analysis (LDA) utilizing linear discriminant effect size (LEfSe) was utilized. Raw data are stored in the Sequence Read Archive (BioProject ID: PRJNA1090409).

2.10. Statistical Analysis

The data are expressed as the mean ± standard error of the mean (SEM). For statistical analysis, GraphPad Prism 8.0 (GraphPad Software Inc., San Diego, CA, USA) was utilized. Duncan's new multiple-range test (MRT) was used in conjunction with one-way ANOVA to evaluate the differences between the mean values for each group. It was deemed statistically significant when $p < 0.05$.

3. Results

3.1. Analysis of the Main Components of BT and EFT

During microbial fermentation, various bioactive components in the tea substrate undergo a series of biochemical reactions [29], so we analyzed the main components in BT and EFT. As shown in Figure 1A, the content of water extract, tea polyphenols, free amino acid, and thearubin in EFT was significantly lower, while the content of flavonoid, soluble sugar, tea polysaccharide, and theabrownin was significantly higher compared to BT. Also, the main phytochemical components of BT and EFT were identified and quantified by HPLC (Figure 1B–E). In comparison to BT, EFT revealed a large drop in epi-gallocatechin-3-gallate and epi-catechin-3-gallate content and a considerable rise in gallic acid, gallocatechin, and epi-catechin content. And there were no significant differences in the theaflavin, catechin, gallocatechin-3-gallate and catechin-3-gallate contents in BT and EFT. It is worth noting that we did not find EGC in EFT, but it was present in BT. Overall, these results indicate significant differences in the main components of BT and EFT after microbial fermentation.

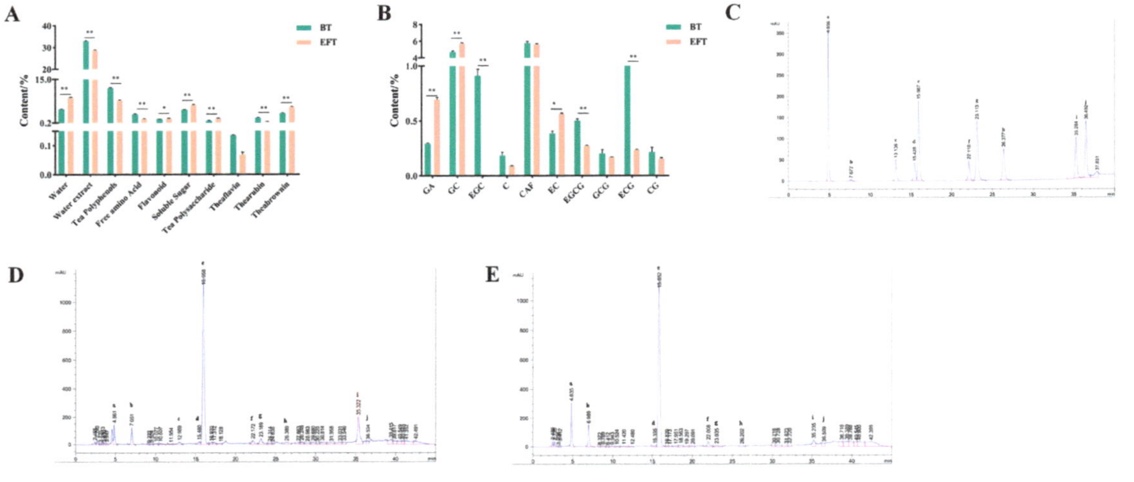

Figure 1. Analysis of the main components of BT and EFT. (**A**) The main compounds of BT and EFT. (**B**) The content of main phytochemicals in the BT and EFT. (**C**) The HPLC chromatograms of the standard samples. (**D**) The HPLC chromatograms of the BT. (**E**) The HPLC chromatograms of the EFT. (a, Gallic acid; b, Gallocatechin; c, Epi-gallocatechin; d, Catechin; e, Caffeine; f, Epi-catechin; g, Epi-gallocatechin-3-gallate; h, Gallocatechin-3-gallate; i, Epi-catechin-3-gallate; j, Catechin-3-gallate). Data are expressed as mean ± SEM; * $p < 0.05$ and ** $p < 0.01$.

3.2. *Effect of BT and EFT on Body Weight and Liver*

C57BL/6J mice were fed with HFFCD and gavage with 800 mg/kg BW BT or EFT for 64 days to investigate the effect of BT and EFT against HFFCD-induced hyperlipidemia. During animal experiments, the amount of food and water consumed by the mice was steady in each group of mice, with the control group taking a greater amount of food (Figure 2A,B), which may be due to the better palatability of the maintenance diet compared to the HFFCD. Also, the body weight of the mice was recorded at three-day intervals, and the Con and Pos groups demonstrated a declining tendency compared to the Mod group, but it was not statistically significant, while the BT and EFT groups displayed a significant decrease in body weight (Figure 2C). Compared to BT, EFT enhanced the inhibition of weight gain as an effect (Figure 2D). Moreover, the Mod group had a higher liver index than the Con group; as expected, the Pos, BT, and EFT groups all had a lower liver index compared with the Mod group, but just the BT group was statistically significant (Figure 2E). The above findings indicated that microbial fermentation not only improved the flavor of black tea, but also enhanced the property of black tea to inhibit weight gain in hyperlipidemic mice.

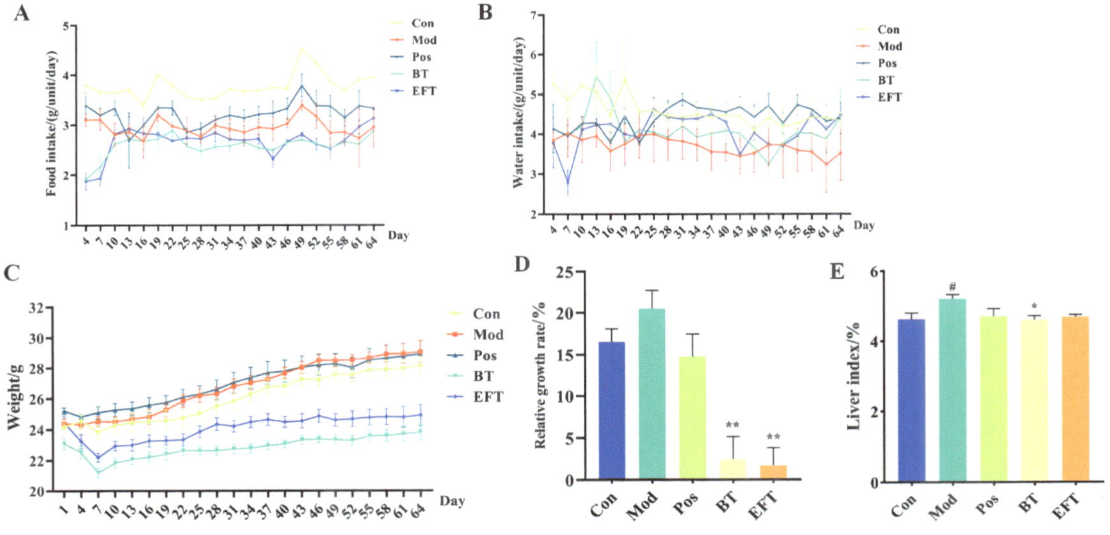

Figure 2. Effect of BT and EFT on body weight and liver weight. (**A**) Food intake. (**B**) Water intake. (**C**) Body weight. (**D**) Relative growth rate. (**E**) Liver index. Data are expressed as mean ± SEM; # $p < 0.05$ versus Con group; * $p < 0.05$ and ** $p < 0.01$ versus the Mod group.

3.3. Effect of BT and EFT on Lipid Level in Serum and Liver

Hyperlipidemia is characterized by elevated levels of TC, TG, and LDL-C in the blood and decreased levels of HDL-C. We measured the levels of TC, TG, LDL-C and HDL-C in the serum, and compared with the Con group, the TC and LDL-C levels of the Mod group were significantly higher than those of the Con group, while the HDL-C levels were significantly lower than those of the Con group, while TG levels also showed a rising trend, but there was no significant difference (Figure 3A–D), which indicated that the hyperlipidemia mouse model was successfully established. In comparison to the Mod group, BT dramatically raised serum levels of HDL-C and decreased serum levels of TC, TG, and LDL-C (Figure 3A–D). Interestingly, compared with BT, EFT enhanced the improvement effect on TC, TG, LDL-C and HDL-C in serum. The liver serves as the metabolic center of the body and is the primary location of lipid synthesis [30], so we also measured the levels of TC, TG, LDL-C and HLD-C in the liver. The Mod group had higher levels of TC, TG, and LDL-C and lower levels of HDL-C compared with the Con group (Figure 3E–H). And this trend was reversed after oral administration of BT and EFT. Compared with the Mod group, BT reduced the levels of TC, TG, and LDL-C in the liver, and increased the HDL-C content. And compared with BT, EFT enhanced the inhibitory effects on TG and LDL-C, as well as the increase in HDL-C levels. In conclusion, our results indicated that BT and EFT were effective in attenuating HFFCD-induced abnormal lipid levels in serum and liver, and more importantly, illustrated that microbial fermentation not only improves the flavor of black tea, but also strengthens the bioactivity of black tea in ameliorating dyslipidemia.

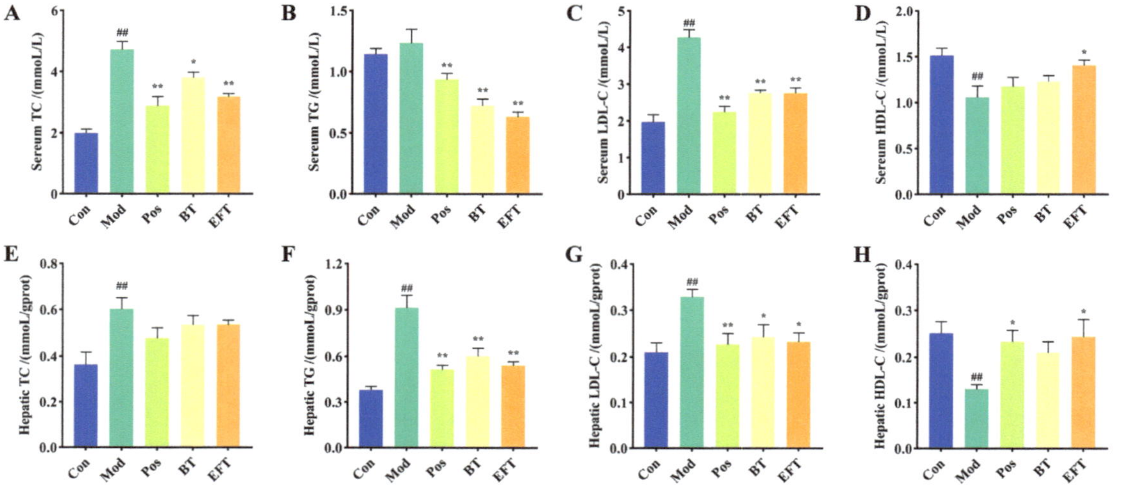

Figure 3. The lipid levels in liver and serum. (**A**) The TC in serum. (**B**) The TG in serum. (**C**) The LDL-C in serum. (**D**) The HDL-C in serum. (**E**) The TC in liver. (**F**) The TG in liver. (**G**) The LDL-C in liver. (**H**) The HDL-C in serum. Data are expressed as mean ± SEM; ## $p < 0.01$ versus Con group; * $p < 0.05$ and ** $p < 0.01$ versus the Mod group.

3.4. Effect of BT and EFT on TBA Level in Serum, Liver, and Feces

A key element of hyperlipidemia is elevated cholesterol levels, and cholesterol is a biosynthetic precursor of bile acids, and bile acid biosynthesis is crucial to the control of systemic cholesterol homeostasis [31]. Therefore, we detected TBA levels in serum, liver and feces. The Mod group showed a lowering trend in TBA levels in blood, liver, and feces when compared to the Con group; however, there was a meaningful difference only in feces. (Figure 4). However, BT and EFT significantly increased TBA levels in liver, serum and feces compared to the Mod group. In addition, EFT enhanced TBA synthesis compared to BT, as we observed higher levels of TBA in liver, serum and feces in the EFT group. Taken together, these results provide important insights that black tea can promote bile acid biosynthesis and microbial fermentation enhances this property of black tea.

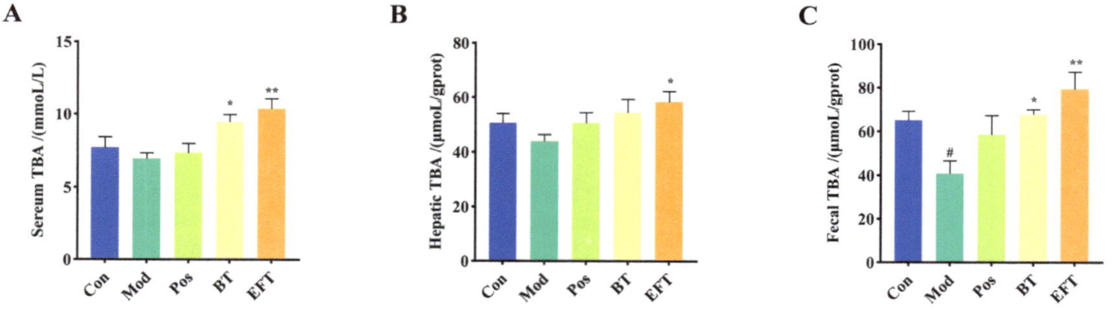

Figure 4. Analysis of TBA levels in serum, liver and feces. (**A**) Serum. (**B**) Liver. (**C**) Feces. Data are expressed as mean ± SEM; # $p < 0.05$ versus Con group; * $p < 0.05$ and ** $p < 0.01$ versus the Mod group.

3.5. Analysis of Liver Histopathology

As mentioned above, HFFCD-induced hyperlipidemia leads to hepatic lipid accumulation. To understand the effects of BT and EFT on the liver structure of HFFCD-induced hyperlipidemic mice, we performed H&E staining of liver sections and scored the liver sections histopathologically with reference to previous studies [28]. From Figure 5, the liver structures in the Con group were neatly arranged and tightly packed, whereas the H&E staining in the Mod group showed infiltration of inflammatory cells and numerous balloon-like changes (as indicated by the green arrows in Figure 5A). However, these pathologic changes improved significantly after simvastatin, BT, and EFT interventions. Our findings indicate that BT and EFT supplementation are effective in ameliorating liver tissue lesions accompanying hyperlipidemia.

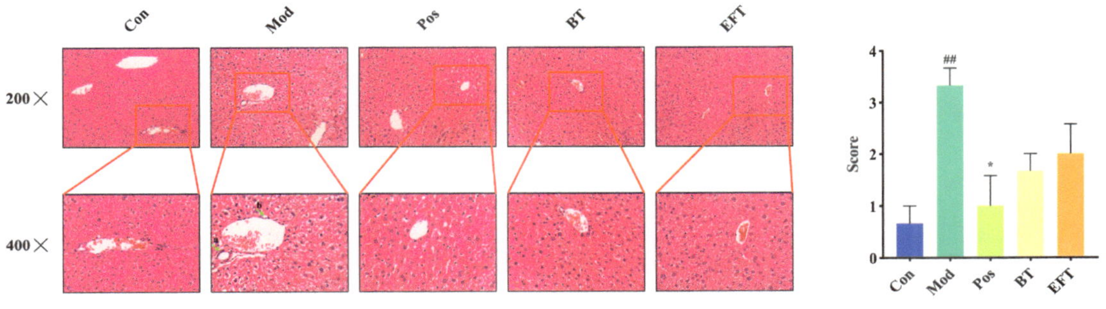

(A) (B)

Figure 5. H&E staining of liver. (**A**) Representative liver section. (**B**) Histopathology score. Data are expressed as mean ± SEM; ## $p < 0.01$ versus Con group; * $p < 0.05$ versus the Mod group.

3.6. Effect of BT and EFT on Expression of Genes and Proteins in Bile Acid Metabolism Signaling Pathway

As demonstrated in the above results, BT and EFT interventions enhanced TBA production in HFFCD-induced hyperlipidemic mice, and thus we examined genes and protein expression of key targets in the bile acid metabolite signaling pathway. The expression of bile acid synthesis-related proteins, including FXR, CYP7A1, and CYP27A1, was significantly reduced in the liver of the Mod group compared with the Con group (Figure 6A–D). Notably, simvastatin did not appear to interfere with protein expression in bile acid metabolism in HFFCD-induced hyperlipidemic mice. However, supplementation with both BT and EFT promoted the expression of these proteins, with a significant increase in protein expression of FXR, CYP7A1, and CYP27A1 in the EFT group compared to the Mod group, but the effect of BT treatment was remarkable only for CYP27A1. With respect to gene expression of FXR, CYP7A1 and CYP27A1, the overall trend was similar to that of the corresponding protein expression. Briefly, the relative expression of these three genes was significantly downregulated in the Mod group, and the trend was significantly reversed by the BT and EFT interventions (Figure 6E–G). Interestingly, both at the protein level and at the gene level, EFT strengthened the expression of FXR, CYP7A1 and CYP27A1 in the liver compared to BT.

After bile acids converted from cholesterol are passed through the intestines, about 95% are reabsorbed (especially at the end of the ileum) and returned to the liver through the portal vein to maintain the bile acid cycle [32,33]. We examined the expression of proteins and genes of FXR and ASBT, which regulate bile acid reuptake. The protein and gene expressions of FXR and ASBT were downregulated in the Mod group compared with the Con group, which indicated abnormal bile acid reabsorption in HFFCD-induced hyperlipidemic mice (Figure 6H–L). However, simvastatin, BT, and EFT interventions reversed this trend and alleviated the abnormal bile acid reabsorption. Taken together, these results indicate that black tea contributes to the conversion of cholesterol into bile acids and that microbial fermentation improves the quality of black tea while also enhancing the property of black tea in promoting bile acid synthesis.

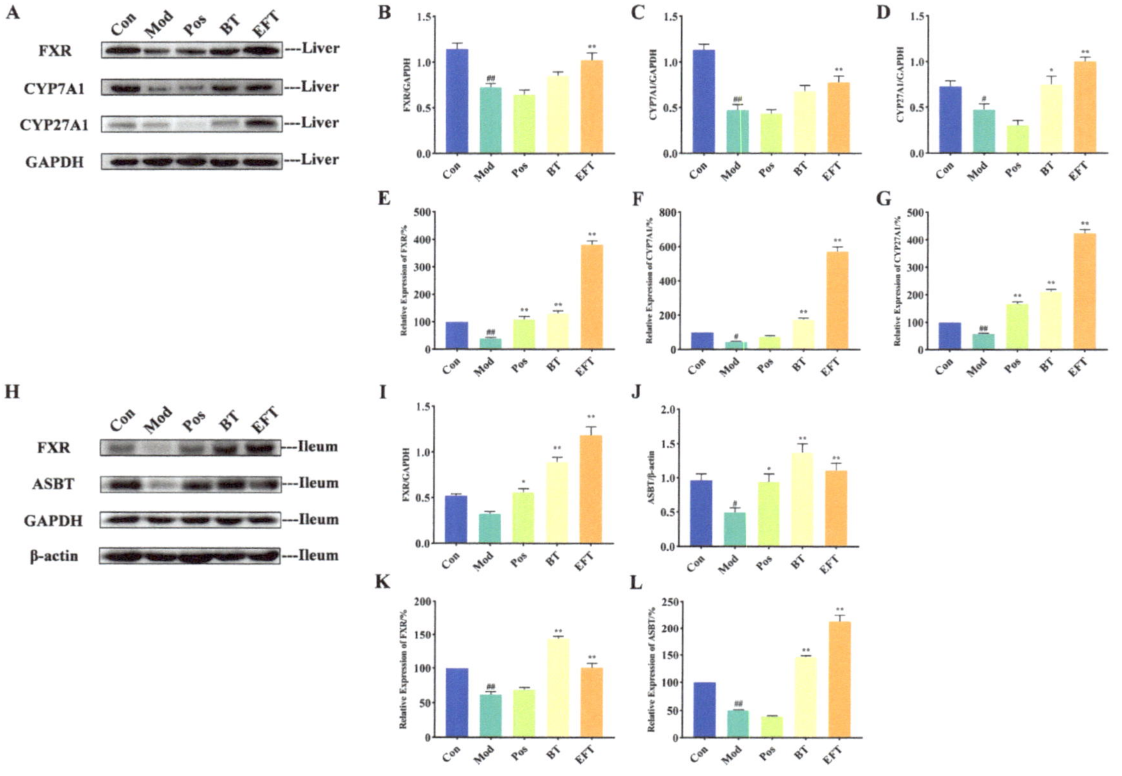

Figure 6. Effect of BT and EFT on expression of genes and proteins in bile acid metabolism signaling pathway. (**A–D**) Protein expression of hepatic FXR, CYP7A1 and CYP27A1. (**E–G**) Gene expression of hepatic FXR, CYP7A1 and CYP27A1. (**H–J**) Protein expression of ileac FXR and ASBT. (**K**,**L**) Gene expression of ileac FXR and ASBT. Data are expressed as mean ± SEM; # $p < 0.05$ and ## $p < 0.01$ versus Con group; * $p < 0.05$ and ** $p < 0.01$ versus the Mod group.

3.7. Effects of BT and EFT on the Gut Microbiota

The second largest genome in humans can be identified in the gut microbes [21], and increasing evidence suggests that it is tightly linked to host metabolism. We carried out 16S rRNA analyses of mice feces to obtain insight into how BT and EFT affected the gut microbial community in HFFCD-induced hyperlipidemic mice.

The Q20 of all samples was greater than 97% and the Q30 was greater than 93%, which indicated that the 16S rRNA sequencing was of excellent quality (Table S3). And the Rarefaction curve indicated that the gut microbial richness detected in each group was saturated. The Abundance-based coverage estimator (ACE) index, Chao 1 index, and observed ASV were reduced in the Mod group compared to the Con group and were not relieved by supplementation with simvastatin, BT, and EFT (Figure 7A–C). However, downward revisions to the Simpson and Shannon indices were reversed (Figure 7D,E). These results suggest that BT and EFT do not improve the microbial community by restoring the number of gut microbial species, but rather by influencing gut microbial species composition and distribution. Further principal coordinates analysis (PCoA) showed that the microbial communities in the Mod group differed significantly from those in the Con, Pos, BT and EFT groups (Figure 7F).

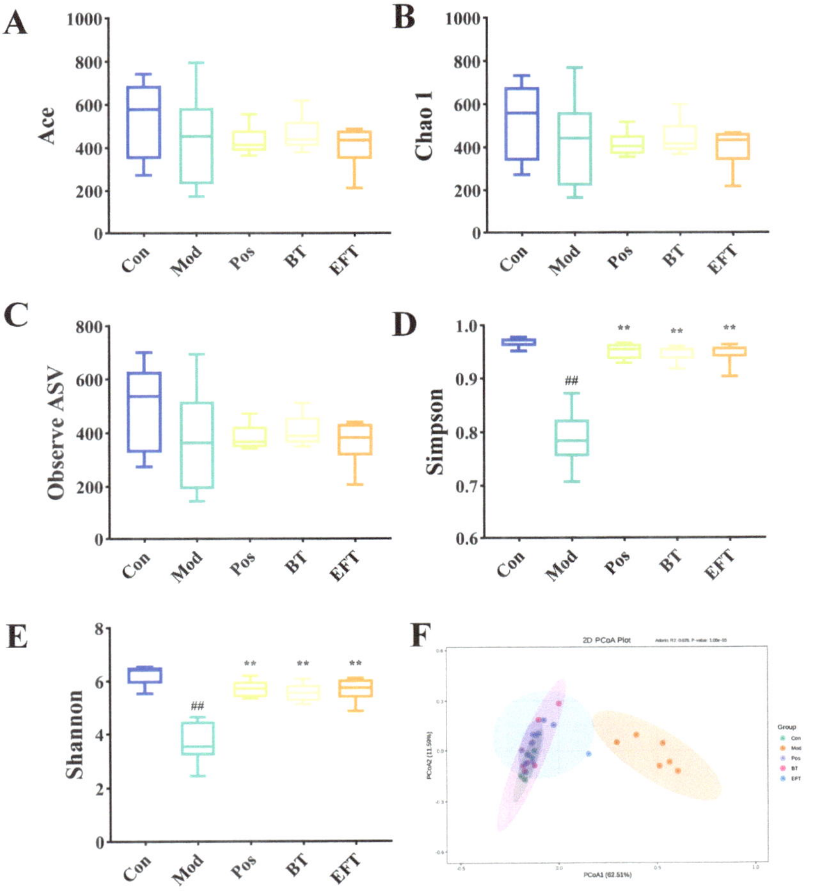

Figure 7. Effects of BT and EFT on gut microbial diversity. (**A**) Ace index; (**B**) Chao 1 index; (**C**) Observe ASV; (**D**) Simpson index; (**E**) Shannon index; (**F**) Principal coordinates analysis (PCoA) based on ASV level. Data are expressed as mean ± SEM; ## $p < 0.01$ versus Con group; ** $p < 0.01$ versus the Mod group.

At the phylum level, elevated Firmicutes, Verrucomicrobiota, Actinobacteriota, and Actinobacteria were significantly downregulated by BT and EFT, and Bacteroidota abundance in hyperlipidemic mice was upregulated by BT and EFT (Figure 8A). The ratio of Firmicutes to Bacteroidetes (F/B) has been reported to be strongly associated with metabolic syndrome, especially obesity [34]. The F/B ratio was significantly higher in the Mod group than in the Con group and was significantly downregulated after the BT and EFT interventions (Figure 8C). At the family level, *Muribaculaceae*, *Rikenellaceae*, and *Bacteroidaceae* abundance decreased in the Mod group, while *Akkermansiaceae*, *Erysipelotrichaceae*, and *Peptostreptococcaceae* abundance increased, and these shifts were reversed by BT and EFT (Figure 8B). To understand the key microorganisms in each group, we performed LefSe analysis of gut microorganisms in five groups. To understand the key microorganisms in each group, we performed LefSe analysis of gut microorganisms in five groups. We identified 58 richly differentiated taxa, including 20 genera. Among them, the predominant taxa in the Con group were *Alistipes* and *Ligilactobacillus*, whereas in the Mod group were *Akkermansia*, *Faecalibaculum*, *Romboutsia*, and *Bifidobacterium*. In addition, the main microorganism in the BT group was *Lactobacillus*, and *Colidextribacter*, *Desulfovibrio* and *Mucispirillum* prevailed in the EFT group (Figure 8D,E). These variations in microorganisms may suggest a potential mechanism for the development of hyperlipidemia. In short, our study demonstrates that BT and EFT may ameliorate hyperlipidemia by restoring the gut microbial community.

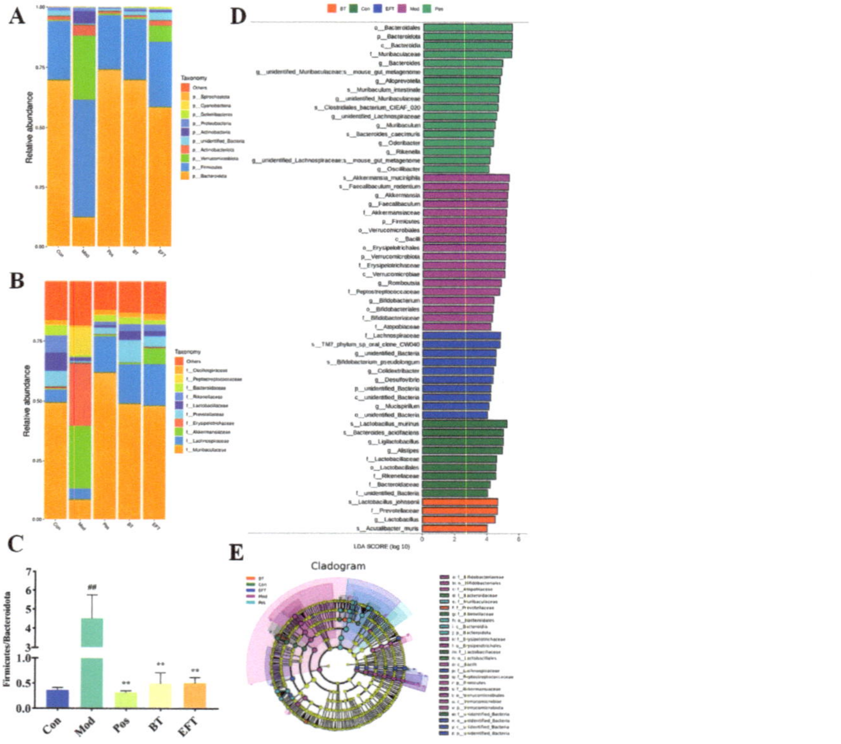

Figure 8. Effect of BT and EFT on gut microorganisms. (**A**) Microbial community composition at phylum level; (**B**) Microbial community composition at family level; (**C**) Ratio of Firmicutes to Bacteroidetes; (**D**) The linear discriminant analysis effect size (LefSe) of LDA score distribution histogram; (**E**) LEfSe analysis of an evolutionary branching diagram. Data are expressed as mean ± SEM; ## $p < 0.01$ versus Con group; ** $p < 0.01$ versus the Mod group.

4. Discussion

Black tea is the most consumed tea in the world and also contributes to fat reduction and weight loss. Liao et al. compared the effects of Keemun black tea and Dianhong black tea on fatty liver and found that both of them could affect the expression of genes related to lipid metabolism [11]. Liu et al. also showed that black tea mediates the inhibition of high-fat diet-induced weight gain and lipid accumulation by intestinal microorganisms [12]. Microbial fermentation is a common method for food flavor improvement and innovation, and has also been reported to produce new active substances. However, previous studies have focused only on the effects of black tea supplementation on hyperlipidemia, and nothing is known about whether microbial fermentation alters the hypolipidemic activity of black tea. In this study, we evaluated the effect of microbial fermentation on the hypolipidemic activity of black tea and found that microbial fermentation significantly reinforced the hypolipidemic efficacy of black tea. Mechanistically, microbial fermentation enhanced the modulatory effects of black tea on the bile acid surrogate pathway and gut microbes.

The essence of microbial fermentation is the conversion of substances with the participation of microbial activity [35]. It is this that leads to differences in the biological activity of fermentation substrates. We observed a decrease in polyphenols, theaflavin, and thearubigins in BT, which could be caused by their oxidative polymerization to theabrownine [36], as we also observed a dramatic increase in theabrownine (Figure 1). Theabrownine has been reported to remodel the gut microbial community, inhibit lipogenesis, and improve hyperlipidemia [36]. In addition, tea polysaccharides as an active substance with potential health benefits have been observed to increase in EFT. A recent report claimed that tea polysaccharides reduce obesity and improve serum TC, TG, LDL-C, and HDL-C abnormalities by modulating gut microbes and their associated short-chain fatty acid and amino acid metabolism [37]. In accordance with other reports, the gallic acid content of tea increases after fermentation, and gallic acid alleviates hypertriglyceridemia and fat accumulation by modulating glycolytic and lipolytic pathways in perirenal adipose tissue [38,39]. Furthermore, we observed an increase in the levels of gallocatechin and epi-catechin in the EFT, while gallocatechin and epi-catechin also seemed to have favorable effects in improving lipid metabolism. gallocatechin positively contributes to the activation of Adenosine 5′-monophosphate-activated protein kinase (AMPK), which is closely related to lipolysis [40,41]. After 12 consecutive weeks of epi-catechin supplementation in high-fat, high-cholesterol-induced hyperlipidemic rats, TC, LDL-C, and TG levels were significantly reduced, and hepatic fat accumulation was alleviated, along with an increase in HDL-C levels, and a decrease in lipid peroxidation and pro-inflammatory cytokine production [42]. It may be that changes in the content of these actives are responsible for the stronger hypolipidemic activity of EFT than of BT, but further research is needed to determine which of these actives play a role.

The primary route for cholesterol metabolism in the liver is the transformation of cholesterol to bile acids [43]. And there are two pathways for the generation of bile acids: the classic pathway, which is controlled by CYP7A1, and the alternative pathway, which is mediated by CYP27A1. The nuclear receptor FXR controls the activity of these two essential enzymes [3,44]. HDL carries cholesterol from peripheral tissues to the liver, where it is transformed into bile acids, and the biliary system then transports the created bile acids to the bile ducts, where they are ultimately expelled into the intestines [44,45]. In order to preserve the bile acid cycle progressing, the majority of the bile acids are reabsorbed by the ileal epithelium through ASBT-mediated processes and then sent back to the liver via the portal vein [44]. Dietary intervention in bile acid metabolism has long been recognized as an effective strategy for alleviating hyperlipidemia. According to Duan et al., whole-grain oat-derived flavonoids were helpful in accelerating the transformation of cholesterol into bile acids, which reduced blood lipid levels [3]. In our research, we found that both BT and EFT promoted the expression of FXR, CYP7A1, and CYP27A1 in the liver, suggesting that the BT and EFT interventions effectively enhanced the conversion of cholesterol to bile acids; on the other hand, BT and EFT also increased the expression of ASBT and FXR in the ileum, further suggesting that BT and EFT facilitated the bile acid cycle. It has also been demonstrated that a chronic high-fat diet disrupts the intestinal barrier function, increases the body's susceptibility to harmful intestinal substances, and leads to a systemic low-inflammatory state [46]. In addition, reduced bile acids can further exacerbate intestinal barrier dysfunction [47]. However, intestinal FXR activation contributes positively to inflammatory suppression and intestinal barrier protection [48]. Compared to the Mod group, we observed elevated bile acid levels and activated FXR expression in the BT and EFT groups, which revealed a protective effect of BT and EFT on the intestinal barrier. It is noteworthy that the effect of EFT was superior to that of BT, both in promoting hepatic bile acid synthesis and intestinal FXR activation, which indicates that hypolipidemic activity and intestinal protection of black tea are enhanced by microbial fermentation.

There is a strong association between Firmicutes and dietary energy intake as well as obesity, and a higher abundance of Firmicutes has been previously reported in populations with higher levels of obesity [49]. We observed that BT and EFT dramatically reduced the abundance of Firmicutes. In addition, Jiang et al. reported that hyperlipidemia led to an extreme reduction in Bacteroidetes [33], which is consistent with our current results, and gratefully BT with EFT reversed this alteration. Bacteroidetes and Firmicutes which have a critical role in energy metabolism in the gut, and the F/B ratio, which is closely related to the metabolic capacity of the microbial community and inflammation, are important indicators of gut ecological dysregulation associated with the metabolic syndrome [50,51]. A significant increase in the ratio of F/B in HFFCD-induced hyperlipidemic mice was also observed in the present research, whereas supplementation with BT and EFT resulted in its essentially return to normal levels. Similar effects are consistent with those reported by Duan et al. who also found a downregulation of both Firmicutes and F/B ratios and an increase in Bacteroidetes abundance after supplementing hyperlipidemic mice with flavonoids [3]. The available evidence suggests that the improvement of hyperlipidemia by BT and EFT may be the result of targeting these microorganisms.

Lactobacillus can alleviate lipid accumulation associated with a high-fat diet, and *Lactobacillus johnsonii* in particular is thought to adhere to intestinal epithelial cells and have serum cholesterol-lowering properties [52,53]. Zhu et al. also demonstrated that bile salt hydrolase from *Lactobacillus johnsonii* could mediate the FXR pathway to exert a hypocholesterolemic effect [54]. *Lactobacillus johnsonii* is a major microbiota in the BT group, perhaps this is a potential mechanism for the hypolipidemic efficacy of BT supplementation. An increasing number of research studies has suggested that a potential mechanism for dietary intervention to alleviate hyperlipidemia is the mediated production of short-chain fatty acids [55]. In the present study, hyperlipidemic mice supplemented with EFT were significantly enriched for *Mucispirillum*, a microorganism that typically produces short-chain fatty acids. Chlorogenic acid supplementation was also previously reported to increase *Mucispirillum* abundance and prevent obesity and related endotoxemia [56]. Ma et al. also observed an increase in *Mucispirillum* abundance as well as a significant elevation in short-chain fatty acid content in their study of *Lyciumbarbarum* Polysaccharide for improvement of type 2 diabetes, which further confirms that *Mucispirillum* may be a key strain in EFT for improvement of hyperlipidemia [57]. Chronic inflammation induces hyperlipidemia [58], and *Colidextribacter*, an anti-inflammatory probiotic [59] which is one of the major microorganisms in the EFT group, may alleviate hyperlipidemia by reducing inflammation in the body and enhancing the intestinal barrier function. Zhao et al. supplemented Sodium Alginate in rats fed a high-fat diet, increased the abundance of *Colidextribacter*, and effectively alleviated non-alcoholic fatty liver and serum lipid levels [60]. In addition, *Faecalibaculum* was positively correlated with serum lipid levels, whereas the reduction in *Romboutsia* was also strongly associated with weight loss [61,62]. The fact that these two strains were observed as the main taxa in the Mod group and not detected in the BT and EFT groups also coincides with this. These microbial modifications at the genus level provide a plausible explanation for the amelioration of HFFCD-induced hyperlipidemia by BT and EFT.

Currently, the vast majority of research has concluded that a reduction in the abundance of *Akkermansia* in the gut is an important marker for a reduction in metabolic syndrome, and a recent review has referred to *Akkermansia muciniphila* as a next-generation probiotic [63,64]. Interestingly, our study demonstrated that *Akkermansia* abundance was significantly increased in hyperlipidemic mice, whereas it was maintained at a lower level in the Con, Pos, BT and EFT groups. Liu et al. also reported higher levels of *Akkermansia* in rats fed a high-fat diet [65]. These phenomena indicate that the effects of *Akkermansia* on organisms are not always positive. *Akkermansia* is mainly found in the intestinal mucosa and uses mucin as a carbon and nitrogen source, and mucin contributes positively to the protection of the intestinal barrier function [66]. Mice fed a fiber-free diet for a protracted period of time have been reported to cause intestinal inflammation, which is associated with *Akkermansia* phagocytosis of intestinal mucin [67]. Since chronic inflammation induces hyperlipidemia [58], amelioration of hyperlipidemia by supplementation with BT and EFT to reduce *Akkermansia* abundance may be mediated by alleviating the level of inflammation. Another explanation for the increase in *Akkermansia* in the Mod group may be that its presence decreases important enzymes that prevent hyperglycemia and hyperlipidemia [68]. Also, this could be another potential mechanism by which BT and EFT alleviate hyperlipidemia. In summary, current results conclude that *Akkermansia* is positively associated with hyperlipidemia and that BT and EFT interventions decreased *Akkermansia* abundance and alleviated hyperlipidemia. However, the role of *Akkermansia* in metabolic syndromes such as hyperlipidemia, obesity, and type 2 diabetes should require further investigation.

5. Conclusions

In conclusion, we indicate that microbial fermentation not only improves the flavor of BT but also enhances the hypolipidemic activity. This is evidenced by the fact that BT reduced serum lipid and liver lipid levels and that EFT enhanced these effects compared to BT. Mechanistic investigations have indicated that BT and EFT promote bile acid cycling, increase gene and protein expression of FXR, CYP7A1, and CYP27A1 in the liver, and promote ileal reabsorption of bile acids, enhancing ileal FXR and ASBT expression. In addition, BT and EFT remodel the gut microbiota. The alleviating effect of BT and EFT on hyperlipidemia may be associated with a decrease in the F/B ratio, a decrease in the abundance of Firmicutes, *Faecalibaculum* and *Romboutsia* and an increase in the abundance of Bacteroidetes. In particular, one of the mechanisms by which BT ameliorates dyslipidemia may be an increase in the abundance of cholesterol-lowering *Lactobacillus johnsonii*. And increased abundance of short-chain fatty acid-producing *Mucispirillum* and anti-inflammatory *Colidextribacter* are potential reasons for the hypolipidemic effect of EFT. In summary, we conclude that BT possesses favorable hypolipidemic efficacy and microbial fermentation which not only improves the flavor of black tea, but also strengthens the hypolipidemic activity of black tea. Our research offers recommendations for scientifically choosing teas and healthy tea consumption for hyperlipidemic individuals, as well as a theoretical underpinning for microbial fermentation to strengthen the hypolipidemic activity of tea and a novel idea for the development of natural tea products to alleviate hyperlipidemia. However, the hypolipidemic active components of tea and the role of *Akkermansia* in hyperlipidemia require further investigation.

Supplementary Materials: The following supporting information can be downloaded at: https://www.mdpi.com/article/10.3390/nu16070998/s1, Figure S1: Rarefaction curve of each group; Table S1: The compositions and energy densities of the diets; Table S2: Sequences of primers for RT-qPCR; Table S3: The sequence quality of 16S rRNA sequencing.

Author Contributions: All authors' contributions as follows—S.S. and F.C. conceived and designed the experiments. L.W. and L.S. performed the experiments and analyzed the data. S.W., R.C., Q.L., J.C., M.H., X.L., Z.L. and Z.Z. contributed reagents. L.W. wrote the paper, and S.S. revised the manuscript. All authors have read and agreed to the published version of the manuscript.

Funding: This study was funded by Food nutrition and health Collaborative Innovation Center of GDAAS [Grant No. XT202229]; Agricultural competitive industry discipline team building project of Guangdong Academy of Agricultural Sciences [Grant No. 202126TD]; the open Project of Guangdong Provincial Key Laboratory of Tea Plant Resources Innovation and Utilization [Grant No. 2020KF05]; Guangdong Basic and Applied Basic Research Foundation [Grant Nos. 2021A1515010958, 2022A1515110190, 2023A1515010264]; Guangzhou Science and Technology Plan Projects [Grant Nos. 20212100005, 202201011455, 2023E04J0781]; Key-Area Research and Development Program of Guangdong Province [Grant No. 2020B0202080003]; Qingyuan Science and Technology Plan Projects [Grant Nos. 2021ZDZX002]; Innovation Fund projects of Guangdong Academy of Agricultural Sciences [Grant Nos. 202115, 202135]; Special fund for scientific innovation strategy-construction of high level Academy of Agriculture Science [Grant Nos. R2019PY-JX004, R2022PY-QY010]; the Innovation Fund projects of Guangdong Key Laboratory of Tea Plant Resources Innovation and Utilization [Grant No. 2021CX02]; Special fund project for introduction of scientific and technological talents of Guangdong Academy of Agricultural Sciences [Grant No. R2021YJ-YB3014]. Funders did not have any role in study design, data collection, and data analysis.

Institutional Review Board Statement: The study was conducted according to the guidelines of the Declaration of Helsinki and approved by the Animal Ethics Committee, Tea Research Institute, Guangdong Academy of Agricultural Sciences (Approval Code: 2019003, Approval Date: 1 June 2019).

Data Availability Statement: The original contributions presented in the study are included in the article/Supplementary Material, further inquiries can be directed to the corresponding author.

Conflicts of Interest: The authors declare no conflict of interest.

Abbreviation

HFFCD: High-cholesterol diet; AMPK, Adenosine 5′-monophosphate-activated protein kinase; BT, Black tea; EFT, Microbial fermented black tea; TC, Total cholesterol; TG, Triglyceride; LDL-C, Low-density lipoprotein cholesterol; LDL-C, High-density lipoprotein cholesterol; TBA, total bile acids; ACE, Abundance-based coverage estimator; ASBT, Apical sodium-dependent bile acid transporter; FXR, Farnesoid X-Activated Receptor; CYP27A1, Cytochrome P450 27A1; CYP7A1, Cholesterol 7alpha-hydroxylase; GAPDH, Glyceraldehyde-3-phosphate dehydrogenase; HPLC, High-Performance Liquid Chromatography.

References

1. Manting, L.; Haihong, Z.; Jing, L.; Shaodong, C.; Yihua, L. The Model of Rat Lipid Metabolism Disorder Induced by Chronic Stress Accompanying High-Fat-Diet. *Lipids Health Dis.* **2011**, *10*, 153. [CrossRef]
2. He, C.; Cheng, D.; Peng, C.; Li, Y.; Zhu, Y.; Lu, N. High-Fat Diet Induces Dysbiosis of Gastric Microbiota Prior to Gut Microbiota in Association with Metabolic Disorders in Mice. *Front. Microbiol.* **2018**, *9*, 639. [CrossRef]
3. Duan, R.; Guan, X.; Huang, K.; Zhang, Y.; Li, S.; Xia, J.; Shen, M. Flavonoids from Whole-Grain Oat Alleviated High-Fat Diet-Induced Hyperlipidemia via Regulating Bile Acid Metabolism and Gut Microbiota in Mice. *J. Agric. Food Chem.* **2021**, *69*, 7629–7640. [CrossRef]
4. Lin, L.-Y.; Hsu, C.-Y.; Lee, H.-A.; Wang, W.-H.; Kurniawan, A.L.; Chao, J.C.-J. Dietary Patterns in Relation to Components of Dyslipidemia and Fasting Plasma Glucose in Adults with Dyslipidemia and Elevated Fasting Plasma Glucose in Taiwan. *Nutrients* **2019**, *11*, 845. [CrossRef]
5. Miao, J.; Zang, X.; Cui, X.; Zhang, J. Autophagy, Hyperlipidemia, and Atherosclerosis. *Adv. Exp. Med. Biol.* **2020**, *1207*, 237–264. [CrossRef]
6. Tang, Q.; Gao, L.; Tong, Z.; Li, W. Hyperlipidemia, COVID-19 and Acute Pancreatitis: A Tale of Three Entities. *Am. J. Med. Sci.* **2022**, *364*, 257–263. [CrossRef]
7. Jia, X.; Xu, W.; Zhang, L.; Li, X.; Wang, R.; Wu, S. Impact of Gut Microbiota and Microbiota-Related Metabolites on Hyperlipidemia. *Front. Cell Infect. Microbiol.* **2021**, *11*, 634780. [CrossRef]
8. Li, T.; Hu, S.; Pang, X.; Wang, J.; Yin, J.; Li, F.; Wang, J.; Yang, X.; Xia, B.; Liu, Y.; et al. The Marine-derived Furanone Reduces Intracellular Lipid Accumulation in Vitro by Targeting LXRα and PPARα. *J. Cell Mol. Med.* **2020**, *24*, 3384–3398. [CrossRef]
9. Pastoriza, S.; Mesías, M.; Cabrera, C.; Rufián-Henares, J.A. Healthy Properties of Green and White Teas: An Update. *Food Funct.* **2017**, *8*, 2650–2662. [CrossRef]
10. van Duynhoven, J.; Vaughan, E.E.; van Dorsten, F.; Gomez-Roldan, V.; de Vos, R.; Vervoort, J.; van der Hooft, J.J.; Roger, L.; Draijer, R.; Jacobs, D.M. Interactions of Black Tea Polyphenols with Human Gut Microbiota: Implications for Gut and Cardiovascular Health1234. *Am. J. Clin. Nutr.* **2013**, *98*, 1631S–1641S. [CrossRef]
11. Liao, W.; Liu, S.; Chen, Y.; Kong, Y.; Wang, D.; Wang, Y.; Ling, T.; Xie, Z.; Khalilova, I.; Huang, J. Effects of Keemun and Dianhong Black Tea in Alleviating Excess Lipid Accumulation in the Liver of Obese Mice: A Comparative Study. *Front. Nutr.* **2022**, *9*, 849582. [CrossRef]
12. Liu, Z.; Chen, Z.; Guo, H.; He, D.; Zhao, H.; Wang, Z.; Zhang, W.; Liao, L.; Zhang, C.; Ni, L. The Modulatory Effect of Infusions of Green Tea, Oolong Tea, and Black Tea on Gut Microbiota in High-Fat-Induced Obese Mice. *Food Funct.* **2016**, *7*, 4869–4879. [CrossRef]
13. Wen, L.; Sun, L.; Chen, R.; Li, Q.; Lai, X.; Cao, J.; Lai, Z.; Zhang, Z.; Li, Q.; Song, G.; et al. Metabolome and Microbiome Analysis to Study the Flavor of Summer Black Tea Improved by Stuck Fermentation. *Foods* **2023**, *12*, 3414. [CrossRef]
14. Moon, K.; Lee, S.; Cha, J. Bacillus Subtilis Fermentation of Malva Verticillata Leaves Enhances Antioxidant Activity and Osteoblast Differentiation. *Foods* **2020**, *9*, 671. [CrossRef]
15. Kim, H.-R.; Kim, S.; Lee, S.-W.; Sin, H.-S.; Kim, S.-Y. Protective Effects of Fermented Paprika (*Capsicum annuum* L.) on Sodium Iodate-Induced Retinal Damage. *Nutrients* **2020**, *13*, 25. [CrossRef]
16. Zhao, Z.-J.; Pan, Y.-Z.; Liu, Q.-J.; Li, X.-H. Exposure Assessment of Lovastatin in Pu-Erh Tea. *Int. J. Food Microbiol.* **2013**, *164*, 26–31. [CrossRef]
17. Zhang, Q.; Wu, Z.-H.; Zhao, S.-S.; Yang, J.; Chen, L.; Wang, X.-Y.; Wang, Z.-Y.; Liu, H.-X. Identification and Spatial Visualization of Dysregulated Bile Acid Metabolism in High-Fat Diet-Fed Mice by Mass Spectral Imaging. *Front. Nutr.* **2022**, *9*, 858603. [CrossRef]

18. Sun, X.; Winglee, K.; Gharaibeh, R.Z.; Gauthier, J.; He, Z.; Tripathi, P.; Avram, D.; Bruner, S.; Fodor, A.; Jobin, C. Microbiota-Derived Metabolic Factors Reduce Campylobacteriosis in Mice. *Gastroenterology* **2018**, *154*, 1751–1763.e2. [CrossRef]
19. Safari, H.; Kaczorowski, N.; Felder, M.L.; Brannon, E.R.; Varghese, M.; Singer, K.; Eniola-Adefeso, O. Biodegradable, Bile Salt Microparticles for Localized Fat Dissolution. *Sci. Adv.* **2020**, *6*, eabd8019. [CrossRef]
20. Ji, J.; Zhang, S.; Yuan, M.; Zhang, M.; Tang, L.; Wang, P.; Liu, Y.; Xu, C.; Luo, P.; Gao, X. Fermented Rosa Roxburghii Tratt Juice Alleviates High-Fat Diet-Induced Hyperlipidemia in Rats by Modulating Gut Microbiota and Metabolites. *Front. Pharmacol.* **2022**, *13*, 883629. [CrossRef]
21. Song, W.; Sun, L.-Y.; Zhu, Z.-J. Effects of Previous Kasai Surgery on Gut Microbiota and Bile Acid in Biliary Atresia with End-Stage Liver Disease. *Front. Med.* **2021**, *8*, 704328. [CrossRef]
22. Wu, C.; Zhao, Y.; Zhang, Y.; Yang, Y.; Su, W.; Yang, Y.; Sun, L.; Zhang, F.; Yu, J.; Wang, Y.; et al. Gut Microbiota Specifically Mediates the Anti-Hypercholesterolemic Effect of Berberine (BBR) and Facilitates to Predict BBR's Cholesterol-Decreasing Efficacy in Patients. *J. Adv. Res.* **2021**, *37*, 197–208. [CrossRef]
23. Liu, C.; Guo, Y.; Sun, L.; Lai, X.; Li, Q.; Zhang, W.; Xiang, L.; Sun, S.; Cao, F. Six Types of Tea Reduce High-Fat-Diet-Induced Fat Accumulation in Mice by Increasing Lipid Metabolism and Suppressing Inflammation. *Food Funct.* **2019**, *10*, 2061–2074. [CrossRef]
24. Wu, Z.; Sun, L.; Chen, R.; Wen, S.; Li, Q.; Lai, X.; Zhang, Z.; Cao, F.; Sun, S. Chinese Tea Alleviates CCl4-Induced Liver Injury through the NF-κBorNrf2Signaling Pathway in C57BL-6J Mice. *Nutrients* **2022**, *14*, 972. [CrossRef]
25. Liu, H.; Chen, R.; Wen, S.; Li, Q.; Lai, X.; Zhang, Z.; Sun, L.; Sun, S.; Cao, F. Tea (*Camellia sinensis*) Ameliorates DSS-Induced Colitis and Liver Injury by Inhibiting TLR4/NF-κB/NLRP3 Inflammasome in Mice. *Biomed. Pharmacother.* **2023**, *158*, 114136. [CrossRef]
26. Wen, S.; Sun, L.; An, R.; Zhang, W.; Xiang, L.; Li, Q.; Lai, X.; Huo, M.; Li, D.; Sun, S. A Combination of Citrus Reticulata Peel and Black Tea Inhibits Migration and Invasion of Liver Cancer via PI3K/AKT and MMPs Signaling Pathway. *Mol. Biol. Rep.* **2020**, *47*, 507–519. [CrossRef]
27. Wu, D.; Chen, R.; Zhang, W.; Lai, X.; Sun, L.; Li, Q.; Zhang, Z.; Cao, J.; Wen, S.; Lai, Z.; et al. Tea and Its Components Reduce the Production of Uric Acid by Inhibiting Xanthine Oxidase. *Food Nutr. Res.* **2022**, *66*. [CrossRef]
28. Xu, H.; Lyu, X.; Guo, X.; Yang, H.; Duan, L.; Zhu, H.; Pan, H.; Gong, F.; Wang, L. Distinct AMPK-Mediated FAS/HSL Pathway Is Implicated in the Alleviating Effect of Nuciferine on Obesity and Hepatic Steatosis in HFD-Fed Mice. *Nutrients* **2022**, *14*, 1898. [CrossRef]
29. Luo, Y.; Zhao, Z.; Chen, H.; Pan, X.; Li, R.; Wu, D.; Hu, X.; Zhang, L.; Wu, H.; Li, X. Dynamic Analysis of Physicochemical Properties and Polysaccharide Composition during the Pile-Fermentation of Post-Fermented Tea. *Foods* **2022**, *11*, 3376. [CrossRef]
30. Shi, X.; Zhou, X.; Chu, X.; Wang, J.; Xie, B.; Ge, J.; Guo, Y.; Li, X.; Yang, G. Allicin Improves Metabolism in High-Fat Diet-Induced Obese Mice by Modulating the Gut Microbiota. *Nutrients* **2019**, *11*, 2909. [CrossRef]
31. Wang, Y.; Gunewardena, S.; Li, F.; Matye, D.J.; Chen, C.; Chao, X.; Jung, T.; Zhang, Y.; Czerwiński, M.; Ni, H.-M.; et al. An FGF15/19-TFEB Regulatory Loop Controls Hepatic Cholesterol and Bile Acid Homeostasis. *Nat. Commun.* **2020**, *11*, 3612. [CrossRef]
32. Hou, G.; Peng, W.; Wei, L.; Li, R.; Yuan, Y.; Huang, X.; Yin, Y. Lactobacillus Delbrueckii Interfere with Bile Acid Enterohepatic Circulation to Regulate Cholesterol Metabolism of Growing-Finishing Pigs via Its Bile Salt Hydrolase Activity. *Front. Nutr.* **2020**, *7*, 617676. [CrossRef]
33. Jiang, T.; Xu, C.; Liu, H.; Liu, M.; Wang, M.; Jiang, J.; Zhang, G.; Yang, C.; Huang, J.; Lou, Z. Linderae Radix Ethanol Extract Alleviates Diet-Induced Hyperlipidemia by Regulating Bile Acid Metabolism Through Gut Microbiota. *Front. Pharmacol.* **2021**, *12*, 627920. [CrossRef]
34. Tao, W.; Liu, W.; Wang, M.; Zhou, W.; Xing, J.; Xu, J.; Pi, X.; Wang, X.; Lu, S.; Yang, Y. Dendrobium Officinale Polysaccharides Better Regulate the Microbiota of Women Than Men. *Foods* **2022**, *11*, 1641. [CrossRef]
35. Feng, Z.; Li, Y.; Li, M.; Wang, Y.; Zhang, L.; Wan, X.; Yang, X. Tea Aroma Formation from Six Model Manufacturing Processes. *Food Chem.* **2019**, *285*, 347–354. [CrossRef]
36. Huang, F.; Zheng, X.; Ma, X.; Jiang, R.; Zhou, W.; Zhou, S.; Zhang, Y.; Lei, S.; Wang, S.; Kuang, J.; et al. Theabrownin from Pu-Erh Tea Attenuates Hypercholesterolemia via Modulation of Gut Microbiota and Bile Acid Metabolism. *Nat. Commun.* **2019**, *10*, 4971. [CrossRef]
37. Zhu, M.; Ouyang, J.; Zhou, F.; Zhao, C.; Zhu, W.; Liu, C.; Huang, P.; Li, J.; Tang, J.; Zhang, Z.; et al. Polysaccharides from Fu Brick Tea Ameliorate Obesity by Modulating Gut Microbiota and Gut Microbiota-Related Short Chain Fatty Acid and Amino Acid Metabolism. *J. Nutr. Biochem.* **2023**, *118*, 109356. [CrossRef]
38. Zhu, M.-Z.; Li, N.; Zhou, F.; Ouyang, J.; Lu, D.-M.; Xu, W.; Li, J.; Lin, H.-Y.; Zhang, Z.; Xiao, J.-B.; et al. Microbial Bioconversion of the Chemical Components in Dark Tea. *Food Chem.* **2020**, *312*, 126043. [CrossRef]
39. Huang, D.-W.; Chang, W.-C.; Yang, H.-J.; Wu, J.S.-B.; Shen, S.-C. Gallic Acid Alleviates Hypertriglyceridemia and Fat Accumulation via Modulating Glycolysis and Lipolysis Pathways in Perirenal Adipose Tissues of Rats Fed a High-Fructose Diet. *Int. J. Mol. Sci.* **2018**, *19*, 254. [CrossRef]
40. Murase, T.; Misawa, K.; Haramizu, S.; Hase, T. Catechin-Induced Activation of the LKB1/AMP-Activated Protein Kinase Pathway. *Biochem. Pharmacol.* **2009**, *78*, 78–84. [CrossRef]

41. Kim, S.-Y.; Lee, M.-S.; Chang, E.; Jung, S.; Ko, H.; Lee, E.; Lee, S.; Kim, C.-T.; Kim, I.-H.; Kim, Y. Tartary Buckwheat Extract Attenuated the Obesity-Induced Inflammation and Increased Muscle PGC-1a/SIRT1 Expression in High Fat Diet-Induced Obese Rats. *Nutrients* **2019**, *11*, 654. [CrossRef]
42. Cheng, H.; Xu, N.; Zhao, W.; Su, J.; Liang, M.; Xie, Z.; Wu, X.; Li, Q. (-)-Epicatechin Regulates Blood Lipids and Attenuates Hepatic Steatosis in Rats Fed High-Fat Diet. *Mol. Nutr. Food Res.* **2017**, *61*, 1700303. [CrossRef]
43. Chiang, J.Y. Recent Advances in Understanding Bile Acid Homeostasis. *F1000Res* **2017**, *6*, 2029. [CrossRef]
44. Yu, L.; Liu, Y.; Wang, S.; Zhang, Q.; Zhao, J.; Zhang, H.; Narbad, A.; Tian, F.; Zhai, Q.; Chen, W. Cholestasis: Exploring the Triangular Relationship of Gut Microbiota-Bile Acid-Cholestasis and the Potential Probiotic Strategies. *Gut Microbes* **2023**, *15*, 2181930. [CrossRef]
45. Chambers, K.F.; Day, P.E.; Aboufarrag, H.T.; Kroon, P.A. Polyphenol Effects on Cholesterol Metabolism via Bile Acid Biosynthesis, CYP7A1: A Review. *Nutrients* **2019**, *11*, 2588. [CrossRef]
46. Larsen, I.S.; Jensen, B.A.H.; Bonazzi, E.; Choi, B.S.Y.; Kristensen, N.N.; Schmidt, E.G.W.; Süenderhauf, A.; Morin, L.; Olsen, P.B.; Hansen, L.B.S.; et al. Fungal Lysozyme Leverages the Gut Microbiota to Curb DSS-Induced Colitis. *Gut Microbes* **2021**, *13*, 1988836. [CrossRef]
47. Zhao, X.; Jiang, L.; Fang, X.; Guo, Z.; Wang, X.; Shi, B.; Meng, Q. Host-Microbiota Interaction-Mediated Resistance to Inflammatory Bowel Disease in Pigs. *Microbiome* **2022**, *10*, 115. [CrossRef]
48. Ijssennagger, N.; van Rooijen, K.S.; Magnúsdóttir, S.; Ramos Pittol, J.M.; Willemsen, E.C.L.; de Zoete, M.R.; Baars, M.J.D.; Stege, P.B.; Colliva, C.; Pellicciari, R.; et al. Ablation of Liver Fxr Results in an Increased Colonic Mucus Barrier in Mice. *JHEP Rep.* **2021**, *3*, 100344. [CrossRef]
49. Dix, C.; Wright, O. Bioavailability of a Novel Form of Microencapsulated Bovine Lactoferrin and Its Effect on Inflammatory Markers and the Gut Microbiome: A Pilot Study. *Nutrients* **2018**, *10*, 1115. [CrossRef]
50. Wang, S.; Liu, S.; Wang, C.; Ye, B.; Lv, L.; Ye, Q.; Xie, S.; Hu, G.; Zou, J. Dietary Antimicrobial Peptides Improve Intestinal Function, Microbial Composition and Oxidative Stress Induced by Aeromonas Hydrophila in Pengze Crucian Carp (Carassius Auratus Var. Pengze). *Antioxidants* **2022**, *11*, 1756. [CrossRef]
51. Du, L.; Zhang, Y.; Ji, S.; Wang, L.; Zhao, X.; Yan, S.; Xiao, X.; Li, S. Mechanisms of Zhenwu Decoction for the Treatment of Renal Fibrosis at Various Stages: What Is the Role of Corynebacterium? *Front. Microbiol.* **2022**, *13*, 913465. [CrossRef]
52. Yang, G.; Hong, E.; Oh, S.; Kim, E. Non-Viable Lactobacillus Johnsonii JNU3402 Protects against Diet-Induced Obesity. *Foods* **2020**, *9*, 1494. [CrossRef]
53. Rodríguez-Lara, A.; Plaza-Díaz, J.; López-Uriarte, P.; Vázquez-Aguilar, A.; Reyes-Castillo, Z.; Álvarez-Mercado, A.I. Fiber Consumption Mediates Differences in Several Gut Microbes in a Subpopulation of Young Mexican Adults. *Nutrients* **2022**, *14*, 1214. [CrossRef]
54. Zhu, H.; Zhao, F.; Zhang, W.; Xia, W.; Chen, Y.; Liu, Y.; Fan, Z.; Zhang, Y.; Yang, Y. Cholesterol-Lowering Effect of Bile Salt Hydrolase from a Lactobacillus Johnsonii Strain Mediated by FXR Pathway Regulation. *Food Funct.* **2022**, *13*, 725–736. [CrossRef]
55. Zheng, Y.; Wang, Y.; Luo, D.; Lin, L.; Lu, X.; Gao, J.; Xiao, C.; Zhao, M. Effect of Bergamot and Laoxianghuang Polysaccharides on Gut Microbiota Derived from Patients with Hyperlipidemia: An Integrative Analysis of Microbiome and Metabolome during In Vitro Fermentation. *Foods* **2022**, *11*, 2039. [CrossRef]
56. Ye, X.; Liu, Y.; Hu, J.; Gao, Y.; Ma, Y.; Wen, D. Chlorogenic Acid-Induced Gut Microbiota Improves Metabolic Endotoxemia. *Front. Endocrinol.* **2021**, *12*, 762691. [CrossRef]
57. Ma, Q.; Zhai, R.; Xie, X.; Chen, T.; Zhang, Z.; Liu, H.; Nie, C.; Yuan, X.; Tu, A.; Tian, B.; et al. Hypoglycemic Effects of Lycium Barbarum Polysaccharide in Type 2 Diabetes Mellitus Mice via Modulating Gut Microbiota. *Front. Nutr.* **2022**, *9*, 916271. [CrossRef]
58. Huang, H.-C.; Hsu, S.-J.; Chang, C.-C.; Chuang, C.-L.; Hou, M.-C.; Lee, F.-Y. Effects of PCSK-9 Inhibition by Alirocumab Treatments on Biliary Cirrhotic Rats. *Int. J. Mol. Sci.* **2022**, *23*, 7378. [CrossRef]
59. Li, R.; Yi, X.; Yang, J.; Zhu, Z.; Wang, Y.; Liu, X.; Huang, X.; Wan, Y.; Fu, X.; Shu, W.; et al. Gut Microbiome Signatures in the Progression of Hepatitis B Virus-Induced Liver Disease. *Front. Microbiol.* **2022**, *13*, 916061. [CrossRef]
60. Zhao, H.; Gao, X.; Liu, Z.; Zhang, L.; Fang, X.; Sun, J.; Zhang, Z.; Sun, Y. Sodium Alginate Prevents Non-Alcoholic Fatty Liver Disease by Modulating the Gut–Liver Axis in High-Fat Diet-Fed Rats. *Nutrients* **2022**, *14*, 4846. [CrossRef]
61. Dang, J.T.; Mocanu, V.; Park, H.; Laffin, M.; Hotte, N.; Karmali, S.; Birch, D.W.; Madsen, K.L. Roux-En-Y Gastric Bypass and Sleeve Gastrectomy Induce Substantial and Persistent Changes in Microbial Communities and Metabolic Pathways. *Gut Microbes* **2022**, *14*, 2050636. [CrossRef]
62. Ruiz-Malagón, A.J.; Rodríguez-Sojo, M.J.; Hidalgo-García, L.; Molina-Tijeras, J.A.; García, F.; Pischel, I.; Romero, M.; Duarte, J.; Diez-Echave, P.; Rodríguez-Cabezas, M.E.; et al. The Antioxidant Activity of Thymus Serpyllum Extract Protects against the Inflammatory State and Modulates Gut Dysbiosis in Diet-Induced Obesity in Mice. *Antioxidants* **2022**, *11*, 1073. [CrossRef]
63. Cani, P.D.; Depommier, C.; Derrien, M.; Everard, A.; de Vos, W.M. Akkermansia Muciniphila: Paradigm for next-Generation Beneficial Microorganisms. *Nat. Rev. Gastroenterol. Hepatol.* **2022**, *19*, 625–637. [CrossRef]
64. Tung, Y.-T.; Zeng, J.-L.; Ho, S.-T.; Xu, J.-W.; Lin, I.-H.; Wu, J.-H. Djulis Hull Improves Insulin Resistance and Modulates the Gut Microbiota in High-Fat Diet (HFD)-Induced Hyperglycaemia. *Antioxidants* **2021**, *11*, 45. [CrossRef]
65. Liu, J.; Song, Y.; Zhao, Q.; Wang, Y.; Li, C.; Zou, L.; Hu, Y. Effects of Tartary Buckwheat Protein on Gut Microbiome and Plasma Metabolite in Rats with High-Fat Diet. *Foods* **2021**, *10*, 2457. [CrossRef]

66. Xiong, Y.; Wu, L.; Shao, L.; Wang, Y.; Huang, Z.; Huang, X.; Li, C.; Wu, A.; Liu, Z.; Fan, X.; et al. Dynamic Alterations of the Gut Microbial Pyrimidine and Purine Metabolism in the Development of Liver Cirrhosis. *Front. Mol. Biosci.* **2022**, *8*, 811399. [CrossRef]
67. Desai, M.S.; Seekatz, A.M.; Koropatkin, N.M.; Kamada, N.; Hickey, C.A.; Wolter, M.; Pudlo, N.A.; Kitamoto, S.; Terrapon, N.; Muller, A.; et al. A Dietary Fiber-Deprived Gut Microbiota Degrades the Colonic Mucus Barrier and Enhances Pathogen Susceptibility. *Cell* **2016**, *167*, 1339–1353.e21. [CrossRef]
68. Chleilat, F.; Schick, A.; Deleemans, J.M.; Ma, K.; Alukic, E.; Wong, J.; Noye Tuplin, E.W.; Nettleton, J.E.; Reimer, R.A. Paternal High Protein Diet Modulates Body Composition, Insulin Sensitivity, Epigenetics, and Gut Microbiota Intergenerationally in Rats. *FASEB J.* **2021**, *35*, e21847. [CrossRef]

nutrients

MDPI

Article

Food-Related Behavioral Patterns in Patients with Inflammatory Bowel Diseases: The Role of Food Involvement and Health Engagement

Lorenzo Palamenghi [1], Dilara Usta [1,*], Salvo Leone [2] and Guendalina Graffigna [1,3]

1 EngageMinds HUB—Consumer, Food & Health Engagement Research Center, Faculty of Agriculture, Food and Environmental Sciences, Catholic University of Sacred Heart, 26100 Cremona, Italy; lorenzo.palamenghi@gmail.com (L.P.); guendalina.graffigna@unicatt.it (G.G.)
2 National Association for Chronic Inflammatory Bowel Diseases (AMICI ETS), 20125 Milan, Italy; salvo.leone@amiciitalia.net
3 Faculty of Psychology, Catholic University of Sacred Heart, 20123 Milan, Italy
* Correspondence: dilara.cengiz@unicatt.it

Abstract: Nutrition has been acknowledged as crucial in IBD and is relevant to patients' motives behind food choices, which are affected by health engagement (HE) and food involvement (FI). This study aimed to profile IBD patients according to their levels of health engagement and food involvement to identify patterns of different motives behind food choices, particularly regarding the use of food to regulate mood. A cross-sectional study was conducted with 890 Italian IBD patients who completed an online survey in April 2021. We measured health engagement, food involvement, motives behind food choices, emotional states, and food-related quality of life (Fr-QoL). K-means cluster analysis was performed to identify participants with similar levels of health engagement and food involvement. Four clusters were identified: "Health-conscious (high HE, low FI)", "Balanced (high HE, high FI)", "Hedonist (high FI, low HE)", and "Careless (low FI, low HE)". Clusters with high FI are inclined toward seeking pleasurable food, but when supported with high health engagement, individuals were less prone to use food to manage mood. Groups with higher health engagement demonstrated lower hospitalization rates and relapses and better Fr-QoL. Profiling IBD patients regarding FI and HE could aid clinicians in identifying individuals at greater risk of maladaptive food-related behaviors.

Keywords: inflammatory bowel disease; emotional regulation; food choices; food involvement; patient engagement; cluster analysis

Citation: Palamenghi, L.; Usta, D.; Leone, S.; Graffigna, G. Food-Related Behavioral Patterns in Patients with Inflammatory Bowel Diseases: The Role of Food Involvement and Health Engagement. *Nutrients* **2024**, *16*, 1185. https://doi.org/10.3390/nu16081185

Academic Editors: Domenico Tafuri and Francesca Latino

Received: 10 March 2024
Revised: 12 April 2024
Accepted: 15 April 2024
Published: 16 April 2024

1. Introduction

Inflammatory bowel disease (IBD), encompassing Crohn's disease (CD) and ulcerative colitis (UC), is a lifelong incurable condition, affecting approximately 0.2% of the European population and around five million worldwide, and represents a substantial and escalating global health challenge [1,2]. IBD manifests with several gastrointestinal symptoms characterized by alternating periods of exacerbation and remission [3], typically emerging as pain, diarrhea, abdominal discomfort, fatigue, reduced appetite, and weight loss, impacting daily living activities and resulting in a significant decrease in health-related quality of life [4]. These symptoms are initiated and triggered by environmental factors, such as diet, in individuals genetically susceptible to the condition, characterized by gut dysbiosis and abnormal immune response [5].

Extensive research demonstrated the impact of diet on IBD onset and prognosis [6,7], and the interplay between nutrition and gut microbiota was acknowledged as a significant factor in the pathogenesis of IBD [8]. To illustrate, the westernized lifestyle has been recognized as a critical contributor to the rising incidence of IBD and exacerbating intestinal symptoms [9,10], supported by the evidence of the effect of saturated fat, cholesterol, and food additive intake on gut inflammation [11]. Also, high protein intake, including red meat and processed meat, is significantly related to the increased risk of IBD development [12] and triggered IBD symptoms [9]. On the contrary, the Mediterranean diet is recognized as a healthy dietary regimen with anti-inflammatory effects and has been linked to a remarkably reduced risk of later onset CD [13] and improvement in inflammation and disease activity [14]. This captivating evidence underscores the significance of diet and nutrition as a regulatory factor in gut inflammation [7].

The scientific community has recognized the significance of dietary behaviors and nutrition in managing IBD [15]. Numerous studies have attempted to establish nutritional guidelines to clarify the impact of diet on the onset and progression of IBD [10,16]. However, evidence indicates a suboptimal adherence of IBD patients to recommendations [17]. So far, various barriers to adapting to a healthy diet have been introduced, such as a lack of knowledge in identifying foods, difficulties correlating disease symptoms with diet, obstacles to accessing IBD-friendly food, challenges with the complexity of diet, and changing habits and lifestyles [18]. Further, adherence to nutritional guidelines and healthy food choices highlights the pivotal role of psychological dynamics in the complex evaluations underpinning dietary decisions [19,20]. Evidence suggests that nutritional restrictions substantially impact psychological well-being by causing distress and negative mood in patients [21]. Frustration, distress, and feelings of deviating from normalcy may contribute to dysregulated food behaviors aimed at alleviating negative emotional states [22]. These coping mechanisms may remarkably influence dietary choices, with certain foods sought for comfort during distress or negative moods, establishing a link between emotional well-being and consumption patterns [23]. The regulatory role of food on mood is also supported by evidence pointing to psychological, hedonistic, and neurochemical pathways that facilitate emotional eating [24,25].

Building on this understanding, food involvement (FI) emerges as a significant psychological construct that interlinks with food utilization as a regulatory mechanism for mood [26]. Food involvement, described as the level of cognitive, behavioral, and emotional commitment a person gives to every aspect of food interaction, has been linked to various dietary patterns [27–29], suggesting that individuals who deeply engage with food, making it a central aspect of their lives, may be more susceptible to the psychological and behavioral tendencies discussed previously. Specifically, the renunciation of food could burden these individuals by exacerbating distress and complicating adherence to dietary guidelines [30]. This is particularly relevant considering their propensity for using food to regulate mood and their preference for hedonic, pleasure-inducing foods [31]. Thus, understanding the nuances of food involvement could illuminate pathways to mitigate the emotional distress associated with dietary restrictions, especially in populations with IBD.

Moreover, the complex relationship between psychological factors underscores the critical need for enhanced patient involvement in managing IBD. In this context, the health engagement (HE) concept appears to be a protective factor in adhering to nutritional guidelines and healthy food choices, elaborating on individuals' emotional and value-based readiness to actively manage their health and lifestyle [32]. Health engagement entails a spectrum of behavioral skills empowering patients to self-manage their health condition, complemented by the requisite knowledge and self-efficacy to implement proactive behaviors [33,34]. Remarkably, health engagement within chronic care contexts has been linked to improved quality of life [33], enhanced health literacy [35,36], and greater adherence to behavioral prescriptions [37].

Unraveling the interplay between food involvement and health engagement could address clinicians' daily practical concerns, including symptom management, monitoring disease progression, planning personalized interventions, providing nutritional guidance, offering psychosocial support, and empowering patients to improve the well-being of individuals managing IBD. Given the limited focus on psychological factors influencing dietary choices among the IBD population and the significant impact of diet on their well-being and symptom management [21,38], there is a need to understand the various factors driving their food choices, including reasons for consuming foods perceived as triggers. Through this study, we aim to uncover the relationship between food involvement and health engagement and assess their utility in delineating distinct groups of IBD patients with varying motives for food choices, particularly regarding the use of food to regulate mood. Furthermore, our study explores whether these identified patient groups exhibit differing food-related quality-of-life levels.

2. Materials and Methods

2.1. Study Design and Participants

A cross-sectional study was carried out between the 3 and the 19 of April 2021 as an online survey on the SurveyMonkey platform. Individuals older than 18 years old with a formal diagnosis of IBD were contacted by sending invitation emails through an Italian IBD patient organization, "AMICI ETS", with a purposive sampling method and completed a survey. Individuals were ineligible to participate if they were unable to give informed consent and had impaired capacity to complete the survey. No incentives were given to the participants. Ethical approval was granted by the Ethics Committee of the Department of Psychology of the Università Cattolica del Sacro Cuore, Italy (protocol code: 36-21). All participants provided written informed consent.

2.2. Measures

The online survey comprised validated self-report questionnaires and ad hoc items. The instruments used in this study can be classified into two groups: (i) for clustering variables, including health engagement and food involvement, and (ii) for dependent variables, involving sociodemographic and disease-related characteristics, food choice motives, emotional states, and food-related quality of life.

2.2.1. Sociodemographic and Disease-Related Characteristics

Sociodemographic and disease-related data included gender, age, diagnosis (Colitis, either ulcerative or indeterminate, or Crohn's disease), duration of diagnosis, level of education (middle school or below, high school, university degree), existing comorbidities, whether had relapses or hospitalizations in the last year, whether had undergone surgery for IBD, and whether currently taking medications for IBD.

2.2.2. Food Involvement

Participants' food involvement was assessed using the Food Involvement Scale (FIS) [27], which addresses the perceived level of importance that a person attributes to food. The FIS consists of 12 items with a 7-point Likert score (1: strongly disagree, 7: strongly agree) regarding food acquisition, preparation, cooking, eating, and disposal and incorporates two dimensions: "set and disposal" (3 items) and "preparation and items" (9 items). An example of an item is "I don't think much about food each day". Total scores ranged between 12 and 84, with higher scores indicating higher food involvement.

2.2.3. Health Engagement

Health engagement was assessed with the Patient Health Engagement Scale (PHE-s®), which was validated among various chronic conditions [32]. The scale is based on the Patient Health Engagement Model, a developmental psychological theory describing the patients' experience of becoming active players in their healthcare pathway. The PHE-s® can grasp the complex psychological experience of health engagement and has an ordinal structure consistent with the PHE model's conceptualization, which, through an algorithm that provides the final score, envisages four different positions along the engagement continuum: Blackout, Alert, Adhesion, and Eudaimonic Project. An example of an item is "When I think about my illness, I feel overwhelmed by emotions". This ordinal scale is measured on a 7-point scoring system to facilitate patient responses and avoid social desirability bias [32].

2.2.4. Food Choice Motives

Food choice motives were evaluated with the revised single-item Food Choice Questionnaire (FCQ) [39]. The FCQ explores people's motivations behind food choices by asking participants to rate each "motive" on a 7-point Likert scale (1: not at all important, 7: very important); it starts with the statement "It is important to me that the food I eat on a typical day is..." and with the following answer options: health, mood, convenience, taste, natural, price, weight control, familiar, environment, animal welfare, and social justice. One additional single-item construct was added to the scale, questioning how important it is for patients that food helps them manage their symptoms.

2.2.5. Emotional States

On a scale from 0 to 100, participants were asked to rate how much they felt three positive emotional states—happiness, hopefulness, and satisfaction—and seven negative emotional states—sadness, fear, anger, disgust, anxiety, distress, and boredom—over the previous 24 h.

2.2.6. Food-Related Quality of Life

Food-related quality of life was measured with the Food-Related Quality of Life Questionnaire (FR-QoL-29), a one-factor self-report measure developed based on qualitative interviews with the IBD population. The FR-QoL, which has good validity and reliability across various characteristics, was used to evaluate participants' food-related quality of life, which refers to their relationship with food and how this impacts their daily lives and psychological well-being [40]. The FR-QoL-29 involves 29 items on a 5-point Likert scale (1: definitely agree, 5: definitely disagree), including four items reversed for scoring. An example of an item is "In the past two weeks, I have regretted eating and drinking things that have made my IBD symptoms worse". Total scores ranged between 29 and 145, with higher scores denoting greater food-related quality of life.

2.3. *Statistical Analysis*

Statistical analyses were conducted using IBM® SPSS 27 (IBM Corp., Armonk, NY, USA). Descriptive statistics of the sample were computed to assess the composition of the sample. Scales were scored according to the literature. Regarding the PHE-s®, a distinct methodology was employed due to limitations posed by its conventional scoring procedure [32], which yields an ordinal score spanning from 1 to 4 based on its theoretical framework [34], rendering it incompatible with the intended cluster analysis (please refer to the subsequent section). Consequently, the 7-point ordinal scales were re-encoded into 4 points to align with the standard scoring protocol, followed by applying a partial credit Rasch model (PCM). The PCM served the dual purpose of assessing the unidimensionality and fit of each ordinal item within the targeted construct while determining the score for each participant [41]. Specifically, the fit mean square (MNSQ) statistics (infit and outfit) were computed to evaluate adherence to the anticipated model; ideally, these statistics should fall within the range of 0.6 to 1.4 to signify a good fit with the Rasch model [42]. Furthermore, analyses of difficulty and step parameters were conducted to ensure adequate discrimination among the various response categories and to maintain the monotonic order.

To identify groups of participants with similar levels of patient health engagement and food involvement, a k-means cluster analysis was conducted utilizing the scores from the PHE-s® and FIS as segmentation variables. Before the k-means clustering, the scores were standardized into z-scores. Outliers, defined as participants with either PHE-s® or FIS z-scores exceeding $|3|$, were systematically excluded from the dataset. Determining the optimal number of clusters was facilitated through a progressive series of analyses, commencing with two clusters and iteratively increasing. Criteria employed to ascertain the most suitable cluster count encompassed the following:

(i) Interpretability of the final clusters' averages and ANOVA's p-values;
(ii) Number of participants in each cluster (closer to homogeneity is better);
(iii) Pseudo-F values, calculated according to the procedure described by Calinski and Harabasz [43]: higher pseudo-Fs are an indication of a better solution;
(iv) Finally, to address the stability of the identified best solution, the Rand index [44] was calculated; the Rand index is considered acceptable above the 0.70 threshold.

A combination of Pearson's χ^2 tests and one-way Welch's ANOVAs was employed to evaluate disparities across the identified clusters concerning the aforementioned dependent variables (as detailed in Section 2.2). Significant χ^2 tests were followed by an inspection of adjusted standardized residuals, where significance was attributed to values exceeding $|2|$ [45]. Additionally, Cramér's V was computed as an indicator of effect size. Games–Howell post hoc tests followed significant ANOVAs. The effect size was quantified using η^2. For variables violating the assumption of normality in the ANOVA (skewness and/or kurtosis > $|1|$), the robustness of findings was assessed by re-running the model post a logarithmic transformation, followed by reassessment of normality parameters. However, for interpretability, results are primarily reported based on non-transformed variables unless otherwise indicated, owing to discrepancies observed between outcomes derived from transformed and non-transformed variables. A p-value of <0.05 was considered statistically significant.

3. Results

3.1. Participant Characteristics

A total of n = 1113 participants responded to the survey. Among them, n = 211 were excluded due to incomplete responses and the premature termination of the online survey. Additionally, n = 12 participants were deemed outliers; consequently, the final analysis included data from n = 890 patients. Table 1 provides an overview of participants' characteristics. The average age of the participants was 47 ± 14 years (min 18, max 85), and on average, patients reported receiving their diagnosis 17 ± 11.5 years ago (min 0.5, max 51).

Table 1. Participants' sociodemographic and disease-related characteristics (n = 890).

Participant Characteristics	n	%
Gender		
Male	355	39.9
Female	535	60.1
Diagnosis		
Colitis (ulcerative or indeterminate)	445	50.0
Crohn's disease	445	50.0
Level of education		
Middle school or lower	87	9.8
High school	422	47.4
University or higher	381	42.8
Existing comorbidities		
Yes	197	22.1
No	693	77.9
IBD relapses/hospitalizations in the last year		
Yes	352	39.6
No	538	60.4
Ever undergone surgery for IBD		
Yes	336	37.8
No	554	62.2
Currently taking drugs/medications for IBD		
Yes	784	88.1
No	106	11.9

3.2. Scale Scoring and Descriptive Statistics

The analysis of the difficulty and the step parameters from the partial credit Rasch model applied to the PHE-s® items indicated a satisfactory ranking of the various response categories, with each scale item adhering to a monotonic order. Furthermore, the infit and outfit statistics fell within acceptable ranges, with values ranging from 0.75 to 0.82 and 0.70 to 0.82, respectively.

Table 2 demonstrates the descriptive statistics of the FIS and PHE-s® before being transformed into z-scores. Both variables showed an acceptably normal distribution, and ordinal scores were calculated to assess the overall distribution of PHE-s® phases by the underlying theoretical framework (Patient Health Engagement Model) [34]. The findings indicate that most participants were in either the Adhesion phase (n = 429), indicative of a moderate degree of health engagement, or the Arousal phase (n = 298). A higher level of health engagement, namely, the "Eudaimonic project", was achieved by only 131 participants, while merely 32 individuals were categorized at the lowest conceivable level, denoted as "Blackout".

Table 2. Descriptive statistics for the scales and single items.

Variables	Min	Max	Mean	SD	Skewness	Kurtosis
Food-related quality of life (Fr-QoL-29)	1.00	4.97	3.10	0.85	−0.344	−0.443
Food choice motives (FCQ) (It is important to me that the food I eat on a typical day...)						
is healthy	1	7	5.96	1.02	−1.275	2.645
is a way of monitoring my mood	1	7	5.03	1.50	−0.810	0.329
is convenient	1	7	4.99	1.42	−0.835	0.568
provides me with pleasurable sensations	1	7	5.57	1.11	−1.105	1.987
is natural	1	7	5.58	1.18	−0.878	1.023
is affordable	1	7	4.93	1.39	−0.618	0.322
helps me control my weight	1	7	4.75	1.66	−0.696	−0.154
is familiar	1	7	4.47	1.41	−0.502	0.134
is environmentally friendly	1	7	5.28	1.29	−0.755	0.626
is animal friendly	1	7	5.18	1.42	−0.628	−0.027
is fairly traded	1	7	4.47	1.14	−0.410	0.038
helps me control my symptoms	1	7	6.18	0.96	−1.224	1.753
Emotional states						
Happiness	0	100	58.66	26.37	−0.571	−0.331
Sadness	0	100	36.46	27.68	0.514	−0.694
Hopefulness	0	100	54.93	29.13	−0.235	−0.865
Fear	0	100	28.72	27.95	0.868	−0.239
Satisfaction	0	100	52.49	27.60	−0.180	−0.860
Anger	0	100	33.49	30.43	0.670	−0.776
Disgust	0	100	19.44	25.92	1.56	1.550
Anxiety	0	100	37.37	30.19	0.423	−0.977
Distress	0	100	46.65	31.28	0.090	−1.170
Boredom	0	100	29.65	29.30	0.778	−0.524
Food involvement (FIS)	2.75	6.33	4.57	0.64	−0.76	−0.308
Health engagement (PHE-s®)	−5.52	8.28	2.40	2.86	−0.148	−0.337

Abbreviations: SD, standard deviation; PHE-s®, Patient Health Engagement Scale; FIS, Food Involvement Scale; Fr-QoL, Food-Related Quality of Life Questionnaire; FCQ, Food Choice Questionnaire.

Regarding the FCQ, the percentages of patients that answered "a lot" (5, 6, or 7 on the Likert scale) were also calculated for each item of the FCQ. Overall, "helps me control my symptoms" was responded to by most patients (78%), as well as "it's healthy" (73%). Being pleasant (59%) and natural (58%) are also reported by most patients as essential drivers.

Remarkably, emotional states and food-related quality of life were correlated with some motives behind food choices and with PHE-s®. Specifically, findings from Spearman's correlation analyses revealed that overall positive emotions were strongly and positively correlated with the PHE-s® score ($\rho = 0.597$; $p < 0.001$) and demonstrated a strong, negative correlation with the Fr-QoL ($\rho = -0.400$; $p < 0.001$). Additionally, positive emotions were found to negatively correlate with the use of food as a means to regulate mood ($\rho = -0.131$; $p < 0.001$) and, intriguingly, with the management of symptoms ($\rho = -0.119$; $p < 0.001$). Finally, positive emotions also showed small correlations with using convenient, easy-to-prepare food ($\rho = -0.096$; $p = 0.004$) and attention to price ($\rho = -0.072$; $p = 0.033$).

On the other hand, negative emotions had a strong, negative correlation with PHE-s® scores ($\rho = -0.575$; $p < 0.001$) and a strong, positive correlation with Fr-QoL-29 scores ($\rho = -0.464$; $p < 0.001$). Coherently with the results reported for positive emotions, the negative emotion score was positively correlated with the use of food as a means to regulate mood ($\rho = 0.173$; $p < 0.001$) and with the preference for convenient food ($\rho = 0.118$; $p < 0.001$). Interestingly, consistent with the described construct, the FIS did not correlate with emotional states but positively correlated with the Fr-QoL ($\rho = 0.143$; $p < 0.001$). Finally, a strong and anticipated correlation emerged between the PHE-s® and the Fr-QoL-29 scores ($\rho = -0.574$; $p < 0.001$).

3.3. Cluster Analysis

Four distinct solutions (comprising two, three, four, and five clusters) were explored to categorize patients based on their PHE-s® and FIS scores. Among these, the four-cluster solution emerged as the most optimal, characterized by clusters with interpretable means and a reasonably balanced distribution of participants. Additionally, Pseudo-F values were computed for each solution, with the four-cluster solution yielding the highest value (473.055, 567.422, 610.208, and 577.528 for the two, three, four, and five-cluster solutions, respectively). Furthermore, the Rand index for the four-cluster solution was calculated to be 0.94, surpassing the acceptability threshold. Table 3 presents the average scores of the final clusters and the outcomes of the ANOVA test, depicting the variations across clusters and the number of subjects in each cluster.

Table 3. Distribution of participants in clusters and results from ANOVA.

Variables	Clusters: 1: Health-Conscious (n = 261)	2: Balanced (n = 241)	3: Hedonist (n = 187)	4: Careless (n = 201)	Welch's F	p	η^2
PHE-s® cluster mean	0.75 ^{a}	0.71 ^{a}	−0.90 ^{b}	−0.94 ^{b}	626.597	<0.001	0.681
FIS cluster mean	−0.80 ^{c}	0.77 ^{b}	0.92 ^{a}	−0.73 ^{c}	594.238	<0.001	0.667

Abbreviations: PHE-s®, Patient Health Engagement Scale; FIS, Food Involvement Scale. Note: p, significance level; η^2, eta-squared; mean scores with the same superscript letter do not differ significantly ($p > 0.05$) from each other (Games–Howell post hoc test). Superscript letters are ordered from the highest value to the lowest.

The four identified clusters exhibited distinct levels of both PHE-s® and FIS scores (F3, 476.93 = 626.597; $p < 0.001$; $\eta^2 = 0.681$; F3, 477.12 = 594.238; $p < 0.001$; $\eta^2 = 0.667$, respectively). The identified clusters are as follows:

(i) A cluster comprising patients with higher PHE-s® scores and lower FIS scores is labeled as the "Health-conscious" group.
(ii) Another group characterized by higher scores on PHE-s® and FIS is the "Balanced" group.
(iii) A cluster with higher FIS scores and lower PHE-s® scores is identified as the "Hedonist" group.
(iv) Lastly, a cluster of patients with lower scores on both PHE-s® and FIS is called the "Careless" group.

3.3.1. Differences between Clusters

Sociodemographic and Disease-Related Characteristics

Overall, the results indicate that the identified clusters are not strongly correlated with sociodemographic characteristics. Specifically, χ^2 tests revealed a moderate association between belonging to a cluster and gender ($\chi^2_{(1, 890)} = 39.529$; $p < 0.001$; V = 0.21). Post hoc examination of adapted standardized residuals revealed that the "Health-conscious" cluster exhibited a higher percentage of males (z = 5.2), while the "Balanced" and "Hedonist" clusters had a higher rate of females (z = 2.5 and z = 4.3, respectively).

Furthermore, the results revealed a weak association between belonging to a cluster and recent hospitalizations ($\chi^2_{(1,\,890)} = 24.616$; $p < 0.001$; V = 0.16) and a moderate association with recent relapses ($\chi^2_{(1,\,890)} = 92.681$; $p < 0.001$; V = 0.32). Specifically, participants in the "Health-conscious" group exhibited less frequent recent hospitalizations and relapses (z = 4.1 and z = 7.0, respectively), while "Balanced" participants experienced fewer recent relapses (z = 3.1). Conversely, "Hedonists" were more likely to have recent hospitalizations and relapses (z = 3.8 and z = 6.9, respectively), and "Careless" participants showed a higher frequency of recent relapses (z = 4.2).

Finally, Welch's ANOVA revealed a significant difference in age across clusters (F3, 476.92 = 13.991; $p < 0.001$; $\eta^2 = 0.045$). Post hoc Games–Howell tests indicated that participants in the "Health-conscious" group had a higher average age (M = 51.07; SD = 14.29) compared to the other clusters (M = 46.76, SD = 12.54; M = 42.8, SD = 12.74; and M = 46.15, SD = 13.89, for clusters 2, 3, and 4, respectively). Additionally, the "Balanced" group had a higher mean age than the "Hedonist" group. However, the differences between Clusters 2 and 4 and Clusters 3 and 4 were not statistically significant. Table 4 presents the results of χ^2 analyses and the percentages of the various group characteristics.

Table 4. Distribution of sociodemographic and clinical characteristics across clusters.

Variables	Clusters				$\chi 2$	*p*	Cramer's V
	1: Health-Conscious	2: Balanced	3: Hedonist	4: Careless			
Gender							
Male	**53.3% (5.2)**	33.2% (−2.5)	26.2% (−4.3)	43.3% (1.1)	39.529	<0.001	0.21
Female	46.7% (−5.2)	**66.8% (2.5)**	**73.8% (4.3)**	56.7% (−1.1)			
Diagnosis							
Colitis	48.7% (−0.5)	48.5% (−0.5)	47.1% (−0.9)	**56.2% (2.0)**			
Crohn's disease	51.3% (0.5)	51.5% (0.5)	53.9% (0.9)	43.8% (−2.0)	4.418	n. s.	
Level of education							
Middle school or lower	11.5% (1.1)	8.7% (−0.6)	5.9% (−2.0)	12.4% (1.4)			
High school	47.9% (0.2)	46.9% (−0.2)	46.5% (−0.3)	48.3% (0.3)	7.525	n. s.	-
University or higher	40.6% (−0.9)	44.4% (0.6)	47.6% (1.5)	39.3% (−1.1)			
Comorbidities							
Yes	22.2% (0.0)	21.6% (−0.2)	23.0% (0.3)	21.9% (−0.1)	0.132	n. s.	-
No	77.8% (0.0)	78.4% (0.2)	77.0% (−0.3)	78.1% (0.1)			
Hospitalizations							
Yes	4.6% (−4.1)	10.8% (−0.4)	**19.3% (3.8)**	13.9% (1.2)	24.616	<0.001	0.16
No	**95.4% (4.1)**	89.2% (0.4)	80.7% (−3.8)	86.1% (−1.2)			
Relapses							
Yes	21.8% (−7.0)	31.1% (−3.1)	**61.5% (6.9)**	**52.2% (4.2)**	92.681	<0.001	0.32
No	**78.2% (7.0)**	**68.9% (3.1)**	38.5% (−6.9)	47.8% (−4.2)			

Note: *p*, significance level; χ^2, chi-square; n. s., not significant; Cramer's V, effect size; cell values represent the percentages of the relative characteristic (e.g., male) in each cluster. Values in brackets represent standardized residuals, and cells highlighted in bold indicate where residuals show a significantly higher percentage relative to the sample at 5% significance (standardized residuals ≥ 2).

Food-Related Quality of Life

Regarding the food-related quality of life, the analysis revealed significant differences in Fr-QoL scores among participants in different clusters (F3, 482.70 = 104.161; $p < 0.001$; $\eta^2 = 0.257$). Specifically, participants in the "Health-conscious" and "Balanced" groups had lower Fr-QoL scores (M = 2.66, SD = 0.79, and M = 2.81, SD = 0.76, respectively) compared to the other groups, indicating a higher quality of life related to food. On the other hand, "Careless" participants had higher scores than those in the "Health-conscious" and "Balanced" groups but lower scores compared to "Hedonists", who exhibited the overall lowest food-related quality of life.

Motives behind Food Choices

Using the single items from the FCQ, the analysis revealed that participants in different clusters attributed varying degrees of importance to different motivations behind their food choices (F tests reported in Table 5 for each item). Specifically, participants in the "Balanced" and "Hedonist" clusters expressed greater concern for the "healthiness" of their food compared to those in the "Health-conscious" and "Careless" groups. Moreover, they exhibited a higher interest in food that provides pleasurable sensations. At the same time, participants in the "Health-conscious" cluster, along with the "Careless" group, showed the slightest interest in this aspect.

Table 5. ANOVA results regarding food-related quality of life and food choice motives.

Variables	Clusters				Welch's F	p	η^2
	1: Health-Conscious	2: Balanced	3: Hedonist	4: Careless			
Food-related quality of life	2.66 ^{c}	2.81 ^{c}	3.68 ^{a}	3.49 ^{b}	104.161	<0.001	0.257
Food choice motives (It is important to me that the food I eat on a typical day...)							
is healthy	5.84 ^{b}	6.16 ^{a}	6.02 ^{a}	5.81 ^{b}	6.749	<0.001	0.020
is a way of monitoring my mood	4.64	4.98 ^{b}	5.42 ^{a}	5.24	12.476	<0.001	0.038
is convenient	-	-	-	-	1.776	n. s.	-
provides me with pleasurable sensations	5.38 ^{b}	5.70	5.73 ^{a}	5.52 ^{a,b}	5.193	0.002	0.017
is natural	-	-	-	-	2.344	n. s.	-
is affordable	4.75 ^{b}	5.03 ^{a,b}	5.13 ^{a}	4.87 ^{a,b}	3.668	0.012	0.011
helps me control my weight	-	-	-	-	0.636	n. s.	-
is familiar	-	-	-	-	0.626	n. s.	-
is environmentally friendly	5.33 ^{a,b}	5.40 ^{a}	5.28 ^{a,b}	5.05 ^{b}	2.604	0.051	0.010
is animal friendly	-	-	-	-	1.013	n. s.	-
is fairly traded	-	-	-	-	1.854	n. s.	-
helps me control my symptoms	5.90 ^{c}	6.18 ^{b}	6.43 ^{a}	6.31 ^{a,b}	13.054	<0.001	0.044

Note: p, significance level; η^2, eta-squared; n. s., not significant, mean scores with the same superscript letter do not differ significantly ($p > 0.05$) from each other (Games–Howell post hoc test). Superscript letters are ordered from the highest value to the lowest.

Individuals with high health engagement (i.e., "Health-conscious" and "Balanced" clusters) were less inclined to use food as a means to manage mood compared to those in the "Hedonist" and "Careless" clusters. Notably, the "Hedonist" cluster was particularly interested in using food to control symptoms, while "Health-conscious" participants showed the slightest interest. Furthermore, "Hedonists" expressed a higher interest in affordability than "Health-conscious" participants. At the same time, a marginally significant difference ($p = 0.051$) indicated that the "Balanced" group exhibited the most interest in environmental friendliness. Conversely, participants in the "Careless" group showed little interest in environmental friendliness. Table 6 summarizes the different identified characteristics of these four groups of patients.

Table 6. Result summary and principal characteristics of participants in each cluster for type of variable.

Variables	Clusters			
	1: Health-Conscious	2: Balanced	3: Hedonist	4: Careless
Defining variables	High health engagement and low food involvement	High health engagement and high food involvement	Low health engagement and high food involvement	Low health engagement and low food involvement
Sociodemographic and clinical characteristics	Males, with no hospitalization or relapses in the last year, higher average age	Females, with no relapses in the last year	Females with relapses and/or hospitalizations in the previous year	Diagnosis of colitis, with relapses in the last year and no surgical treatment in the past

Table 6. *Cont.*

Variables	Clusters			
	1: Health-Conscious	2: Balanced	3: Hedonist	4: Careless
Food-related quality of life and emotional state	High food-related quality of life	High food-related quality of life	Reported food-related quality of life is the lowest among groups	Reported food-related quality of life is low
Food choice drivers	This group does not show a particular driver regarding food. Compared to the other groups, these participants reported less interest in healthiness, mood modulation, sensations, affordability, and friendliness toward the environment. Food is not deemed essential for symptom management either	More interest in food's healthiness and its pleasurable sensations. However, seeking pleasure is unrelated to mood modulation, as they reported this driver as less critical. Attention toward the environment and symptom management are also present (the latter is not as high as for other clusters)	Higher interest in food's healthiness and its pleasurable sensations. In this case, however, there is also a reported use of food as a means to regulate mood, which is an essential driver for this group. Affordability and symptom control are also important drivers	This group shows a low interest in the healthiness of food and a high use of food for mood regulation. Inconsistently, they also report the importance of managing symptoms through food. Environmental friendliness is not essential

4. Discussion

The results of the current study, derived from a comprehensive clustering analysis based on the level of low-high health engagement and food involvement, uncovered intriguing insights into the multifaceted nature of food choices among individuals living with IBD.

In particular, food involvement holds a significant influence on the dietary preferences of individuals with IBD, as evidenced by our findings, which resonate with the existing literature [46], particularly regarding the pivotal role of food involvement in influencing food preferences. Our participants had a moderate level of food involvement, and consistent with prior research [47], individuals with higher levels of food involvement showed a predisposition toward healthier food choices, which underscores the significance of food-related attitudes and interests in shaping dietary behaviors. This inclination toward healthier eating patterns may stem from heightened nutritional awareness and appreciation among individuals with elevated food involvement, as they prioritize foods that align with health goals and dietary guidelines. Moreover, the multifaceted nature of food involvement, encompassing cognitive, emotional, and sensory dimensions, may have shaped food preferences and consumption patterns in this population [48]. Despite the lack of association with emotional states, the positive correlation between the Fr-QoL and food involvement implies that while emotional experiences may not directly influence food involvement, individuals who are more engaged with food-related activities and decisions may perceive a greater sense of satisfaction or fulfillment in their overall quality of life, particularly in the context of dietary aspects. This finding highlights the multifaceted nature of quality of life in chronic illness, where factors such as food involvement play a distinct role in shaping individuals' perceptions of well-being.

The current study elucidated associations between emotional states and health engagement among IBD patients. Contrary to Barello and colleagues' findings [33], our participants had a moderate level of health engagement, indicating a noteworthy interest in health-related behaviors despite the challenges posed by their medical condition. Notably, a robust positive correlation between overall positive emotions and health engagement underscores the influence of emotional well-being on individuals' inclination toward health-conscious behaviors [34,49]. This finding aligns with another study conducted on a diabetes population that has manifested the synergy among these subjective experiences in terms of sustainable and effective disease management [50]. Therefore, individuals experiencing more positive emotions may exhibit greater motivation and commitment toward actively engaging in health-promoting activities, including dietary choices, despite the challenges posed by their medical condition.

Further exploration into the motivations underlying food choices reveals a complex interplay between nutritional considerations and emotional elaboration. The identification of four distinct patient clusters—"Health-conscious (high HE, low FI)", "Balanced (high HE, high FI)", "Hedonist (high FI, low HE)", and "Careless (low FI, low HE)"—led to a nuanced understanding of the underlying motivations and behavioral patterns guiding dietary decisions. Regarding the individuals with high food involvement, an interesting difference emerged: "Hedonist" patients attributed higher importance to the capacity of food to help regulate mood and alleviate distress compared to "Balanced" patients. Indeed, individuals with high food involvement are inclined toward seeking pleasurable food [51]; however, supported by high health engagement, individuals demonstrated improved symptomatology and emotional well-being, thereby reducing the necessity to resort to food as a mechanism for emotional regulation [21]. Recent research by Wardle et al. (2018) revealed that low mood and high anxiety in patients with Crohn's disease were associated with more frequent binge eating and decreased control of food cravings [52]. We claim that when individuals with high food involvement lack positive adaptation to their condition and experience worsened emotional states, they may become more prone to abuse comfort food as a means of mood regulation. Furthermore, it is interesting that the "Health-conscious" group appears to lack a prevalent driver for food selection compared to the other three clusters. This phenomenon could be attributed to the superior clinical status of these patients in comparison to the two groups demonstrating low health engagement.

While the four clusters did not represent particular differences regarding their sociodemographic characteristics (except the gender distribution), the groups with high health engagement levels demonstrated lower hospitalization rates and relapses. A recent study reported that health-engaged individuals are more likely to adhere to their treatment plans, including medication regimens and lifestyle modifications, leading to better disease control and reducing the likelihood risk of hospitalizations [53]. Furthermore, our results revealed a significant positive correlation between health engagement and food-related quality of life, indicating that a higher health engagement level is associated with a better perception of food-related factors on the overall quality of life among IBD patients. This implies that individuals actively engaged in managing their health may perceive fewer limitations or disruptions in their daily lives stemming from food-related issues and tend to have improved psychological well-being, potentially due to better coping strategies or adaptive behaviors. Similarly, higher levels of quality of life have been linked with higher health engagement among individuals with IBD [33,54]. In addition, individuals with the lowest food-related quality of life levels were classified into the "Hedonist" cluster, and their high food involvement may have contributed to increased cravings. In this sense, these cravings may be exacerbated by dietary restrictions imposed upon individuals in the "Hedonist" group, thereby impacting their overall quality of life related to food.

4.1. Limitations

The current study is subject to some limitations. Initially, the k-means cluster analysis serves as an exploratory technique. While the derived solution comprising four clusters proved to be interpretable and stable, exhibiting favorable metrics such as the Rand index, it is imperative to acknowledge the potential existence of alternative solutions. Secondly, the influence of symptoms such as abdominal pain, diarrhea, and bloating may prompt individuals to modify their diets in an attempt to alleviate discomfort and manage their condition effectively. Then, all participants in this study were voluntarily recruited from a patient organization, potentially introducing bias into our sample and limiting the generalizability of findings to all patients with IBD. While the study was conducted in 2021, it is important to note that the fundamental principles underlying dietary behaviors and health management strategies are unlikely to undergo significant changes over a relatively short period. Also, this investigation focused exclusively on Italian patients, whose unique food culture may hinder the reproducibility of results in populations where food holds different cultural significance. Finally, the instruments used in this study relied on self-reported data; however, it is worth mentioning that the instruments, apart from the measure of emotional state, are validated to reduce potential biases. The choice to use a scale from 0 to 100 for measuring emotional state in our study was deliberate and aligned with our research objectives; nonetheless, we aim to incorporate alternative measures in future research endeavors. To mitigate these limitations, future research endeavors should address these constraints and validate and extend these findings across diverse cultural contexts.

4.2. Relevance for Clinical Practice

By integrating strategies to mitigate the impact of emotional eating on symptom severity and disease progression, healthcare providers can optimize patient outcomes and enhance overall quality of life. For instance, patients with low health engagement coupled with high food involvement may benefit from interventions targeting two key objectives: firstly, enhancing their health engagement to foster sustained behavioral change and, secondly, assisting them in exploring alternative mood modulation strategies beyond reliance on food. Conversely, individuals characterized by high levels of both health engagement and food involvement necessitate interventions primarily focused on maintaining health engagement and cultivating a positive mindset rather than solely educational endeavors. Such patients have either initiated or are in the process of implementing positive dietary changes, which are not inherently at odds with their keen interest in food. Nevertheless, due to the inherent challenges associated with adhering to behavioral modifications, ongoing monitoring and support from healthcare professionals are essential to facilitate the maintenance of these changes.

From a clinical perspective, our findings highlight the need for personalized and holistic approaches to dietary management in IBD patients. Furthermore, implementing psychological profiling for patients could facilitate the derivation of robust inferences regarding their motivational styles and the underlying motivations for their dietary selections. Developing approaches to education and adherence support specifically tailored to cater to the distinct motivational dimensions characteristic of particular profiles makes it feasible to leverage these aspects strategically. Such differentiated, bespoke approaches should be co-designed considering the psychological and motivational profiles of diverse patients. This strategy may promote a more patient-centered approach to engaging individuals in self-management and adherence practices, enhancing the likelihood of achieving sustained and positive behavioral change.

5. Conclusions

In conclusion, our study offers valuable insights into the intricate motivations guiding food choices among individuals with IBD. Based on our findings, food involvement significantly impacted the dietary preferences of the participants, with higher levels of food involvement associated with the inclination to seek pleasurable food. However, individuals with high health engagement levels reflected a better emotional state which resulted in less need for food to regulate their mood. Despite the lack of direct association with emotional states, individuals highly engaged with health experienced a greater sense of satisfaction and fulfillment in their overall quality of life, particularly concerning dietary aspects. This might be due to the fact that patients with a higher PHE-s® score have less severe symptomatology and thus, potentially, fewer problems with food. Moreover, given the importance of understanding patients' subjective experiences, utilizing measurement tools could be useful to identify patterns in the dietary habits of IBD patients and design personalized interventions to promote disease management. Future research exploring the relationship between psychological-behavioral variables and clinical outcomes is strongly recommended.

Author Contributions: Conceptualization, L.P. and G.G.; Methodology, L.P., S.L. and G.G.; Software, L.P.; Validation, L.P., S.L. and G.G.; Formal Analysis, L.P., D.U., S.L. and G.G.; Investigation, L.P., D.U., S.L. and G.G.; Data Curation, L.P., D.U. and S.L.; Interpretation, L.P., D.U., S.L. and G.G.; Writing—Original Draft Preparation, L.P. and D.U.; Writing—Review and Editing, D.U. and G.G.; Visualization, D.U.; Supervision, S.L. and G.G.; Project Administration, L.P. and G.G.; Funding Acquisition, L.P. All authors have read and agreed to the published version of the manuscript.

Funding: This work was part of a broader project regarding the relationship between food, IBD, and the psychosocial well-being of patients (project "Crohnviviamo"), funded by Nestlé Health Science with grant number R2094400310. Also, this work is partly supported by the EU Horizon Europe Program grant miGut-Health: personalized blueprint of intestinal health (101095470). Views and opinions expressed are however those of the authors only and do not necessarily reflect those of the European Union. Neither the European Union nor the granting authority can be held responsible for them.

Institutional Review Board Statement: All subjects gave their informed consent for inclusion before they participated in the study. The study was conducted according to the guidelines of the Declaration of Helsinki and approved by the Ethics Committee of the Università Cattolica del Sacro Cuore, Italy (protocol code: 36-21, approval date: 30 April 2021).

Informed Consent Statement: Informed consent was obtained from all subjects involved in the study.

Data Availability Statement: The data supporting this study's findings are available on request from the corresponding author. The data are not publicly available due to privacy or ethical restrictions.

Acknowledgments: The authors would like to acknowledge the IBD patient organization AMICI ETS for supporting the Crohnviviamo project.

Conflicts of Interest: The authors declare no conflicts of interest. This funding source had no role in the design of this study, in its execution, in the writing of this report, or in the decision to submit the results.

References

1. Zhao, M.; Gönczi, L.; Lakatos, P.L.; Burisch, J. The burden of inflammatory bowel disease in Europe in 2020. *J. Crohn's Colitis* **2021**, *15*, 1573–1587. [CrossRef] [PubMed]
2. Wang, R.; Li, Z.; Liu, S.; Zhang, D. Global, regional and national burden of inflammatory bowel disease in 204 countries and territories from 1990 to 2019: A systematic analysis based on the Global Burden of Disease Study 2019. *BMJ Open* **2023**, *13*, e065186. [CrossRef] [PubMed]
3. Johansen, I.; Småstuen, M.C.; Løkkeberg, S.T.; Kristensen, V.A.; Høivik, M.L.; Lund, C.; Olsen, B.; Strande, V.; Huppertz-Hauss, G.; Aabrekk, T.B. Symptoms and symptom clusters in patients newly diagnosed with inflammatory bowel disease: Results from the IBSEN III Study. *BMC Gastroenterol.* **2023**, *23*, 255. [CrossRef] [PubMed]
4. Jones, J.L.; Nguyen, G.C.; Benchimol, E.I.; Bernstein, C.N.; Bitton, A.; Kaplan, G.G.; Murthy, S.K.; Lee, K.; Cooke-Lauder, J.; Otley, A.R. The impact of inflammatory bowel disease in Canada 2018: Quality of life. *J. Can. Assoc. Gastroenterol.* **2019**, *2* (Suppl. 1), S42–S48. [CrossRef] [PubMed]
5. Ananthakrishnan, A.N.; Bernstein, C.N.; Iliopoulos, D.; Macpherson, A.; Neurath, M.F.; Ali, R.A.R.; Vavricka, S.R.; Fiocchi, C. Environmental triggers in IBD: A review of progress and evidence. *Nat. Rev. Gastroenterol. Hepatol.* **2018**, *15*, 39–49. [CrossRef] [PubMed]
6. Campmans-Kuijpers, M.J.; Dijkstra, G. Food and food groups in inflammatory bowel disease (IBD): The design of the groningen anti-inflammatory diet (GrAID). *Nutrients* **2021**, *13*, 1067. [CrossRef] [PubMed]
7. Yan, J.; Wang, L.; Gu, Y.; Hou, H.; Liu, T.; Ding, Y.; Cao, H. Dietary patterns and gut microbiota changes in inflammatory bowel disease: Current insights and future challenges. *Nutrients* **2022**, *14*, 4003. [CrossRef] [PubMed]
8. Levine, A.; Boneh, R.S.; Wine, E. Evolving role of diet in the pathogenesis and treatment of inflammatory bowel diseases. *Gut* **2018**, *67*, 1726–1738. [CrossRef] [PubMed]
9. Peters, V.; Spooren, C.E.; Pierik, M.J.; Weersma, R.K.; van Dullemen, H.M.; Festen, E.A.; Visschedijk, M.C.; Masclee, A.A.; Hendrix, E.M.; Almeida, R.J. Dietary intake pattern is associated with occurrence of flares in IBD patients. *J. Crohn's Colitis* **2021**, *15*, 1305–1315. [CrossRef]
10. Rizzello, F.; Spisni, E.; Giovanardi, E.; Imbesi, V.; Salice, M.; Alvisi, P.; Valerii, M.C.; Gionchetti, P. Implications of the westernized diet in the onset and progression of IBD. *Nutrients* **2019**, *11*, 1033. [CrossRef]

11. Bolte, L.A.; Vila, A.V.; Imhann, F.; Collij, V.; Gacesa, R.; Peters, V.; Wijmenga, C.; Kurilshikov, A.; Campmans-Kuijpers, M.J.; Fu, J. Long-term dietary patterns are associated with pro-inflammatory and anti-inflammatory features of the gut microbiome. *Gut* **2021**, *70*, 1287–1298. [CrossRef] [PubMed]
12. Tayyem, R.F.; Qalqili, T.R.; Ajeen, R.; Rayyan, Y.M. Dietary patterns and the risk of inflammatory bowel disease: Findings from a case-control study. *Nutrients* **2021**, *13*, 1889. [CrossRef] [PubMed]
13. Khalili, H.; Håkansson, N.; Chan, S.S.; Chen, Y.; Lochhead, P.; Ludvigsson, J.F.; Chan, A.T.; Hart, A.R.; Olén, O.; Wolk, A. Adherence to a Mediterranean diet is associated with a lower risk of later-onset Crohn's disease: Results from two large prospective cohort studies. *Gut* **2020**, *69*, 1637–1644. [CrossRef]
14. Papada, E.; Amerikanou, C.; Forbes, A.; Kaliora, A.C. Adherence to Mediterranean diet in Crohn's disease. *Eur. J. Nutr.* **2020**, *59*, 1115–1121. [CrossRef] [PubMed]
15. Marsh, A.; Kinneally, J.; Robertson, T.; Lord, A.; Young, A.; Radford–Smith, G. Food avoidance in outpatients with Inflammatory Bowel Disease–Who, what and why. *Clin. Nutr. ESPEN* **2019**, *31*, 10–16. [CrossRef] [PubMed]
16. Bischoff, S.C.; Escher, J.; Hébuterne, X.; Kłęk, S.; Krznaric, Z.; Schneider, S.; Shamir, R.; Stardelova, K.; Wierdsma, N.; Wiskin, A.E. ESPEN practical guideline: Clinical Nutrition in inflammatory bowel disease. *Clin. Nutr.* **2020**, *39*, 632–653. [CrossRef] [PubMed]
17. Marsh, A.; Rindfleish, S.; Bennett, K.; Croft, A.; Chachay, V. Outcomes of dietary management approaches in active ulcerative colitis: A systematic review. *Clin. Nutr.* **2022**, *41*, 298–306. [CrossRef]
18. Szeto, J.; Noejovich, C.; Verma, R.; Miranda, P.; Pinto-Sanchez, M.; Verdu, E.; Armstrong, D. Barriers to dietary modification in Inlammatory Bowel Disease (IBD): A mixed-methods assessment of patient perceptions. *J. Can. Assoc. Gastroenterol.* **2024**, *7* (Suppl. 1), 213–214. [CrossRef]
19. Richard, A.; Meule, A.; Blechert, J. Implicit evaluation of chocolate and motivational need states interact in predicting chocolate intake in everyday life. *Eat. Behav.* **2019**, *33*, 1–6. [CrossRef]
20. Yin, T.T.; Tu, W.-J.; Li, Y.-T.; Xu, W.-J.; Xu, G.-H. 'Eating is like experiencing a gamble': A qualitative study exploring the dietary decision-making process in adults with inflammatory bowel disease. *Health Expect.* **2023**, *27*, e13873. [CrossRef]
21. Palamenghi, L.; Figliuc, P.; Leone, S.; Graffigna, G. Food and Inflammatory Bowel Diseases: A scoping review on the impact of food on patients' psychosocial quality of life. *Health Soc. Care Community* **2022**, *30*, 1695–1712. [CrossRef]
22. Chuong, K.H.; Haw, J.; Stintzi, A.; Mack, D.R.; O'Doherty, K.C. Dietary strategies and food practices of pediatric patients, and their parents, living with inflammatory bowel disease: A qualitative interview study. *Int. J. Qual. Stud. Health Well-Being* **2019**, *14*, 1648945. [CrossRef] [PubMed]
23. Rines, J.; Daley, K.; Loo, S.; Safari, K.; Walsh, D.; Gill, M.; Moayyedi, P.; Fernandes, A.; Marlett, N.; Marshall, D. A patient-led, peer-to-peer qualitative study on the psychosocial relationship between young adults with inflammatory bowel disease and food. *Health Expect.* **2022**, *25*, 1486–1497. [CrossRef] [PubMed]
24. Coccurello, R.; Maccarrone, M. Hedonic eating and the "delicious circle": From lipid-derived mediators to brain dopamine and back. *Front. Neurosci.* **2018**, *12*, 271. [CrossRef] [PubMed]
25. Barnhart, W.R.; Braden, A.L.; Price, E. Emotion regulation difficulties interact with negative, not positive, emotional eating to strengthen relationships with disordered eating: An exploratory study. *Appetite* **2021**, *158*, 105038. [CrossRef]
26. Castellini, G.; Graffigna, G. Assessing involvement with food: A systematic review of measures and tools. *Food Qual. Prefer.* **2022**, *97*, 104444. [CrossRef]
27. Bell, R.; Marshall, D.W. The construct of food involvement in behavioral research: Scale development and validation. *Appetite* **2003**, *40*, 235–244. [CrossRef] [PubMed]
28. Mak, A.H.; Lumbers, M.; Eves, A.; Chang, R.C. The effects of food-related personality traits on tourist food consumption motivations. *Asia Pac. J. Tour. Res.* **2017**, *22*, 1–20. [CrossRef]
29. Vorage, L.; Wiseman, N.; Graca, J.; Harris, N. The association of demographic characteristics and food choice motives with the consumption of functional foods in emerging adults. *Nutrients* **2020**, *12*, 2582. [CrossRef]
30. Dziewa, M.; Bańka, B.; Herbet, M.; Piątkowska-Chmiel, I. Eating disorders and diabetes: Facing the dual challenge. *Nutrients* **2023**, *15*, 3955. [CrossRef] [PubMed]
31. Leeds, J.; Keith, R.; Woloshynowych, M. Food and Mood: Exploring the determinants of food choices and the effects of food consumption on mood among women in Inner London. *World Nutr.* **2020**, *11*, 68–96. [CrossRef]
32. Graffigna, G.; Barello, S.; Bonanomi, A.; Lozza, E. Measuring patient engagement: Development and psychometric properties of the Patient Health Engagement (PHE) Scale. *Front. Psychol.* **2015**, *6*, 274. [CrossRef] [PubMed]
33. Barello, S.; Guida, E.; Leone, S.; Previtali, E.; Graffigna, G. Does patient engagement affect IBD patients' health-related quality of life? Findings from a cross-sectional study among people with inflammatory bowel diseases. *Health Qual. Life Outcomes* **2021**, *19*, 77. [CrossRef] [PubMed]
34. Graffigna, G.; Barello, S. Spotlight on the Patient Health Engagement model (PHE model): A psychosocial theory to understand people's meaningful engagement in their own health care. *Patient Prefer. Adherence* **2018**, *2018*, 1261–1271. [CrossRef] [PubMed]
35. Barello, S.; Palamenghi, L.; Graffigna, G. The mediating role of the patient health engagement model on the relationship between patient perceived autonomy supportive healthcare climate and health literacy skills. *Int. J. Environ. Res. Public Health* **2020**, *17*, 1741. [CrossRef]
36. Cengiz, D.; Agrali, H.; Korkmaz, F.; Akyar, I. Mediation role of patient engagement on the association between health literacy and self-care behaviours in individuals with type 2 diabetes. *Int. J. Nurs. Pract.* **2022**, *28*, e13051. [CrossRef] [PubMed]

37. Graffigna, G.; Barello, S.; Bonanomi, A. The role of Patient Health Engagement Model (PHE-model) in affecting patient activation and medication adherence: A structural equation model. *PLoS ONE* **2017**, *12*, e0179865. [CrossRef] [PubMed]
38. Day, A.S.; Yao, C.K.; Costello, S.P.; Andrews, J.M.; Bryant, R.V. Food avoidance, restrictive eating behaviour and association with quality of life in adults with inflammatory bowel disease: A systematic scoping review. *Appetite* **2021**, *167*, 105650. [CrossRef] [PubMed]
39. Onwezen, M.; Reinders, M.; Verain, M.; Snoek, H. The development of a single-item Food Choice Questionnaire. *Food Qual. Prefer.* **2019**, *71*, 34–45. [CrossRef]
40. Hughes, L.D.; King, L.; Morgan, M.; Ayis, S.; Direkze, N.; Lomer, M.C.; Lindsay, J.O.; Whelan, K. Food-related quality of life in inflammatory bowel disease: Development and validation of a questionnaire. *J. Crohn's Colitis* **2016**, *10*, 194–201. [CrossRef]
41. Andrich, D. A structure of index and causal variables. *Rasch Meas. Transac.* **2014**, *28*, 1475–1477.
42. Linacre, J. *A User's Guide to WINSTEPS Ministeps*. Rasch-Model Computer Program; Program Manual 3.71. 70. 2011. Available online: https://www.winsteps.com/winman/copyright.htm (accessed on 9 April 2024).
43. Caliński, T.; Harabasz, J. A dendrite method for cluster analysis. *Commun. Stat. Theory Methods* **1974**, *3*, 1–27. [CrossRef]
44. Rand, W.M. Objective criteria for the evaluation of clustering methods. *Am. Stat. Assoc. Bull.* **1971**, *66*, 846–850. [CrossRef]
45. Sharpe, D. Chi-square test is statistically significant: Now what? *Pract. Assess. Res. Eval.* **2019**, *20*, 8. [CrossRef]
46. Farragher, T.; Wang, W.C.; Worsley, A. The associations of vegetable consumption with food mavenism, personal values, food knowledge and demographic factors. *Appetite* **2016**, *97*, 29–36. [CrossRef] [PubMed]
47. Medoro, C.; Cianciabella, M.; Magli, M.; Daniele, G.M.; Lippi, N.; Gatti, E.; Volpe, R.; Longo, V.; Nazzaro, F.; Mattoni, S. Food involvement, food choices, and bioactive compounds consumption correlation during COVID-19 pandemic: How food engagement influences consumers' food habits. *Nutrients* **2022**, *14*, 1490. [CrossRef] [PubMed]
48. Castellini, G.; Graffigna, G. "Food is more than just a source of nutrients": A qualitative phenomenological study on Food Involvement. *Appetite* **2022**, *178*, 106179. [CrossRef] [PubMed]
49. Hadgkiss, E.J.; Jelinek, G.A.; Taylor, K.L.; Marck, C.H.; van der Meer, D.M.; Pereira, N.G.; Weiland, T.J. Engagement in a program promoting lifestyle modification is associated with better patient-reported outcomes for people with MS. *Neurol. Sci.* **2015**, *36*, 845–852. [CrossRef] [PubMed]
50. Graffigna, G.; Barello, S.; Libreri, C.; Bosio, C.A. How to engage type-2 diabetic patients in their own health management: Implications for clinical practice. *BMC Public Health* **2014**, *14*, 648. [CrossRef]
51. Bédard, A.; Lamarche, P.-O.; Grégoire, L.-M.; Trudel-Guy, C.; Provencher, V.; Desroches, S.; Lemieux, S. Can eating pleasure be a lever for healthy eating? A systematic scoping review of eating pleasure and its links with dietary behaviors and health. *PLoS ONE* **2020**, *15*, e0244292. [CrossRef]
52. Wardle, R.A.; Thapaliya, G.; Nowak, A.; Radford, S.; Dalton, M.; Finlayson, G.; Moran, G.W. An examination of appetite and disordered eating in active Crohn's disease. *J. Crohn's Colitis* **2018**, *12*, 819–825. [CrossRef] [PubMed]
53. Ngorsuraches, S.; Da Rosa, P.; Ge, X.; Djira, G.; Michael, S.; Wey, H. Patient engagement as a predictor for health outcomes and costs in multiple chronic conditions. *Value Health* **2018**, *21*, S88–S89. [CrossRef]
54. Graffigna, G.; Bosio, C.; Pagnini, F.; Volpato, E.; Previtali, E.; Leone, S.; D'Amico, F.; Armuzzi, A.; Danese, S. Promoting psycho-social wellbeing for engaging inflammatory bowel disease patients in their care: An Italian consensus statement. *BMC Psychol.* **2021**, *9*, 186. [CrossRef] [PubMed]

nutrients

Article

Cardiovascular Disease Burden Attributable to High Sodium Intake in China: A Longitudinal Study from 1990 to 2019

Liying Jiang [1,2], Wanying Shen [3], Anqi Wang [4], Haiqin Fang [5], Qihe Wang [5], Huzhong Li [5], Sana Liu [5], Yi Shen [3,*] and Aidong Liu [6,*]

1 Jiading Central Hospital, Shanghai University of Medicine & Health Sciences, Shanghai 201899, China; j_meili@126.com
2 Department of Prevention Medicine, College of Public Health, Shanghai University of Medicine & Health Sciences, Shanghai 201318, China
3 Department of Epidemiology, School of Public Health, Nantong University, Nantong 226019, China; shenwanying-@outlook.com
4 Graduate School, Shanghai University of Traditional & Chinese Medicine, Shanghai 201203, China; wanganqi202309@163.com
5 Department of Nutrition Division I, China National Center for Food Safety Risk Assessment, Beijing 100022, China; fanghaiqin@cfsa.net.cn (H.F.); wangqihe@cfsa.net.cn (Q.W.); lihuzhong@cfsa.net.cn (H.L.); liusana@cfsa.net.cn (S.L.)
6 National Institute for Nutrition and Health, Chinese Center for Disease Control and Prevention, Beijing 100050, China
* Correspondence: sunny@ntu.edu.cn (Y.S.); liuad@ninh.chinacdc.cn (A.L.); Tel.: +86-010-52165568 (A.L.)

Abstract: Background: Overconsumption of sodium has been identified as a key driving factor for diet-related cardiovascular diseases (CVDs). China, being a country bearing a hefty burden of CVD, has a large population with diverse cultural traditions and ethnic beliefs, which complicates the patterns of dietary sodium intake, necessitating a systematic investigation into the profile of the high sodium intake (HSI)-related burden of CVD within its subregions. This study aims to estimate the evolving patterns of HSI-induced CVD burden across China from 1990 to 2019. Methods: The methodology used in the Global Burden of Disease Study was followed to assess deaths and disability-adjusted life years (DALYs) by age, sex, region, and socio-demographic index (SDI). The estimated annual percentage change (EAPC) was calculated to quantify the secular changes in the age-standardized mortality rate (ASMR) and age-standardized DALY rate (ASDR). Results: In 2019, 0.79 million deaths and 1.93 million DALYs of CVD were attributed to HSI, an increase of 53.91% and 39.39% since 1990, respectively. Nevertheless, a downward trend in ASMR (EAPC: −1.45, 95% CI: −1.55, −1.35) and ASDR (EAPC: −1.61, 95% CI: −1.68, −1.53) was detected over time. ASMR and ASDR were higher for males, individuals aged ≥60 years, and regions with low-middle SDI. A markedly negative association between the EAPC in both ASMR and ASDR and the SDI was found in 2019 (ρ = −0.659, $p < 0.001$ and ρ = −0.558, $p < 0.001$, respectively). Conclusions: The HSI-induced CVD burden is gender-, age-, and socioeconomic-dependent. Integrated and targeted strategies for CVD prevention are anticipated in the future throughout China.

Keywords: high sodium intake; cardiovascular disease; burden of disease; China

Citation: Jiang, L.; Shen, W.; Wang, A.; Fang, H.; Wang, Q.; Li, H.; Liu, S.; Shen, Y.; Liu, A. Cardiovascular Disease Burden Attributable to High Sodium Intake in China: A Longitudinal Study from 1990 to 2019. *Nutrients* **2024**, *16*, 1307. https://doi.org/10.3390/nu16091307

Academic Editors: Domenico Tafuri and Francesca Latino

Received: 28 March 2024
Revised: 21 April 2024
Accepted: 23 April 2024
Published: 26 April 2024

1. Introduction

Cardiovascular diseases (CVDs), representing a group of circulatory system disorders that mainly include heart and vascular diseases, have been the leading cause of morbidity and mortality worldwide over the past decades, and this trend has carried over to now. The newly updated Global Burden of Disease (GBD) Study 2019 assessed that an estimated 18.6 million deaths and 393 million disability-adjusted life years (DALYs) resulted from CVD, accounting for 33% and 15.5% of total deaths and DALYs around the world, respectively [1]. Despite all the endeavors to bring the growing burden of CVD into greater relief, the future trend of it seems still not optimistic. Currently, the pace of decline in CVD prevalence has already decelerated in high-income countries, while the upward trend still persists in low- and middle-income countries [2]. China bears the world's heaviest burden of CVD [3]. The estimated number of deaths resulting from the disease doubled from 2.42 million in 1990 to 4.58 million in 2019, chalking up rapid and substantial growth in this country [4]. In 2019, CVD was responsible for 46.74% and 44.26% of deaths in rural and urban regions of China, being the cause of death in 2 out of every 5 cases [5]. In the face of population growth and aging, China's burden of CVD is forecast to rise in the years ahead. A study indicated that, taking only the demographic shifts into account, the annual CVD incidents in China are projected to rise by over 50% between 2010 and 2030, and the current trend of risk factors will further the burden during this period [6].

Primary prevention, focusing on modifiable risk factors, was considered a promising way to reduce the incidence of CVD and promote its subsequent prognosis. Risk factors confirmed to be related to CVD include diabetes, hypertension, dyslipidemia, obesity, smoking, a lack of physical activity, and an unhealthy diet [7]. Unhealthy diets are one of the most relevant modifiable risk factors for CVD. High sodium intake (HSI), as an unhealthy dietary practice, has shown a consistent association with this disease in previous studies [8–10]. Limiting daily sodium intake can effectively lower the risk of elevated blood pressure and subsequent CVD, while HSI may have the opposite effect through pathways including stiffened endothelial cells, thickened resistance arteries, and blocked nitric oxide synthesis [11,12]. The risk of fatal and non-fatal CVD is reckoned at 17% for every 1000 mg/d increase in sodium excretion, according to a cohort analysis [13]. Patients with established CVD may be particularly vulnerable to excess sodium, as the risk seems to be greater for those with prior cardiovascular events, heart failure, and diabetes [14]. Therefore, it is important for people, especially those patients, to consume a safe amount of sodium. The World Health Organization (WHO) advises adults to limit their daily sodium intake to below 2000 mg [15]. The American Heart Association proposes that the general public should aim for a sodium consumption of less than 2400 mg/d, whereas individuals with or at a high risk of CVD should aim for less than 1500 mg/d [16].

Although interventions such as legislative policies and awareness campaigns aimed at reducing sodium intake have been widely implemented, current sodium consumption remains beyond ideal, as evidenced by the estimated global average sodium intake of 4310 mg/d.15 Worse still, according to the China National Nutrition and Health Survey (CNNHS), the average daily sodium intake of Chinese adults was 5013 mg/d, far exceeding the recommended level [17]. The sodium intake among Chinese residents varies greatly on temporal and spatial scales, and the dynamic changes in the CVD burden attributable to HSI in China remain to be elucidated. On this basis, this study evaluated the spatiotemporal trends in HSI-related CVD burden in China from 1990 to 2019. Recognizing these epidemiological characteristics and identifying the potential driving factors could lead to a better understanding of the disease burden resulting from HSI in China and serve to formulate evidence-based solutions and control policies tailored to regional-specific needs, thus responding to the call to reduce the CVD burden globally.

2. Materials and Methods

2.1. Study Data

The Global Burden of Diseases, Injuries, and Risk Factors Study provided a systematic assessment of health loss for 369 diseases and injuries, age-, and sex-specific mortality for 286 causes, and 87 risk factors in 204 countries and territories from 1990 to 2019, with a wide array of standardized analytical procedures, including data screenings, data adjustments, and DisMod-MR (a Bayesian meta-regression tool) estimations to improve the quality and comparability of the data [1]. Through the Global Health Data Exchange Tool (https://vizhub.healthdata.org/gbd-results/ (accessed on 1 August 2023)), we obtained data on deaths, DALYs, age-standardized mortality rates (ASMR), and age-standardized DALY rates (ASDR) of CVD attributable to HSI from 1990 to 2019, which were processed primarily based on raw data from original sources, including the Disease Surveillance Points system in China, the cause-of-death reporting system collected by the Chinese Center for Disease Control and Prevention, medical certification of causes of death for Macao and Hong Kong, and nationwide surveys [18]. Details regarding sodium intake and the incidence and mortality of CVD across China, spanning from January 1990 to December 2019, were obtained from a database compiled by the China National Center for Food Safety Risk Assessment (NFSA) from January 1990 to December 2019 to quantify the estimates of HSI-related CVD burden by age, sex, year, and region and assess its relationship with the sociodemographic index (SDI). This analysis was conducted on 33 administrative units at the provincial level, encompassing 31 provinces, municipalities, and autonomous regions, as well as the Special Administrative Regions (SAR) of Hong Kong and Macao.

As a compound indicator, SDI is measured by the equally weighted geometric mean of lag-distributed income per capita, average years of education for those aged 15 and above, and total fertility rate among women below the age of 25, thus reflecting the socioeconomic level relevant to the health of a certain geographic area [1,19]. It ranges from 0 to 1, representing the varying level of health development from minimum to maximum. The reference SDI quartile as well as the SDI values estimated for China were retrieved in GBD 2017. The national SDI in China was 0.71 in 2017 and its provincial SDI, ranging from 0.47 to 0.86, can be categorized into 4 levels: high SDI region (Beijing, Shanghai, Hong Kong, and Macao), high-middle SDI region (Tianjin, Hebei, Fujian, Guangdong, Jilin, Liaoning, Shandong, Jiangsu, Zhejiang, Heilongjiang, and Inner Mongolia), low-middle SDI region (Tibet, Gansu, and Guizhou), and middle SDI region (all remaining regions) [20].

Since the database is publicly accessible and does not include identifiable information, there is no need for approval by the Research Ethics Committee.

2.2. Definitions of Dietary Risk Factors and Associated Outcomes

Sodium intake assessment among the Chinese population in GBD 2019 has been described previously [19,21]. Briefly, DisMod-MR was utilized to convert sodium intake acquired from 24 h dietary recalls and food frequency questionnaires in nine representative Chinese studies, including the China Nutrition Survey in 1992 and 2002 and the KaiLuan Study during 2006–2016, into 24 h urinary sodium to reduce the bias (the gold standard data source for GBD 2019). A spatiotemporal Gaussian process regression framework was employed to estimate the mean sodium intake by age, sex, year, and region. HSI was defined as sodium intake (in grams per day) greater than 3 g/d (95% UI: 1 g, 5 g). The International Statistical Classification of Diseases and Related Health Issues-10 (ICD-10) was used by GBD to identify the cases of CVDs, and codes for 11 CVD subcategories were presented in Supplementary Materials (Table S1).

2.3. Estimation of High Sodium Intake-Attributed CVD Burden

Methods employed in GBD 2019 to estimate the risk-attributable disease burden have been expounded upon elsewhere [1,19]. In brief, the proportion of preventable disease burden was quantified for each risk factor had the level of exposure been maintained at the level linked to minimum risk, which was defined as the theoretical minimum-risk exposure level (TMREL). Founded on exposure data of each risk factor released in the large-scale population surveys and reports, DisMod-MR and spatiotemporal Gaussian process regression were utilized to synthesize data and determine the exposure level of risk and its TMREL. The relative risk (RR) for the HSI-CVD pair was summarized as a function of exposure based on the meta-analysis of prospective observational studies published globally. Population-attributable fractions (PAFs), presenting the proportion of CVD burden that might be reduced within a given timeframe and population if the exposure to HSI in the past was lowered to the counterfactual level of the TMREL, were calculated for the HSI-CVD pair by age, sex, and year by using exposure level estimates of RR and TMREL. The standard GBD PAF formula is outlined as follows:

$$PAF_{asgt} = \frac{\sum_{x=l}^{u} RR_{asg}(x)P_{asgt}(x) - RR_{asg}(TMREL_{as})}{\sum_{x=l}^{u} RR_{as}(x)P_{asgt}(x)}$$

where PAF_{asgt} represents the PAF for CVD burden attributable to HSI for age group a, sex s, geographic area g, and year t. RR_{ast} (x) denotes the *RR* between exposure level x (from l to u) of HSI and CVD for age group a, sex s, and year t; and P_{asgt} (x) stands for the proportion of the population exposed to HSI at the level x for age group a, sex s, geographic area g, and year t. $TMREL_{as}$ refers to the TMREL for age group a, and sex s.

DALYs were calculated with a sum of years of life lost (YLLs, multiplying observed deaths from each cause in each age group by the age-specific standard life expectancy estimated using life table methods) and years lived with disability (YLDs, years lived with any short-term or long-term health loss weighted for severity by the disability weights). Then, the burden of CVD attributable to HSI was estimated by multiplying the corresponding PAF for each age group, sex, year, and geographic area with the total number of CVD deaths and DALYs.

2.4. Statistical Analyses

To remove the effects of the differences in age structure within a population, age-standardized rates (ASRs) would be necessary. Hence, data on age-standardized mortality and DALY rates were reported in numerical form with 95% uncertainty intervals (UIs) based on the interquartile range of the 25th and 75th ranked estimates across 1000 samples. The PAF was calculated to measure the contribution of HSI exposure to the subsequent occurrence of CVD-related burden using a comparative risk assessment framework. The estimated annual percentage change (EAPC), which indicates the trajectory of ASRs over a specified interval, was applied to reflect the secular trends in ASMR and ASDR of CVD between 1990 and 2019.

A linear regression model was fitted to the logarithmic age-standardized indicators: $\ln(ASR) = \alpha + \beta x + \varepsilon$, where x was the calendar year. After that, EAPC and its 95% confidence interval (CI) could be derived on a log scale from the model $100 \times [\exp(\beta) - 1]$. It is deemed indicative of a rising trend in ASRs of deaths and DALYs if the lower limit of the 95% CI for the corresponding EAPC estimation was higher than 0, and the reverse is true when the upper limit of the 95% CI was lower than 0. Otherwise, the ASRs were considered to be stable.

To investigate the association between the CVD burden attributable to HSI and the socioeconomic development across various provinces, relationships between the ASRs of deaths and DALYs and the SDI from 1990 to 2019 were examined by smoothing spline models. Spearman's test was used to explore the potential factors influencing the EAPC of the CVD burden caused by HSI. $p < 0.05$ was considered significant. The R program (version 4.0.2) was utilized to perform all the statistical analyses.

3. Results

3.1. Deaths and ASMR of CVD Attributable to HSI

Globally, the number of CVD deaths attributable to high sodium intake increased by 41.08% over the past thirty years, from 1,215,851 in 1990 to 1,715,381 in 2019. However, the ASMR decreased from 33.16 (95% CI: 10.39, 69.96) to 21.51 (95% CI: 5.58, 47.2) per 100,000 population at the same time, with an average annual decline of 1.53% (95% CI: −1.57, −1.50).

In China, the estimated number of CVD deaths attributable to HSI in 2019 was 788,587.8, increasing by 53.91% compared with 512,346.4 in 1990. Meanwhile, the ASMR decreased from 68.52 (95% CI: 27.97, 119.98) to 42.46 (95% CI: 15.4, 76.86) per 100,000 population, and the annual trends of changes in ASMR were slightly slower than the global level, with an EAPC of −1.45 (95% CI: −1.55, −1.35). Notably, there was considerable gender inequality in the CVD burden due to HSI. The sheer numbers of CVD deaths in males were 1.54 times higher in 1990 and 1.98 times higher in 2019 than those in females, respectively. The ASMR between 1990 and 2019 was lessened from 52.16 (95% CI: 17.36, 100.5) to 26.86 (95% CI: 6.52, 55.72) per 100,000 population for females and reduced from 89.96 (95% CI: 39.84, 150.1) to 62.57 (95% CI: 24.39, 109.97) per 100,000 population for males. The percentage change in ASMR displayed a coincident downward trend in both genders, with EAPC being more pronounced in females than in males.

At the SDI region level, those with high-middle SDI experienced the highest number of CVD deaths attributable to HSI in both 1990 and 2019, while regions with high SDI had the complete opposite. At the same point in the timeline, after adjusting for the effects of differences in age structure, the ASMR was largest in regions with low-to-middle SDI and lowest in those with high SDI. EACP in ASMR differed by SDI quintile but presented a downward trend in all SDI regions in unison between 1990 and 2019, and it decreased the most in the high SDI regions and the least in the low SDI regions (Table 1).

Table 1. Deaths and ASMR of cardiovascular disease due to high sodium intake in 1990 and 2019 and their percentage changes from 1990 to 2019.

Characteristics	1990		2019		1990–2019	
	Deaths Cases, No. (95% UI)	ASMR per 100,000 No. (95% UI)	Deaths Cases, No. (95% UI)	ASMR per 100,000 No. (95% UI)	PAFs % (95% UI)	EAPC (%) in ASMR No. (95% CI)
Global	1,215,850.67 (397,663.19 to 2,505,379.89)	33.16 (10.39 to 69.96)	1,715,381.13 (454,452.62 to 3,713,240.37)	21.51 (5.58 to 47.20)	0.09 (0.02 to 0.19)	−1.53 (−1.57 to −1.50)
China	512,346.4 (233,687.7 to 853,907.61)	68.52 (27.97 to 119.98)	788,587.8 (307,537.43 to 1,385,711.49)	42.46 (15.40 to 76.86)	0.15 (0.06 to 0.28)	−1.45 (−1.55 to −1.35)
Gender (China)						
Male	311,061.16 (151,828.83 to 490,285.48)	89.96 (39.84 to 150.10)	524,270.43 (225,211.66 to 876,595.25)	62.57 (24.39 to 109.97)	0.17 (0.07 to 0.29)	−0.96 (−1.08 to −0.83)
Female	201,285.24 (73,041.74 to 371,881.92)	52.16 (17.36 to 100.50)	264,317.36 (67,441.92 to 539,511.81)	26.86 (6.52 to 55.72)	0.12 (0.03 to 0.24)	−2.18 (−2.29 to −2.07)
Socio-demographic Index (SDI)						
High SDI	3772.18 (370.52 to 4293.03)	52.22 (22.38 to 88.22)	4575.81 (1609.17 to 6936.32)	19.16 (6.52 to 36.28)	19.16 (0.05 to 0.28)	−3.56 (−3.71 to −3.02)
High-middle SDI	20,309.93 (9732.94 to 28,932.90)	76.44 (30.92 to 132.48)	31,137.83 (12,152.52 to 49,705.38)	42.89 (14.48 to 84.42)	0.15 (0.05 to 0.27)	−1.88 (−2.23 to −1.63)
Middle SDI	16,908.47 (7177.02 to 28,677.93)	67.49 (25.61 to 121.99)	26,349.53 (9721.99 to 54,485.84)	48.46 (17.31 to 91.24)	5.98 (0.05 to 0.27)	−0.93 (−1.22 to −0.66)
Low-middle SDI	6740.50 (2819.47 to 7834.64)	87.39 (38.45 to 154.12)	10,841.83 (3600.87 to 20,705.91)	61.33 (23.33 to 111.58)	0.16 (0.06 to 0.28)	−0.74 (−0.88 to −0.60)

ASMR = age-standardized mortality rate; PAF = population attributable fraction; EAPC = estimated annual percentage change.

At the provincial level, the CVD deaths attributable to HSI mainly occurred in Shandong, Henan, Hebei, and Sichuan, sharing more than 30% of the deaths in China in both 1990 and 2019. The fractional contribution of total CVD deaths caused by his ranged from 0.09% to 0.23% in 2019. Specifically, Tibet, Xinjiang, and Zhejiang were among the top three, while Gansu, Guangxi, and Hainan were ranked in the bottom three in terms of PAFs of HSI-related CVD deaths. The highest ASMR of CVD attributable to HSI was observed in Tibet at 105.86 (95% CI: 73.14, 265.75), Xinjiang at 82.32 (95% CI: 35.03, 142.37), and Qinghai at 71.14 (95% CI: 30.04, 124.28) per 100,000 population in 2019. In contrast, the lowest ASMR was found in Hong Kong at 13.9 (95% CI: 4.64, 27.92), Macau at 16.9 (95% CI: 5.64, 31.78), and Shanghai at 16.91 (95% CI: 5.31, 33.04) per 100,000 population in 2019. What is more, the heat map indicated that in 2019, the high values of ASMR were mostly concentrated in the provinces located in western and northern inland areas. On the contrary, the eastern and southern coastal regions of China exhibited generally lower ASMR values. During 1990 to 2019, the EAPC in ASMR of HSI-induced CVD decreased significantly in almost all provinces, and the most conspicuous decline was seen in Beijing (EAPC: −4.47, 95% UI: −4.78, −4.17), followed by Shanghai (EAPC: −3.85, 95% UI: −4.19, −3.5) and Jilin (EAPC: −3.23, 95% UI: −3.52, −2.94), while an upward tendency for ASMR was only observed in Gansu (EAPC: 0.56, 95% UI: 0.37, 0.75) and Xinjiang (EAPC: 0.38, 95% UI: 0.26, 0.5) (Supplementary Table S2; Figure 1A,C).

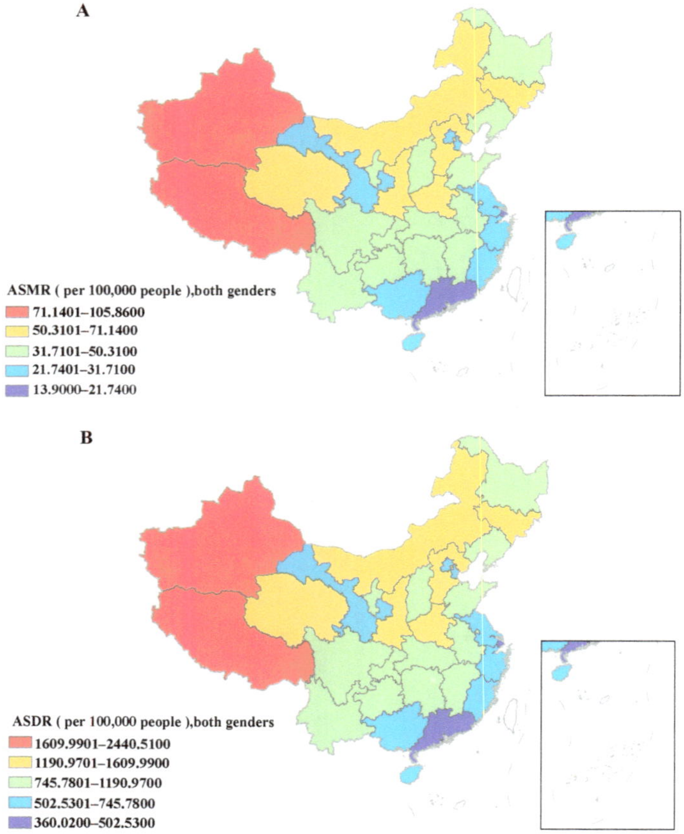

Figure 1. *Cont.*

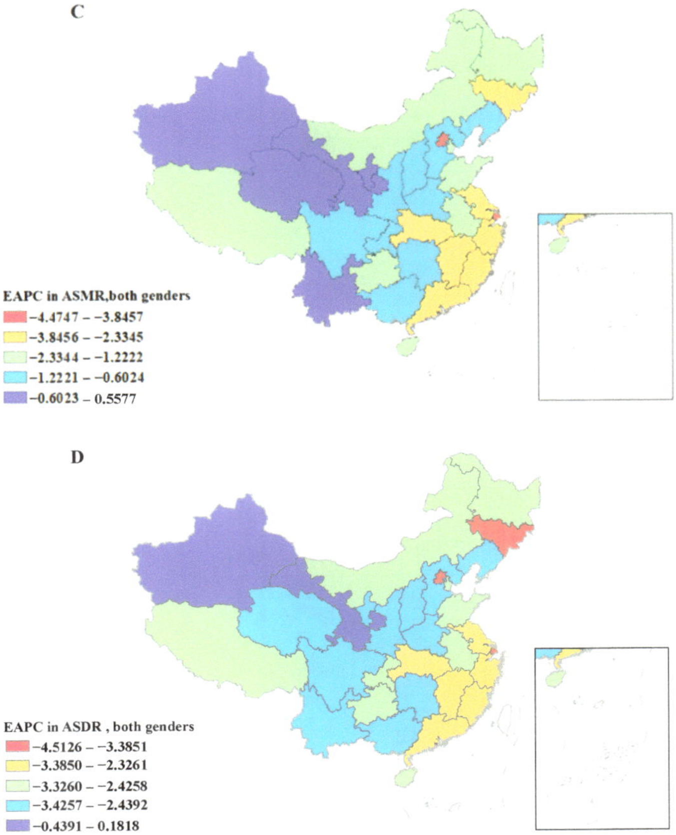

Figure 1. The burden of cardiovascular disease due to high sodium intake for both genders. (**A**) The spatial distribution of ASMR in 2019. (**B**) The spatial distribution of ASDR in 2019. (**C**) The EAPC in ASMR from 1990 to 2019. (**D**) The EAPC in ASDR from 1990 to 2019. ASMR, age-standardized mortality rate; DALYs, disability-adjusted life years; ASDR, age-standardized DALYs rate; EAPC, estimated annual percentage change.

3.2. DALYs and ASDR of CVD Attributable to HSI

The global DALYs of high sodium-related CVD had an increase of 33.06% during the last three decades, from 304,684.79 in 1990 to 405,406.75 in 2019. Meanwhile, the ASDR decreased from 756.66 (95% CI: 262.0, 1495.6) per 100,000 person-years to 490.68 (95% CI: 146.3, 1010.7) per 100,000 person-years, with an average annual decline of 1.54% (95% CI: −1.57, −1.50).

In China, the estimated number of CVD DALYs attributable to HSI was 138,523.85 in 2019, marking a 39.39% rise from 193,086.28 in 1990. Still, the ASDR witnessed a decline from 1582.78 (95% CI: 751.7, 2585.9) per 100,000 person-years to 954.80 (95% CI: 425.7, 1601.0) per 100,000 person-years, and the annual trends of changes in ASDR were somewhat faster than the global level, with an EAPC of −1.61 (95% CI: −1.68, −1.53). Similar patterns in deaths and ASMR could also be observed in DALYs and ASDR in terms of gender. That is, males experienced a greater load of DALYs and ASDR and had a much slower decline in ASDR compared with females, with an EAPC of −1.13 (95% CI: −1.22, −1.03) in males versus an EAPC of −2.44 (95% CI: −2.53, −2.35) in females from 1990 to 2019.

At the SDI region level, the situation in DALYs, ASDR, and EAPC in ASDR was similar to the case in deaths, ASMR, and EAPC in ASMR of CVD caused by HSI we described before (Table 2).

Table 2. DALYs and ASDR of cardiovascular disease due to high sodium intake in 1990 and 2019 and their percentage changes from 1990 to 2019.

Characteristics	1990		2019		1990–2019	
	DALYs, No. (95% UI)	ASDR per 100,000 No. (95% UI)	DALYs, No. (95% UI)	ASDR per 100,000 No. (95% UI)	PAFs % (95% UI)	EAPC (%) in ASDR No. (95% CI)
Global	30,468,479.33 (10,829,442.92 to 59,419,298)	756.66 (262.03 to 1495.60)	40,540,674.51 (12,262,912.95 to 83,022,361.27)	490.68 (146.27 to 1010.73)	0.10 (0.03 to 0.21)	−1.54 (−1.57 to −1.50)
China	13,852,385.09 (6,916,321.04 to 22,011,605.74)	1582.78 (751.65 to 2585.91)	19,308,628.39 (8,894,786.58 to 31,852,222.15)	954.80 (425.86 to 1600.97)	0.19 (0.09 to 0.32)	−1.61 (−1.68 to −1.53)
Gender						
Male	8,608,041.4 (4,574,858.95 to 13,276,567.48)	2008.91 (1004.58 to 3173.03)	13,190,360.85 (6,551,116.67 to 21,235,740.84)	1358.90 (646.75 cto 2217.05)	0.22 (0.11 to 0.34)	−1.13 (−1.22 to −1.03)
Female	5,244,343.69 (2,247,167.81 to 9,012,808.73)	1196.65 (482.56 to 2093.97)	6,118,267.54 (1,979,661.12 to 11,147,594.56)	587.74 (185.00 to 1082.24)	0.15 (0.05 to 0.28)	−2.44 (−2.53 to −2.35)
Socio-demographic Index (SDI)						
High SDI	96,492.83 (88,944.24 to 198,188.75)	1185.88 (415.09 to 1959.03)	109,499.06 (62,723.97 to 218,135.01)	456.96 (205.66 to 826.28)	0.19 (0.09 to 0.31)	−3.33 (−3.58 to −3.07)
High-middle SDI	542,598.62 (286,974.19 to 869,579.72)	1820.67 (876.75 to 3011.52)	770,508.40 (350,208.69 to 1,320,252.39)	971.97 (378.71 to 1660.43)	0.19 (0.08 to 0.32)	−2.02 (−2.28 to −1.84)
Middle SDI	461,382.41 (220,529.43 to 851,330.99)	1591.61 (778.84 to 2691.05)	638,164.44 (290,164.88 to 989,275.51)	1074.08 (437.44 to 1876.64)	0.19 (0.09 to 0.32)	−1.20 (−1.41 to −0.95)
Low-middle SDI	192,364.27 (118,846.97 to 325,071.57)	2185.11 (1025.61 to 3633.15)	274,191.05 (106,753.79 to 491,301.09)	1412.95 (638.08 to 2380.01)	0.20 (0.09 to 0.32)	−1.10 (−1.23 to −0.98)

ASMR = age-standardized mortality rate; PAF = population attributable fraction; EAPC = estimated annual percentage change. At the provincial level, it was still Shandong, Henan, Hebei, and Sichuan that had the majority of CVD DALYs attributable to HSI, accounting for over 30% of the DALYs in China in both 1990 and 2019. The proportion of total CVD DALYs resulting from HSI ranged from 0.12% to 0.27% in 2019. Tibet, Xinjiang, and Zhejiang maintained their positions as the top three, while Gansu, Guangxi, and Hainan remained in the bottom three for PAFs. The highest ASDR of CVD attributable to HSI was observed in Tibet at 2440.51 (95% CI: 1274.33, 3772.27), Xinjiang at 1877.72 (95% CI: 934.52, 3041.33), and Hebei at 1609.99 (95% CI: 721.61, 2644.41) per 100,000 person-years in 2019. The lowest ASDR was detected in Hong Kong at 360.02 (95% CI: 149.85, 648.54), Shanghai at 381.45 (95% CI: 159.6, 662.46), and Macau at 442.08 (95% CI: 186.17, 730.45) per 100,000 person-years in 2019. Additionally, the heat map showed that China's regional distribution variations in ASDR of CVD due to HSI were basically consistent with those in ASMR. What is also similar to the findings in ASMR is that almost all provinces had significant drops in ASDR of CVD attributable to HSI. Between 1990 and 2019, the top three still were Beijing (EAPC: −4.51, 95% UI: −4.81, −4.21), Shanghai (EAPC: −3.59, 95% UI: −3.89, −3.29), and Jilin (EAPC: −3.39, 95% UI: −3.65, −3.12), while ASDR in Gansu (EAPC: 0.18, 95% UI: 0.02, 0.34) continuously rose and remained stable in Xinjiang (EAPC: 0.07, 95% UI: −0.08, 0.23) (Supplementary Table S3; and Figure 1B,D).

3.3. Deaths and ASMR of CVD Attributable to HSI by Age and Gender

In 2019, the number of CVD deaths attributable to HSI peaked in groups aged 70–74 in both sexes, and it remained higher in males than in females until the age of 89, whereas the opposite was true for those aged 90 and above. The ASMR of CVD attributable to HSI approximated linear growth with age in groups younger than 90–94 years for both sexes (Ptrend < 0.05). In all age groups, the ASMR for males is higher than that for females, and the disparity between genders intensifies with age (Figure 2A). Between 1990 and 2019, the EAPC in ASMR showed a "V" relationship with age in each SDI region. Specifically, in low-middle and middle SDI regions, the trend of ASMR was downward in individuals aged below 74 and 84 and upward in individuals aged above 75 and 85, respectively, but the sharpest increase in those aged 85–89 and the steepest decrease in those aged 50–54 were commonly observed in both regions; in high-middle and high SDI regions, the EAPC in ASMR was below 0 across all age groups, and the most decline occurred in those aged 55–59 in both regions, with the latter regions having a larger decrease (Figure 3A).

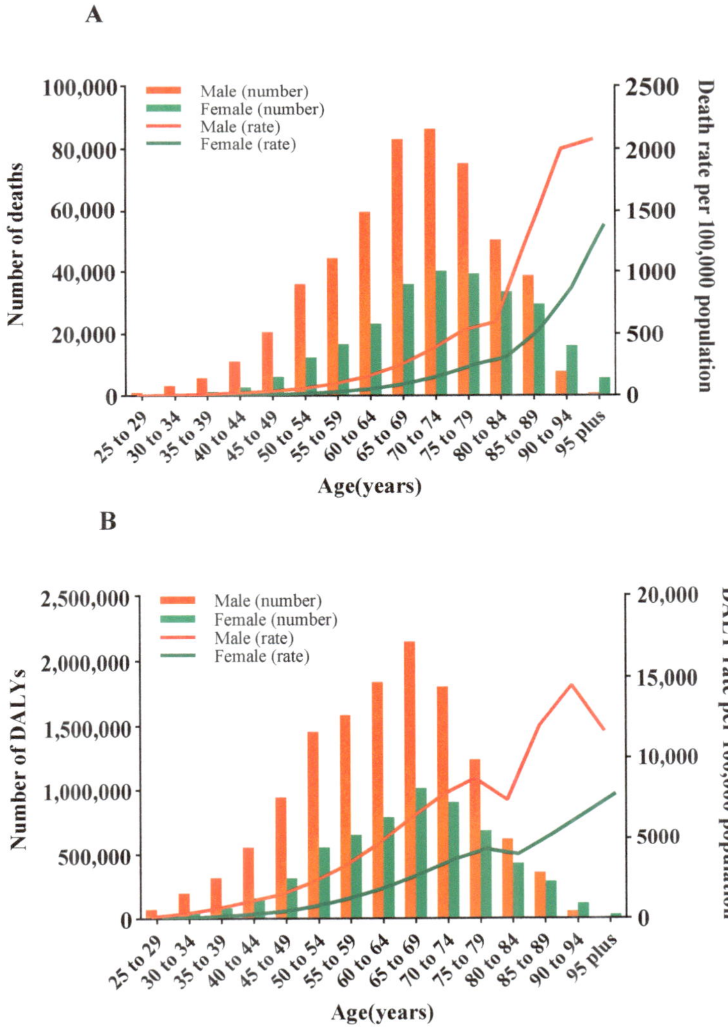

Figure 2. Numbers and rates of deaths and DALYs of cardiovascular disease due to high sodium intake across age groups by gender in 2019. (**A**) Deaths. (**B**) DALYs. DALYs are disability-adjusted life years.

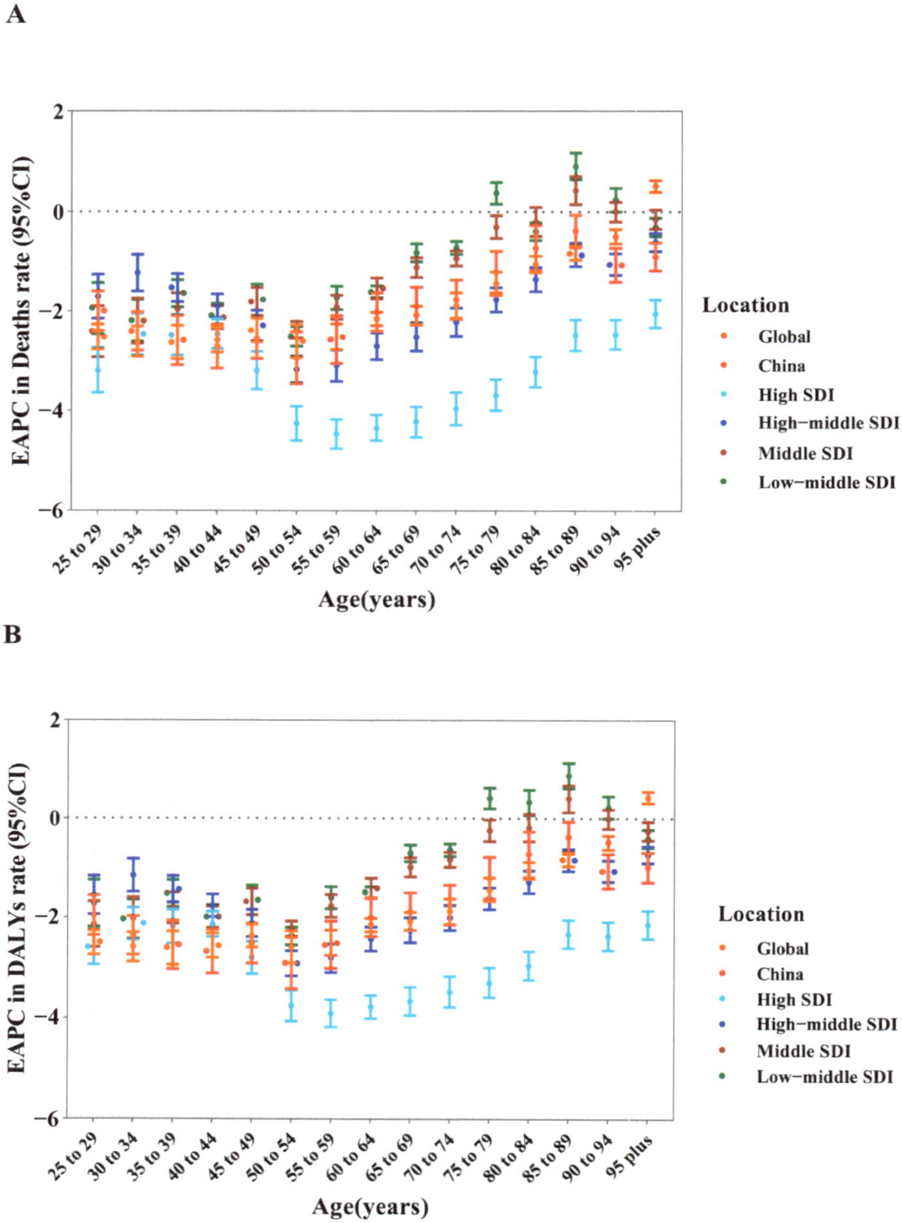

Figure 3. The temporal trends in burden of cardiovascular disease due to high sodium intake across age groups by location, 1990–2019. (**A**) EAPC in mortality rate. (**B**) EAPC in DALYs rate. EAPC, estimated annual percentage change. DALYs are disability-adjusted life years.

3.4. DALYs and ASDR of CVD Attributable to HSI by Age and Gender

In 2019, the number of CVD DALYs attributable to HSI peaked in groups aged 65–69, regardless of gender, and it followed the pattern in deaths for males and females aged under and over 90. The ASDR of CVD attributable to HSI exhibited an elevated trend with age in groups younger than 90–94 years for both sexes and then decreased in males but continually increased in females (Ptrend < 0.05) (Figure 2B). At the SDI regional level, the temporal trend of ASDR in age groups resembled the pattern we described earlier for ASMR (Figure 3B).

3.5. The Association between ASMR and ASDR of CVD Attributable to HSI and SDI

In general, the ASMR of CVD attributable to HSI had non-linear associations with SDI from 1990 to 2019. The ASMR declined with the increment of SDI over time when the index was below 0.4 or above 0.7, while it tended to level off when SDI was between 0.4 and 0.7. The ASMR in the provinces of Xinjiang and Gansu steadily increased with the growth of SDI during the observation period. In spite of the overall downward trend of ASMR in Beijing, Tianjin, Guizhou, Hebei, Henan, Heilongjiang, Jilin, Liaoning, Shaanxi, Shandong, Inner Mongolia, Tibet, Xinjiang, and Qinghai, these regions still showed higher ASMR than expected based on comparisons of SDI for all years. Conversely, the ASMR in other provinces was below or at the same level as the expected values for all years. In addition, ASMR exceeded the expected values in examined the early stages of the time series for Hubei, Jiangxi, and Anhui, but it decreased as SDI increased. The relationship between ASDR and SDI shared the same patterns as that between ASMR and SDI (Figure 4).

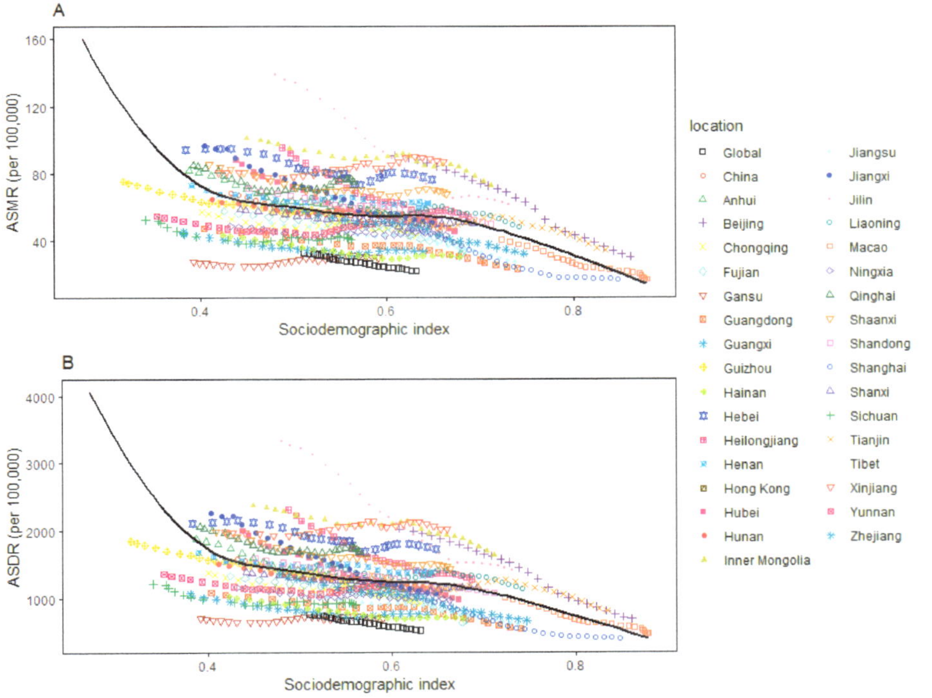

Figure 4. The burden of cardiovascular disease due to high sodium intake across 33 provinces by SDI for both genders, 1990–2019. (**A**) The correlation between ASMR and SDI. (**B**) The correlation between ASDR and SDI. SDI, sociodemographic index; ASMR, age-standardized mortality rate; ASDR, age-standardized DALY rate.

A strong negative correlation was detected between the EAPC in ASRs of deaths and DALYs of and the SDI in 2019 ($\rho = -0.659$, $p < 0.001$ and $\rho = -0.558$, $p < 0.001$, respectively) (Figure 5C,D). However, no significant correlation was found between the EAPC in ASRs of deaths and DALYs and the ASMR or ASDR of CVD in 1990 (ρ = −0.123, *p* = 0.494 and ρ = −0.161, *p* = 0.370, respectively) (Figure 5A,B).

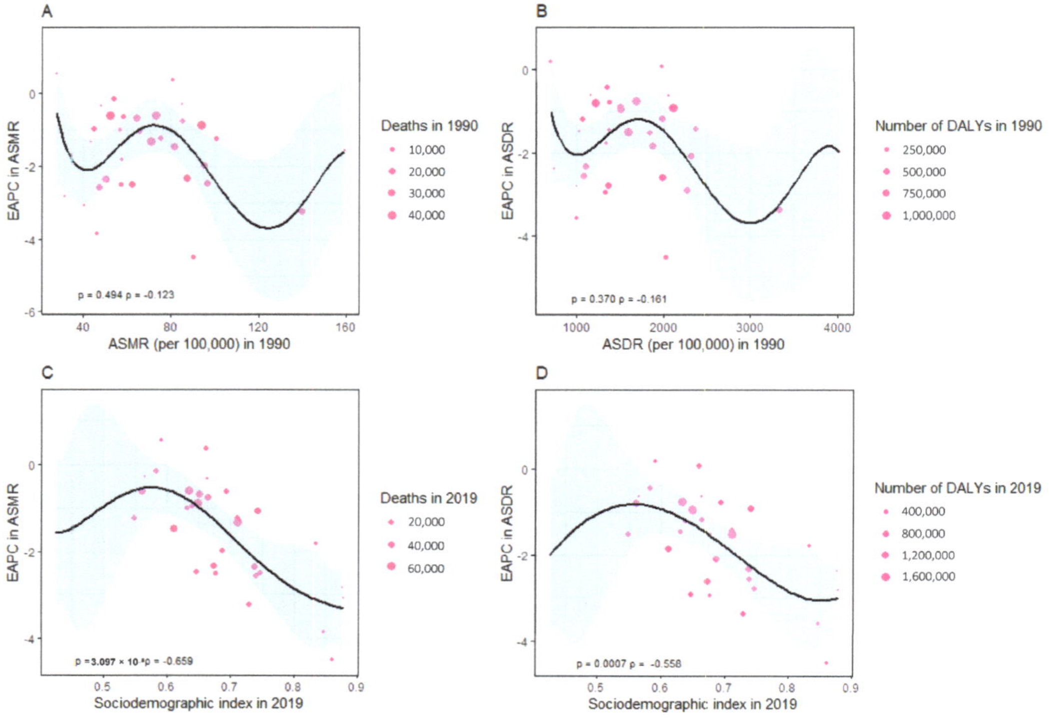

Figure 5. Factors influencing the EAPC. (**A**) The association between EAPC and ASMR in 1990. (**B**) The association between EAPC in ASDR and ASDR in 1990. (**C**) The association between EAPC in ASMR and SDI in 2019. (**D**) The association between EAPC in ASDR and SDI in 2019. Provinces were depicted by circles, with the size of each increasing according to the number of cardiovascular disease-related deaths and DALYs due to high sodium intake. EAPC, estimated annual percentage change; ASMR, age-standardized mortality rate; ASDR, age-standardized DALYs rate; DALYs, disability-adjusted life years.

4. Discussion

Given the prevalent high-sodium diets and the observed patterns of exposure, this article exerts the first large-scale effort to conduct a comprehensive assessment of the epidemiological trends of CVD burden caused by HSI in China's 33 provincial administrative units from 1990 to 2019, with a view to providing scientific evidence and thoughtful insights needed for efficiency in preventing the disease at a low cost.

In 2019, China presented a higher proportion of all CVD-related deaths and DALYs attributed to HSI compared to the global average. Salt has always maintained a difficult-to-shake presence in the dietary habits of Chinese residents, with nearly 70% of their intake of sodium originating from salt added during daily food preparation [17]. Despite a decline in salt intake from 16 g/d in 1991 to 10.4 g/d in 2015, credited to salt reduction efforts and health education campaigns, the issue of excessive-sodium diets remains problematic across China, considering the recommended upper limit given by the WHO [22]. China may need to make necessary adaptations based on current disease epidemiological trends associated with HSI exposure, seeking strategies that benefit the most widespread and vulnerable populations.

Our finding that, across all age groups, the CVD burden attributable to HSI was heavier in males than in females supports the observations in previous studies [23]. Males have higher intake and excretion of sodium compared to females, as reported in presumed studies, suggesting that levels of sodium exposure in the internal environment between males and females may well account for gender differences in HSI-related burdens [24–26]. Generally, men are more likely to engage in unhealthy lifestyles. A study conducted aboard deduced that males are more inclined to choose convenient dining options such as fast food and takeout [27]. In the domestic context of the evolving modern food system and modern marketing, males are also found to have a higher frequency of dining out for business and social purposes [28]. In fact, food eaten out is usually characterized by an unfavorable dietary profile. The sodium levels in Chinese restaurant dishes were revealed to be extremely high, with a single serving dish containing an average of 3331.2 $\pm$ 4156.9 mg sodium, which is almost 1.7 times the proposed intake for preventing non-communicable chronic diseases for Chinese adults [29]. In addition, social smoking and drinking, an entrenched part of Chinese culture, are also more prevalent among males, possibly as a result of social role differentiation. Cigarette and alcohol consumption, as identified risk factors, could also exacerbate the pathogenesis of CVD when combined with unhealthy diets. Another reason for lower HSI-related burdens in females is that they seem to be more efficient at maintaining Na^+ homeostasis during acclimatization to dietary Na^+ challenges, thereby reducing their vulnerability to the adverse effects of HSI exposure [30]. Premenopausal estrogen levels in females were also found to have a protective effect against CVD [31]. Therefore, gender-based sodium reduction policies in this regard should be urgently needed.

Although the ASMR and ASDR suggested a softened trend of CVD attributable to HSI from 1990 to 2019, the absolute number of its deaths and DALYs has increased substantially throughout the years in China. The growth and accelerated aging of the population may partly explain the findings. The ratio of China's elderly, aged 60 and above, was reported to continually expand in the context of the country's population growing from 1.1 billion to 1.4 billion, accounting for approximately 11.5% of the total in 2019, with expectations of reaching 16.9% by 2030 [18,32]. The aging population could act as a pivotal factor in the overall burden of diseases, including CVD caused by high-salt diets. We further explored that there was a greater burden of HSI-related CVD in the elderly compared to the youngsters, the explanations of which may be boiled down to several factors. Weakened blood vessel elasticity and heart function with aging could potentially predispose the elderly to the negative effects of HSI. Additionally, the implemented sodium reduction measures in China's select counties and provinces may produce more pronounced effects on younger individuals. As for older adults, who have a consistent diet and less food consumption over a long period of time, these interventions could not significantly reduce their cumulative burden [33]. When the mitigation of the HSI-induced CVD burden was insufficient to counteract the impact of the worsening graying, a net increase in the burden would not be surprising.

Variations in the trends of ASMR and ASDR during 1990 to 2019 at the provincial level have been observed, showing complex associations with factors regarding nature and social setting. In 2019, Tibet was hardest hit by the HSI-induced CVD burden. The Tibet autonomous region is located on the Qinghai-Tibet Plateau, the highest plateau in the world with an average altitude of more than 4500 m, where most of the Tibetans are concentrated. A previous study among Tibetans exhibited a 2% increase in the prevalence of hypertension for every 100 m elevation, indicating such an environment is inherently to the detriment of subsequent cardiovascular outcomes [34]. Based on the unique climatic and geographical conditions there, the local inhabitants have dietary patterns peculiar to their group that feature a high proportion of meat, dairy products, and grain-based foods but a low proportion of leafy green vegetables and fruits, usually involving high levels of sodium but low levels of potassium [35]. This heightened sodium-to-potassium ratio is particularly concerning given the evidence that the synergistic effect of diets high in Na^+

and low in K^+ is greater than that of each per se on the pathogenesis of hypertension or CVD [36]. The lowest burden was found in Hong Kong and Macau. In fact, these places are also noted for people's longevity. In addition to a sound healthcare system and strong public health measures, the populace is relatively well-educated, enabling them to have a full awareness of CVD and its risk factors and make informed behavioral changes [37]. This can be mirrored in their health-oriented diet to some extent. Residents there prefer to cook the ingredients in a more simplified way, like stewing or steaming, minimizing excessive processing and nutrient loss. Fresh fish, vegetables, and fruits, as essentials of local dietary preferences, have shown significant beneficial effects in reducing CVD risks in Hong Kong Chinese samples [38]. The result that the burden of CVD attributable to HSI was strongly contrasting between northern and Southern China confirmed the same finding in previous studies [39]. In terms of lifestyle issues, levels of sodium intake appear to play a dominant role in shaping the north–south differential for disease, with residents' average daily sodium intake in Northern China being 2.32 g higher than that in Southern China [40]. In terms of environmental factors, ambient temperature and air quality could also influence CVD. While both high and low extremes increase mortality, the low-temperature effects on the disease persist for a longer duration [41]. The temperature difference between the two sides of the north–south boundary can be much greater in winter than in summer, fostering the contrasted disease burden. Additionally, due to extensive coal consumption driven by heavy industry development or winter indoor heating, the northern regions of China experience higher concentrations of PM 2.5 particles [42]. Interactions between PM 2.5 and HSI may co-exist, whereby prolonged exposure to PM 2.5 exacerbates the effect of HSI on CVD.

Wide disparities existed in HSI-caused CVD burden among regions with different SDI quintiles, reflecting disequilibrium in sodium intake control and medical healthcare. In our study, the ASRs of deaths and DALYs were comparatively the lowest in high SDI regions from 1990 to 2019. That makes sense, as residents in these regions have greater chances to avail themselves of quality education, well-established medical systems, and policy priorities, all of which are conducive to easing the disease burden. Interestingly, the correlation between the CVD burden and SDI was not monotonic or linear. In particular, the downturn in declines in ASRs of deaths and DALYs in low-middle SDI regions was more striking than that in middle SDI and high-middle SDI regions. This indicates that low-income regions, being the regions saddled with the heaviest health burden, are more likely to make a real difference by investing in diet enhancement. From 1990 to 2019, the annual decreasing trends in ASMR and ASDR were observed across all age groups for high-middle and high SDI regions, while the increasing trends were found at the end of the scale for low-middle and middle SDI regions. However, EAPC in ASMR or ASDR in different SDI regions uniformly showed a linear positive correlation with age $\geq$60 years, which may be due in part to the prolonged life expectancy by virtue of the reduction in deaths from common infectious diseases, leaving the number of elderly individuals much higher. Concerted efforts are required to address the outstanding CVD burden across China, especially in regions with low SDI and high sodium diets.

Aggregated information provided by representative datasets instead of the details of the total number and geographical distribution of those individuals allowed us to understand the broader epidemiological patterns and their overall impact of HSI on cardiovascular health outcomes across China. The interpretation of our study is still limited in several ways. Firstly, the analysis was reliant on the quantity and quality of the acquired data. Even though a good-quality vital registration system in China has provided adequate data on CVD mortality and morbidity in provinces, data for non-fatal outcomes of CVD may still be underreported for remote and poorer areas. Secondly, the profile of CVD burden in the urban–rural stratification has not been evaluated and examined due to data gaps. Thirdly, only individuals aged 25 and older were included in this study, while the details for younger groups are unavailable. Fourthly, the sodium-to-potassium ratio, which is detrimental for CVD onset, might potentially influence our findings, given its decrease from 4.1 to 3.1 between 1991 and 2015 in China [43]. Nevertheless, this decline was predominantly driven by a reduction in sodium intake, while potassium intake exhibited relative stability. Consequently, its effect on our results is expected to be slight. Fifthly, this study adopted a global perspective to analyze the CVD burden caused by HSI in China and underscored the benefits of nationwide salt reduction. Ecological studies may not establish causality and provide guidance at the individual level. Lastly, the statistical significance of the relationship between HSI and CVD might be affected by potential confounders, such as medication use and accessibility to medical resources, and interactions among dietary components that could not be adjusted promptly on account of data availability.

5. Conclusions

In summary, this study is grounded in a sizable and varied population hailing from different regions, ensuring sufficient statistical power for the analysis. This study demonstrated the profile of the CVD burden attributable to a high-sodium diet among Chinese residents between 1990 and 2019. Although a pronounced reduction in HSI-related CVD burden was achieved, the increasing number of deaths and DALYs still makes it an incremental public health issue over the past three decades. Males and the elderly bore a heavier disease burden in the context of the demographic transition. Diet management holds enormous potential to enhance public health, particularly in low-income regions. These findings call for prioritized public health interventions that are evidence-based and data-driven.

Supplementary Materials: The following supporting information can be downloaded at: https://www.mdpi.com/article/10.3390/nu16091307/s1, Table S1: Codes in ICD-10 for major CVD subcategories; Table S2: Deaths and ASMR of cardiovascular disease attributable to high sodium intake in 1990 and 2019 and the temporal trends from 1990 to 2019; Table S3: DALYs and ASDR of cardiovascular disease attributable to high sodium intake in 1990 and 2019 and the temporal trends from 1990 to 2019.

Author Contributions: L.J.: protocol draft and original paper preparation; W.S. and A.W.: statistical analysis and interpretation of results. H.F., Q.W., H.L. and S.L.: data collection and basic analysis. Y.S. and A.L.: revisions of the manuscript. All authors have read and agreed to the published version of the manuscript.

Funding: We acknowledge funding from the National Key R&D Program of China (No. 2022YFF1100400), Nantong Municipal Science and Technology Bureau, Jiangsu Province (MS12021054), and Local High-level University (cultivation) project in Shanghai (E1-2602-21-201006-6).

Institutional Review Board Statement: The study protocol was reviewed and approved by the Ethics Committee of the China National Center for Food Safety Risk Assessment.

Informed Consent Statement: Informed consent was waived because no identifiable information was included in the analyses.

Data Availability Statement: The authors declare that the data and material of this study are available for reasonable request. The data are not publicly available due to information restriction.

Acknowledgments: The writing team thanks the Institute of Health Metrics and Evaluation for providing the available deidentified data and the China National Center for Food Safety Risk Assessment for assistance.

Conflicts of Interest: The authors declare no conflict of interest.

Abbreviations

Cardiovascular disease (CVD); Global Burden of Disease Study (GBD); High sodium intake (HSI); disability-adjusted life years (DALYs); Years of life lost (YLLs); Years lived with disability (YLD); age-standardized mortality rate (ASMR); age-standardized DALYs rate (ASDR); age-standardized rates (ASRs); population-attributable fraction (PAFs); estimated annual percentage change (EAPC); socio-demographic index (SDI); Confidence Interval

(CI); Uncertainty Intervals (UIs); World Health Organization (WHO); China National Nutrition and Health Survey (CNNHS); National Center for Food Safety Risk Assessment (NFSA); Special Administrative Regions (SAR); International Statistical Classification of Diseases and Related Health Issues (ICD).

References

1. Vos, T.; Lim, S.S.; Abbafati, C.; Abbas, K.M.; Abbasi, M.; Abbasifard, M.; Abbasi-Kangevari, M.; Abbastabar, H.; Abd-Allah, F.; Abdelalim, A.; et al. Global burden of 369 diseases and injuries in 204 countries and territories, 1990–2019: A systematic analysis for the Global Burden of Disease Study 2019. *Lancet* **2020**, *396*, 1204–1222. [CrossRef] [PubMed]
2. Zou, Z.; Cini, K.; Dong, B.; Ma, Y.; Ma, J.; Burgner, D.P.; Patton, G.C. Time Trends in Cardiovascular Disease Mortality across the BRICS: An Age-Period-Cohort Analysis of Key Nations with Emerging Economies Using the Global Burden of Disease Study 2017. *Circulation* **2020**, *141*, 790–799. [CrossRef] [PubMed]
3. Du, X.; Patel, A.; Anderson, C.S.; Dong, J.; Ma, C. Epidemiology of Cardiovascular Disease in China and Opportunities for Improvement. *J. Am. Coll. Cardiol.* **2019**, *73*, 3135–3147. [CrossRef] [PubMed]
4. Zhang, J.; Tong, H.; Jiang, L.; Zhang, Y.; Hu, J. Trends and disparities in China's cardiovascular disease burden from 1990 to 2019. *Nutr. Metab. Cardiovasc. Dis.* **2023**, *33*, 2344–2354. [CrossRef] [PubMed]
5. Hu, S.-S. Report on cardiovascular health and diseases in China 2021: An updated summary. *J. Geriatr. Cardiol.* **2023**, *20*, 399–430. [CrossRef] [PubMed]
6. Moran, A.; Gu, D.; Zhao, D.; Coxson, P.; Wang, Y.C.; Chen, C.-S.; Liu, J.; Cheng, J.; Bibbins-Domingo, K.; Shen, Y.-M.; et al. Future Cardiovascular Disease in China: Markov Model and Risk Factor Scenario Projections from the Coronary Heart Disease Policy Model—China. *Circ. Cardiovasc. Qual. Outcomes* **2010**, *3*, 243–252. [CrossRef] [PubMed]
7. Mozaffarian, D.; Wilson, P.W.F.; Kannel, W.B. Beyond Established and Novel Risk Factors: Lifestyle Risk Factors for Cardiovascular Disease. *Circulation* **2008**, *117*, 3031–3038. [CrossRef] [PubMed]
8. Mozaffarian, D.; Fahimi, S.; Singh, G.M.; Micha, R.; Khatibzadeh, S.; Engell, R.E.; Lim, S.; Danaei, G.; Ezzati, M.; Powles, J. Global Sodium Consumption and Death from Cardiovascular Causes. *N. Engl. J. Med.* **2014**, *371*, 624–634. [CrossRef] [PubMed]
9. O'Donnell, M.J.; Yusuf, S.; Mente, A.; Gao, P.; Mann, J.F.; Teo, K.; McQueen, M.; Sleight, P.; Sharma, A.M.; Dans, A.; et al. Urinary Sodium and Potassium Excretion and Risk of Cardiovascular Events. *Jama* **2011**, *306*, 2229–2238. [CrossRef]
10. O'Donnell, M.; Mente, A.; Rangarajan, S.; McQueen, M.J.; Wang, X.; Liu, L.; Yan, H.; Lee, S.F.; Mony, P.; Devanath, A.; et al. Urinary Sodium and Potassium Excretion, Mortality, and Cardiovascular Events. *N. Engl. J. Med.* **2014**, *371*, 612–623. [CrossRef]
11. Neal, B.; Wu, Y.; Feng, X.; Zhang, R.; Zhang, Y.; Shi, J.; Zhang, J.; Tian, M.; Huang, L.; Li, Z.; et al. Effect of Salt Substitution on Cardiovascular Events and Death. *N. Engl. J. Med.* **2021**, *385*, 1067–1077. [CrossRef] [PubMed]
12. Grillo, A.; Salvi, L.; Coruzzi, P.; Salvi, P.; Parati, G. Sodium Intake and Hypertension. *Nutrients* **2019**, *11*, 1970. [CrossRef] [PubMed]
13. Cook, N.R.; Appel, L.J.; Whelton, P.K. Lower Levels of Sodium Intake and Reduced Cardiovascular Risk. *Circulation* **2014**, *129*, 981–989. [CrossRef] [PubMed]
14. Thomas, M.C.; Moran, J.; Forsblom, C.; Harjutsalo, V.; Thorn, L.; Ahola, A.; Wadén, J.; Tolonen, N.; Saraheimo, M.; Gordin, D.; et al. The Association Between Dietary Sodium Intake, ESRD, and All-Cause Mortality in Patients with Type 1 Diabetes. *Diabetes Care* **2011**, *34*, 861–866. [CrossRef] [PubMed]
15. WHO Global Report on Sodium Intake Reduction. Available online: https://www.who.int/publications-detail-redirect/9789240069985 (accessed on 1 August 2023).
16. Eckel, R.H.; Jakicic, J.M.; Ard, J.D.; De Jesus, J.M.; Houston Miller, N.; Hubbard, V.S.; Lee, I.-M.; Lichtenstein, A.H.; Loria, C.M.; Millen, B.E.; et al. 2013 AHA/ACC Guideline on Lifestyle Management to Reduce Cardiovascular Risk. *J. Am. Coll. Cardiol.* **2014**, *63*, 2960–2984. [CrossRef]
17. Fang, K.; He, Y.; Fang, Y.; Lian, Y. Dietary Sodium Intake and Food Sources among Chinese Adults: Data from the CNNHS 2010–2012. *Nutrients* **2020**, *12*, 453. [CrossRef] [PubMed]
18. Wang, H.; Abbas, K.M.; Abbasifard, M.; Abbasi-Kangevari, M.; Abbastabar, H.; Abd-Allah, F.; Abdelalim, A.; Abolhassani, H.; Abreu, L.G.; Abrigo, M.R.M.; et al. Global age-sex-specific fertility, mortality, healthy life expectancy (HALE), and population estimates in 204 countries and territories, 1950–2019: A comprehensive demographic analysis for the Global Burden of Disease Study 2019. *Lancet* **2020**, *396*, 1160–1203. [CrossRef] [PubMed]
19. Murray, C.J.L.; Aravkin, A.Y.; Zheng, P.; Abbafati, C.; Abbas, K.M.; Abbasi-Kangevari, M.; Abd-Allah, F.; Abdelalim, A.; Abdollahi, M.; Abdollahpour, I.; et al. Global burden of 87 risk factors in 204 countries and territories, 1990–2019: A systematic analysis for the Global Burden of Disease Study 2019. *Lancet* **2020**, *396*, 1223–1249. [CrossRef] [PubMed]
20. Li, M.; Li, X.; Zhao, Y.; Zhang, L.; Yang, J.; Zhou, M.; Wang, Z. The burden of ischemic heart disease and type 2 diabetes mellitus attributable to diet high in sugar-sweetened beverages in China: An analysis for the Global Burden of Disease Study 2017. *J. Diabetes* **2021**, *13*, 482–493. [CrossRef]
21. Chen, X.; Du, J.; Wu, X.; Cao, W.; Sun, S. Global burden attributable to high sodium intake from 1990 to 2019. *Nutr. Metab. Cardiovasc. Dis.* **2021**, *31*, 3314–3321. [CrossRef]

22. Liu, Z.; Man, Q.; Li, Y.; Yang, X.; Ding, G.; Zhang, J.; Zhao, W. Estimation of 24-hour urinary sodium and potassium excretion among Chinese adults: A cross-sectional study from the China National Nutrition Survey. *Am. J. Clin. Nutr.* **2024**, *119*, 164–173. [CrossRef]
23. Wang, K.; Jin, Y.; Wang, M.; Liu, J.; Bu, X.; Mu, J.; Lu, J. Global cardiovascular diseases burden attributable to high sodium intake from 1990 to 2019. *J. Clin. Hypertens.* **2023**, *25*, 868–879. [CrossRef] [PubMed]
24. Brown, I.J.; Tzoulaki, I.; Candeias, V.; Elliott, P. Salt intakes around the world: Implications for public health. *Int. J. Epidemiol.* **2009**, *38*, 791–813. [CrossRef]
25. Trieu, K.; Ospanova, F.; Tazhibayev, S.; Jewell, J.; Breda, J.; Santos, J.A.; Webster, J. Sodium and potassium intakes in the Kazakhstan population estimated using 24-h urinary excretion: Evidence for national action. *Eur. J. Nutr.* **2021**, *60*, 1537–1546. [CrossRef]
26. Afroza, U.; Abrar, A.K.; Nowar, A.; Akhtar, J.; Mamun, M.A.A.; Sobhan, S.M.M.; Cobb, L.; Ide, N.; Choudhury, S.R. Salt Intake Estimation from Urine Samples in South Asian Population: Scoping Review. *Nutrients* **2023**, *15*, 4358. [CrossRef] [PubMed]
27. McGuire, S.; Todd, J.E.; Mancino, L.; Lin, B.-H. The Impact of Food Away from Home on Adult Diet Quality. ERR-90, U.S. Department of Agriculture, Econ. Res. Serv., February 2010. *Adv. Nutr.* **2011**, *2*, 442–443. [CrossRef]
28. Ju, L.; Yu, D.; Guo, Q.; Fang, H.; Xu, X.; Li, S.; Cheng, X.; Piao, W.; Cai, S.; Zhao, L. Eating out behavior and its impact on obesity among Chinese residents aged 18–59 in 2015. *Wei Sheng Yan Jiu* **2021**, *50*, 395–400. [CrossRef]
29. Du, W.; Wang, H.; Zhang, J.; Zhang, X.; Wei, N.; Li, Y.; Tan, M.; Zhang, P.; He, F.J. Sodium content of restaurant dishes in China: A cross-sectional survey. *Nutr. J.* **2022**, *21*, 10. [CrossRef] [PubMed]
30. Johnston, J.G.; Speed, J.S.; Jin, C.; Pollock, D.M. Loss of endothelin B receptor function impairs sodium excretion in a time- and sex-dependent manner. *Am. J. Physiol.-Ren. Physiol.* **2016**, *311*, F991–F998. [CrossRef]
31. Kannel, W.B.; Levy, D. Menopause, Hormones, and Cardiovascular Vulnerability in Women. *Arch. Intern. Med.* **2004**, *164*, 479. [CrossRef]
32. World Population Prospects—Population Division—United Nations. Available online: https://population.un.org/wpp/ (accessed on 15 August 2023).
33. Chen, S.; Shan, L.C.; Tao, W.; Lu, T.; Regan, Á.; Han, H.; Guo, L.; Deng, T.; Wall, P. A survey of Chinese consumers' knowledge, beliefs and behavioural intentions regarding salt intake and salt reduction. *Public Health Nutr.* **2020**, *23*, 1450–1459. [CrossRef]
34. Mingji, C.; Onakpoya, I.J.; Perera, R.; Ward, A.M.; Heneghan, C.J. Relationship between altitude and the prevalence of hypertension in Tibet: A systematic review. *Heart* **2015**, *101*, 1054–1060. [CrossRef] [PubMed]
35. Li, K.; Zhang, Q.; Cai, H.; He, R.; Nima, Q.; Li, Y.; Suolang, D.; Cidan, Z.; Wangqing, P.; Zhao, X.; et al. Association of Tibetan Habitual Food and Metabolic Syndrome Among Tibetan People in China: A Cross-Sectional Study. *Front. Nutr.* **2022**, *9*, 888317. [CrossRef] [PubMed]
36. Cook, N.R.; Obarzanek, E.; Cutler, J.A.; Buring, J.E.; Rexrode, K.M.; Kumanyika, S.K.; Appel, L.J.; Whelton, P.K. Joint effects of sodium and potassium intake on subsequent cardiovascular disease: The trials of hypertension prevention follow-up study. *Arch. Intern. Med.* **2009**, *169*, 32–40. [CrossRef] [PubMed]
37. Lee, V.W.Y.; Fong, F.Y.H.; Cheng, F.W.T.; Yan, B.P.Y. Life's simple 7 and cardiovascular disease risk knowledge in Hong Kong. *BMC Cardiovasc. Disord.* **2019**, *19*, 185. [CrossRef] [PubMed]
38. Loke, A.Y.; Chan, K.N. Dietary habits of patients with coronary atherosclerosis: Case–control study. *J. Adv. Nurs.* **2005**, *52*, 159–169. [CrossRef]
39. Ma, L.-Y.; Chen, W.-W.; Gao, R.-L.; Liu, L.-S.; Zhu, M.-L.; Wang, Y.-J.; Wu, Z.-S.; Li, H.-J.; Gu, D.-F.; Yang, Y.-J.; et al. China cardiovascular diseases report 2018: An updated summary. *J. Geriatr. Cardiol. JGC* **2020**, *17*, 1.
40. Wang, M.; Huang, Y.; Song, Y.; Chen, J.; Liu, X. Study on Environmental and Lifestyle Factors for the North–South Differential of Cardiovascular Disease in China. *Front. Public Health* **2021**, *9*, 615152. [CrossRef]
41. Braga, A.L.F.; Zanobetti, A.; Schwartz, J. The time course of weather-related deaths. *Epidemiology* **2001**, *12*, 662–667. [CrossRef]
42. Zhang, S.; Routledge, M.N. The contribution of $PM_{2.5}$ to cardiovascular disease in China. *Environ. Sci. Pollut. Res.* **2020**, *27*, 37502–37513. [CrossRef]
43. Du, S.; Wang, H.; Zhang, B.; Popkin, B.M. Dietary potassium intake remains low and sodium intake remains high, and most sodium is derived from home food preparation for Chinese adults, 1991–2015 trends. *J. Nutr.* **2020**, *150*, 1230–1239. [CrossRef] [PubMed]

nutrients

Article

Selenium May Be Involved in Esophageal Squamous Cancer Prevention by Affecting GPx3 and FABP1 Expression: A Case-Control Study Based on Bioinformatic Analysis

Niannian Wang [1], Da Pan [1], Xiaopan Zhu [1], Xingyuan Ren [1], Xingyi Jin [1], Xiangjun Chen [1,2], Yuanyuan Wang [1], Ming Su [3], Guiju Sun [1] and Shaokang Wang [1,2,*]

[1] Key Laboratory of Environmental Medicine and Engineering, Ministry of Education, Department of Nutrition and Food Hygiene, School of Public Health, Southeast University, Nanjing 210009, China; wangnian-nian@foxmail.com (N.W.); pantianqi92@foxmail.com (D.P.); zhuxiaopandhr@163.com (X.Z.); mn13451537048@126.com (X.R.); xingyijin@foxmail.com (X.J.); cxjll910@163.com (X.C.); 230218460@seu.edu.cn (Y.W.); gjsun@seu.edu.cn (G.S.)

[2] Department of Public Health, School of Medicine, Xizang Minzu University, Xianyang 712000, China

[3] Huai'an District Center for Disease Control and Prevention, Huai'an 223001, China; suminglala@163.com

* Correspondence: shaokangwang@seu.edu.cn; Tel.: +86-153-1206-0480

Citation: Wang, N.; Pan, D.; Zhu, X.; Ren, X.; Jin, X.; Chen, X.; Wang, Y.; Su, M.; Sun, G.; Wang, S. Selenium May Be Involved in Esophageal Squamous Cancer Prevention by Affecting GPx3 and FABP1 Expression: A Case-Control Study Based on Bioinformatic Analysis. *Nutrients* **2024**, *16*, 1322. https://doi.org/10.3390/nu16091322

Academic Editors: Domenico Tafuri and Francesca Latino

Received: 17 March 2024
Revised: 17 April 2024
Accepted: 25 April 2024
Published: 28 April 2024

Correction Statement: This article has been republished with a minor change. The change does not affect the scientific content of the article and further details are available within the backmatter of the website version of this article.

Abstract: The role of selenium in the developmental process of esophageal cancer (EC) requires further investigation. To explore the relationship between selenium-related factors and EC through bioinformatic analysis, a case-control study was conducted to verify the results. Utilizing the GEPIA and TCGA databases, we delineated the differential expression of glutathione peroxidase 3 (GPx3) in EC and normal tissues, identified differentially expressed genes (DEGs), and a performed visualization analysis. Additionally, 100 pairs of dietary and plasma samples from esophageal precancerous lesions (EPLs) of esophageal squamous cancer (ESCC) cases and healthy controls from Huai'an district, Jiangsu, were screened. The levels of dietary selenium, plasma selenium, and related enzymes were analyzed using inductively coupled plasma mass spectrometry (ICP-MS) or ELISA kits. The results showed lower GPx3 expression in tumor tissues compared to normal tissues. Further analysis revealed that DEGs were mainly involved in the fat digestion and absorption pathway, and the core protein fatty acid binding protein 1 (FABP1) was significantly upregulated and negatively correlated with GPx3 expression. Our case-control study found that selenium itself was not associated with EPLs risk. However, both the decreased concentration of GPx3 and the increase in FABP1 were positively correlated with the EPLs risk (p for trend = 0.035 and 0.046, respectively). The different expressions of GPx3 and FABP1 reflect the potential of selenium for preventing ESCC at the EPLs stage. GPx3 may affect EC through FABP1, which remains to be further studied.

Keywords: esophagus cancer; esophageal precancerous lesions; selenium; GPx3; FABP1

1. Introduction

Esophageal cancer (EC) has garnered significant attention within the global cancer landscape due to its high incidence and mortality rates [1]. The absence of early and mid-stage neoplasm diagnosis, owing to its inconspicuous nature, poses significant challenges to achieving a favorable prognosis or managing mortality rates [2]. In 2022, the World Health Organization (WHO) reported that, globally, approximately 511,000 cases of EC were recorded, placing it as the 11th most common cancer worldwide. Additionally, about 445,000 deaths were attributed to EC, ranking it seventh in terms of mortality rates [3]. China, a country with a high incidence of EC, reported over 187,000 deaths in 2022, making it the 5th leading cause of cancer mortality within the country [4]. Given these statistics, it becomes evident that the primary prevention of EC is crucial. Widespread recognition has been accorded to the reduction of EC risk through dietary and living-environment improvements [5].

Selenium is an essential trace element that can be detected in soil, water, and air [6]. It enters the food chain through bioaccumulation and serves as a primary route for selenium supplementation in humans [7,8]. Many studies have demonstrated the significant effect of selenium on EC prevention. Ahsan et al. [9] found that selenium exhibits anti-inflammatory and antioxidant effects, inhibiting the proliferation of EC cell lines and retarding its development in mice. In addition, long-term selenium supplementation has shown positive effects on gene expression associated with inflammation, cell proliferation, and apoptosis in normal esophageal tissues of rats [10]. Cohort studies also can be found. Several separate studies from Iran found that serum selenium levels were significantly lower in EC patients than in controls. Meanwhile, the concentration of selenium was lower in EC tissues compared to non-cancerous tissues [11–13]. A meta-analysis by Cai et al. [14], which comprised five studies, showed that high selenium exposure may reduce the EC risk.

Selenium is normally bound to proteins in vivo and is involved in physiological activities in the form of selenoproteins [15]. A deficiency of selenium leads to low levels of antioxidant selenoproteins [16,17]. Glutathione peroxidase (GPxs) is a family of antioxidant enzymes with eight members, GPx1–GPx8 [18]. The human GPx3 gene is located on chromosome 5 and consists of five exons. GPx3 is an extracellular glycoprotein, and therefore, it can be considered an efficient extracellular device against oxidative stress [19]. As the only extracellular isoform of the family, GPx3, can circulate and be detected in plasma [20]. Some studies have found that selenium levels in vivo have a significant positive correlation with GPx3, leading to its widespread utilization as a biomarker [21,22].

The downregulation of GPx3 expression is closely associated with cancers. Several studies have shown reduced serum GPx3 activity in cancer patients [23,24]. In addition, studies utilizing GPx3 KO mice have shown accelerated dysplasia in tumor lesions and an increased incidence of cancer [25,26]. However, studies focusing on GPx3 and its relationship with EC remain limited. A case-control study showed that the expression level of GPx3 in normal esophageal epithelial tissues was three times higher than that in EC tissues, indicating that the downregulation of GPx3 serves as a significant indicator of EC development [27]. However, the bioinformatic analysis involved in this study did not delve into the possible causes of GPx3 downregulation to support the experimental results. Another bioinformatic analysis also clarified that GPx3 expression was downregulated in EC tissues. An attempt was made to find differentially expressed genes (DEGs), but no experimental data were available to further validate [28].

We hope to clarify the role of selenium and related metabolites in reducing EC incidence through bioinformatic analyses to support health promotion. However, EC patients have lost the opportunity to undergo primary and secondary prevention. Therefore, studies targeting esophageal precancerous lesions (EPLs) should be highlighted. EPLs often progress to EC in situ following deterioration, but appropriate intervention can slow down or even reverse this process [29,30]. Meanwhile, a study found that genetic mutations and copy-number alterations observed in EPL tissues closely resemble those found in EC tissues [31]. Therefore, it is feasible to validate the expression of dietary selenium intake and related proteins in patients with EPLs based on bioinformatics analyses, which, at the same time, has greater relevance for the prevention of EC.

EC comprises two primary subtypes: esophageal squamous cell carcinoma (ESCC) and esophageal adenocarcinoma (EAC). ESCC is prevalent in Asia, South Africa, and South America, with major contributing factors including smoking, alcohol consumption, food choices, and eating patterns [32]. EAC is the main histological type of EC in the West, including North America, Europe, and Australia. Huai'an district in Jiangsu province is one of the high-risk areas of ESCC in China. The total number of new cases of ESCC in this region from 1998 to 2016 was 20,892, accounting for 38.01% of the new cases of cancer during the same period, both of which ranked first in the ranking of malignant tumors in the region [33].

Our study explored the possible association of GPx3 with EC and its underlying mechanisms systematically through DEG screening, GO and KEGG pathway enrichment analysis, and PPI visualization. Subsequently, a 1:1 case-control study about EPLs in ESCC from Huai'an district, Jiangsu, was used to validate the results of the bioinformatic analysis and to explore the effects of selenium and related proteins on the risk of EPLs and to explore the effects of selenium and related proteins on the risk of EPLs.

2. Materials and Methods

2.1. Bioinformation Analysis

The Gene Expression Profiling Interactive Analysis (GEPIA) database (http://gepia.cancer-pku.cn/, accessed on 28 April 2022) was used to analyze the differential expression of GPx3 between tumor tissues and normal tissues. The EC dataset was selected, and we set the $|Log_2FC|$ cutoff to 2 and the q-value cutoff to 0.01. GPx3 was retrieved and more detailed gene expression was obtained.

The RNA sequence data and clinical information on esophagus cancer were downloaded from The Cancer Genome Atlas (TCGA) database (https://www.cancer.gov/ccg/research/genome-sequencing/tcga, accessed on 30 April 2022). The exclusion criteria were as follows: (a) without clinical or prognostic information, (b) other malignancies in addition to EC, and (c) patients without follow-up data or overall survival < 30 days.

The sample data were analyzed using the "edgeR", "limma" and "gplot" packages in R. The fold change (FC) and the corresponding p value were considered indicators for DEGs. We set $\log|FC| > 2.0$ and $p < 0.01$ as significant. The DAVID database can provide biofunctional annotation and analysis for large-scale genes. All selected DEGs were ID-transformed by DAVID, and GO and KEGG enrichment analysis and mapping were performed based on the "clusterProfiler" and "ggplot2" packages in R. STRING was used to pool selected DEGs for protein–protein interaction (PPI) network identification. The results were exported as images to Cytoscape v3.8.2 for visual analysis, and the top-ten hub genes were clustered using the CytoHubba plugin for display.

GEPIA was used to analyze the correlation of GPx3 and FABP1. $p < 0.05$ (two tailed) were considered statistically significant.

2.2. The Case-Control Study

2.2.1. Population-Sample Collection

The recruitment of subjects came from The Early Diagnosis and Early Treatment Project of Esophageal Cancer (EDETPEC) [34,35]. The study was conducted according to the guidelines of the Declaration of Helsinki and approved by the Institutional Review Board of Southeast University Zhongda Hospital (Nanjing, China) with an approval number of 2012ZDllKY19.0. Beginning in 2011, EDETREC recruited subjects through face-to-face questionnaire interviews by village healthcare workers. The subjects were in the age range of 35 to 75 years and did not have a history of any kind of cancer. Participants underwent a routine endoscopy examination, tissues were biopsied, and diagnosed according to the histologic criteria of dysplasia. Based on this, 1731 residents who had lived in the towns of Jibiao, Jingkou, and Qiuqiao for more than 5 years were recruited between January 2015 and June 2017 for this study. The routine endoscopy examination revealed that 148 of them were diagnosed with EPLs.

The recognized EPL of ESCC is esophageal squamous dysplasia, which usually occurs in the mucosa of the esophagus. The WHO considers EPLs in ESCC to be esophageal squamous cells characterized by cytological atypia (nuclear atypia: enlargement, pleomorphism, hyperpigmentation, loss of polarity, and overlapping) and structural atypia (abnormal epithelial maturation). The WHO classifies EPLs as low and high grade. Involvement of only the lower half of the epithelium, with only mild cytological atypia was classified as a low-grade EPL, whereas high grade meant involvement of more than half of the epithelium or severe cytological atypia (regardless of the extent of epithelial involvement) [36].

A total of 144 residents were diagnosed with low-grade EPLs, of which 100 cases were randomly selected. One hundred cases were selected as healthy controls from 1583 healthy residents, matched by sex, age (± 2 years), and village. The epidemiological data and dietary information obtained from the survey have been published [34]. We have also previously described in detail the collection and preservation of the dietary and plasma samples from the 200 subjects [35].

2.2.2. Laboratory Measurements

The dietary and plasma samples need to be filtered and purified before being tested for indicators (Supplementary Materials: Figure S1). Inductively coupled plasma-mass spectrometry (ICP-MS) (Agilent Technologies (China) Co., Beijing, China) was used to determine the levels of selenium in the dietary and plasma samples. Under the optimal operating conditions of the instrument, the blank control sample, the mixed standard working solutions, and the diluted sample solutions for measurement were introduced into the ICP-MS by the injection system and peristaltic pump. The selenium contents were calculated from the measured abundance values of the samples corresponding to the standard curve automatically plotted by the instrument.

The expression concentrations of GPx3 and FABP1 in plasma samples were determined by ELISA kits (Nanjing Jiancheng Bioengineering Institute, Nanjing, China). The assay was performed in strict accordance with the manufacturer's recommended procedure. The OD values were measured at 450 nm using an enzyme marker (Tecan Trading Co., Ltd., Shanghai, China).

2.2.3. Statistical Analysis

A Wilcoxon rank-sum test was conducted to analyze the results of indicator measurements. A Spearman correlation analysis was used to understand the correlation of plasma selenium levels with dietary selenium content, GPx3, and FABP1 in patients with EPLs and healthy subjects. Conditional logistic regression was performed to evaluate the association between selenium-related variables and the risk of EPLs, with adjustment for age, gender, body mass index (BMI), education level, annual income, tobacco smoking, alcohol drinking, tea drinking, meals on time, spicy taste acceptance, family history of esophagus cancer, and history of digestive diseases. Selenium-related variables were triangulated to complete subsequent analyses. All covariates were categorical except age and BMI, which were continuous variables. For better analysis, BMI was set as a dummy variable. It was classified based on the criteria from the WHO (http://apps.who.int/gho/data/node.main.BMIANTHROPOMETRY?lang=en, accessed on 30 April 2022), with underweight (BMI < 18.5), normal (BMI 18.5–24.9), overweight (BMI 25–29.9) and obese (BMI $\geq$ 30). Each variable was analyzed by way of "enter". $p < 0.05$ (two-tailed) were considered statistically significant. All statistical analyses were performed on SPSS (version 22.0). The statistical review of the study was performed by a biomedical statistician.

3. Results

3.1. Expression of GPx3 in Tumor and Normal Tissues

A total of 182 tumor tissues of EC and 286 normal tissues were retrieved from GEPIA to analyze the differential expression of GPx3. The median gene expression of GPx3 in cancer tissues was 38.489, which was significantly lower compared to 401.971 in normal tissues. As shown in Figure 1, the expression of GPx3 was significantly lower in the tumors than in the controls.

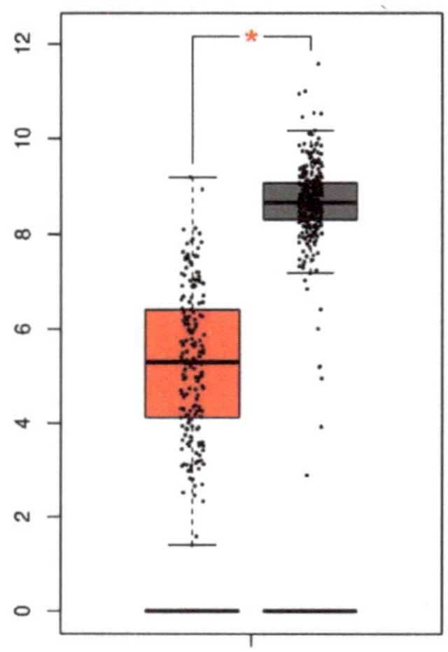

Figure 1. Expression of GPx3 in tumor and normal tissues. For better visualization, GPx3 expressions are log-transformed. We use $\log_2(\text{TPM} + 1)$ for log scale. TPM, transcripts per million. * The difference is statistically significant ($p < 0.05$).

3.2. Visualization of DEGs in the Tumor Tissues with Low Expressed GPx3

A total of 162 tumor tissues were searched from the TCGA. The samples were divided into two groups using the median of the GPx3 expression levels as a limit. As shown in Table 1, there were no significant differences in age and EC stage between the two sample groups of GPx3.

Table 1. Characteristics of patients in high and low expression of GPx3.

	Stage	GPx3		
		Low	**High**	***p* Value**
n		81	81	
Age, $\overline{X} \pm s$		62.49 ± 11.79	61.99 ± 12.31	0.790
T, n (%)	T1	11 (7.6)	16 (11)	0.478
	T2	19 (13.1)	18 (12.4)	
	T3	42 (29)	35 (24.1)	
	T4	1 (0.7)	3 (2.1)	
N, n (%)	N0	35 (24.3)	31 (21.5)	0.272
	N1	30 (20.8)	33 (22.9)	
	N2	3 (2.1)	6 (4.2)	
	N3	5 (3.5)	1 (0.7)	
M, n (%)	M0	58 (45)	63 (48.8)	0.486
	M1	5 (3.9)	3 (2.3)	

A total of 165 DEGs were demonstrated, of which 121 were downregulated genes and 44 were upregulated ones. (Supplementary Materials: Table S1 and Figure 2A). GO and KEGG were responsible for the enrichment analysis of the DEGs. As shown in Figure 2B, most of the DEGs are located at the apical part of the cell in tissues with low expression of GPx3. In addition, their main function is to participate in receptor ligand activity and the process of epidermal cell differentiation. The DEGs are mainly involved in fat digestion and the absorption pathway.

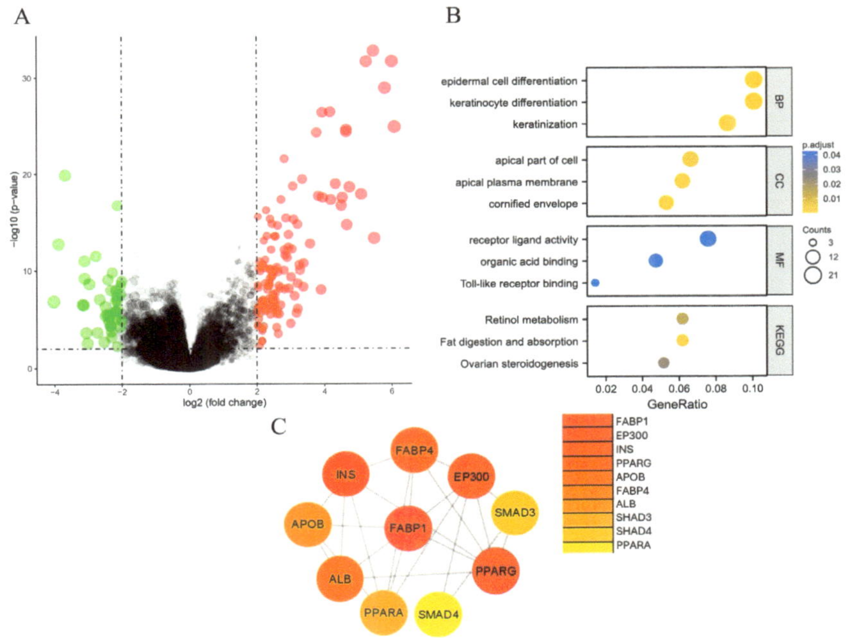

Figure 2. The visualization of the DEGs in low-expression GPx3 vs. high-expression GPx3. (**A**) The volcano plot of the DEGs, in which red dots represent downregulated genes, green dots represent upregulated genes, and black dots show no differences in gene expression; (**B**) GO and KEGG enrichment analysis. (**C**) The top-10 hub DEGs.

The top-10 hub genes ranked were screened based on String online analysis and visualized by Cytoscape. The most central protein, fatty acid-binding protein 1 (FABP1), was found to be upregulated in cancer tissues with low GPx3 expression (Figure 2C).

We further analyzed the correlation between GPx3 and FABP1 using GEPIA to validate the DEG visualization results. Figure 3 shows that there was a significant negative correlation between GPx3 and FABP1 (p = 0.012).

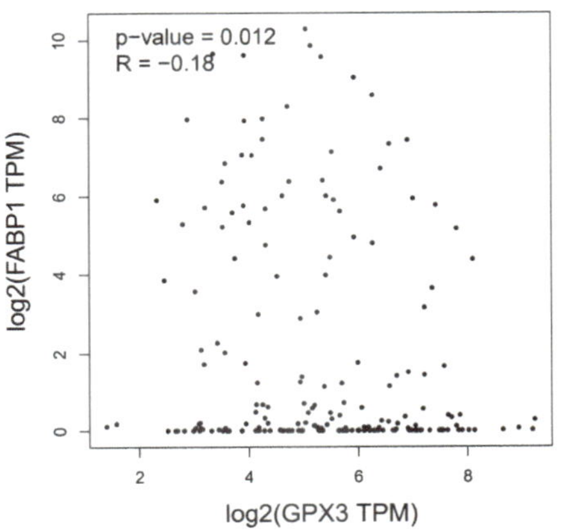

Figure 3. The correlation of GPx3 and FABP1.

3.3. Selenium-Related Variables of the Subjects

As shown in Table 2, no statistical difference in the daily dietary intake of selenium was found between the two groups. However, the plasma selenium level of EPL cases was significantly lower than that of healthy controls ($p < 0.001$). The protein concentrations of GPx3 did not differ between the two groups. A further detection, however, revealed that the EPLs group had higher concentrations of FABP1.

Table 2. Comparison of selenium-related variables between two groups.

Median (25th–75th)	EPLs	Controls	*p* Value
Dietary samples (n)	100	100	
Dietary Selenium (μg/d)	136.72 (94.77–180.06)	147.20 (99.20–177.65)	0.498
Blood samples (n)	100	100	
Plasma Selenium (μg/L)	49.46 (41.75–57.71)	55.97 (43.07–68.23)	<0.001 **
GPx3 (pmol/mL)	42.78 (35.68–51.18)	44.82 (34.48–56.56)	0.561
FABP1 (ng/L)	1433.61 (1337.08–1512.72)	1388.40 (1297.61–1481.82)	0.011 *

* The difference is statistically significant ($p < 0.05$). ** The difference is statistically significant ($p < 0.01$).

We found that dietary selenium intake was significantly positively correlated with plasma selenium, suggesting that exogenous supplementation of selenium can change selenium levels in vivo. In addition, there was also a positive correlation between plasma selenium levels and GPx3 concentrations, which adds to the evidence that GPx3 can serve as a biomarker of selenium status in vivo. Although FABP1 expression was negatively correlated with plasma selenium in both groups, neither was significant (Table 3).

Table 3. Correlation between plasma selenium and related variables between the two groups.

	Plasma Selenium	Dietary Selenium	GPx3	FABP1
EPLs	1.00	0.931 **	0.141 *	−0.76
Controls	1.00	0.924 **	0.239 *	−0.82

* The difference is statistically significant ($p < 0.05$). ** The difference is statistically significant ($p < 0.01$).

3.4. Correlation between Selenium-Related Variables and Risk of EPLs

As shown in Table 4, our study did not find an association between a daily dietary intake of selenium and EPLs risk (p for trend = 0.545). The concentration of plasma selenium located in the highest tertile was associated with a reduced risk of EPLs compared to that in the lowest tertile (OR (95% CI) = 0.34 (0.13–0.93)). In addition, the concentration of GPx3 was negatively associated with the EPLs risk (p for trend = 0.035).

Table 4. ORs (and 95% CIs) for selenium-related variables with EPLs.

	Q1	Q2	Q3	*p* for Trend
	OR (95% CI)	OR (95% CI)	OR (95% CI)	
Dietary samples				
Dietary Selenium (μg/d)	11.44–114.27	114.28–165.68	165.69–379.39	
EPLs cases (%)	48.48	52.24	49.25	
crude model	1.00	1.22 (0.56–2.67)	1.05 (0.49–2.24)	0.846
adjusted model #	1.00	1.91 (0.69–5.29)	1.45 (0.52–4.01)	0.545
Blood samples				
Plasma Selenium (μg/L)	5.03–45.53	45.54–59.26	59.27–115.92	
EPLs cases (%)	56.06	58.82	34.85	
crude model	1.00	1.29 (0.63–2.63)	0.39 (0.19–0.82) *	0.021 *
adjusted model	1.00	2.04 (0.79–5.25)	0.34 (0.13–0.93) *	0.103
GPx3 (pmol/mL)	17.00–37.79	37.80–47.99	48.00–125.96	
EPLs cases (%)	52.24	61.19	36.36	

Table 4. *Cont.*

	Q1	Q2	Q3	*p* for Trend
	OR (95%CI)	OR (95%CI)	OR (95%CI)	
crude model	1.00	1.64 (0.78–3.47)	0.56 (0.27–1.13)	0.291
adjusted model	1.00	1.56 (0.60–4.04)	0.27 (0.10–0.70) *	0.035 *
FABP1 (ng/L)	1192.84–1346.47	1346.48–1465.23	1465.24–1649.36	
EPLs cases (%)	73.68	52.24	55.22	
crude model	1.00	1.41 (0.73–2.71)	1.54 (0.82–2.90)	0.180
adjusted model	1.00	1.77 (0.77–4.09)	2.30 (0.99–5.31)	0.046 *

* The difference is statistically significant ($p < 0.05$). # The model adjusted for age, gender, BMI, education level, annual income, tobacco smoking, alcohol drinking, tea drinking, meals on time, spicy taste acceptance, family history of esophagus cancer, and history of digestive diseases.

However, the risk of EPLs increases with higher protein concentrations of FABP1 (*p* for trend = 0.04). This is consistent with the results of our previous bioinformatic analyses.

4. Discussion

GPx3 expression was lower in tumor tissues compared to normal tissues. DEGs were found to act mainly in the fat digestion and absorption pathway, and the upregulation of the most central hub protein, FABP1, was also associated with low GPx3 expression. Subsequent case-control studies supported these results. We further found that although selenium was not associated with EPLs, the reduced concentration of GPx3 and the upregulation of FABP1 increased the risk of EPLs.

The reduced GPx3 expression in the plasma of EPL cases may be related to epigenetic regulation. Nirgude et al. [37] highlighted GPx3 mutations and copy-number alterations in cancers through TCGA analysis. In addition, there was hypermethylation of the CpG island within the GPx3 promoter in ESCC cell lines. Moreover, the GPx3 promoter exhibited more frequent methylation in ESCC tissues compared to adjacent normal tissues [38,39]. Consequently, this methylation pattern leads to the downregulation of both mRNA and protein expression of GPx3 in EC [27]. Secondly, GPx3 was shown to be effective in slowing down the proliferation, migration, and invasion of lung cancer cells under oxidative stress and in significantly reducing the production of ROS [40,41]. Meanwhile, GPx3 silencing resulted in ROS production, DNA damage, and increased apoptosis in Caco2 cells [25]. This suggests that GPx3 deficiency is associated with enhanced lipid peroxidation [42,43].

Lipid peroxides serve as pivotal agents in cancer by disrupting the body's homeostasis through REDOX balance disturbances [44]. As we analyzed with KEGG, FABP1 played an important role in the fat digestion and absorption pathway. This is due to the presence of fatty acid-binding cavities within the FABP1 protein that facilitate fatty acid uptake and transport [45]. Wu et al. [46] found that the overexpression of FABP1 significantly increased fatty acid uptake in hepatocytes. Therefore, FABP1 is considered a protective antioxidant molecule that controls the accumulation of peroxidation by inducing the inactivation of active lipids [47]. The thiobarbituric acid-reactive substances (TBARS) are the marker of lipid peroxidation [48]. FABP1 KO mice were found to have significantly reduced levels of TBARS in the liver, along with a decreased expression of markers of inflammation and oxidative stress [49]. The absence of GPx3 in EPLs may lead to the increase of TBARS, thus stimulating the increase of FABP1 expression to coordinate the oxidative stress state of the body. Therefore, our database analysis revealed a significant upregulation of FABP1 in EC tissues with low GPx3 expression. Meanwhile, our case control also found a significant increase in FABP1 expression in the plasma of EPLs with downregulated GPx3 expression.

In addition to the manifestations of epigenetic regulation of GPx3 in ESCC, it is possible that selenium status in vivo may also influence its expression. However, neither dietary selenium nor plasma selenium levels were associated with the risk of EPLs, despite demonstrating a positive correlation between selenium and GPx3. A meta-analysis conducted by Vinceti et al. [50] concluded that the incidence and mortality of several cancers, including

EC, were not associated with selenium supplementation or its in vivo status. This relatively strong evidence supports our results.

The possible contribution of selenium to reducing the incidence of ESCC was mentioned previously, and our non-predicted results may be related to the differences in environmental selenium concentrations. The concentration of selenium in soils varies with the type, texture, and organic matter content of the soil, as well as with local rainfall. Like plants, it is influenced by the physicochemical properties of the soil (redox state, pH, and microbial activity). The average concentration of selenium in soils ranges from 0.1 to 0.7 mg/kg. It ranges from 0.8 to 2 mg/kg in clay soils and up to 2 to 4.5 mg/kg in tropical soils. In addition, the presence of selenium in the atmosphere is related to natural and anthropogenic activities. Soil erosion, volcanic activity, and waste incineration all contribute to changes in atmospheric selenium concentrations [6]. It is not difficult to reason that differences in selenium levels in soil and atmospheric deposits can lead to different selenium concentrations in water. The selenium present in diverse soils and water sources can be partially absorbed by crops and ingested by bred animals, subsequently entering the human body. Consequently, the selenium content of plants used as food or feed in different environments may vary significantly. Health problems associated with selenium may be altered due to differences in the body's tolerance to selenium. A U-shaped nonlinear dose–response relationship has been demonstrated, indicating that both selenium-deficient and selenium-enriched states can be detrimental to health [51]. Therefore, disparities in selenium status due to natural exposure in different habitations might influence overall health conditions [52–55].

In addition to natural levels of selenium exposure, Hurst et al. [39] found that cancer risk is related to the body's ability to absorb and metabolize selenium. Interactions between vitamins and minerals may mask the association between selenium and ESCC [40,41]. For instance, trivalent iron forms complexes with selenium, leading to precipitation and preventing its absorption by intestinal cells. When sulfur concentrations exceed 2.4 g/kg DM, selenium absorption is reduced due to competition [56,57].

Furthermore, the specificity of subjects may lead to differences in EC subtypes, which makes it challenging to study the relationship between selenium and ESCC. While China is a region with a high prevalence of ESCC, the current evidence linking selenium and EC is mostly from European and American populations [58]. In addition, the evidence suggests that, despite consistent nationality, ethnic heterogeneity still produces large differences in both the incidence and subtypes of EC [59]. This diversity may be related to variations in dietary patterns, lifestyle habits, and the genetic genes related to selenium absorption and metabolism across different ethnic groups [5,60,61].

Our study also has limitations. First, the fluctuation of GPx3 and FABP1 levels during the development of EPLs into EC was not investigated. Barrett's esophagus (BE) is a kind of EPL that may progress into EAC [62,63]. Evidence highlights that FABP1 was significantly highly expressed in BE, while the expression gradually decreased as the disease worsened until the occurrence of EAC [64,65]. This is despite the fact that a study proved rational drug therapy resulted in the downregulation of FABP1 expression in BE patients [66]. We detected a significant increase in the FABP1 expression level in EPL cases. However, we remain uncertain whether FABP1 expression also gradually decreases until the diagnosis of EC is established. Second, EC is mostly classified as ESCC in China, leading to a lack of sufficient data pertaining to EAC for the validation of bioinformatics results. We acknowledge the necessity for more extensive international collaboration to overcome these limitations. In addition, a deeper exploration is necessary. Although we have found an association between GPx3 and FABP1 based on the fat digestion and absorption pathway, further experiments and analyses, such as investigating their in situ expression, are still needed to clarify the underlying mechanistic reasons for this association.

5. Conclusions

Our study did not establish a direct association between selenium itself and ESCC. However, the significant difference observed between GPx3 and FABP1 in the two groups suggested the potential significance of selenium in EC prevention. In addition, FABP1 may serve as a novel biomarker for EC and EPLs in ESCC, but this still needs to be explored by more in-depth studies.

Supplementary Materials: The following supporting information can be downloaded at https://www.mdpi.com/article/10.3390/nu16091322/s1, Table S1: Differential genes screened from tissues with high expression of GPx3; Figure S1: Flowchart of sample pretreatment for detection of A, dietary selenium and B, plasma selenium level.

Author Contributions: N.W. and D.P. designed research; N.W., X.Z., X.R., X.J. and X.C. conducted research; S.W., M.S. and G.S. provided essential materials; Y.W. analyzed data; N.W. wrote the paper; S.W. had primary responsibility for final content. All authors have read and agreed to the published version of the manuscript.

Funding: This research was supported by the National Natural Science Foundation of China, No. 81673147, the Danone Dietary Nutrition Research and Education Foundation, DIC2020-08, and the Postgraduate Research and Practice Innovation Program of Jiangsu Province, KYCX23_0329.

Institutional Review Board Statement: The study was conducted according to the guidelines of the Declaration of Helsinki and approved by the Institutional Review Board of Southeast University Zhongda Hospital (Jiangsu Province, China) with approval number 2012ZDllKY19.0.

Informed Consent Statement: Informed consent was obtained from all subjects involved in the study. Written informed consent has been obtained from the patient(s) to publish this paper.

Data Availability Statement: All data are available with the consent of the corresponding author.

Acknowledgments: We thank all the participants, researchers, and collaborators who were involved in the study for their efforts and contributions.

Conflicts of Interest: The authors declare no conflicts of interest.

References

1. Morgan, E.; Soerjomataram, I.; Rumgay, H.; Coleman, H.G.; Thrift, A.P.; Vignat, J.; Laversanne, M.; Ferlay, J.; Arnold, M. The Global Landscape of Esophageal Squamous Cell Carcinoma and Esophageal Adenocarcinoma Incidence and Mortality in 2020 and Projections to 2040: New Estimates From GLOBOCAN 2020. *Gastroenterology* **2022**, *163*, 649–658.e2. [CrossRef] [PubMed]
2. Wang, W.W.; Zhao, Z.H.; Wang, L.; Li, P.; Chen, K.S.; Zhang, J.Y.; Li, W.C.; Jiang, G.Z.; Li, X.N. MicroRNA-134 prevents the progression of esophageal squamous cell carcinoma via the PLXNA1-mediated MAPK signalling pathway. *eBioMedicine* **2019**, *46*, 66–78. [CrossRef] [PubMed]
3. Bray, F.; Laversanne, M.; Sung, H.Y.A.; Ferlay, J.; Siegel, R.L.; Soerjomataram, I.; Jemal, A. Global cancer statistics 2022: GLOBOCAN estimates of incidence and mortality worldwide for 36 cancers in 185 countries. *CA Cancer J. Clin.* **2024**. [CrossRef] [PubMed]
4. Zheng, R.S.; Chen, R.; Han, B.F.; Wang, S.M.; Li, L.; Sun, K.X.; Zeng, H.M.; Wei, W.W.; He, J. Cancer incidence and mortality in China, 2022. *Zhonghua Zhong Liu Za Zhi [Chin. J. Oncol.]* **2024**, *46*, 221–231. [CrossRef] [PubMed]
5. Arnal, M.J.D.; Arenas, A.F.; Arbeloa, A.L. Esophageal cancer: Risk factors, screening and endoscopic treatment in Western and Eastern countries. *World J. Gastroenterol.* **2015**, *21*, 7933–7943. [CrossRef] [PubMed]
6. Mehdi, Y.; Hornick, J.L.; Istasse, L.; Dufrasne, I. Selenium in the environment, metabolism and involvement in body functions. *Molecules* **2013**, *18*, 3292–3311. [CrossRef] [PubMed]
7. Mendieta, J.M.; Fields, C.L.; Ossorio, M.A.; Roy, T.M. Respiratory failure associated with the post-polio syndrome. *J. Ky. Med. Assoc.* **1989**, *87*, 63–65. [PubMed]
8. Kershaw, J.L.; Hall, A.J. Mercury in cetaceans: Exposure, bioaccumulation and toxicity. *Sci. Total Environ.* **2019**, *694*, 133683. [CrossRef]
9. Ahsan, A.; Liu, Z.; Su, R.; Liu, C.; Liao, X.; Su, M. Potential Chemotherapeutic Effect of Selenium for Improved Canceration of Esophageal Cancer. *Int. J. Mol. Sci.* **2022**, *23*, 5509. [CrossRef]
10. Younesian, O.; Sheikh Arabi, M.; Jafari, S.M.; Joshaghani, H. Long-Term Excessive Selenium Supplementation Affects Gene Expression in Esophageal Tissue of Rats. *Biol. Trace Elem. Res.* **2023**, *201*, 3387–3394. [CrossRef]

11. Nozadi, F.; Azadi, N.; Mansouri, B.; Tavakoli, T.; Mehrpour, O. Association between trace element concentrations in cancerous and non-cancerous tissues with the risk of gastrointestinal cancers in Eastern Iran. *Environ. Sci. Pollut. Res. Int.* **2021**, *28*, 62530–62540. [CrossRef]
12. Hashemi, S.M.; Mashhadi, M.; Moghaddam, A.A.; Yousefi, J.; Mofrad, A.D.; Sadeghi, M.; Allahyari, A. The Relationship between Serum Selenium and Zinc with Gastroesophageal Cancers in the Southeast of Iran. *Indian J. Med. Paediatr. Oncol.* **2017**, *38*, 169–172.
13. Nouarie, M.; Pourshams, A.; Kamangar, F.; Sotoudeh, M.; Derakhshan, M.H.; Akbari, M.R.; Fakheri, H.; Zahedi, M.J.; Caldwell, K.; Abnet, C.C.; et al. Ecologic study of serum selenium and upper gastrointestinal cancers in Iran. *World J. Gastroenterol.* **2004**, *10*, 2544–2546. [CrossRef] [PubMed]
14. Cai, X.; Wang, C.; Yu, W.; Fan, W.; Wang, S.; Shen, N.; Wu, P.; Li, X.; Wang, F. Selenium Exposure and Cancer Risk: An Updated Meta-analysis and Meta-regression. *Sci. Rep.* **2016**, *6*, 19213. [CrossRef]
15. Kang, D.; Lee, J.; Wu, C.; Guo, X.; Lee, B.J.; Chun, J.S.; Kim, J.H. The role of selenium metabolism and selenoproteins in cartilage homeostasis and arthropathies. *Exp. Mol. Med.* **2020**, *52*, 1198–1208. [CrossRef]
16. Tan, L.C.; Nancharaiah, Y.V.; van Hullebusch, E.D.; Lens, P.N.L. Selenium: Environmental significance, pollution, and biological treatment technologies. *Biotechnol. Adv.* **2016**, *34*, 886–907. [CrossRef]
17. Touat-Hamici, Z.; Legrain, Y.; Bulteau, A.L.; Chavatte, L. Selective up-regulation of human selenoproteins in response to oxidative stress. *J. Biol. Chem.* **2014**, *289*, 14750–14761. [CrossRef] [PubMed]
18. Pei, J.; Pan, X.; Wei, G.; Hua, Y. Research progress of glutathione peroxidase family (GPX) in redoxidation. *Front. Pharmacol.* **2023**, *14*, 1147414. [CrossRef]
19. Brigelius-Flohé, R.; Flohé, L. Regulatory Phenomena in the Glutathione Peroxidase Superfamily. *Antioxid. Redox Signal.* **2020**, *33*, 498–516. [CrossRef]
20. Kipp, A.P. Selenium-Dependent Glutathione Peroxidases During Tumor Development. *Adv. Cancer Res.* **2017**, *136*, 109–138. [CrossRef]
21. Sun, Q.; Oltra, E.; Dijck-Brouwer, D.A.J.; Chillon, T.S.; Seemann, P.; Asaad, S.; Demircan, K.; Espejo-Oltra, J.A.; Sánchez-Fito, T.; Martín-Martínez, E.; et al. Autoantibodies to selenoprotein P in chronic fatigue syndrome suggest selenium transport impairment and acquired resistance to thyroid hormone. *Redox Biol.* **2023**, *65*, 102796. [CrossRef] [PubMed]
22. Ashton, K.; Hooper, L.; Harvey, L.J.; Hurst, R.; Casgrain, A.; Fairweather-Tait, S.J. Methods of assessment of selenium status in humans: A systematic review. *Am. J. Clin. Nutr.* **2009**, *89*, 2025S–2039S. [CrossRef]
23. Psathakis, D.; Wedemeyer, N.; Oevermann, E.; Krug, F.; Siegers, C.P.; Bruch, H.P. Blood selenium and glutathione peroxidase status in patients with colorectal cancer. *Dis. Colon Rectum* **1998**, *41*, 328–335. [CrossRef]
24. Oh, I.J.; Kim, H.E.; Song, S.Y.; Na, K.J.; Kim, K.S.; Kim, Y.C.; Lee, S.W. Diagnostic value of serum glutathione peroxidase 3 levels in patients with lung cancer. *Thorac. Cancer* **2014**, *5*, 425–430. [CrossRef]
25. Barrett, C.W.; Ning, W.; Chen, X.; Smith, J.J.; Washington, M.K.; Hill, K.E.; Coburn, L.A.; Peek, R.M.; Chaturvedi, R.; Wilson, K.T.; et al. Tumor suppressor function of the plasma glutathione peroxidase gpx3 in colitis-associated carcinoma. *Cancer Res.* **2013**, *73*, 1245–1255. [CrossRef]
26. Chang, S.N.; Lee, J.M.; Oh, H.; Park, J.H. Glutathione Peroxidase 3 Inhibits Prostate Tumorigenesis in TRAMP Mice. *Prostate* **2016**, *76*, 1387–1398. [CrossRef] [PubMed]
27. Lin, Y.; Zhang, Y.; Chen, Y.; Liu, Z. Promoter methylation and clinical significance of GPX3 in esophageal squamous cell carcinoma. *Pathol. Res. Pract.* **2019**, *215*, 152676. [CrossRef] [PubMed]
28. Zhang, J.; Zhang, N.; Yang, X.; Xin, X.; Jia, C.H.; Li, S.; Lu, Q.; Jiang, T.; Wang, T. Machine Learning and Novel Biomarkers Associated with Immune Infiltration for the Diagnosis of Esophageal Squamous Cell Carcinoma. *J. Oncol.* **2022**, *2022*, 6732780. [CrossRef]
29. Wang, N.; Pan, D.; Wang, X.; Su, M.; Wang, X.; Yan, Q.; Sun, G.; Wang, S. NAPRT, but Not NAMPT, Provides Additional Support for NAD Synthesis in Esophageal Precancerous Lesions. *Nutrients* **2022**, *14*, 4916. [CrossRef]
30. Qin, J.; Zhu, Y.; Ding, Y.; Niu, T.; Zhang, Y.; Wu, H.; Zhu, L.; Yuan, B.; Qiao, Y.; Lu, J.; et al. DNA polymerase beta deficiency promotes the occurrence of esophageal precancerous lesions in mice. *Neoplasia* **2021**, *23*, 663–675. [CrossRef]
31. Liu, X.; Zhang, M.; Ying, S.M.; Zhang, C.; Lin, R.H.; Zheng, J.X.; Zhang, G.H.; Tian, D.P.; Guo, Y.; Du, C.W.; et al. Genetic Alterations in Esophageal Tissues From Squamous Dysplasia to Carcinoma. *Gastroenterology* **2017**, *153*, 166–177. [CrossRef]
32. Abnet, C.C.; Arnold, M.; Wei, W.Q. Epidemiology of Esophageal Squamous Cell Carcinoma. *Gastroenterology* **2018**, *154*, 360–373. [CrossRef]
33. Wang, S.K.; Pan, D.; Chen, Z.T.; Song, G.; Han, R.Q.; Sun, G.J.; Su, M. Trends in Incidence and Mortality of Esophageal Cancer in Huai'an District, a High-Risk Area in Northern Jiangsu Province, China. *Cancer Control* **2022**, *29*. [CrossRef]
34. Pan, D.; Su, M.; Zhang, T.; Miao, C.; Fu, L.; Yang, L.; Song, G.; Raine, P.J.; Wang, S.; Sun, G. A Distinct Epidemiologic Pattern of Precancerous Lesions of Esophageal Squamous Cell Carcinoma in a High-risk Area of Huai'an, Jiangsu Province, China. *Cancer Prev. Res.* **2019**, *12*, 449–462. [CrossRef]
35. Pan, D.; Wang, S.; Su, M.; Sun, G.; Zhu, X.; Ghahvechi Chaeipeima, M.; Guo, Z.; Wang, N.; Zhang, Z.; Cui, M. Vitamin B(12) may play a preventive role in esophageal precancerous lesions: A case-control study based on markers in blood and 3-day duplicate diet samples. *Eur. J. Nutr.* **2021**, *60*, 3375–3386. [CrossRef]

36. WHO Classification of Tumours Editorial Board (Ed.) *WHO Classification of Tumours, 5th Edition: Digestive System Tumours*; International Agency for Research on Cancer: Lyon, France, 2019; Volume 1, p. 37.
37. Nirgude, S.; Choudhary, B. Insights into the role of GPX3, a highly efficient plasma antioxidant, in cancer. *Biochem. Pharmacol.* **2021**, *184*, 114365. [CrossRef]
38. He, Y.; Wang, Y.; Li, P.; Zhu, S.; Wang, J.; Zhang, S. Identification of GPX3 epigenetically silenced by CpG methylation in human esophageal squamous cell carcinoma. *Dig. Dis. Sci.* **2011**, *56*, 681–688. [CrossRef]
39. Salta, S.; Macedo-Silva, C.; Miranda-Goncalves, V.; Lopes, N.; Gigliano, D.; Guimaraes, R.; Farinha, M.; Sousa, O.; Henrique, R.; Jeronimo, C. A DNA methylation-based test for esophageal cancer detection. *Biomark. Res.* **2020**, *8*, 68. [CrossRef]
40. An, B.C.; Choi, Y.D.; Oh, I.J.; Kim, J.H.; Park, J.I.; Lee, S.W. GPx3-mediated redox signaling arrests the cell cycle and acts as a tumor suppressor in lung cancer cell lines. *PLoS ONE* **2018**, *13*, e0204170. [CrossRef]
41. Chang, C.; Worley, B.L.; Phaeton, R.; Hempel, N. Extracellular Glutathione Peroxidase GPx3 and Its Role in Cancer. *Cancers* **2020**, *12*, 2197. [CrossRef]
42. Guven, M.; Ozturk, B.; Sayal, A.; Ozeturk, A.; Ulutin, T. Lipid peroxidation and antioxidant system in the blood of cancerous patients with metastasis. *Cancer Biochem. Biophys.* **1999**, *17*, 155–162.
43. He, S.X.; Wu, B.; Chang, X.M.; Li, H.X.; Qiao, W. Effects of selenium on peripheral blood mononuclear cell membrane fluidity, interleukin-2 production and interleukin-2 receptor expression in patients with chronic hepatitis. *World J. Gastroenterol.* **2004**, *10*, 3531–3533. [CrossRef]
44. Gaschler, M.M.; Stockwell, B.R. Lipid peroxidation in cell death. *Biochem. Biophys. Res. Commun.* **2017**, *482*, 419–425. [CrossRef]
45. Hotamisligil, G.S.; Bernlohr, D.A. Metabolic functions of FABPs—Mechanisms and therapeutic implications. *Nat. Rev. Endocrinol.* **2015**, *11*, 592–605. [CrossRef]
46. Wu, Y.L.; Peng, X.E.; Zhu, Y.B.; Yan, X.L.; Chen, W.N.; Lin, X. Hepatitis B Virus X Protein Induces Hepatic Steatosis by Enhancing the Expression of Liver Fatty Acid Binding Protein. *J. Virol.* **2016**, *90*, 1729–1740. [CrossRef]
47. Yamamoto, T.; Noiri, E.; Ono, Y.; Doi, K.; Negishi, K.; Kamijo, A.; Kimura, K.; Fujita, T.; Kinukawa, T.; Taniguchi, H.; et al. Renal L-type fatty acid–binding protein in acute ischemic injury. *J. Am. Soc. Nephrol.* **2007**, *18*, 2894–2902. [CrossRef]
48. Jablonska, E.; Gromadzinska, J.; Peplonska, B.; Fendler, W.; Reszka, E.; Krol, M.B.; Wieczorek, E.; Bukowska, A.; Gresner, P.; Galicki, M.; et al. Lipid peroxidation and glutathione peroxidase activity relationship in breast cancer depends on functional polymorphism of GPX1. *BMC Cancer* **2015**, *15*, 657. [CrossRef]
49. Mukai, T.; Egawa, M.; Takeuchi, T.; Yamashita, H.; Kusudo, T. Silencing of FABP1 ameliorates hepatic steatosis, inflammation, and oxidative stress in mice with nonalcoholic fatty liver disease. *FEBS Open Bio* **2017**, *7*, 1009–1016. [CrossRef]
50. Vinceti, M.; Filippini, T.; Del Giovane, C.; Dennert, G.; Zwahlen, M.; Brinkman, M.; Zeegers, M.P.; Horneber, M.; D'Amico, R.; Crespi, C.M. Selenium for preventing cancer. *Cochrane Database Syst. Rev.* **2018**, *1*, CD005195. [CrossRef]
51. Sun, Y.; Wang, Z.; Gong, P.; Yao, W.; Ba, Q.; Wang, H. Review on the health-promoting effect of adequate selenium status. *Front. Nutr.* **2023**, *10*, 1136458. [CrossRef]
52. Dinh, Q.T.; Cui, Z.; Huang, J.; Tran, T.A.T.; Wang, D.; Yang, W.; Zhou, F.; Wang, M.; Yu, D.; Liang, D. Selenium distribution in the Chinese environment and its relationship with human health: A review. *Environ. Int.* **2018**, *112*, 294–309. [CrossRef]
53. Fordyce, F.M.; Johnson, C.C.; Navaratna, U.R.; Appleton, J.D.; Dissanayake, C.B. Selenium and iodine in soil, rice and drinking water in relation to endemic goitre in Sri Lanka. *Sci. Total Environ.* **2000**, *263*, 127–141. [CrossRef]
54. Favorito, J.E.; Grossl, P.R.; Davis, T.Z.; Eick, M.J.; Hankes, N. Soil-plant-animal relationships and geochemistry of selenium in the Western Phosphate Resource Area (United States): A review. *Chemosphere* **2021**, *266*, 128959. [CrossRef]
55. Soderlund, M.; Virkanen, J.; Holgersson, S.; Lehto, J. Sorption and speciation of selenium in boreal forest soil. *J. Environ. Radioact.* **2016**, *164*, 220–231. [CrossRef]
56. Spears, J.W.; Weiss, W.P. Role of antioxidants and trace elements in health and immunity of transition dairy cows. *Vet. J.* **2008**, *176*, 70–76. [CrossRef]
57. Thomson, C.D.; Robinson, M.F. Urinary and Fecal Excretions and Absorption of a Large Supplement of SELENIUM—Superiority of Selenate over Selenite. *Am. J. Clin. Nutr.* **1986**, *44*, 659–663. [CrossRef]
58. Pearce, N. The globalization of epidemiology: Introductory remarks. *Int. J. Epidemiol.* **2004**, *33*, 1127–1131. [CrossRef]
59. Cook, M.B.; Chow, W.H.; Devesa, S.S. Oesophageal cancer incidence in the United States by race, sex, and histologic type, 1977–2005. *Br. J. Cancer* **2009**, *101*, 855–859. [CrossRef]
60. Matejcic, M.; Iqbal Parker, M. Gene-environment interactions in esophageal cancer. *Crit. Rev. Clin. Lab. Sci.* **2015**, *52*, 211–231. [CrossRef]
61. van Blankenstein, M.; Looman, C.W.; Hop, W.C.; Bytzer, P. The incidence of adenocarcinoma and squamous cell carcinoma of the esophagus: Barrett's esophagus makes a difference. *Am. J. Gastroenterol.* **2005**, *100*, 766–774. [CrossRef]
62. Gilbert, S.; Jobe, B.A. Surgical therapy for Barrett's esophagus with high-grade dysplasia and early esophageal carcinoma. *Surg. Oncol. Clin. N. Am.* **2009**, *18*, 523–531. [CrossRef] [PubMed]
63. Wheeler, J.B.; Reed, C.E. Epidemiology of esophageal cancer. *Surg. Clin. N. Am.* **2012**, *92*, 1077–1087. [CrossRef] [PubMed]
64. Srivastava, S.; Kern, F.; Sharma, N.; McKeon, F.; Xian, W.; Yeoh, K.G.; Ho, K.Y.; Teh, M. FABP1 and Hepar expression levels in Barrett's esophagus and associated neoplasia in an Asian population. *Dig. Liver Dis.* **2017**, *49*, 1104–1109. [CrossRef] [PubMed]

65. Dum, D.; Ocokoljic, A.; Lennartz, M.; Hube-Magg, C.; Reiswich, V.; Hoflmayer, D.; Jacobsen, F.; Bernreuther, C.; Lebok, P.; Sauter, G.; et al. FABP1 expression in human tumors: A tissue microarray study on 17,071 tumors. *Virchows Arch.* **2022**, *481*, 945–961. [CrossRef]
66. Abrams, J.A.; Del Portillo, A.; Hills, C.; Compres, G.; Friedman, R.A.; Cheng, B.; Poneros, J.; Lightdale, C.J.; De La Rue, R.; di Pietro, M.; et al. Randomized Controlled Trial of the Gastrin/CCK(2) Receptor Antagonist Netazepide in Patients with Barrett's Esophagus. *Cancer Prev. Res.* **2021**, *14*, 675–682. [CrossRef]

nutrients

MDPI

Article

Association of Sleep Quality and General, Mental, and Oral Health with Lifestyle Traits (Dietary Intake, Smoking Status) in Arthritis: A Cross-Sectional Study from the Canadian Community Health Survey (CCHS)

Zohre Gheisary [1,†], Ibrahim Hoja [2,3,†], Juxin Liu [4], Petros Papagerakis [2], Lynn P. Weber [5], Mark Fenton [6], George S. Katselis [7], Jessica R. L. Lieffers [8,*] and Silvana Papagerakis [2,*]

1 Department of Biochemistry, Microbiology, and Immunology, College of Medicine, University of Saskatchewan, 107 Wiggins Road, Saskatoon, SK S7N 5E5, Canada; zog389@usask.ca
2 Laboratory of Precision Oral Health and Chronobiology, Faculty of Dentistry, Laval University, Dental Medicine Pavilion, 2420, rue de la Terrasse, Quebec City, QC G1V 0A6, Canada; ibhoj@ulaval.ca (I.H.); petros.papagerakis@fmd.ulaval.ca (P.P.)
3 Health Sciences, College of Medicine, University of Saskatchewan, 107 Wiggins Road, Saskatoon, SK S7N 5E5, Canada
4 Department of Mathematics and Statistics, College of Arts and Science, University of Saskatchewan, 106 Wiggins Road, Saskatoon, SK S7N 5E6, Canada; jul086@mail.usask.ca
5 Department of Veterinary Biomedical Sciences, Western College of Veterinary Medicine, University of Saskatchewan, 52 Campus Drive, Saskatoon, SK S7N 5B4, Canada; lynn.weber@usask.ca
6 Division of Respirology, Critical Care, and Sleep Medicine, College of Medicine, University of Saskatchewan, Saskatoon, SK S7N 5E5, Canada; mef132@mail.usask.ca
7 Department of Medicine, Canadian Centre for Rural and Agricultural Health, College of Medicine, University of Saskatchewan, 104 Clinic Place, Saskatoon, SK S7N 2Z4, Canada; george.katselis@usask.ca
8 College of Pharmacy and Nutrition, University of Saskatchewan, 107 Wiggins Road, Saskatoon, SK S7N 5E5, Canada
* Correspondence: jessica.lieffers@usask.ca (J.R.L.L.); silvana.papagerakis@fmd.ulaval.ca (S.P.)
† These authors contributed equally to this work.

Citation: Gheisary, Z.; Hoja, I.; Liu, J.; Papagerakis, P.; Weber, L.P.; Fenton, M.; Katselis, G.S.; Lieffers, J.R.L.; Papagerakis, S. Association of Sleep Quality and General, Mental, and Oral Health with Lifestyle Traits (Dietary Intake, Smoking Status) in Arthritis: A Cross-Sectional Study from the Canadian Community Health Survey (CCHS). *Nutrients* **2024**, *16*, 2091. https://doi.org/10.3390/nu16132091

Academic Editors: Domenico Tafuri and Francesca Latino

Received: 27 May 2024
Revised: 23 June 2024
Accepted: 24 June 2024
Published: 29 June 2024

Abstract: Arthritis is associated with health challenges. Lifestyle traits are believed to influence arthritis development and progression; however, data to support personalized treatment regimens based on holistic lifestyle factors are missing. This study aims to provide a comprehensive list of associations between lifestyle traits and the health status of individuals with arthritis in the Canadian population, using binary logistic regression analysis on data from the Canadian Community Health Survey, which includes 104,359 respondents. Firstly, we explored the association between arthritis and various aspects of health status including self-reported lifestyle factors. Secondly, we examined the associations between self-reported dietary intake and smoking status with general, mental, and oral health, and sleep disturbance among individuals both with and without arthritis. Our analysis revealed that individuals with arthritis reported considerably poorer general, mental, and oral health, and poorer sleep quality compared to those without arthritis. Associations were also found between self-reported dietary intake and various measures of health status in individuals with arthritis. Smoking and exposure to passive smoking were associated not only with arthritis but also with compromised sleep quality and poorer general, mental, and oral health in people with and without arthritis. This study highlights the need for personalized and holistic approaches that may include a combination of dietary interventions, oral health improvements, sleep therapies, and smoking cessation for improved arthritis prevention and care.

Keywords: arthritis; lifestyle; diet; smoking; sleep; health status; logistic regression

1. Introduction

According to the latest update from the Arthritis Community Research and Evaluation Unit (ACREU) on the status of arthritis in Canada released in 2019, around six million (~1 in 5) Canadians have been impacted by arthritis [1]; these authors further estimate this number will grow to over 24% by 2040 [2]. Arthritis is a chronic disease that comprises over 100 different diseases affecting joints, surrounding and other connective tissues and is associated with pain, discomfort, major functional disability, and activity restriction [1–3]. Moreover, it is associated with poor general health and a higher prevalence of other chronic diseases/comorbidities including cardiovascular diseases, diabetes, cancer, metabolic syndrome, sleep disturbance, and mental health issues including depression, mood disorders, and anxiety [4–7]. Arthritis is also more prevalent in women, older adults, in individuals who are overweight, and in people who have lower income and/or lower education levels [1].

Studies have also shown that people with arthritis are more likely to report sleep disturbances compared to people without arthritis [8–13]. More specifically, individuals with arthritis report various sleep issues including short sleep duration (i.e., less than 6 h per night), daytime sleepiness, difficulty staying awake and going to sleep, non-refreshing sleep, poor to fair quality of sleep, and insomnia [11,13]. Also, the frequency and type of sleep disturbance varies depending on the type of arthritis [13]. In both rheumatoid arthritis (RA) and osteoarthritis (OA), greater sleep disturbance is correlated with disease activity and increased arthritic joints and joint pain [12,14]. Also, sleep problems in people with arthritis are more frequently reported in younger individuals, women, individuals with lower income, and those with a higher body mass index (BMI) [3,15]. Numerous reasons could explain poor sleep outcomes in people with arthritis, including joint pain and discomfort. Also, another possible reason for a relationship between arthritis and sleep quality is the reciprocal connection between sleep and the immune system [16]. Sleep has an important role in regulating the immune system dynamically through affecting immune cells, including antigen-presenting cell and T-cell distributions and the production of inflammatory cytokines [16]. Therefore, sufficient and high-quality sleep is fundamental to maintaining overall health as it plays a critical role in enhancing immune function. Insufficient sleep and untreated sleep disorders have also been shown to be associated with several medical epidemics and health issues, including poorer general health, oral health, and mental health [17–19].

Another common health concern in individuals with arthritis is mental wellness. According to the Centers for Disease Control and Prevention (CDC), 22.5% and 12.1% of adults with arthritis report symptoms of anxiety and depression, respectively, compared to 10.7% and 4.7% of adults without arthritis [20]. Also, a multi-national cross-sectional study using eighteen population-based surveys from different countries that assessed the association between arthritis and anxiety and mood disorders found that the prevalence of mental disorders is higher in persons with arthritis compared to persons without arthritis [21]. This result was also confirmed by another muti-national study that investigated the relationship between arthritis and mental health outcomes including depression, anxiety, and stress across 46 low- and middle-income countries. This work also showed a higher frequency of arthritis in women compared to men and a lower frequency of arthritis among individuals with more education [22]. Also, the importance of mental health issues in people with arthritis is not limited to its association with poorer RA outcomes and quality of life; mental health issues have also been found to predict a reduced response to treatment due to poor medication compliance [22,23].

In addition to the mentioned health complications, arthritis is associated with a higher probability of struggling with oral health including a higher prevalence of periodontitis [24,25]. In 2008, Pischon et al. found that the association between RA and periodontitis was independent of demographic and lifestyle factors such as sex and age [26].

In people with arthritis, lifestyle factors (e.g., smoking, physical activity, alcohol use, and diet) may also have a role in the development and progression of the disease.

Numerous studies have found a relationship between RA and smoking. For example, a meta-analysis study of 18 case–control and cohort studies in the United States and Europe found that current male and female smokers have a higher risk of developing RA compared to non-smokers. Additionally, the meta-analysis study revealed that heavy smokers, defined as individuals with a smoking history of 20 pack-years or more, have an elevated risk of developing RA regardless of gender [27]. Furthermore, a meta-analysis of six case–control and prospective cohort studies conducted in the United States and Europe reported that individuals exposed to passive smoking have a 12% higher prevalence of RA versus individuals who are not exposed [28].

Several studies have consistently shown that individuals with RA also tend to have lower levels of physical activity compared to those without this issue [29,30]. However, multiple studies have consistently found that exercise and increased physical activity are linked to reduced disease activity and systemic inflammation in individuals with RA [31–33]. Furthermore, various studies have provided evidence that exercise interventions in individuals with RA have had a positive impact on disease symptoms, specifically in reducing pain levels [34], alleviating fatigue [35], and enhancing overall functioning [36].

In studies that examined the effect of alcohol consumption on RA, different prospective and cross-sectional studies have suggested that alcohol consumption reduces both the risk of developing RA and the severity of this disease [37,38]. In a study conducted by Maxwell et al., they observed that alcohol consumption was linked to a significant and dose-dependent decrease in susceptibility to RA [39]. Moreover, a prospective study from Sweden found that alcohol consumption was associated with reduced RA disease activity in females but not in males [40]. Moreover, in 2009, Källberg et al. assessed the interaction between alcohol consumption and smoking in influencing RA risk and found that drinking alcohol reduced risk more pronouncedly among smokers than non-smokers [41]. This reduction in risk could be explained by the anti-inflammatory effects associated with moderate alcohol consumption [42,43].

Diet is another lifestyle factor that has been shown to have an impact on RA, including the development and progression of this disease. High-fat diets (HFDs) are an important factor leading to obesity [44]. This relationship is important as a meta-analysis study concluded that obesity was associated with a higher disease activity score 28 in people with RA [45]. A meta-analysis of observational studies also found evidence that the association between obesity and an increased risk of RA is stronger among females than males [46]. Moreover, the intake of high-fructose beverages at least five times a week increases the risk of developing arthritis [47]. In 2014, Hu et al. also found that the consumption of sugar-sweetened soda increases the risk of RA among women [48]. On the other hand, certain dietary components could play a role in reducing the risk and activity of RA. For example, a case-control study has demonstrated that the Mediterranean diet may lower the risk of RA among males with seropositive RA [49]. Kostoglou-Athanassiou et al. have also suggested that omega-3 fatty acids may reduce RA activity [50]. Moreover, in 2018, Tedeschi et al. proposed that consuming fish more than twice a week reduces RA activity [51].

Lifestyle factors not only contribute to the risk and progression of RA but are also linked to the occurrence of sleep disorders. For example, several studies have found that smoking reduces sleep quality, and it is a significant risk factor associated with sleep disorders [52–54]. Also, a Japanese study found that women who were current smokers had more difficulty initiating sleep than women who had never smoked [55]. The association of alcohol use with sleep is a complex phenomenon. Many studies have also shown an increased prevalence of insomnia among alcohol-dependent individuals [56,57]. Furthermore, alcohol dependence is associated with obstructive sleep apnea and insufficient sleep duration [58,59]. The association between sleep disorders and alcohol consumption may also vary according to gender and race [60]. Also, physical activity is a lifestyle factor that contributes to improved health and reduces the risk of chronic diseases [61]. Diet is also a lifestyle factor that has a complex relationship with sleep. For example, St-Onge et al. found that a high-carbohydrate intake reduces sleep onset latency and a high-fat intake decreases sleep efficiency [62]. A cross-sectional study also found an association between an unhealthy diet and poor sleep quality [63], and Campanini et al. suggested that following a Mediterranean diet decreases the risk of poor sleep quality [64]. Also, an observational study found that individuals with insomnia reported a higher intake of fat than individuals without sleep disorders [65]. However, a conspicuous knowledge gap persists regarding understanding the association of lifestyle factors with sleep quality in people with arthritis. Closing this scientific gap is imperative for the development of targeted interventions specifically tailored to address sleep disturbances in this population, considering their distinctive lifestyle challenges.

Lifestyle factors also have an important impact on mental health. Various studies have found a strong association between smoking and mental health issues such as anxiety and affective disorders [66,67]. Similarly, individuals consuming higher amounts of alcohol are more likely to have mental health issues, such as anxiety and depression, compared to those who do not [68,69]. Mental health is also affected by physical activity. Physical activity has been shown to have positive psychosocial outcomes and a beneficial effect on reducing anxiety in young people [70]. Dietary behaviors have also been shown to have an influence on mental health. A healthy diet is associated with a reduced risk of mental health issues [71]. Also, a cross-sectional study showed that high fruit and vegetable intake are important for mental and physical health [72]; moreover, diets rich in crucial nutrients such as omega-3 fatty acids and B vitamins have an important role in promoting good mental health [73]. However, there is a notable gap in understanding the association of these lifestyle factors with mental health in individuals with arthritis; addressing this scientific gap helps to better understand and address the intricate relationships between lifestyle choices and mental well-being in people with arthritis.

Given the aforementioned associations, we hypothesize that positive lifestyle factors, such as modifying dietary intake and smoking status, could be positively associated with various measures of health status in people with arthritis. These health status measures include sleep disturbances, mental health issues, general health problems, and oral health issues, which often co-occur with the main symptoms of arthritis (such as joint pain and swelling). To confirm this hypothesis, we used data from the Canadian Community Health Survey (CCHS), which is a large Canadian annual health survey that captures information on chronic diseases, health status, and lifestyle factors in a representative sample of Canadian residents. To the best of our knowledge, this is the first study to examine the associations between lifestyle traits and arthritis and also association of various lifestyle traits on different health status measures in people with and without arthritis in a large sample of Canadians.

The current study design consists of two primary aims. The first aim of this study is to a) investigate associations between arthritis and demographic characteristics as well as various aspects of self-reported health status (i.e., sleep quality, general health, mental health, and oral health) and b) investigate associations between arthritis and self-reported lifestyle traits (i.e., dietary intake, smoking status, drinking status, and physical activity). The second aim of this study is to a) assess associations between self-reported lifestyle traits (i.e., dietary intake and smoking status) and health status (i.e., sleep quality and general, mental, and oral health) and b) investigate the associations between sleep quality and duration and health status (i.e., general, mental, and oral health) in individuals with and without arthritis, separately (Figure 1).

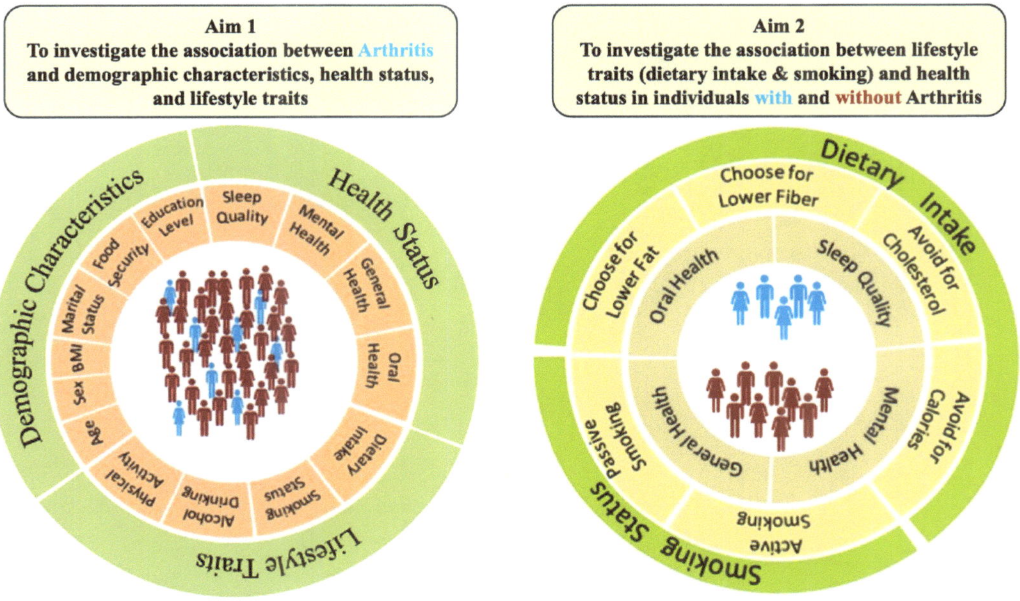

Figure 1. Study design and research question. The study design consists of two primary aims. The first aim of this study is to investigate associations between arthritis and demographic characteristics, health status, and lifestyle traits. The second aim is to investigate the association of lifestyle traits including dietary intake (choose lower fat/higher fiber and avoid cholesterol/calories) and smoking status (active/passive) with health status including sleep quality and general, mental, and oral health in individuals with and without arthritis. "Created with BioRender.com.".

2. Materials and Methods

2.1. Study Design, Data Source and Population

Data for this cross-sectional study were obtained from the 2017–2018 Canadian Community Health Survey (CCHS) public use microdata file. The CCHS is an annual cross-sectional survey with the central objective of gathering information to monitor health status, the use of health care services, and health determinants in the general population of Canadians aged 12 years and older. The 2017–2018 CCHS collected data from approximately 113,000 respondents. This survey provides information on a sample of representative individuals in the ten provinces and three territories of Canada. This survey excluded persons living on reserves and other Aboriginal settlements, youth aged 12–17 years living in foster homes, full-time members of Canadian Forces, institutionalized populations, and persons living in the Quebec health regions of Nunavik and Terres-Cries-de-la-Baie-James, which altogether encompassed less than 3% of the target population. Further details on sampling strategies and questionnaires are described elsewhere [74]. The CCHS microdata file is publicly available and was therefore deemed exempt from review by the University of Saskatchewan Research Ethics Board.

In this study, we selected individuals aged 18 years and older who answered a question in the survey regarding arthritis status. Figure 2 shows participant flow.

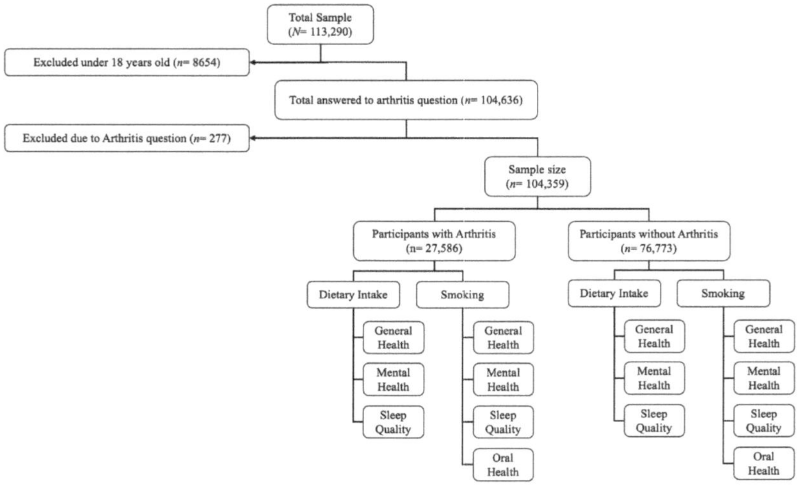

Figure 2. Flow diagram showing the final analytic sample for each outcome from the 2017–2018 CCHS.

2.2. Measures

2.2.1. Outcome Variables

1. For the first research aim, the outcome of interest was self-reported arthritis status derived from the question: "Do you have arthritis, for example osteoarthritis, rheumatoid arthritis, gout or any other type, excluding fibromyalgia?" ("Yes", "No").
2. For the second research aim, we separated the population based on their answer to the question on arthritis status mentioned above. With these two separate groupings (i.e., those who self-reported having arthritis; those who self-reported not having arthritis), we assessed the associations of a) self-reported lifestyle traits (i.e., dietary intake and smoking status) on several health status measures of interest including sleep quality, general health, mental health, and oral health and b) the associations of sleep quality and duration on health status measures of interest including general health, mental health, and oral health. It should be noted that in the second part of the second research question (i.e., 2b), "sleep quality" and "duration" are treated as independent variables, whereas for the first study aim they are considered outcome variables, along with general health, mental health, and oral health. These outcomes are described in more detail below.

Sleep quality: Responses to three questions related to sleep quality were used as separate outcomes. The three questions were as follows: "How often do you find your sleep refreshing?"; "How often do you find it difficult to stay awake when you want to?"; "How often do you have trouble going to sleep or staying asleep?". Responses for all three questions included "Never", "Rarely", "Sometimes", "Most of the time", and "All of the time." Respondents who answered "Never" or "Rarely" were classified as not refreshed, not troubled, and no difficulty sleeping, respectively.

General health, mental health, and oral health: The overall self-reported status of general health, mental health, and oral health was derived from the questions asking "In general, would you say your health/mental health/oral health is...?". Responses included "Excellent", "Very Good", "Good", "Fair", and "Poor". Responses were grouped into the following categories: "Poor" (fair/poor) and "Good" (excellent/very good/good).

Self-reported chronic conditions were also included, including diabetes, cancer, high blood cholesterol, and high blood pressure. The self-reported status of these conditions was derived from the questions asking "Do you have diabetes?", "Do you have cancer?",

"Do you have high blood pressure?", and "Do you have high blood cholesterol or lipids?". Response options included "Yes" and "No" and were grouped into those two categories.

Mental health (i.e., mood and anxiety disorders): Mood and anxiety disorders were derived from the following questions: "Do you have a mood disorder such as depression, bipolar disorder, mania or dysthymia?" and "Do you have an anxiety disorder such as a phobia, obsessive-compulsive disorder or a panic disorder?". Response options included "Yes" and "No" and were grouped into those two categories.

Oral health: In addition to overall oral health, several self-reported oral health outcomes were also included, including bleeding gums, mouth pain, mouth dryness, and uncomfortable eating. The status of bleeding gums, mouth pain, mouth dryness, and uncomfortable eating were derived from the questions asking "How often have you had bleeding gums while brushing or flossing your teeth?", "How often have you had any other persistent or ongoing pain anywhere in your mouth?", "How often have you had persistent dry mouth?", and "How often have you found it uncomfortable to eat any food because of problems with your mouth?". Responses included "Often", "Sometimes", "Rarely", and "Never". Responses were grouped into the following categories: "Yes" (often/sometimes) and "No" (rarely/never).

2.2.2. Independent Variables

Several independent variables of interest were included, which are described in detail below:

Sociodemographic variables: The independent variables of interest for both study aims included respondent self-reported demographic information (sex (male/female); age (years, grouped as follows: 65 years and older, 50–64 years, 35–49 years, and 18–34 years); marital status (married/common law, widowed/divorced/separated, and single); education (less than secondary school graduation, secondary school graduation, and post-secondary certificate diploma or university degree); food security (moderate/severe food insecurity and food secure).

Sleep quality variables: For the first research aim, the sleep variables included as independent variables were the same as listed for the outcome variables described earlier for the second research aim.

General health/mental health/oral health: For the first research aim, the general health, mental health, and oral health variables that were included as independent variables were the same as described for the outcome variables discussed earlier for the second research aim.

Lifestyle variables: Several independent variables related to self-reported lifestyle traits were also included. These variables are described in more detail below.

- Body mass index (BMI): BMI (kg/m^2) was calculated using self-reported height and weight according to international standard classification: underweight (<18.5); normal weight (18.5–24.9); overweight (25–29.9); obese class I, II, and III ($\geq$30).
- Self-reported food choices: Information on the self-reported intake of fat, fiber, cholesterol, and calories was also included. The following questions were used to derive this information: "Do you choose certain foods because of the lower fat content?" (responses included "Yes" and "No"), "Do you choose certain foods because of the fibre content?" (responses included "Yes" and "No"), "Do you avoid certain foods because of the cholesterol content?" (responses included "Yes" and "No"), and "Do you avoid certain foods because of the calorie content?" (responses included "Yes" and "No"). Responses to each of these questions were grouped into two categories: "Yes" and "No."
- Fruit and vegetable consumption: The total self-reported daily consumption of fruits and vegetables was categorized as "Eats fruits and vegetables less than 5 times per day", "Eats fruits and vegetables between 5 and 10 times per day", and "Eats fruits and vegetables more than 10 times per day." We recategorized the total fruit and vegetable

consumption as "Eats fruits and vegetables less than 5 times per day" and "Eats fruits and vegetables equal or more than 5 times per day."

- Smoking status: Smoking status was categorized based on the respondents' self-reported smoking habits. The response options included "Current daily smoker and current occasional smoker", "Former daily smoker (non-smoker now) and former occasional smoker (non-smoker now)", "Lifetime abstainer (never smoked a whole cigarette", and "Experimental smoker (at least 1 cig, non-smoker now)." We recategorized these response options into "Current daily/occasional smoker (current smoker)", "Former daily/occasional smoker (former smoker)" and "Experimental smoker or lifetime abstainer (never smoker)."
- Exposure to second-hand smoke: Information on self-reported second-hand smoking exposure was also included by using information captured from the following question: "Including both household members and regular visitors, does anyone smoke inside your home, every day or almost every day?" Responses were grouped into two groups: "Yes" and "No".
- Alcohol use: Self-reported alcohol drinking habits were categorized as "Regular drinkers who drink at least once a month to every day", "Occasional drinkers who drink less than once a month", and "Never drinkers who never had a drink or did not drink lifetime or during the past 12 months."
- Physical activity: According to Canadian Physical Activity Guideline (CPAG), adults are recommended to have at least 150 min of moderate-to-vigorous-intensity aerobic activity per week. Self-reported physical activity was categorized according to the CPAG, which included "Physically active at/above recommended level from CPAG" for individuals with equal to or more than 150 min of activity, "Physically active below recommended level from CPAG" for individuals with less than 150 min of activity, and "No physical activity" for individuals with 0 min of activity reported.

2.3. Data Analysis

Descriptive statistics were used to determine the frequencies of the independent variables cross-tabulated with the outcome variables of interest.

For the first research aim, we examined the association between arthritis and each demographic characteristic (sex, age, marital status, education, and food security), self-reported lifestyle trait (i.e., dietary intake, smoking status, drinking status, BMI, and physical activity), and health status measure (general health, mental health, oral health, and sleep quality) using univariate logistic regression. For the second research aim, first we identified the five most important and well-known arthritis confounders amongst the demographic characteristics (age and sex) and lifestyle traits (BMI, smoking status and drinking status). We then conducted univariate logistic regression analysis to assess the crude association between each independent variable of interest (lifestyle traits including dietary intake and smoking status) and the outcome variables (sleep quality, general health, mental health, and oral health) in people with and without arthritis separately (Model Type 1). For Model Type 2, we included demographic characteristics (age and sex). The final model (Model Type 3) included BMI, smoking status and drinking status as covariates in addition to age and sex. The significant contribution of adding BMI, smoking status (where relevant), and drinking status was confirmed by the forward selection based on likelihood ratio tests. The Hosmer Lemeshow goodness-of-fit test was used to assess the goodness of fit with a p-value > 0.05 indicating the model is a good fit.

Participants were weighted by the sampling weight provided in the CCHS microdata file allowing for the findings to be attributed to the general Canadian population [75]; these sample weights were applied in all analyses. All analyses were carried out using SPSS software, Version 28.0 (IBM, Armonk, NY, USA).

3. Results

3.1. Association of Sociodemographic Factors, Lifestyle Traits, and Other Aspects of Health Status with Arthritis

Regression statistics were conducted to examine the association of arthritis with various factors, including sociodemographic factors, lifestyle traits, and other aspects of health status.

3.1.1. Sociodemographic Information

A total of 104,359 participants (54.2% females and 45.8% males) constituted the analytical sample; 27,586 (26.4%) participants self-reported arthritis. In total, females consisted of 61.5% of all participants who self-reported arthritis. Females were 50% more likely to self-report arthritis compared to males (odds ratio (OR) 1.50; 95% confidence interval (CI) 1.46–1.55). The analysis showed that adults ≥65 years of age and 50–64 years of age were 31.09 (95% CI: 28.69–33.70) and 14.57 (95% CI: 13.43–15.80) times more likely to self-report arthritis compared to those who were 18–34 years of age, respectively. Married/common law participants were 1.95 (95% CI: 1.88–2.04) times more likely to self-report arthritis compared to single participants. Moreover, participants with less than a secondary school education were 2.42 (95% CI: 2.34–2.51) times more likely to self-report arthritis compared to those who had post-secondary education degrees. Additionally, we found that individuals experiencing moderate to severe food insecurity were 1.18 (95% CI: 1.12–1.24) times more likely to self-report arthritis when compared to individuals who were food secure. More details information are presented in Table 1.

Table 1. Sociodemographic characteristics of participants with and without arthritis.

Variable (*N* = 104,359)	**W/O Arthritis (%) 76,773 (73.6%)**	**With Arthritis (%) 27,586 (26.4%)**	**OR † (95% CI)**	***p*-Value**
Sex, *n* = 104,359 (100% out of *N*)				
Male	37,226 (48.5%)	10,620 (38.5%)	1.00 (ref)	
Female	39,547 (51.5%)	16,966 (61.5%)	1.50 (1.46–1.55)	<0.001
Age (years), *n* = 104,359 (100% out of *N*)				
18–34	21,427 (27.9%)	659 (2.4%)	1.00 (ref)	
35–49	19,427 (25.3%)	2449 (8.9%)	4.10 (3.75–4.48)	<0.001
50–64	19,419 (25.3%)	8700 (31.5%)	14.57 (13.43–15.80)	<0.001
≥65	16,500 (21.5%)	15,778 (57.2%)	31.09 (28.69–33.70)	<0.001
Marital status, *n* = 104,118 (99.8% out of *N*)				
Single	20,381 (26.6%)	3460 (12.6%)	1.00 (ref)	
Married/common law	43,011 (56.2%)	14,265 (51.8%)	1.95 (1.88–2.04)	<0.001
Widowed/divorced/separated	13,210 (17.2%)	9791 (35.6%)	4.37 (4.18–4.57)	<0.001
Education, *n* = 102,832 (98.5% out of *N*)				
Post-secondary, diploma, or university degree	48,134 (63.5%)	14,129 (52.2%)	1.00 (ref)	
Secondary school graduation	18,309 (24.2%)	6299 (23.3%)	1.17 (1.13–1.21)	<0.001
Less than secondary school graduation	9327 (12.3%)	6634 (24.5%)	2.42 (2.34–2.51)	<0.001
Food Security, *n* = 102,848 (98.6% out of *N*)				
Food secure	69,251 (91.5%)	24,437 (90.1%)	1.00 (ref)	
Moderately/severely food insecure	6470 (8.5%)	2690 (9.9%)	1.18 (1.12–1.24)	<0.001

Sociodemographic characteristics of participants with and without arthritis of the Canadian cohort aged 18 years and older (Canadian Community Health Survey, 2017–2018). Data presented as odds ratios and 95% confidence intervals. † Logistic regression. Data considered statistically significant when *p*-value ≤ 0.05. *N* = total participants; *n* = participants within variable; W/O = without; OR = odds ratio; CI = confidence interval; ref = reference.

3.1.2. Lifestyle Traits

Table 2 summarizes the association between self-reported lifestyle traits (such as BMI, food intake, physical activity, alcohol intake, and smoking) and arthritis. Participants who self-reported a BMI that was overweight or obese were 1.45 (95% CI: 1.40–1.50) and 2.11

(95% CI: 2.04–2.19) times more self-report to have arthritis compared to those who were normal weight, respectively. Participants who were not physically active were 2.22 (95% CI: 2.15–2.29) times more likely to self-report arthritis compared to those who had physical activity levels meeting or exceeding the recommended level from CPAG. However, regular alcohol drinkers were 46% less likely to have arthritis compared to those who did not drink.

Table 2. Lifestyle traits of participants with and without arthritis.

Variable	W/O Arthritis (%)	With Arthritis (%)	OR † (95% CI)	*p*-Value
Self-reported BMI, *n* = 97,768 (93.7% out of *N*)				
Normal weight	30,784 (42.6%)	7821 (30.7%)	1.00 (ref)	
Underweight	1480 (2.1%)	410 (1.6%)	1.09 (0.98–1.22)	0.130
Overweight	25,028 (34.6%)	9241 (36.2%)	1.45 (1.40–1.50)	<0.001
Obese—Class I, II, III	14,974 (20.7%)	8030 (31.5%)	2.11 (2.04–2.19)	<0.001
Physical activity indicator, *n* = 102,350 (98.1% out of *N*)				
≥CPAG level	42,713 (56.6%)	11,466 (42.7%)	1.00 (ref)	
<CPAG level	17,384 (23.0%)	6265 (23.3%)	1.34 (1.30–1.39)	<0.001
No physical activity	15,373 (20.4%)	9149 (34.0%)	2.22 (2.15–2.29)	<0.001
Alcohol Drinking, *n* = 103,926 (99.6% out of *N*)				
No drinking in the last year	13,869 (18.1%)	7542 (27.5%)	1.00 (ref)	
Occasional	12,643 (16.6%)	5209 (19.0%)	0.76 (0.73–0.79)	<0.001
Regular	49,959 (65.3%)	14,704 (53.5%)	0.54 (0.52–0.56)	<0.001
Food Choice—Choosing lower fat content, *n* = 13,740 (13.1% of *N*)				
Not choosing	4376 (42.3%)	1175 (34.6%)	1.00 (ref)	
Choosing	5969 (57.7%)	2220 (65.4%)	1.39 (1.28–1.50)	<0.001
Food Choice—Choosing fiber content, *n* = 13,737 (13.1% of *N*)				
Not choosing	4186 (40.5%)	1075 (31.6%)	1.00 (ref)	
Choosing	6152 (59.5%)	2324 (68.4%)	1.47 (1.36–1.60)	<0.001
Food Choice—Avoiding cholesterol content, *n* = 13,714 (13.1% of *N*)				
Not avoiding	6148 (59.5%)	1802 (53.2%)	1.00 (ref)	
Avoiding	4177 (40.5%)	1587 (46.8%)	1.30 (1.20–1.40)	<0.001
Food Choice—Avoiding calorie content, *n* = 13,734 (13.1% of *N*)				
Not avoiding	4962 (48.0%)	1595 (47.0%)	1.00 (ref)	
Avoiding	5378 (52.0%)	1799 (53.0%)	1.04 (0.96–1.13)	0.315
Total daily consumption of fruits and vegetables, *n* = 2241 (2.1% out of *N*)				
≥5 times	442 (24.4%)	101 (23.7%)	1.00 (ref)	
<5 times	1373 (75.6%)	325 (76.3%)	1.04 (0.81–1.33)	0.780
Smoking, *n* = 103,935 (99.6% out of *N*)				
Never	40,974 (53.6%)	11,328 (41.3%)	1.00 (ref)	
Current	14,638 (19.1%)	4852 (17.7%)	1.20 (1.15–1.25)	<0.001
Former	20,914 (27.3%)	11,229 (41.0%)	1.94 (1.88–2.00)	<0.001
Exposure to second hand smoke, *n* = 27,528 (26.3% of *N*)				
Not passive smoker	18,691 (90.8%)	6155 (88.5%)	1.00 (ref)	
Passive smoker	1885 (9.2%)	797 (11.5%)	1.28 (1.18–1.40)	<0.001

Lifestyle traits of participants with and without arthritis of the Canadian cohort aged 18 years and older (Canadian Community Health Survey, 2017–2018). Data presented as odds ratios and 95% confidence intervals. † Logistic regression. Data considered statistically significant when *p*-value ≤ 0.05. CPAG = Canada's Physical Activity Guide; BMI = body mass index; *N* = total participants; *n* = participants within variable; W/O = without; OR = odds ratio; CI = confidence interval; ref = reference.

The number of respondents who answered both questions about arthritis and dietary intake was around 13% of the total population. Our results found there was a statistically significant association between self-reported arthritis and food choice. Participants who choose lower fat foods, choose foods because of fiber content and avoid foods because of cholesterol content were more likely to self-report arthritis compared to individuals who did not follow those behaviors (lower fat content: OR 1.39, 95% CI: 1.28–1.50; fiber content: OR 1.47, 95% CI: 1.36–1.60; avoiding cholesterol content: OR 1.30, 95% CI: 1.20–1.40). However, the association was not statistically significant between arthritis and avoiding calorie content (*p*-value: 0.315). In addition, we found no statistically significant association between self-reported arthritis and fruit and vegetable intake (*p*-value: 0.780).

For smoking, the number of respondents who answered both questions about arthritis and smoking status was 99.6% of the total population. Current and former smokers were more likely to self-report arthritis (current smoker: OR 1.20, 95% CI: 1.15–1.25; former smoker: OR 1.94, 95% CI: 1.88–2.00) compared to never smokers, respectively. Regarding exposure to second-hand smoke, those with exposure to second hand smoke at home were significantly more likely to self-report arthritis compared to those who did not have those exposures (OR 1.28, 95% CI: 1.18–1.40) (Table 2).

3.1.3. Sleep Quality

The analysis of the association between sleep quality and arthritis revealed that participants who found their sleep refreshing had a 32% lower likelihood of self-reporting arthritis compared to those who did not find their sleep refreshing. In addition, participants who experienced difficulty staying awake (OR 1.21, 95% CI: 1.16–1.27) and had trouble going to sleep (OR 1.63, 95% CI: 1.57–1.70) were more likely to self-report arthritis compared to participants who did not report these sleep issues. Furthermore, participants who slept ≥7 h per night were 21% less likely to self-report having arthritis compared to those who slept <7 h per night (OR 0.79, 95% CI: 0.76–0.82) (Table 3).

Table 3. Health status of participants with and without arthritis.

Variable		W/O Arthritis (%)	With Arthritis (%)	OR † (95% CI)	*p*-Value
Sleep Quality					
Refreshing sleep	No	5311 (13.7%)	2336 (19.0%)	1.00 (ref)	
n = 50,918 (48.8% out of *N*)	Yes	33,327 (86.3%)	9944 (81.0%)	0.68 (0.64–0.72)	<0.001
Difficulty staying awake	No	28,346 (73.5%)	8528 (69.6%)	1.00 (ref)	
n = 50,849 (48.7% out of *N*)	Yes	10,245 (26.5%)	3730 (30.4%)	1.21 (1.16–1.27)	<0.001
Having trouble going to sleep	No	20,454 (52.8%)	5018 (40.7%)	1.00 (ref)	
n = 51,018 (48.9% out of *N*)	Yes	18,248 (47.2%)	7298 (59.3%)	1.63 (1.57–1.70)	<0.001
Spent ≥7 h per night sleeping	No	15,801 (40.9%)	5734 (46.8%)	1.00 (ref)	
n = 50,834 (48.7% out of *N*)	Yes	22,788 (59.1%)	6511 (53.2%)	0.79 (0.76–0.82)	<0.001
General Health Condition					
General health	Good	69,377 (90.6%)	19,584 (71.2%)	1.00 (ref)	
n = 104,198 (99.8% out of *N*)	Poor	7303 (9.5%)	7934 (28.8%)	3.85 (3.71–3.99)	<0.001
Diabetes	No	71,314 (93.0%)	22,923 (83.2%)	1.00 (ref)	
n = 104,254 (99.9% out of *N*)	Yes	5393 (7.0%)	4624 (16.8%)	2.67 (2.56–2.78)	<0.001
Cancer	No	75,375 (98.3%)	26,450 (96.2%)	1.00 (ref)	
n = 104,177 (99.8% out of *N*)	Yes	1304 (1.7%)	1048 (3.8%)	2.29 (2.11–2.49)	<0.001
High total cholesterol	No	67,143 (88.4%)	19,890 (23.5%)	1.00 (ref)	
n = 102,983 (98.7% out of *N*)	Yes	8793 (11.6%)	7157 (26.5%)	2.75 (2.65–2.85)	<0.001
High blood pressure	No	63,297 (82.7%)	16,294 (59.3%)	1.00 (ref)	
n = 104,009 (99.7% out of *N*)	Yes	13,237 (17.3%)	11,181 (40.7%)	3.28 (3.18–3.38)	<0.001
Mental Health Condition					
Mental health	Good	69,782 (93.0%)	23,618 (89.2%)	1.00 (ref)	
n = 101,540 (97.3% out of *N*)	Poor	5272 (7.0%)	2868 (10.8%)	1.61 (1.53–1.69)	<0.001
Mood disorder	No	69,915 (91.2%)	23,669 (86.0%)	1.00 (ref)	
n = 104,170 (99.8% out of *N*)	Yes	6730 (8.8%)	3856 (14.0%)	1.69 (1.62–1.77)	<0.001
Anxiety disorder	No	69,963 (91.3%)	24,151 (87.8%)	1.00 (ref)	
n = 104,148 (99.8% out of *N*)	Yes	6674 (8.7%)	3360 (12.2%)	1.46 (1.40–1.52)	<0.001
Oral Health Condition					
Oral health	Good	35,190 (90.3%)	11,580 (85.5%)	1.00 (ref)	
n = 52,510 (50.3% out of *N*)	Poor	3769 (9.7%)	1971 (14.5%)	1.59 (1.50–1.69)	<0.001
Bleeding gums	No	26,893 (74.0%)	8947 (80.4%)	1.00 (ref)	
n = 47,468 (45.5% out of *N*)	Yes	9445 (26.0%)	2183 (19.6%)	0.70 (0.66–0.73)	<0.001
Mouth pain	No	34,837 (89.4%)	11,554 (85.2%)	1.00 (ref)	
n = 52,520 (50.1% out of *N*)	Yes	4126 (10.6%)	2003 (14.8%)	1.46 (1.38–1.55)	<0.001
Mouth dryness	No	31,876 (81.9%)	8491 (62.7%)	1.00 (ref)	
n = 52,463 (50.3% out of *N*)	Yes	7041 (18.1%)	5055 (37.3%)	2.70 (2.58–2.82)	<0.001
Uncomfortable eating	No	33,025 (84.8%)	10,649 (78.6%)	1.00 (ref)	
n = 52,515 (50.3% out of *N*)	Yes	5938 (15.2%)	2903 (21.4%)	1.52 (1.44–1.59)	<0.001

Health condition of participants with and without arthritis of the Canadian cohort (Canadian Community Health Survey, 2017–2018). Data presented as odds ratios (ORs) and 95% confidence intervals. † Logistic regression. Data considered statistically significant when *p*-value ≤ 0.05. *N* = total participants; *n* = participants within variable; W/O = without; OR = odds ratio; CI = confidence interval; ref = reference.

3.1.4. General Health

We found a significant association between self-reported general health and arthritis. More specifically, we found that the participants with poor general health were more likely to self-report arthritis compared to those with good general health (OR 3.85, 95% CI: 3.71–3.99) (Table 3). Moreover, our findings demonstrated a strong association between chronic health status and the occurrence of arthritis. Participants with chronic health conditions, including diabetes (OR 2.67, 95% CI: 2.56–2.78), cancer (OR 2.29, 95% CI: 2.11–2.49), high total cholesterol (OR 2.75, 95% CI: 2.65–2.85), and high blood pressure (OR 3.28, 95% CI: 3.18–3.38) were more likely to self-report arthritis compared to those without a chronic health condition (Table 3).

3.1.5. Mental Health

When looking at the association between mental health and arthritis, participants with poor mental health were 61% more likely to self-report arthritis (OR 1.61, 95% CI: 1.53–1.69) compared to those self-reporting good mental health (Table 3). Additionally, our findings revealed that individuals who self-reported mood disorders (OR 1.69, 95% CI: 1.62–1.77) and anxiety disorders (OR 1.46, 95% CI: 1.40–1.52) were more likely to self-report arthritis compared to those without mood disorders and anxiety disorders, respectively (Table 3).

3.1.6. Oral Health

Our results identified that the participants with poor oral health were more likely to self-report arthritis compared to those with good oral health (OR 1.59, 95% CI: 1.50–1.69). Furthermore, our findings revealed that participants who self-reported mouth pain (OR 1.46, 95% CI: 1.38–1.55) and mouth dryness (OR 2.70, 95% CI: 2.58–2.82) were more likely to self-report arthritis compared to those without mouth pain and mouth dryness, respectively. Moreover, we found that the participants who felt uncomfortable eating (OR 1.52, 95% CI: 1.44–1.59) were more likely to self-report arthritis compared to those who did not. On the other hand, the participants who had bleeding gums were 30% less likely to self-report arthritis compared to those who had healthy gums (Table 3).

3.2. The Association between Lifestyle Traits and Health Status in Participants with and without Arthritis

A binary logistic regression analysis was used to generate different models to examine the association between lifestyle traits (i.e., various self-reported dietary intake behaviors and smoking) and health status (sleep quality, general health, mental health, and oral health) in two different groups (i.e., (a) people with arthritis and (b) people without arthritis). Three logistic regression models were run for each association:

- Model 1 examined the (crude) unadjusted association;
- Model 2 examined the association, while adjusting for the effects of age and sex;
- Model 3 examined the association, while adjusting for age, sex, BMI, smoking status (where relevant) and drinking status.

Tables 4 and 5 provide the association information (OR) for each relationship for Model 3; this information will also be described in the following sections. Information about the associations for Models 1 and 2 is provided in Supplementary Materials (Tables S1 and S2).

Table 4. Association between sleep quality, general health and mental health with dietary intake in participants with and without arthritis.

Variables	Population	OR (95% CI)	*p*-Value
Refreshing sleep			
Individuals who choose lower fat content (ref: not choosing)	Arthritis (n = 3227) W/O arthritis (n = 9900)	1.51 (1.49–1.52) 1.00 (0.99–1.01)	<0.001 0.708

Table 4. *Cont.*

Variables	Population	OR (95% CI)	*p*-Value
Individuals who choose fiber content	Arthritis (*n* = 3230)	1.15 (1.14–1.17)	<0.001
(ref: not choosing)	W/O arthritis (*n* = 9892)	1.18 (1.18–1.19)	<0.001
Individuals who avoid cholesterol content	Arthritis (*n* = 3219)	1.24 (1.23–1.26)	<0.001
(ref: not avoiding)	W/O arthritis (*n* = 9882)	1.19 (1.18–1.20)	<0.001
Individuals who avoid calorie content	Arthritis (*n* = 3227)	1.23 (1.22–1.25)	<0.001
(ref: not avoiding)	W/O arthritis (*n* = 9896)	1.03 (1.03–1.04)	<0.001
Difficulty staying awake			
Individuals who choose lower fat content	Arthritis (*n* = 3215)	1.21 (1.19–1.22)	<0.001
(ref: not choosing)	W/O arthritis (*n* = 9876)	1.24 (1.24–1.25)	<0.001
Individuals who choose fiber content	Arthritis (*n* = 3217)	1.30 (1.28–1.31)	<0.001
(ref: not choosing)	W/O arthritis (*n* = 9868)	1.25 (1.25–1.26)	<0.001
Individuals who avoid cholesterol content	Arthritis (*n* = 3207)	1.26 (1.24 –1.27)	<0.001
(ref: not choosing)	W/O arthritis (*n* = 9859)	1.25 (1.24–1.26)	<0.001
Individuals who avoid calorie content	Arthritis (*n* = 3214)	1.14 (1.13–1.16)	<0.001
(ref: not avoiding)	W/O arthritis (*n* = 9873)	1.17 (1.16–1.17)	<0.001
Trouble going to sleep			
Individuals who choose lower fat content	Arthritis (*n* = 3228)	0.95 (0.94–0.96)	<0.001
(ref: not choosing)	W/O arthritis (*n* = 9908)	1.09 (1.08–1.09)	<0.001
Individuals who choose fiber content	Arthritis (*n* = 3231)	1.12 (1.11–1.14)	<0.001
(ref: not choosing)	W/O arthritis (*n* = 9900)	1.04 (1.04–1.05)	<0.001
Individuals who avoid cholesterol content	Arthritis (*n* = 3221)	1.14 (1.13–1.16)	<0.001
(ref: not avoiding)	W/O arthritis (*n* = 9890)	1.13 (1.12–1.13)	<0.001
Individuals who avoid calorie content	Arthritis (*n* = 3228)	1.07 (1.06–1.08)	<0.001
(ref: not avoiding)	W/O arthritis (*n* = 9904)	1.33 (1.32–1.33)	<0.001
Spent ≥7 h per night sleeping			
Individuals who choose lower fat content	Arthritis (*n* = 3204)	1.00 (0.99–1.01)	0.785
(ref: not choosing)	W/O arthritis (*n* = 9878)	0.96 (0.95–0.96)	<0.001
Individuals who choose fiber content	Arthritis (*n* = 3206)	1.01 (1.00–1.02)	0.064
(ref: not choosing)	W/O arthritis (*n* = 9870)	1.06 (1.05–1.06)	<0.001
Individuals who avoid cholesterol content	Arthritis (*n* = 3196)	0.88 (0.87–0.89)	<0.001
(ref: not avoiding)	W/O arthritis (*n* = 9860)	0.83 (0.83–0.84)	<0.001
Individuals who avoid calorie content	Arthritis (*n* = 3203)	1.00 (0.98–1.01)	0.301
(ref: not avoiding)	W/O arthritis (*n* = 9874)	0.94 (0.93–0.94)	<0.001
Perceived general health			
Individuals who choose lower fat content	Arthritis (*n* = 3236)	1.12 (1.10–1.13)	<0.001
(ref: not choosing)	W/O arthritis (*n* = 9908)	1.13 (1.12–1.14)	<0.001
Individuals who choose fiber content	Arthritis (*n* = 3240)	1.06 (1.04–1.07)	<0.001
(ref: not choosing)	W/O arthritis (*n* = 9901)	1.11 (1.10–1.12)	<0.001
Individuals who avoid cholesterol content	Arthritis (*n* = 3229)	1.00 (0.99–1.02)	0.519
(ref: not avoiding)	W/O arthritis (*n* = 9889)	0.91 (0.91–0.92)	<0.001
Individuals who avoid calorie content	Arthritis (*n* = 3235)	1.20 (1.18–1.21)	<0.001
(ref: not avoiding)	W/O arthritis (*n* = 9903)	1.38 (1.37–1.39)	<0.001
Perceived mental health			
Individuals who choose lower fat content	Arthritis (*n* = 3228)	1.18 (1.16–1.20)	<0.001
(ref: not choosing)	W/O arthritis (*n* = 9905)	1.30 (1.29–1.31)	<0.001
Individuals who choose fiber content	Arthritis (*n* = 3231)	0.96 (0.95–0.98)	<0.001
(ref: not choosing)	W/O arthritis (*n* = 9896)	1.54 (1.52–1.55)	<0.001
Individuals who avoid cholesterol content	Arthritis (*n* = 3221)	0.99 (0.97–1.00)	0.063
(ref: not avoiding)	W/O arthritis (*n* = 9885)	1.25 (1.24–1.26)	<0.001
Individuals who avoid calorie content	Arthritis (*n* = 3227)	1.28 (1.26–1.30)	<0.001
(ref: not avoiding)	W/O arthritis (*n* = 9900)	1.38 (1.37–1.40)	<0.001
Mood disorder			
Individuals who choose lower fat content	Arthritis (*n* = 3233)	1.04 (1.02–1.05)	<0.001
(ref: not choosing)	W/O arthritis (*n* = 9904)	0.88 (0.87–0.88)	<0.001

Table 4. *Cont.*

Variables	Population	OR (95% CI)	p-Value
Individuals who choose fiber content	Arthritis (n = 3236)	1.22 (1.20–1.24)	<0.001
(ref: not choosing)	W/O arthritis (n = 9896)	0.88 (0.88–0.89)	<0.001
Individuals who avoid cholesterol content	Arthritis (n = 3226)	0.97 (0.95–0.98)	<0.001
(ref: not avoiding)	W/O arthritis (n = 9886)	0.85 (0.84–0.86)	<0.001
Individuals who avoid calorie content	Arthritis (n = 3232)	0.91 (0.90–0.93)	<0.001
(ref: not avoiding)	W/O arthritis (n = 9899)	0.83 (0.83–0.84)	<0.001
Anxiety disorder			
Individuals who choose lower fat content	Arthritis (n = 3231)	0.94 (0.93–0.96)	<0.001
(ref: not choosing)	W/O arthritis (n = 9904)	0.92 (0.91–0.93)	<0.001
Individuals who choose fiber content	Arthritis (n = 3234)	1.03 (1.01–1.05)	<0.001
(ref: not choosing)	W/O arthritis (n = 9896)	0.94 (0.93–0.95)	<0.001
Individuals who avoid cholesterol content	Arthritis (n = 3224)	1.32 (1.30–1.34)	<0.001
(ref: not avoiding)	W/O arthritis (n = 9886)	0.92 (0.91–0.93)	<0.001
Individuals who avoid calorie content	Arthritis (n = 3230)	1.17 (1.15–1.19)	<0.001
(ref: not avoiding)	W/O arthritis (n = 9899)	0.87 (0.86–0.87)	<0.001

Association of sleep quality, general health, and mental health with dietary intake in participants with and without arthritis. Binary logistic regression adjusted for age, sex, BMI, smoking status, and drinking status (Model 3). Data presented as odds ratios (ORs) and 95% confidence intervals; all analyses were weighted. Data considered statistically significant when p-value ≤ 0.05. W/O = without; OR = odds ratio; CI = confidence interval; n = number of respondents; ref = reference.

Table 5. Association between sleep quality, general health, mental health, and oral health with smoking/passive smoking in participants with and without arthritis.

Variables	Population		OR (95% CI)	p–Value
Smokers (ref: never smoker)				
Refreshing sleep in individuals who are a current or former smoker	Arthritis	Current	0.54 (0.53–0.54)	<0.001
	n = 11,818	Former	0.79 (0.78–0.80)	<0.001
	W/O arthritis	Current	0.63 (0.63–0.63)	<0.001
	n = 37,124	Former	0.85 (0.84–0.85)	<0.001
Difficulty staying awake in individuals who are a current or former smoker	Arthritis	Current	1.14 (1.13–1.15)	<0.001
	n = 11,798	Former	0.94 (0.93–0.94)	<0.001
	W/O arthritis	Current	1.08 (1.08–1.08)	<0.001
	n = 37,097	Former	0.92 (0.92–0.92)	<0.001
Trouble going to sleep in individuals who are a current or former smoker	Arthritis	Current	0.99 (0.98–0.99)	<0.001
	n = 11,851	Former	1.13 (1.12–1.14)	<0.001
	W/O arthritis	Current	1.33 (1.32–1.33)	<0.001
	n = 37,184	Former	1.18 (1.17–1.18)	<0.001
Spent $\geq$7 h per night sleeping in individuals who are a current or former smoker	Arthritis	Current	0.91 (0.90–0.92)	<0.001
	n = 11,790	Former	1.05 (1.04–1.05)	<0.001
	W/O arthritis	Current	0.80 (0.79–0.80)	<0.001
	n = 37,104	Former	0.97 (0.97–0.98)	<0.001
Perceived General health in individuals who are a current or former smoker	Arthritis	Current	0.41 (0.40–0.41)	<0.001
	n = 25,223	Former	0.76 (0.76–0.77)	<0.001
	W/O arthritis	Current	0.35 (0.34–0.35)	<0.001
	n = 71,814	Former	0.68 (0.68–0.68)	<0.001
Perceived mental health in individuals who are a current or former smoker	Arthritis	Current	0.41 (0.40–0.41)	<0.001
	n = 25,209	Former	0.84 (0.83–0.84)	<0.001
	W/O arthritis	Current	0.40 (0.40–0.41)	<0.001
	n = 71,801	Former	0.84 (0.84–0.84)	<0.001

Table 5. *Cont.*

Variables	Population		OR (95% CI)	*p*–Value
Mood disorder in individuals who are a current or former smoker	Arthritis *n* = 25,236	Current Former	2.88 (2.86–2.89) 1.43 (1.42–1.44)	<0.001 <0.001
	W/O arthritis *n* = 71,798	Current Former	2.73 (2.72–2.74) 1.58 (1.57–1.59)	<0.001 <0.001
Anxiety disorder in individuals who are a current or former smoker	Arthritis *n* = 25,221	Current Former	2.87 (2.85–2.89) 1.30 (1.29–1.30)	<0.001 <0.001
	W/O arthritis *n* = 71,801	Current Former	2.62 (2.61–2.63) 1.63 (1.63–1.64)	<0.001 <0.001
Perceived oral health in individuals who are a current or former smoker	Arthritis *n* = 12,908	Current Former	0.36 (0.36–0.36) 0.93 (0.92–0.94)	<0.001 <0.001
	W/O arthritis *n* = 37,275	Current Former	0.27 (0.27–0.28) 0.68 (0.68–0.68)	<0.001 <0.001
Mouth pain in individuals who are a current or former smoker	Arthritis *n* = 12,914	Current Former	1.47 (1.46–1.48) 1.11 (1.10–1.11)	<0.001 <0.001
	W/O arthritis *n* = 37,275	Current Former	1.81 (1.80–1.82) 1.22 (1.22–1.23)	<0.001 <0.001
Bleeding gums in individuals who are a current or former smoker	Arthritis *n* = 10,597	Current Former	0.71 (0.71–0.72) 0.97 (0.97–0.98)	<0.001 <0.001
	W/O arthritis *n* = 34,750	Current Former	1.01 (1.00–1.01) 1.03 (1.02–1.03)	0.009 <0.001
Mouth dryness in individuals who are a current or former smoker	Arthritis *n* = 12,905	Current Former	2.03 (2.02–2.04) 1.24 (1.23–1.25)	<0.001 <0.001
	W/O arthritis *n* = 37,235	Current Former	2.08 (2.07–2.08) 1.39 (1.39–1.40)	<0.001 <0.001
Uncomfortable eating in individuals who are a current or former smoker	Arthritis *n* = 12,908	Current Former	1.65 (1.63–1.66) 1.15 (1.14–1.15)	<0.001 <0.001
	W/O arthritis *n* = 37,274	Current Former	1.94 (1.93–1.95) 1.39 (1.39–1.40)	<0.001 <0.001
Passive smokers (ref: not passive smoker)				
Refreshing sleep in individuals who report passive smoke exposure	Arthritis (*n* = 5533) W/O arthritis (*n* = 17,327)		0.85 (0.83–0.86) 0.94 (0.93–0.95)	<0.001 <0.001
Difficulty staying awake in individuals who report passive smoke exposure	Arthritis (*n* = 5524) W/O arthritis (*n* = 17,327)		1.12 (1.10–1.13) 1.07 (1.06–1.08)	<0.001 <0.001
Trouble going to sleep in individuals who report passive smoke exposure	Arthritis (*n* = 5551) W/O arthritis (*n* = 17,361)		1.15 (1.13–1.17) 1.06 (1.05–1.06)	<0.001 <0.001
Spent ≥7 h per night sleeping in individuals who report passive smoke exposure	Arthritis (*n* = 5532) W/O arthritis (*n* = 17,346)		1.02 (1.01–1.04) 0.91 (0.91–0.92)	0.003 <0.001
Perceived general health in individuals who report passive smoke exposure	Arthritis (*n* = 6534) W/O arthritis (*n* = 19,543)		0.69 (0.68–0.70) 0.69 (0.68–0.69)	<0.001 <0.001
Perceived mental health in individuals who report passive smoke exposure	Arthritis (*n* = 6533) W/O arthritis (*n* = 19,537)		0.64 (0.62–0.65) 0.76 (0.75–0.77)	<0.001 <0.001
Mood disorder in individuals who report passive smoke exposure	Arthritis (*n* = 6535) W/O arthritis (*n* = 19,545)		1.18 (1.16–1.20) 1.54 (1.53–1.56)	<0.001 <0.001
Anxiety disorder in individuals who report passive smoke exposure	Arthritis (*n* = 6528) W/O arthritis (*n* = 19,546)		1.61 (1.58–1.64) 1.36 (1.35–1.38)	<0.001 <0.001
Perceived oral health in individuals who report passive smoke exposure	Arthritis (*n* = 5098) W/O arthritis (*n* = 16,205)		0.73 (0.71–0.74) 0.75 (0.74–0.76)	<0.001 <0.001
Mouth pain in individuals who report passive smoke exposure	Arthritis (*n* = 5093) W/O arthritis (*n* = 16,202)		1.84 (1.80–1.87) 1.22 (1.21–1.23)	<0.001 <0.001

Table 5. *Cont.*

Variables	Population	OR (95% CI)	*p*–Value
Bleeding gums in individuals who report passive smoke exposure	Arthritis (*n* = 3785) W/O arthritis (*n* = 14,692)	1.26 (1.24–1.28) 1.03 (1.02–1.04)	<0.001 <0.001
Mouth dryness in individuals who report passive smoke exposure	Arthritis (*n* = 5093) W/O arthritis (*n* = 16,188)	1.46 (1.44–1.48) 1.36 (1.35–1.37)	<0.001 <0.001
Uncomfortable eating in individuals who report passive smoke exposure	Arthritis (*n* = 5091) W/O arthritis (*n* = 16,198)	1.27 (1.25–1.29) 1.33 (1.32–1.34)	<0.001 <0.001

Association of sleep quality, general health, mental health, and oral health with smoking and passive smoking in participants with and without arthritis.. Binary logistic regression adjusted for age, sex, BMI, smoking status, and drinking status. Data presented as odds ratios (ORs) and 95% confidence intervals; all analyses were weighted. Data considered statistically significant when p-value ≤ 0.05. W/O = without; OR = odds ratio; CI = confidence interval; n = number of respondents; ref = reference.

3.2.1. Self-Reported Health Status and Dietary Intake Behaviors

Sleep Quality and Dietary Intake Behaviors

The number of respondents with arthritis who answered questions about self-reported dietary intake behaviors (i.e., choosing lower fat foods, choosing high fiber foods, avoiding high cholesterol foods, and avoiding high calorie foods) and sleep quality (i.e., refreshing sleep, difficulty staying wake, trouble going to sleep, and sleep duration) was around 12% of the sample who reported having arthritis. The number of respondents without arthritis who met these criteria was around 13% of the sample without arthritis (Table 4).

When examining the relationship between sleep quality and self-reported dietary behaviors while adjusting for age, sex, BMI, and smoking and drinking status, our results showed that participants with arthritis who self-reported choosing foods with lower fat content were more likely to report refreshing sleep (OR = 1.51, 95% CI: 1.49–1.52) compared to participants who did not choose these types of foods. However, it is important to note that the association between choosing foods with a lower fat content and refreshing sleep was not found to be statistically significant among participants without arthritis (p-value 0.708). Both participants with and without arthritis who reported choosing foods with a higher fiber content exhibited a higher likelihood of having refreshing sleep (participants with arthritis: OR = 1.15, 95% CI: 1.14–1.17; participants without arthritis: OR = 1.18, 95% CI: 1.18–1.19) compared to those who did not prioritize these foods. Additionally, participants who self-reported avoiding foods for cholesterol content were more likely to experience refreshing sleep (participants with arthritis: OR = 1.24, 95% CI: 1.23–1.26; participants without arthritis: OR = 1.19, 95% CI: 1.18–1.20). Interestingly, those who avoid foods for calorie content were more likely to report refreshing sleep (participants with arthritis: OR = 1.23, 95% CI:1.22–1.25; participants without arthritis: OR = 1.03, 95% CI: 1.03–1.04) compared to participants who did not avoid these foods (Table 4).

When examining the association between dietary choices and staying awake, our results find that participants who self-reported choosing foods with lower fat content were more likely to report difficulty staying awake (participants with arthritis: OR = 1.21, 95% CI:1.19–1.22; participants without arthritis: OR = 1.24, 95% CI: 1.24–1.25) compared to participants who did not. In the same way, the participants who self-reported choosing food with a higher fiber content were more likely to report difficulties staying awake (participants with arthritis: OR = 1.30, 95% CI: 1.28–1.31; participants without arthritis: OR = 1.25, 95% CI: 1.25–1.26) compared to participants who did not. The participants who self-reported avoiding foods for cholesterol content were more likely to have difficulties staying awake (participants with arthritis: OR = 1.26, 95% CI: 1.24–1.27; participants without arthritis: OR = 1.25, 95% CI: 1.24–1.26). Similarly, participants who were avoiding food for calorie content were more likely to report difficulties staying awake (participants with arthritis: OR = 1.14, 95% CI: 1.13–1.16; participants without arthritis: OR = 1.17, 95% CI: 1.16–1.17) compared to participants who were not avoiding such food (Table 4).

When assessing the association between dietary choices and trouble going to sleep, we found that participants with arthritis who self-reported choosing foods with a lower fat content were less likely to have trouble going to sleep (OR = 0.95, 95% CI: 0.94–0.96). However, participants without arthritis who self-reported choosing foods with a lower fat content were more likely to have trouble going to sleep (OR = 1.09, 95% CI: 1.08–1.09) compared to participants who did not. Furthermore, both participants with and without arthritis who self-reported choosing food for fiber content were more likely to report trouble going to sleep (participants with arthritis OR = 1.12, 95% CI: 1.11–1.14; participants without arthritis OR = 1.04, 95% CI: 1.04–1.05) compared to participants who did not. Additionally, participants who self-reported avoiding foods for cholesterol content were more likely to have trouble going to sleep (participants with arthritis: OR = 1.14, 95% CI: 1.13–1.16; participants without arthritis: OR = 1.13, 95% CI: 1.12–1.13). Similarly, those who were avoiding food for calorie content were more likely to report trouble going to sleep (participants with arthritis: OR = 1.07, 95% CI: 1.06–1.08; participants without arthritis: OR = 1.33, 95% CI: 1.32–1.33) compared to participants who were not avoiding such food (Table 4).

We also examined the association between sleep duration and self-reported dietary intake behaviors while adjusting for age, sex, BMI, and smoking and drinking status. Overall, participants without arthritis who self-reported choosing foods with lower fat content were less likely to sleep ≥7 h per night (OR = 0.96, 95% CI: 0.95–0.96) compared to participants who did not make such food choices. However, there was no statistically significant association for this relationship in participants with arthritis (p-value = 0.785). Regarding choosing food for fiber content, there was no statistically significant association observed between sleeping for ≥7 h per night and dietary choices based on fiber content among participants with arthritis (p-value = 0.064), but the participants without arthritis were more likely to sleep ≥7 h per night (OR = 1.06, 95% CI: 1.05–1.06) compared to participants who did not make such food choices. Regarding avoiding cholesterol-containing foods, participants with arthritis and those without arthritis who self-reported avoiding cholesterol-containing foods had lower odds of sleeping ≥7 h per night compared to participants who did not avoid such foods (participants with arthritis: OR = 0.88, 95% CI: 0.87–0.89; participants without arthritis: OR = 0.83, 95% CI: 0.83–0.84). Among participants without arthritis, those who self-reported avoiding foods for calorie content were 6% less likely to spend ≥7 h per night sleeping than those who did not avoid these foods (OR = 0.94, 95% CI: 0.93–0.94); however, this relationship was not significant in participants with arthritis (p-value = 0.301) (Table 4).

General Health and Dietary Intake Behaviors

When examining the association between self-reported general health and dietary intake behaviors, while adjusting for age, sex, BMI, and smoking and drinking status, the results indicated higher odds of perceiving general health as excellent/very good/good in both participants with and without arthritis who were choosing food with a lower fat content (with arthritis OR = 1.12, 95% CI: 1.10–1.13; without arthritis OR = 1.13, 95% CI: 1.12–1.14), choosing fiber content (with arthritis OR = 1.06, 95% CI:1.04–1.07; without arthritis OR = 1.11, 95% CI: 1.10–1.12), and avoiding calorie content (with arthritis OR = 1.20, 95% CI: 1.18–1.21; without arthritis OR = 1.38, 95% CI: 1.37–1.39) compared to those who did not choose those foods.. The participants without arthritis who were avoiding cholesterol containing foods were 9% less likely to perceive excellent/very good/good general health (OR = 0.91, 95% CI: 0.91–0.92); however, this association was not statistically significant in people with arthritis (p-value = 0.519) (Table 4).

Mental Health and Dietary Intake Behaviors

In investigating the associations between self-reported dietary intake behaviors and mental health (i.e., self-reported mental health, mood disorder, and anxiety disorder), adjustments were made for age, sex, BMI, and smoking and drinking status. The subsequent analysis revealed that participants with and without arthritis who were choosing food with a lower fat content were more likely to self-report excellent/very good/good mental health compared to participants who did not choose food with a lower fat content (with arthritis OR = 1.18, 95% CI: 1.16–1.20; without arthritis OR = 1.30, 95% CI: 1.29–1.31). The participants with arthritis who were choosing food with a higher fiber content were less likely to report that they perceived their mental health as excellent/very good/good (OR = 0.96, 95% CI: 0.95–0.98) compared to those who did not choose those foods, but the participants without arthritis who were choosing food with a higher fiber content were 54% more likely to self-report excellent/very good/good mental health (OR = 1.54, 95% CI: 1.52–1.55) compared to participants who did not choose those foods. The association between avoiding food for cholesterol content and excellent/very good/good mental health was not statistically significant among participants with arthritis (p-value: 0.063). However, the participants without arthritis who were avoiding food for cholesterol content were more likely to report that they perceived their mental health as excellent/very good/good (OR = 1.25, 95% CI: 1.24–1.26). Participants both with and without arthritis who were avoiding food for calorie content had higher odds of perceiving their mental health as excellent/very good/good (with arthritis OR = 1.28, 95% CI: 1.26–1.30; without arthritis OR = 1.38, 95% CI: 1.37–1.40) compared to those who did not avoid those foods.

When we examined the relation between self-reported dietary intake behaviors and mood disorder, our results found that the participants with arthritis who were choosing food with a lower fat content were more likely to report mood disorders (OR = 1.04, 95% CI: 1.02–1.05), but participants without arthritis were less likely to report mood disorders (OR = 0.88, 95% CI: 0.87–0.88) than participants who did not choose. Participants with arthritis who were choosing food with a higher fiber content were more likely to report that they suffered from a mood disorder (OR = 1.22, 95% CI: 1.20–1.24), but the participants without arthritis who were choosing food with a higher fiber content were less likely to have mood disorders (OR = 0.88, 95% CI: 0.88–0.89) compared to participants who did not make those food choices. The participants who were avoiding food for cholesterol content were less likely to experience mood disorders (with arthritis OR = 0.97, 95% CI: 0.95–0.98; without arthritis OR = 0.85, 95% CI: 0.84–0.86) compared to participants who were not. Furthermore, the participants who were avoiding food for calorie content were less likely to experience mood disorders (with arthritis OR = 0.91, 95% CI: 0.90–0.93; without arthritis OR = 0.83, 95% CI: 0.83–0.84) compared to participants who were not.

When we examined the association between self-reported dietary intake behaviors and anxiety disorders, our results showed that the participants with and without arthritis who were choosing food with a lower fat content were less likely to report anxiety disorders (with arthritis OR = 0.94, 95% CI: 0.93–0.96; without arthritis OR = 0.92, 95% CI: 0.91–0.93) compared to participants who were not. Participants with arthritis who were choosing food with a higher fiber content were more likely to report anxiety disorders (OR = 1.03, 95% CI: 1.01–1.05). However, the participants without arthritis who were choosing food with a higher fiber content were less likely to have anxiety disorders (OR = 0.94, 95% CI: 0.93–0.95). The participants with arthritis who were avoiding food for cholesterol content had higher odds of experiencing anxiety disorders (OR = 1.32, 95% CI: 1.30–1.34), but participants without arthritis who avoided food for cholesterol content were 8% less likely to report anxiety disorders (OR = 0.92, 95% CI: 0.91–0.93). Anxiety disorders were more likely to occur in participants with arthritis who avoided high-calorie-content foods (OR = 1.17, 95% CI: 1.15–1.19), but participants without arthritis who avoided high-calorie-content foods were 13% less likely to report that they had an anxiety disorder (OR = 0.87, 95% CI: 0.86–0.87) compared to participants who were not avoiding such foods (Table 4). It is important to highlight that that there was no overlapping population within the arthritis cohort that responded to both the dietary intake and oral health questions. Consequently, a comprehensive analysis in this context was not feasible.

3.2.2. Self-Reported Health Status and Smoking

Sleep Quality and Smoking

When examining the association between sleep quality (refreshing sleep, difficulty staying awake, trouble going to sleep, and sleep duration) and smoking while adjusting for age, sex, BMI, and drinking status, the results showed that in comparison to never smokers, the participants with arthritis who were current smokers were less likely to have refreshing sleep (OR = 0.54, 95% CI: 0.53–0.54) and trouble going to sleep (OR = 0.99, 95% CI: 0.98–0.99), and more likely to report difficulties staying awake (OR = 1.14, 95% CI: 1.13–1.15) compared to non-smokers. The current smoker participants without arthritis were less likely to have refreshing sleep (OR = 0.63, 95% CI: 0.63–0.63) and more likely to report difficulty staying awake and trouble going to sleep (OR = 1.08, 95% CI: 1.08–1.08 and OR = 1.33, 95% CI: 1.32–1.33), respectively, compared to non-smokers. Current smokers were less likely to sleep $\geq$7 h per night (with arthritis OR = 0.91, 95% CI: 0.90–0.92; without arthritis OR = 0.80, 95% CI: 0.79–0.80), respectively, compared with non-smokers (Table 5).

Among participants with and without arthritis who were passive smokers, refreshing sleep was less likely to occur by 15% and 6%, respectively (with arthritis OR = 0.85, 95% CI: 0.83–0.86; without arthritis OR = 0.94, 95% CI: 0.93–0.95). Passive smokers with and without arthritis were both more likely to report difficulties staying awake (with arthritis OR = 1.12, 95% CI: 1.10–1.13; without arthritis OR = 1.07, 95% CI: 1.06–1.08) and more likely to have trouble going to sleep (with arthritis OR = 1.15, 95% CI: 1.13–1.17; without arthritis OR = 1.06, 95% CI: 1.05–1.06) compared to individuals who were not passive smokers (Table 5).

General Health and Smoking

When examining the association between general health and smoking among participants with and without arthritis, while adjusting for age, sex, BMI, and drinking status, the results found that participants with and without arthritis who were current smokers were less likely to self-report general health as excellent/very good/good (with arthritis OR = 0.41, 95% CI: 0.40–0.41; without arthritis OR = 0.35, 95% CI: 0.34–0.35) compared to non-smokers. Also, both groups of participants who smoked passively were less likely to self-report general health as excellent/very good/good (with arthritis OR = 0.69, 95% CI: 0.68–0.70; without arthritis OR = 0.69, 95% CI: 0.68–0.69) compared to individuals who were not passive smokers (Table 5).

Mental Health and Smoking

When examining the association between mental health and smoking while adjusting for age, sex, BMI, and drinking status, our results showed that individuals with and without arthritis who were current smokers were less likely to self-report excellent/very good/good mental health compared to individuals who never smoked (with arthritis OR = 0.41, 95% CI: 0.40–0.41; without arthritis OR = 0.40, 95% CI: 0.40–0.41). Both arthritis and non-arthritis groups who were current smokers were more likely to report mood disorders (with arthritis OR = 2.88, 95% CI: 2.86–2.89; without arthritis OR = 2.73, 95% CI: 2.72–2.74) and anxiety disorders (with arthritis OR = 2.87, 95% CI: 2.85–2.89; without arthritis OR = 2.62, 95% CI: 2.61–2.63) compared to non-smokers. Among passive smokers, participants with arthritis and those without arthritis were less likely to self-report mental health as excellent/very good/good by 36% and 24%, respectively (with arthritis OR = 0.64, 95% CI: 0.62–0.65; without arthritis OR = 0.76, 95% CI: 0.75–0.77) than those who did not passively smoke. Participants with and without arthritis who were passive smokers were more likely to have mood disorders (with arthritis OR = 1.18, 95% CI: 1.16–1.20; without arthritis OR = 1.54, 95% CI: 1.53–1.56) and anxiety disorders (with arthritis OR = 1.61, 95% CI: 1.58–1.64; without arthritis OR = 1.36, 95% CI: 1.35–1.38) compared to participants who were not passive smokers (Table 5).

Oral Health and Smoking

When examining the association between oral health and smoking among participants with and without arthritis, while adjusting for age, sex, BMI, and drinking status, the results showed that current smokers were less likely to have perceived oral health as excellent/very good/good (with arthritis OR = 0.36, 95% CI: 0.36–0.36; without arthritis OR = 0.27, 95% CI: 0.27–0.28).

Individuals who were current smokers were significantly more likely to have mouth pain (with arthritis OR = 1.47, 95% CI: 1.46–1.48; without arthritis OR = 1.81, 95% CI: 1.80–1.82), mouth dryness (with arthritis OR = 2.03, 95% CI: 2.02–2.04; without arthritis OR = 2.08, 95% CI: 2.07–2.08), and be uncomfortable while eating (with arthritis OR = 1.65, 95% CI: 1.63–1.66; without arthritis OR = 1.94, 95% CI: 1.93–1.95) compared to individuals who reported never smoking. However, participants with arthritis who were current and former smokers were less likely to have bleeding gums (OR = 0.71, 95% CI: 0.71–0.72 and OR = 0.97, 95% CI: 0.97–0.98, respectively). However, participants without arthritis who were current and former smokers were more likely to have bleeding gums (OR = 1.01, 95% CI: 1.00–1.01; OR = 1.03, 95% CI: 1.02–1.03, respectively) (Table 5).

Regarding passive smoking, respondents who reported this exposure were less likely to have perceived oral health as excellent/very good/good (with arthritis OR = 0.73, 95% CI: 0.71–0.74; without arthritis OR = 0.75, 95% CI: 0.74–0.76) compared to non-passive smokers. Also, these groups were significantly more likely to have mouth pain (with arthritis 84%, without arthritis 22%), bleeding gums (with arthritis 26%, without arthritis 3%), mouth dryness (with arthritis 46%, without arthritis 36%), and be uncomfortable while eating (with arthritis 27%, without arthritis 33%) in comparison to individuals who were not passive smokers (Table 5).

3.3. The Association between Sleep and Health Status in Participants with and without Arthritis

In examining the association between arthritis and refreshing sleep, our analysis revealed that individuals with arthritis were 32% less likely to report experiencing refreshing sleep (OR = 0.68, 95% CI 0.64–0.72) compared to those without arthritis (Table 3). When examining the association between refreshing sleep and general health, mental health, and oral health while adjusting for age, sex, BMI, and smoking and drinking status, our results showed that participants with and without arthritis who found their sleep refreshing were more likely to report good perceived general health (with arthritis OR = 2.96, 95% CI: 2.94–2.98; without arthritis OR = 3.57, 95% CI: 3.55–3.58), good perceived mental health (with arthritis OR = 4.32, 95% CI: 4.28–4.36; without arthritis OR = 3.76, 95% CI: 3.74–3.78), and good perceived oral health (with arthritis OR = 2.19, 95% CI: 2.16–2.22; without arthritis OR = 2.48, 95% CI: 2.46–2.50) compared to those who did not find their sleep refreshing (Table 6).

Table 6. Association of sleep quality with general, mental, and oral health in participants with and without arthritis.

Variables	Population	OR (95% CI)	*p*-Value
Refreshing sleep (ref: Sleep unrefreshing)			
Perceived General Health	Arthritis (*n* = 11,800)	2.96 (2.94–2.98)	<0.001
	W/O arthritis (*n* = 37,039)	3.57 (3.55–3.58)	<0.001
Perceived Mental Health	Arthritis (*n* = 11,791)	4.32 (4.28–4.36)	<0.001
	W/O arthritis (*n* = 37,080)	3.76 (3.74 –3.78)	<0.001
Perceived Oral Health	Arthritis (*n* = 5147)	2.19 (2.16–2.22)	<0.001
	W/O arthritis (*n* = 16,650)	2.48 (2.46–2.50)	<0.001
Difficulty staying awake (ref: No difficulty)			
Perceived General Health	Arthritis (*n* = 11,778)	0.53 (0.53–0.54)	<0.001
	W/O arthritis (*n* = 37,066)	0.55 (0.55–0.56)	<0.001

Table 6. *Cont.*

Variables	Population	OR (95% CI)	*p*-Value
Perceived Mental Health	Arthritis (n = 11,772) W/O arthritis (n = 37,054)	0.41 (0.40–0.41) 0.47 (0.46–0.47)	<0.001 <0.001
Perceived Oral Health	Arthritis (n = 5139) W/O arthritis (n = 16,657)	0.61 (0.60–0.62) 0.64 (0.64–0.65)	<0.001 <0.001
Trouble going to sleep (ref: No trouble)			
Perceived General Health	Arthritis (n = 11,830) W/O arthritis (n = 37,151)	0.58 (0.58–0.59) 0.39 (0.38–0.39)	<0.001 <0.001
Perceived Mental Health	Arthritis (n = 11,825) W/O arthritis (n = 37,138)	0.41 (0.41–0.42) 0.33 (0.33–0.33)	<0.001 <0.001
Perceived Oral Health	Arthritis (n = 5163) W/O arthritis (n = 16,687)	0.72 (0.72–0.73) 0.54 (0.53–0.54)	<0.001 <0.001
Spent ≥7 h per night sleeping (ref: <7 h)			
Perceived General Health	Arthritis (n = 11,770) W/O arthritis (n = 37,069)	1.59 (1.58–1.60) 1.37 (1.36–1.38)	<0.001 <0.001
Perceived Mental Health	Arthritis (n = 11,765) W/O arthritis (n = 37,059)	1.60 (1.59 –1.62) 1.54 (1.54–1.55)	<0.001 <0.001
Perceived Oral Health	Arthritis (n = 5147) W/O arthritis (n = 16,676)	1.69 (1.67–1.71) 1.66 (1.65–1.67)	<0.001 <0.001

Association of sleep quality with general, mental, and oral health in participants with and without arthritis. Binary logistic regression adjusted for age, sex, BMI, smoking status, and drinking status. Data presented as odds ratios and 95% confidence intervals; all analyses were weighted. Data considered statistically significant when p-value ≤ 0.05. W/O = without; OR = odds ratio; CI = confidence interval; n = number of respondents; ref = reference.

Furthermore, our results showed that people with arthritis were more likely to report difficulty staying awake (OR = 1.21, 95% CI 1.16–1.27) than people without arthritis (Table 3). When examining the association between difficulty staying awake and general health, mental health, and oral health while adjusting for age, sex, BMI, and smoking and drinking status, our results revealed that the people with and without arthritis who had difficulty staying awake were less likely to report good perceived general health (with arthritis OR = 0.53, 95% CI: 0.53–0.54; without arthritis OR = 0.55, 95% CI: 0.55–0.56), good perceived mental health (with arthritis OR = 0.41, 95% CI: 0.40–0.41; without arthritis OR = 0.47, 95% CI: 0.46–0.47), and good perceived oral health (with arthritis OR = 0.61, 95% CI: 0.60–0.62; without arthritis OR = 0.64, 95% CI: 0.64–0.65) compared to those who did not have difficulty staying awake (Table 6).

When exploring the association between arthritis and trouble going to sleep, our findings indicated that individuals with arthritis were 63% more likely to report trouble going to sleep (OR = 1.63, 95% CI 1.57–1.70) compared to those without arthritis (Table 3). Moreover, our results showed that the people with and without arthritis who had trouble going to sleep were less likely to report good perceived general health (with arthritis OR = 0.58, 95% CI: 0.58–0.59; without arthritis OR = 0.39, 95% CI: 0.38–0.39), good perceived mental health (with arthritis OR = 0.41, 95% CI: 0.41–0.42; without arthritis OR = 0.33, 95% CI: 0.33–0.33), and good perceived oral health (with arthritis OR = 0.72, 95% CI: 0.72–0.73; without arthritis OR = 0.54, 95% CI: 0.53–0.54) than those who did not have trouble going to sleep, while adjusting for age, sex, BMI, and smoking and drinking status (Table 6).

In examining the association between arthritis and sleep duration, our analysis revealed that people with arthritis were less likely to spend ≥7 h sleeping per night (OR = 0.79, 95% CI 0.76–0.82) than people without arthritis (Table 3). Additionally, in examining the association between sleep duration and general health, mental health, and oral health while adjusting for age, sex, BMI, and smoking and drinking status, our results showed that the participants with and without arthritis who spent ≥7 h sleeping per night were more likely to report good perceived general health (with arthritis OR = 1.59, 95% CI: 1.58–1.60; without arthritis OR = 1.37, 95% CI: 1.36–1.38), good perceived mental health (with arthritis OR = 1.60, 95% CI: 1.59–1.62; without arthritis OR = 1.54, 95% CI: 1.54–1.55), and good perceived oral health (with arthritis OR = 1.69, 95% CI: 1.67–1.71; without arthritis OR = 1.66, 95% CI: 1.65–1.67) than those spending <7 h per night sleeping (Table 6).

Supplementary Material S3 provides information about the associations between arthritis and sleep quality and duration for Models 1 and 2.

4. Discussion

The first objective of this study was to examine associations between arthritis and lifestyle traits, demographic characteristics, and self-reported health status. Additionally, we aim to explore associatios of lifestyle traits on the health-related conditions commonly associated with arthritis, including sleep quality, general health, mental health, and oral health in people with and without arthritis. This analysis considers previously studied individual-level covariates, such as age, sex, BMI, and alcohol intake to provide a comprehensive understanding of the complex relationships at play.

4.1. Arthritis, Demographic Characteristics, and Lifestyle Traits

The results of our study found statistically significant associations between arthritis and both sociodemographic characteristics (age, sex, education, marital status, and food security) and lifestyle traits (BMI, physical activity, smoking status, alcohol drinking status, and dietary behaviors) (Tables 1 and 2). Consistent with previous studies, we found that women, individuals aged 65 years and older, and those with a low education level were more likely to self-report arthritis [1]. Additionally, people who were food insecure were more likely to self-report arthritis, which is in line with previous studies showing an association between food insecurity and health problems that cause functional limitation including arthritis [76–78]. This relationship may be due to limited access to healthcare, leading to poor disease management and exacerbation [78]. Also, functional disability (which can occur with arthritis) may affect the ability to work, resulting in less financial resources to purchase groceries, leading to food insecurity [78]. Our findings highlight an association between demographic characteristics and arthritis, emphasizing the need for targeted interventions and public health strategies to address these factors for arthritis management and prevention.

Our findings also showed a negative association between physical activity and arthritis. These findings are possibly due to lower activity levels among individuals who may be experiencing arthritis symptoms and/or who have concerns about exacerbating the condition. Nevertheless, health providers often encourage individuals with arthritis to be more engaged in physical activity [79], as research has shown that physical activity can have numerous benefits for people with arthritis including improved pain management, increased mobility, and enhanced overall quality of life [79,80]. According to the CDC, adults with arthritis should aim for 150 min per week of moderately intense activities or 75 min per week of vigorously intense aerobic activities [81]. This underscores the necessity for a public health policy aimed at guiding healthcare providers in effectively advising people with arthritis on the positive impact of physical activity for both disease management and overall well-being.

Our analysis also found a positive association between BMI and arthritis which is consistent with previous research [82,83]. This association is possibly because excess body weight places additional stress on the joints, which can lead to increased wear and tear over time. Moreover, adipose tissue produces cytokines and other inflammatory molecules that can contribute to the development of arthritis [84]. Thus, maintaining a healthy weight through regular physical activity and a balanced diet may reduce the risk of developing arthritis, and can also help manage symptoms in individuals who already have the condition.

In parallel with previous studies, our results demonstrated a negative association between alcohol consumption and arthritis; individuals with arthritis who consume alcohol tend to experience lower disease activity and a better quality of life [38,85]. Furthermore, Larsson et al. (2018) found that individuals with arthritis who stopped drinking had worse physical functionality and health-related quality of life, and more pain and fatigue compared to individuals who did not stop drinking alcohol [86]. However, relying on alcohol as a coping mechanism can lead to other long-term health problems including addiction, liver damage, cardiovascular issues, and mental health concerns [87]. Additionally, alcohol can interact with certain arthritis medications like Methotrexate, potentially reducing their effectiveness or causing adverse effects [88].

As in previous studies showing that smokers are at higher risk for arthritis, we found associations between both current and former smokers and arthritis compared to never smokers. We also found that this association was higher among former smokers compared to current smokers. There are a few possible explanations for why former smoking status may have a higher association with arthritis. Smoking cessation can lead to weight gain, and this extra weight can put additional stress on the joints, which may increase the association between arthritis and former smoking status [89]. On the other hand, in current smokers, the anti-inflammatory effect of nicotine may mask the symptoms of arthritis [90].

4.2. *Health Status and Lifestyle Traits*

4.2.1. Sleep and Arthritis

In the CCHS, the quality of sleep is assessed through questions asking about the frequency of refreshing sleep, difficulty staying awake, and trouble going to sleep. The last two questions are used to assess insomnia symptoms in Canada through assessing daytime dysfunction and sleep onset latency, respectively [91,92]. Insomnia is defined as a complaint of difficulty going to sleep, trouble to maintaining sleep, or waking up too early, leading to non-restorative or poor-quality sleep [93]. Our analysis showed a negative association between arthritis and quality of sleep. We found that arthritis was negatively associated with refreshing sleep and an adequate sleep duration ($\geq$7 h) and positively associated with difficulty staying awake and trouble going to sleep. These findings are in line with other studies showing a high prevalence of sleep disorders among individuals with rheumatoid arthritis and significant associations between disease activity and poor sleep quality [94,95]. The associations identified in our study underscore the importance of integrating sleep assessments and interventions into routine care for individuals with arthritis.

4.2.2. Dietary Intake and Sleep in Arthritis

Our study showed that people with arthritis were more likely to self-report making healthier food choices including choosing food with a lower fat content and higher fiber content and avoiding food with a high cholesterol content compared to those without arthritis. This finding is in line with other studies showing a lower intake of discretionary foods in people with RA compared to the general population, which might be due to the fact that individuals with RA believe their food impacts their disease symptoms, and therefore they avoid these foods [51,96].

In this study, we hypothesized that the types of food chosen might not only affect arthritis but also have an association on other aspects of health status commonly associated with arthritis, such as sleep disturbance. To the best of our knowledge, this study is the first to assess the association between sleep quality and self-reported dietary behaviors in people with arthritis in Canada. In general, research suggests that there is a bidirectional association between sleep quality/quantity and dietary intake. Poor sleep quality can affect food choices and the desire to eat, and on the other hand, the type of nutrients we consume can influence sleep patterns in healthy people [97,98]. Our study found that choosing food with a lower fat content and avoiding cholesterol were associated with refreshing sleep in individuals with arthritis. This finding aligns with the results of a cross-sectional study that utilized data from a national survey in the United States which investigated the link between health, dietary intake, and sleep patterns. This study found that lower fat and cholesterol intake was correlated with better quality sleep and reduced daytime sleepiness in the general population [99]. Furthermore, consistent with our findings that indicated the association between a high fiber content and refreshing sleep, observational and interventional studies have found associations between a high fiber intake and deeper sleep (longer slow-wave sleep), which leads to better quality sleep [100,101]. Our study also found an association between avoiding food for calorie content and refreshing sleep in people with and without arthritis. However, some other studies have shown that caloric amounts, either high energy intake or caloric deprivation, have no significant impact on sleep quality or duration [102–104]. In general, studies on nutrition and sleep often exhibit high diversity and complexity in terms of participants and experimental conditions. These variations can contribute to conflicting study results, highlighting the need for more targeted and controlled investigations to draw definitive conclusions in these areas. Furthermore, sleep health is complex and comprises many dimensions, and not all dimensions are necessarily impacted by dietary intake at the same time.

4.2.3. Mental Health and Arthritis

Our analysis found associations between self-reported mental health, anxiety, and mood disorder and arthritis. Studies have shown that sleep disturbance in people with arthritis is associated with both disease intensity and poor mental health issues, including anxiety and mood disorders [105–107]. Furthermore, chronic pain, regardless of the anatomical location, has an important relationship with anxiety and depression [108,109]. Therefore, it is not surprising that psychiatric disorders, including anxiety, mood disorders, and depression, occur with greater frequency among people with arthritis, as our study also found [21,108–110]. The important role of mental health for quality of life highlights the need to take measures to address or prevent mental health problems in this population. These findings emphasize the need for healthcare approaches that address both physical and mental well-being in individuals with arthritis.

4.2.4. Mental Health and Dietary Intake in Arthritis

Over the last decade, we witnessed an exponential rise in the number of both prospective and cross-sectional studies in "nutritional psychiatry" documenting an association between diet quality and mental disorders [111,112]. Evidence supporting this association has also been steadily growing through discovering the important biological mediators affecting neurotransmission and enhancing neural signaling [113–117]. Our findings also found an association between mental health and self-reported dietary intake in individuals with and without arthritis. We found a positive association between choosing a low-fat diet and good general and mental health, as well as a negative association between choosing a low-fat diet and anxiety disorders. This finding is consistent with the study that indicated the long-term beneficial effect of a low-fat diet on mood state in participants with obesity and metabolic syndrome [118].

Cholesterol may have an impact on mental health due to its effect on neurotransmissions in the central nervous system [119]. Our study found a negative association between avoiding food for cholesterol content and mood disorders in people with and without arthritis which is consistent with studies showing a higher prevalence of depressive moods in individuals with higher serum total cholesterol levels [120,121]. Considering the dispute over the main contributory role of dietary cholesterol in elevated serum cholesterol and the complexities surrounding the relationship between dietary cholesterol and its impact on serum cholesterol [122], the association between serum cholesterol level and mood disorders cannot be concluded from our study.

Our study also revealed an association between choosing foods with high fiber and mental health issues, including self-reported mental health, anxiety, and mood disorders in individuals with arthritis. This aligns with existing research that documented the association of fiber intake or compliance to a high-fiber dietary pattern, like the Mediterranean Diet, with anxiety and depression [123–126]. It is proposed that dietary fiber has a beneficial effect on mental health through gut microbiota modulation that finally leads to a balanced gut microbiota–brain axis or a microbiota composition effect on neurotransmission [127–129]. However, it is noteworthy that a few other studies have reported inconsistent findings, indicating that the association may vary based on the study population or the source of dietary fiber [128,130].

Our analysis also found an association between avoiding food for calorie content and mental health, including self-reported mental health and mood disorders in individuals with and without arthritis. This finding is reinforced by evidence from both animal- and population-based studies underscoring the beneficial effects of a caloric restriction design on mood improvement [131,132]. One possible mechanism by which caloric restriction is believed to alleviate depressive syndrome is the activation of the hypothalamic–pituitary–adrenal axis and the increase in glucocorticoids, ultimately leading to depressive symptom alleviation [131,133].

4.2.5. Smoking and Arthritis

Our study revealed a significant association between smoking (current or former) and arthritis in comparison to never smokers, which is consistent with prior research on smoking and various types of arthritis. Multiple studies have demonstrated the association between smoking and RA [134–136]. A meta-analysis study revealed that smokers are more likely to develop RA than non-smokers [27]. On the other hand, meta-analysis studies have indicated a negative association between smoking and osteoarthritis [137,138]. The variation in the association between smoking and arthritis may depend on the type of arthritis.

4.2.6. Smoking/Passive Smoking and Sleep in Arthritis

Previous studies have shown that smokers were more likely to experience reduced sleep quality compared to non-smokers [54,139]. Our study found that smoking was associated with a poorer sleep quality among participants with and without arthritis. Specifically, both current and former smokers with arthritis were more likely to report poor sleep quality compared to never smokers. This association extends beyond active smoking to passive smoking, with numerous studies linking passive smoking to poorer sleep quality [140–142]. Our findings align with other studies showing that passive smokers with arthritis are more likely to report poor sleep quality compared to non-exposed individuals. It is well-established that sleep disturbance has been associated with different types of arthritis such as osteoarthritis (OA) [143] and RA [144]. Several studies have revealed an increase in pain, depression, and anxiety in people with RA following sleep disturbance [105,145]. In 2006, Luqmani et al. indicated that the early assessment and management of sleep disturbance is recommended to stabilize and monitor care for individuals with RA [146]. Given the association between sleep disturbance and smoking among participants with arthritis as shown in our study, addressing this link is crucial in developing effective treatment strategies for this group. However, further research is needed to better understand the underlying mechanisms and establish a robust causal relationship for targeted interventions.

4.2.7. Smoking/Passive Smoking and Mental Health in Arthritis

The current study demonstrated a significant association between mental health and arthritis. Individuals with arthritis experienced a higher rate of mood and anxiety disorders than those without arthritis, consistent with previous research [147,148]. Moreover, several studies have shown a link between smoking and poor mental health [149–151]. In our investigation into the association of smoking and mental health in individuals with arthritis, current and former smokers with arthritis were more likely to have mood and anxiety disorders compared to non-smokers with arthritis. This association was less strong among former smokers with arthritis. These results are consistent with studies showing improved mental health following smoking cessation [152,153].

In addition, we found that passive smokers with arthritis were more likely to report mood and anxiety disorders than those not exposed to passive smoke. These results are in line with existing research that has demonstrated a higher risk of mood and anxiety disorders among smokers compared to non-smokers [149,154]. Canadian studies have specifically linked passive smoking to anxiety disorders, while other research has associated depression with passive smoking [155–157]. These findings underscore the importance of addressing smoking cessation and providing mental health support for arthritis management. Future studies should confirm the association between smoking and mental disorders among people with arthritis and investigate the underlying mechanisms that link smoking, mental health, and arthritis.

4.2.8. Oral Health and Arthritis

In the current study, we found that participants with arthritis were 59% more likely to report poor oral health compared to participants without arthritis. These results are in accordance with previous studies indicating that people with RA have a poorer oral-health-related quality of life and have a higher risk of periodontal disease when compared to individuals without RA [158–160]. Also, various studies indicated that the risk of deteriorating oral health increased for people with RA and osteoarthritis [161,162].

4.2.9. Smoking/Passive Smoking and Oral Health in Arthritis

Nearly every system and organ in the body is adversely affected by smoking, including the oral cavity. It is not surprising that smoking has the same adverse effect on oral health among people with arthritis. In 2007, Millar and Locker analyzed cycle 2.1 of the CCHS, and they showed that both current and former smokers were more likely to have oral health problems than those who had never smoked [163]. Many studies have indicated that smoking is associated with poor oral health, including the prevalence and severity of periodontal diseases [164–166]. Our study confirmed that current and former smokers with arthritis were more likely to exhibit poor oral health than non-smokers with arthritis. Additionally, individuals with arthritis who had been exposed to passive smoking were more likely to report poor oral health compared to those with arthritis who had not been exposed to passive smoking. Our study's findings align with studies that demonstrated an association between passive smoking and oral health issues including oral cancer [167–169]. Interestingly, despite the detrimental effect of smoking on oral health, our study showed that participants with arthritis who smoked were less likely to experience bleeding gums than non-smokers. Also, former smokers with arthritis were less likely to report bleeding gums than non smoker, aligning with research indicating smoking can mask signs of periodontitis and reduces gum bleeding [170]. Also, the study showed that smoking cessation increases gum bleeding [171].

Arthritis negatively affects oral health, contributing to dental caries and gingivitis, and temporomandibular joint problems, often resulting in mouth pain in individuals with arthritis [172–174]. Moreover, it has been well established that smoking increases the risk of mouth pain [163,175]. To our knowledge, no study has been conducted that reports the association between smoking and mouth pain among people with arthritis. Our study revealed that people with arthritis who were current or former smokers as well as passive smokers suffered from mouth pain more than people with arthritis who had never smoked.

Our study revealed a significant association between arthritis and dry mouth. Specifically, participants with arthritis were 170% more likely to report dry mouth compared to individuals without arthritis. This aligns with studies linking autoimmune diseases to salivary gland dysfunction and xerostomia [176–178]. Moreover, our current study showed that current and former smokers with arthritis were more likely to report dry mouth than non-smokers with arthritis, consistent with previous research linking dry mouth to smoking [179,180]. A previous study indicated that individuals suffering from dry mouth experience oral symptoms such as dental caries, gingivitis, and mastication problems [181], which were more pronounced among smoker participants with arthritis in our study.

The association between smoking and oral health problems among people with arthritis that was shown in this study underscores the importance for dentists to create proper treatment plans for these individuals. Future studies should focus on investigate underlying mechanisms and evaluate smoking cessation interventions. These studies can enhance our understanding and inform targeted interventions to improve oral health in individuals with arthritis.

4.3. Sleep and Health Status

The current study underscores the association between arthritis and poor sleep quality and duration. Moreover, in examining the association of sleep quality and duration with perceived general health, mental health, and oral health among people with arthritis, we found that people with arthritis who reported experiencing refreshing sleep were significantly more likely to report positive perceptions across all health statuses than those who did not find their sleep refreshing. Conversely, people with arthritis facing difficulties staying awake or trouble going to sleep were less likely to report positive perceived general health, mental health, and oral health than those who did not face these issues. Additionally, an adequate sleep duration ($\geq$7 h per night) was associated with positive perceived health status compared to an inadequate sleep duration (<7 h per night). The identified associations between sleep and health status elucidated in our study underscores the importance of adopting comprehensive and personalized approaches in the prevention and care of arthritis. The circadian clock plays a crucial role in the regulation of various physiological process, including those that can be affected by arthritis [182]. There is a consensus that the disruption of circadian rhythms can have a profound effect on human health status and well-being [183–185]. Epidemiological studies have shown that individuals who work night/irregular shifts or have irregular sleep patterns are at increased risk for metabolic disorders, such as obesity and type 2 diabetes, cardiovascular disease, cancer, and neurodegenerative and autoimmune conditions [186–188]. Emerging studies have suggested a key role of the circadian clock in the onset and disease progression of arthritis, along with nascent evidence suggesting that optimizing individual circadian rhythms through lifestyle interventions may benefit those patients. This study can contribute to an increased understanding on the links between circadian disruption, susceptibility to disease, and disease development and progression in support of potential therapeutic opportunities for a range of medical conditions including arthritis, as well as strategies for preventing circadian-related health conditions and improving clinical outcomes through lifestyle changes.

4.4. Strengths and Limitations

A significant strength of this study lies in its robust study design, which meticulously encompasses various variables, effectively addressing two pivotal research aims within the framework of one distinctive population study. Moreover, this study is the first of its kind, not only in examining associations between lifestyle traits and arthritis, but also their association with health status commonly associated with arthritis among the Canadian population. This research utilizes a large, reliable survey, contributing to the existing body of knowledge by providing novel insights into the intricate associations between lifestyle traits and various types of health status measures in individuals with arthritis.

The primary limitation of this study is its cross-sectional design, limiting the capacity to determine a temporal link or cause-and-effect relationship between outcome and exposure. One constraint of this survey stems from a lack of detailed information about the specific type and stage of arthritis, the duration of the disease, and the treatments used. Because of this lack of information, there is likely a wide variation in disease characteristics among respondents who have confirmed arthritis. As a result, the survey's reliability and the clarity of the results are affected by this significant diversity among individuals who responded positively to the arthritis-related question. Another limitation of our study the lack of participants who completed both questions on oral health status and dietary intake. This lack of data prevents a comprehensive analysis of these important associations. This study also relies on self-reported data from the CCHS, which can introduce biases and inaccuracies. The absence of objective health indicators is a limitation. Future research should include objective health markers alongside self-reported data for a more comprehensive understanding. Furthermore, one challenge for surveys using self-reported questionnaires is answer reliability, as participants may make more socially acceptable choices rather than providing the true answers. In addition, the questions asking about self-reported dietary intake only ask participants to report information about self-reported choices of foods based on single dietary components and, and therefore do not provide a well-recognized, valid estimation of intake. Furthermore, we expect that some respondents may have had difficulties answering some of these questions about food choices as many members of the public have limited awareness of food sources of different nutrients. In addition, the examination of the association between a single macronutrient and health status may generate inconsistent results, as often the intake of one macronutrient leads to changes in other macronutrients. These limitations highlight the need for a cautious interpretation of our results and emphasize the importance of addressing methodological limitations in future research on dietary influences.

5. Conclusions

This study represents an in-depth exploration of possible associations between lifestyle traits and health status in Canadian participants with arthritis. Employing a meticulous analysis, we have unveiled noteworthy associations that carry substantial potential implications for both the prevention and management of arthritis.

Our study brought to light the associations between arthritis and demographic characteristics, revealing the heightened susceptibility of women, older individuals, and those with lower educational levels. The investigation extended to health statuses commonly associated with arthritis, uncovering the association between arthritis and poor sleep quality and poor general health, mental health, and oral health. Furthermore, our study explored associations between dietary choices and general health, mental health, and sleep quality in individuals with arthritis. Additionally, our findings highlight the association between smoking (active and passive) and poor sleep quality and poor general health, mental health, and oral health in people with arthritis.

Supplementary Materials: The following supporting information can be downloaded at: https://www.mdpi.com/article/10.3390/nu16132091/s1, Supplemental Table S1: Association between sleep quality, general and mental health with dietary intake in arthritis/non-arthritis participants, Model 1, 2, and 3; Supplemental Table S2: Association between sleep quality, general, mental, and oral health with smoking/passive smoking in arthritis/non-arthritis participants, Model 1, 2, and 3; Supplemental Table S3: Association between Perceived General and Mental and Oral Health with Sleep Quality in Participants with Arthritis and without Arthritis, Model 1, 2, and 3.

Author Contributions: Conceptualization, Z.G., S.P. and J.R.L.L.; formal analysis, Z.G. and I.H.; investigation, Z.G. and I.H.; writing—original draft preparation, Z.G. and I.H.; writing—review and editing, Z.G., I.H., J.L., P.P., L.P.W., M.F., G.S.K., J.R.L.L. and S.P.; visualization, Z.G. and I.H.; supervision, J.L., P.P., J.R.L.L. and S.P.; project administration, S.P. All authors have read and agreed to the published version of the manuscript.

Funding: This research received no external funding.

Institutional Review Board Statement: Not applicable.

Informed Consent Statement: Not applicable.

Data Availability Statement: Canadian Community Health Survey microdata file is publicly available on the Statistics Canada website.

Conflicts of Interest: The authors declare no conflict of interest.

Abbreviations

ACREU	Arthritis Community Research and Evaluation Unit
BMI	Body Mass Index
CCHS	Canadian Community Health Survey
CDC	Center for Disease Control and Prevention
CI	Confidence Interval
HFD	High-fat Diet
OA	Osteoarthritis
OR	Odd Ratio
RA	Rheumatoid Arthritis

References

1. Badley, E.M.; Wilfong, J.M.; Zahid, S.; Perruccio, A.V. *The Status of Arthritis in Canada: National Report*; The Canadian Arthritis Society, Arthritis Community Research and Evaluation Unit: Toronto, ON, Canada, 2019; pp. 1–34.
2. Badley, E.M.; Goulart, C.M.; Millstone, D.B.; Perruccio, A.V. An Update on Arthritis in Canada—National and Provincial Data Regarding the Past, Present, and Future. *J. Rheumatol.* **2019**, *46*, 579–586. [CrossRef] [PubMed]
3. Lyne, L.; Åkerstedt, T.; Alfredsson, L.; Lehtonen, T.; Saevarsdottir, S.; Klareskog, L.; Westerlind, H. Sleep Problems in Rheumatoid Arthritis over 12 Years from Diagnosis: Results from the Swedish EIRA Study. *RMD Open* **2022**, *8*, e001800. [CrossRef] [PubMed]
4. Badley, E.M. *Arthritis in Canada: An Ongoing Challenge*; Arthritis Community Research And Evaluation Unit (ACREU): Ottawa, ON, Canada, 2003.
5. Rao, Y.; Xu, X.; Liu, D.; Reis, C.; Newman, I.; Qin, L.; Sharma, M.; Shen, J.; Zhao, Y. Health-Related Quality of Life in Patients with Arthritis: A Cross-Sectional Survey among Middle-Aged Adults in Chongqing, China. *Int. J. Environ. Res. Public. Health* **2018**, *15*, 768. [CrossRef] [PubMed]
6. Yang, L.; Sibbritt, D.; Adams, J. A Critical Review of Complementary and Alternative Medicine Use among People with Arthritis: A Focus upon Prevalence, Cost, User Profiles, Motivation, Decision-Making, Perceived Benefits and Communication. *Rheumatol. Int.* **2017**, *37*, 337–351. [CrossRef] [PubMed]
7. Beydon, M.; Pinto, S.; De Rycke, Y.; Fautrel, B.; Mariette, X.; Seror, R.; Tubach, F. Risk of Cancer for Patients with Rheumatoid Arthritis versus General Population: A National Claims Database Cohort Study. *Lancet Reg. Health. Eur.* **2023**, *35*, 100768. [CrossRef] [PubMed]
8. Lee, Y.C.; Chibnik, L.B.; Lu, B.; Wasan, A.D.; Edwards, R.R.; Fossel, A.H.; Helfgott, S.M.; Solomon, D.H.; Clauw, D.J.; Karlson, E.W. The Relationship between Disease Activity, Sleep, Psychiatric Distress and Pain Sensitivity in Rheumatoid Arthritis: A Cross-Sectional Study. *Arthritis Res. Ther.* **2009**, *11*, R160. [CrossRef] [PubMed]
9. Louie, G.H.; Tektonidou, M.G.; Caban--Martinez, A.J.; Ward, M.M. Sleep Disturbances in Adults with Arthritis: Prevalence, Mediators, and Subgroups at Greatest Risk. Data from the 2007 National Health Interview Survey. *Arthritis Care Res.* **2011**, *63*, 247–260. [CrossRef] [PubMed]
10. Power, J.D.; Perruccio, A.V.; Badley, E.M. Pain as a Mediator of Sleep Problems in Arthritis and Other Chronic Conditions. *Arthritis Rheum.* **2005**, *53*, 911–919. [CrossRef]
11. Taylor-Gjevre, R.M.; Gjevre, J.A.; Nair, B.; Skomro, R.; Lim, H.J. Components of Sleep Quality and Sleep Fragmentation in Rheumatoid Arthritis and Osteoarthritis. *Musculoskelet. Care* **2011**, *9*, 152–159. [CrossRef]
12. Wolfe, F.; Michaud, K.; Li, T. Sleep Disturbance in Patients with Rheumatoid Arthritis: Evaluation by Medical Outcomes Study and Visual Analog Sleep Scales. *J. Rheumatol.* **2006**, *33*, 1942–1951.
13. Abad, V.C.; Sarinas, P.S.A.; Guilleminault, C. Sleep and Rheumatologic Disorders. *Sleep. Med. Rev.* **2008**, *12*, 211–228. [CrossRef] [PubMed]
14. Wilcox, S.; Brenes, G.A.; Levine, D.; Sevick, M.A.; Shumaker, S.A.; Craven, T. Factors Related to Sleep Disturbance in Older Adults Experiencing Knee Pain or Knee Pain with Radiographic Evidence of Knee Osteoarthritis. *J. Am. Geriatr. Soc.* **2000**, *48*, 1241–1251. [CrossRef]
15. Hawker, G.A.; French, M.R.; Waugh, E.J.; Gignac, M.A.M.; Cheung, C.; Murray, B.J. The Multidimensionality of Sleep Quality and Its Relationship to Fatigue in Older Adults with Painful Osteoarthritis. *Osteoarthr. Cartil.* **2010**, *18*, 1365–1371. [CrossRef]
16. Irwin, M.R.; Opp, M.R. Sleep Health: Reciprocal Regulation of Sleep and Innate Immunity. *Neuropsychopharmacology* **2017**, *42*, 129–155. [CrossRef] [PubMed]
17. Freeman, D.; Sheaves, B.; Waite, F.; Harvey, A.G.; Harrison, P.J. Sleep Disturbance and Psychiatric Disorders. *Lancet Psychiatry* **2020**, *7*, 628–637. [CrossRef] [PubMed]
18. Ramar, K.; Malhotra, R.K.; Carden, K.A.; Martin, J.L.; Abbasi-Feinberg, F.; Aurora, R.N.; Kapur, V.K.; Olson, E.J.; Rosen, C.L.; Rowley, J.A.; et al. Sleep Is Essential to Health: An American Academy of Sleep Medicine Position Statement. *J. Clin. Sleep Med.* **2021**, *17*, 2115–2119. [CrossRef] [PubMed]
19. Carra, M.C.; Schmitt, A.; Thomas, F.; Danchin, N.; Pannier, B.; Bouchard, P. Sleep Disorders and Oral Health: A Cross-Sectional Study. *Clin. Oral. Investig.* **2017**, *21*, 975–983. [CrossRef]

20. Guglielmo, D.; Hootman, J.M.; Boring, M.A.; Murphy, L.B.; Theis, K.A.; Croft, J.B.; Barbour, K.E.; Katz, P.P.; Helmick, C.G. Symptoms of Anxiety and Depression Among Adults with Arthritis—United States, 2015–2017. *MMWR Morb. Mortal. Wkly Rep.* **2018**, *67*, 1081–1087. [CrossRef]
21. He, Y.; Zhang, M.; Lin, E.H.B.; Bruffaerts, R.; Posada-Villa, J.; Angermeyer, M.C.; Levinson, D.; de Girolamo, G.; Uda, H.; Mneimneh, Z.; et al. Mental Disorders among Persons with Arthritis: Results from the World Mental Health Surveys. *Psychol. Med.* **2008**, *38*, 1639–1650. [CrossRef]
22. Stubbs, B.; Veronese, N.; Vancampfort, D.; Thompson, T.; Kohler, C.; Schofield, P.; Solmi, M.; Mugisha, J.; Kahl, K.G.; Pillinger, T.; et al. Lifetime Self-Reported Arthritis Is Associated with Elevated Levels of Mental Health Burden: A Multi-National cross Sectional Study across 46 Low- and Middle-Income Countries. *Sci. Rep.* **2017**, *7*, 7138. [CrossRef]
23. Lwin, M.N.; Serhal, L.; Holroyd, C.; Edwards, C.J. Rheumatoid Arthritis: The Impact of Mental Health on Disease: A Narrative Review. *Rheumatol. Ther.* **2020**, *7*, 457–471. [CrossRef]
24. Jung, E.S.; Choi, Y.Y.; Lee, K.H. Relationship between Rheumatoid Arthritis and Periodontal Disease in Korean Adults: Data from the Sixth Korea National Health and Nutrition Examination Survey, 2013 to 2015. *J. Periodontol.* **2019**, *90*, 350–357. [CrossRef] [PubMed]
25. Shim, H.; Koo, J.; Ahn, J. Association between Rheumatoid Arthritis and Poor Self-Perceived Oral Health in Korean Adults. *Healthcare* **2022**, *10*, 427. [CrossRef] [PubMed]
26. Pischon, N.; Pischon, T.; Kröger, J.; Gülmez, E.; Kleber, B.-M.; Bernimoulin, J.-P.; Landau, H.; Brinkmann, P.-G.; Schlattmann, P.; Zernicke, J.; et al. Association Among Rheumatoid Arthritis, Oral Hygiene, and Periodontitis. *J. Periodontol.* **2008**, *79*, 979–986. [CrossRef] [PubMed]
27. Sugiyama, D.; Nishimura, K.; Tamaki, K.; Tsuji, G.; Nakazawa, T.; Morinobu, A.; Kumagai, S. Impact of Smoking as a Risk Factor for Developing Rheumatoid Arthritis: A Meta-Analysis of Observational Studies. *Ann. Rheum. Dis.* **2010**, *69*, 70–81. [CrossRef]
28. Zhang, X.; Zhang, X.; Yang, Y.; Zhi, K.; Chen, Y.; Zhao, J.; Cui, W.; Zhao, X.; Zhang, Z.; An, Y.; et al. Association between Passive Smoking and the Risk of Rheumatoid Arthritis: A Systematic Review and Meta-Analysis. *Clin. Rheumatol.* **2023**, *42*, 663–672. [CrossRef]
29. Shih, M.; Hootman, J.M.; Kruger, J.; Helmick, C.G. Physical Activity in Men and Women with Arthritis. *Am. J. Prev. Med.* **2006**, *30*, 385–393. [CrossRef]
30. Tierney, M.; Fraser, A.; Kennedy, N. Physical Activity in Rheumatoid Arthritis: A Systematic Review. *J. Phys. Act. Health* **2012**, *9*, 1036–1048. [CrossRef]
31. van den Ende, C.H.; Breedveld, F.C.; le Cessie, S.; Dijkmans, B.A.; de Mug, A.W.; Hazes, J.M. Effect of Intensive Exercise on Patients with Active Rheumatoid Arthritis: A Randomised Clinical Trial. *Ann. Rheum. Dis.* **2000**, *59*, 615–621. [CrossRef]
32. Stavropoulos-Kalinoglou, A.; Metsios, G.S.; Veldhuijzen van Zanten, J.J.J.C.S.; Nightingale, P.; Kitas, G.D.; Koutedakis, Y. Individualised Aerobic and Resistance Exercise Training Improves Cardiorespiratory Fitness and Reduces Cardiovascular Risk in Patients with Rheumatoid Arthritis. *Ann. Rheum. Dis.* **2013**, *72*, 1819–1825. [CrossRef]
33. Wadley, A.J.; Veldhuijzen van Zanten, J.J.C.S.; Stavropoulos-Kalinoglou, A.; Metsios, G.S.; Smith, J.P.; Kitas, G.D.; Aldred, S. Three Months of Moderate-Intensity Exercise Reduced Plasma 3-Nitrotyrosine in Rheumatoid Arthritis Patients. *Eur. J. Appl. Physiol.* **2014**, *114*, 1483–1492. [CrossRef] [PubMed]
34. Löfgren, M.; Opava, C.H.; Demmelmaier, I.; Fridén, C.; Lundberg, I.E.; Nordgren, B.; Kosek, E. Long-Term, Health-Enhancing Physical Activity Is Associated with Reduction of Pain but Not Pain Sensitivity or Improved Exercise-Induced Hypoalgesia in Persons with Rheumatoid Arthritis. *Arthritis Res. Ther.* **2018**, *20*, 262. [CrossRef] [PubMed]
35. Durcan, L.; Wilson, F.; Cunnane, G. The Effect of Exercise on Sleep and Fatigue in Rheumatoid Arthritis: A Randomized Controlled Study. *J. Rheumatol.* **2014**, *41*, 1966–1973. [CrossRef] [PubMed]
36. Baillet, A.; Vaillant, M.; Guinot, M.; Juvin, R.; Gaudin, P. Efficacy of Resistance Exercises in Rheumatoid Arthritis: Meta-Analysis of Randomized Controlled Trials. *Rheumatology* **2012**, *51*, 519–527. [CrossRef]
37. Lu, B.; Solomon, D.H.; Costenbader, K.H.; Karlson, E.W. Alcohol Consumption and Risk of Incident Rheumatoid Arthritis in Women: A Prospective Study. *Arthritis Rheumatol.* **2014**, *66*, 1998–2005. [CrossRef] [PubMed]
38. Turk, J.N.; Zahavi, E.R.; Gorman, A.E.; Murray, K.; Turk, M.A.; Veale, D.J. Exploring the Effect of Alcohol on Disease Activity and Outcomes in Rheumatoid Arthritis through Systematic Review and Meta-Analysis. *Sci. Rep.* **2021**, *11*, 10474. [CrossRef] [PubMed]
39. Maxwell, J.R.; Gowers, I.R.; Moore, D.J.; Wilson, A.G. Alcohol Consumption Is Inversely Associated with Risk and Severity of Rheumatoid Arthritis. *Rheumatology* **2010**, *49*, 2140–2146. [CrossRef] [PubMed]
40. Bergman, S.; Symeonidou, S.; Andersson, M.L.; Söderlin, M.K. Alcohol Consumption Is Associated with Lower Self-Reported Disease Activity and Better Health-Related Quality of Life in Female Rheumatoid Arthritis Patients in Sweden: Data from BARFOT, a Multicenter Study on Early RA. *BMC Musculoskelet. Disord.* **2013**, *14*, 218. [CrossRef] [PubMed]
41. Källberg, H.; Jacobsen, S.; Bengtsson, C.; Pedersen, M.; Padyukov, L.; Garred, P.; Frisch, M.; Karlson, E.W.; Klareskog, L.; Alfredsson, L. Alcohol Consumption Is Associated with Decreased Risk of Rheumatoid Arthritis: Results from Two Scandinavian Case–Control Studies. *Ann. Rheum. Dis.* **2009**, *68*, 222–227. [CrossRef]
42. Barr, T.; Helms, C.; Grant, K.; Messaoudi, I. Opposing Effects of Alcohol on the Immune System. *Prog. Neuropsychopharmacol. Biol. Psychiatry* **2016**, *65*, 242. [CrossRef]

43. Mangnus, L.; Van Steenbergen, H.W.; Nieuwenhuis, W.P.; Reijnierse, M.; Van Der Helm-van Mil, A.H.M. Moderate Use of Alcohol Is Associated with Lower Levels of C Reactive Protein but Not with Less Severe Joint Inflammation: A Cross-Sectional Study in Early RA and Healthy Volunteers. *RMD Open* **2018**, *4*, e000577. [CrossRef] [PubMed]
44. Teodoro, J.S.; Varela, A.T.; Rolo, A.P.; Palmeira, C.M. High-Fat and Obesogenic Diets: Current and Future Strategies to Fight Obesity and Diabetes. *Genes Nutr.* **2014**, *9*, 406. [CrossRef]
45. Liu, Y.; Hazlewood, G.S.; Kaplan, G.G.; Eksteen, B.; Barnabe, C. Impact of Obesity on Remission and Disease Activity in Rheumatoid Arthritis: A Systematic Review and Meta-Analysis. *Arthritis Care Res.* **2017**, *69*, 157–165. [CrossRef]
46. Feng, J.; Chen, Q.; Yu, F.; Wang, Z.; Chen, S.; Jin, Z.; Cai, Q.; Liu, Y.; He, J. Body Mass Index and Risk of Rheumatoid Arthritis. *Medicine* **2016**, *95*, e2859. [CrossRef] [PubMed]
47. DeChristopher, L.R.; Uribarri, J.; Tucker, K.L. Intake of High-Fructose Corn Syrup Sweetened Soft Drinks, Fruit Drinks and Apple Juice Is Associated with Prevalent Arthritis in US Adults, Aged 20–30 Years. *Nutr. Diabetes* **2016**, *6*, e199. [CrossRef]
48. Hu, Y.; Costenbader, K.H.; Gao, X.; Al-Daabil, M.; Sparks, J.A.; Solomon, D.H.; Hu, F.B.; Karlson, E.W.; Lu, B. Sugar-Sweetened Soda Consumption and Risk of Developing Rheumatoid Arthritis in Women. *Am. J. Clin. Nutr.* **2014**, *100*, 959–967. [CrossRef] [PubMed]
49. Johansson, K.; Askling, J.; Alfredsson, L.; Di Giuseppe, D. Mediterranean Diet and Risk of Rheumatoid Arthritis: A Population-Based Case-Control Study. *Arthritis Res. Ther.* **2018**, *20*, 175. [CrossRef] [PubMed]
50. Kostoglou-Athanassiou, I.; Athanassiou, L.; Athanassiou, P. The Effect of Omega-3 Fatty Acids on Rheumatoid Arthritis. *Mediterr. J. Rheumatol.* **2020**, *31*, 190–194. [CrossRef] [PubMed]
51. Tedeschi, S.K.; Bathon, J.M.; Giles, J.T.; Lin, T.-C.; Yoshida, K.; Solomon, D.H. Relationship between Fish Consumption and Disease Activity in Rheumatoid Arthritis. *Arthritis Care Res.* **2018**, *70*, 327–332. [CrossRef]
52. Htoo, A.; Talwar, A.; Feinsilver, S.H.; Greenberg, H. Smoking and Sleep Disorders. *Med. Clin. N. Am.* **2004**, *88*, 1575–1591. [CrossRef]
53. Hu, N.; Wang, C.; Liao, Y.; Dai, Q.; Cao, S. Smoking and Incidence of Insomnia: A Systematic Review and Meta-Analysis of Cohort Studies. *Public Health* **2021**, *198*, 324–331. [CrossRef]
54. Jaehne, A.; Unbehaun, T.; Feige, B.; Lutz, U.C.; Batra, A.; Riemann, D. How Smoking Affects Sleep: A Polysomnographical Analysis. *Sleep. Med.* **2012**, *13*, 1286–1292. [CrossRef]
55. Nakata, A.; Takahashi, M.; Haratani, T.; Ikeda, T.; Hojou, M.; Fujioka, Y.; Araki, S. Association of Active and Passive Smoking with Sleep Disturbances and Short Sleep Duration among Japanese Working Population. *Int. J. Behav. Med.* **2008**, *15*, 81–91. [CrossRef]
56. Chaudhary, N.S.; Kampman, K.M.; Kranzler, H.R.; Grandner, M.A.; Debbarma, S.; Chakravorty, S. Insomnia in Alcohol Dependent Subjects Is Associated with Greater Psychosocial Problem Severity. *Addict. Behav.* **2015**, *50*, 165–172. [CrossRef]
57. Zhabenko, N.; Wojnar, M.; Brower, K.J. Prevalence and Correlates of Insomnia in a Polish Sample of Alcohol-Dependent Patients. *Alcohol. Clin. Exp. Res.* **2012**, *36*, 1600–1607. [CrossRef] [PubMed]
58. Chakravorty, S.; Chaudhary, N.S.; Brower, K.J. Alcohol Dependence and Its Relationship with Insomnia and Other Sleep Disorders. *Alcohol. Clin. Exp. Res.* **2016**, *40*, 2271–2282. [CrossRef] [PubMed]
59. Taveira, K.V.M.; Kuntze, M.M.; Berretta, F.; de Souza, B.D.M.; Godolfim, L.R.; Demathe, T.; De Luca Canto, G.; Porporatti, A.L. Association between Obstructive Sleep Apnea and Alcohol, Caffeine and Tobacco: A Meta-Analysis. *J. Oral Rehabil.* **2018**, *45*, 890–902. [CrossRef] [PubMed]
60. Jackson, C.L.; Gaston, S.A.; Liu, R.; Mukamal, K.; Rimm, E.B. The Relationship between Alcohol Drinking Patterns and Sleep Duration among Black and White Men and Women in the United States. *Int. J. Environ. Res. Public Health* **2018**, *15*, 557. [CrossRef]
61. Warburton, D.E.R.; Charlesworth, S.; Ivey, A.; Nettlefold, L.; Bredin, S.S.D. A Systematic Review of the Evidence for Canada's Physical Activity Guidelines for Adults. *Int. J. Behav. Nutr. Phys. Act.* **2010**, *7*, 39. [CrossRef]
62. St-Onge, M.-P.; Mikic, A.; Pietrolungo, C.E. Effects of Diet on Sleep Quality. *Adv. Nutr.* **2016**, *7*, 938–949. [CrossRef]
63. Katagiri, R.; Asakura, K.; Kobayashi, S.; Suga, H.; Sasaki, S. Low Intake of Vegetables, High Intake of Confectionary, and Unhealthy Eating Habits Are Associated with Poor Sleep Quality among Middle-Aged Female Japanese Workers. *J. Occup. Health* **2014**, *56*, 359–368. [CrossRef] [PubMed]
64. Campanini, M.Z.; Guallar-Castillón, P.; Rodríguez-Artalejo, F.; Lopez-Garcia, E. Mediterranean Diet and Changes in Sleep Duration and Indicators of Sleep Quality in Older Adults. *Sleep* **2017**, *40*, zsw083. [CrossRef] [PubMed]
65. Tan, X.; Alén, M.; Cheng, S.M.; Mikkola, T.M.; Tenhunen, J.; Lyytikäinen, A.; Wiklund, P.; Cong, F.; Saarinen, A.; Tarkka, I.; et al. Associations of Disordered Sleep with Body Fat Distribution, Physical Activity and Diet among Overweight Middle-Aged Men. *J. Sleep. Res.* **2015**, *24*, 414–424. [CrossRef] [PubMed]
66. Lawrence, D.; Mitrou, F.; Zubrick, S.R. Smoking and Mental Illness: Results from Population Surveys in Australia and the United States. *BMC Public Health* **2009**, *9*, 285. [CrossRef] [PubMed]
67. Plurphanswat, N.; Kaestner, R.; Rodu, B. The Effect of Smoking on Mental Health. *Am. J. Health Behav.* **2017**, *41*, 471–483. [CrossRef] [PubMed]
68. Pereira, G.; Wood, L.; Foster, S.; Haggar, F. Access to Alcohol Outlets, Alcohol Consumption and Mental Health. *PLoS ONE* **2013**, *8*, e53461. [CrossRef] [PubMed]
69. Weitzman, E.R. Poor Mental Health, Depression, and Associations With Alcohol Consumption, Harm, and Abuse in a National Sample of Young Adults in College. *J. Nerv. Ment. Dis.* **2004**, *192*, 269–277. [CrossRef] [PubMed]

70. Biddle, S.J.H.; Asare, M. Physical Activity and Mental Health in Children and Adolescents: A Review of Reviews. *Br. J. Sports Med.* **2011**, *45*, 886–895. [CrossRef] [PubMed]
71. Lassale, C.; Batty, G.D.; Baghdadli, A.; Jacka, F.; Sánchez-Villegas, A.; Kivimäki, M.; Akbaraly, T. Healthy Dietary Indices and Risk of Depressive Outcomes: A Systematic Review and Meta-Analysis of Observational Studies. *Mol. Psychiatry* **2019**, *24*, 965–986. [CrossRef]
72. Blanchflower, D.G.; Oswald, A.J.; Stewart-Brown, S. Is Psychological Well-Being Linked to the Consumption of Fruit and Vegetables? *Soc. Indic. Res.* **2013**, *114*, 785–801. [CrossRef]
73. Parletta, N.; Milte, C.M.; Meyer, B.J. Nutritional Modulation of Cognitive Function and Mental Health. *J. Nutr. Biochem.* **2013**, *24*, 725–743. [CrossRef] [PubMed]
74. Surveys and Statistical Programs—Canadian Community Health Survey—Annual Component (CCHS). Available online: https://www23.statcan.gc.ca/imdb/p2SV.pl?Function=getSurvey&SDDS=3226 (accessed on 6 January 2024).
75. Elamy, A.-H.H.; Shuaib, A.; Carriere, K.C.; Jeerakathil, T. Common Comorbidities of Stroke in the Canadian Population. *Can. J. Neurol. Sci.* **2020**, *47*, 314–319. [CrossRef] [PubMed]
76. Cai, Q.; Pesa, J.; Wang, R.; Fu, A.Z. Depression and Food Insecurity among Patients with Rheumatoid Arthritis in NHANES. *BMC Rheumatol.* **2022**, *6*, 6. [CrossRef]
77. Seligman, H.K.; Laraia, B.A.; Kushel, M.B. Food Insecurity Is Associated with Chronic Disease among Low-Income NHANES Participants. *J. Nutr.* **2010**, *140*, 304–310. [CrossRef] [PubMed]
78. Venci, B.J.; Lee, S.-Y. Functional Limitation and Chronic Diseases Are Associated with Food Insecurity among U.S. Adults. *Ann. Epidemiol.* **2018**, *28*, 182–188. [CrossRef]
79. Katz, P.; Andonian, B.J.; Huffman, K.M. Benefits and Promotion of Physical Activity in Rheumatoid Arthritis. *Curr. Opin. Rheumatol.* **2020**, *32*, 307–314. [CrossRef]
80. Dunlop, D.D.; Song, J.; Semanik, P.A.; Sharma, L.; Chang, R.W. Physical Activity Levels and Functional Performance in the Osteoarthritis Initiative: A Graded Relationship. *Arthritis Rheum.* **2011**, *63*, 127–136. [CrossRef]
81. CDC Physical Activity for Arthritis. Available online: https://www.cdc.gov/arthritis/basics/physical-activity/index.html (accessed on 9 January 2024).
82. Qin, B.; Yang, M.; Fu, H.; Ma, N.; Wei, T.; Tang, Q.; Hu, Z.; Liang, Y.; Yang, Z.; Zhong, R. Body Mass Index and the Risk of Rheumatoid Arthritis: A Systematic Review and Dose-Response Meta-Analysis. *Arthritis Res. Ther.* **2015**, *17*. [CrossRef]
83. Lu, B.; Hiraki, L.T.; Sparks, J.A.; Malspeis, S.; Chen, C.Y.; Awosogba, J.A.; Arkema, E.V.; Costenbader, K.H.; Karlson, E.W. Being Overweight or Obese and Risk of Developing Rheumatoid Arthritis among Women: A Prospective Cohort Study. *Ann. Rheum. Dis.* **2014**, *73*, 1914–1922. [CrossRef]
84. Feng, X.; Xu, X.; Shi, Y.; Liu, X.; Liu, H.; Hou, H.; Ji, L.; Li, Y.; Wang, W.; Wang, Y.; et al. Body Mass Index and the Risk of Rheumatoid Arthritis: An Updated Dose-Response Meta-Analysis. *Biomed. Res. Int.* **2019**, *2019*, 1–12. [CrossRef]
85. Baker, J.F.; England, B.R.; Mikuls, T.R.; Hsu, J.Y.; George, M.D.; Pedro, S.; Sayles, H.; Michaud, K. Changes in Alcohol Use and Associations With Disease Activity, Health Status, and Mortality in Rheumatoid Arthritis. *Arthritis Care Res.* **2020**, *72*, 301–308. [CrossRef] [PubMed]
86. Larsson, I.; Andersson, M.L.E. Reasons to Stop Drinking Alcohol among Patients with Rheumatoid Arthritis in Sweden: A Mixed-Methods Study. *BMJ Open* **2018**, *8*, 24367. [CrossRef] [PubMed]
87. Cargiulo, T. Understanding the Health Impact of Alcohol Dependence. *Am. J. Health-Syst. Pharm.* **2007**, *64*, S5–S11. [CrossRef]
88. Humphreys, J.H.; Warner, A.; Costello, R.; Lunt, M.; Verstappen, S.M.M.; Dixon, W.G. Quantifying the Hepatotoxic Risk of Alcohol Consumption in Patients with Rheumatoid Arthritis Taking Methotrexate. *Ann. Rheum. Dis.* **2017**, *76*, 1509–1514. [CrossRef] [PubMed]
89. Filozof, C.; Fernandez Pinilla, M.C.; Fernandez-Cruz, A. Smoking Cessation and Weight Gain. *Obes. Rev.* **2004**, *5*, 95–103. [CrossRef] [PubMed]
90. Zhang, W.; Lin, H.; Zou, M.; Yuan, Q.; Huang, Z.; Pan, X.; Zhang, W. Nicotine in Inflammatory Diseases: Anti-Inflammatory and Pro-Inflammatory Effects. *Front. Immunol.* **2022**, *13*, 826889. [CrossRef] [PubMed]
91. Chaput, J.-P.; Yau, J.; Rao, D.P.; Morin, C.M. Prevalence of Insomnia for Canadians Aged 6 to 79. *Health Rep.* **2018**, *29*, 16–20. [PubMed]
92. Vézina-Im, L.-A.; Lebel, A.; Gagnon, P.; Nicklas, T.A.; Baranowski, T. Association between Sleep and Overweight/Obesity among Women of Childbearing Age in Canada. *Can. J. Public Health* **2018**, *109*, 516–526. [CrossRef] [PubMed]
93. Bonnet, M.H.; Burton, G.G.; Arand, D.L. Physiological and Medical Findings in Insomnia: Implications for Diagnosis and Care. *Sleep Med. Rev.* **2014**, *18*, 111–122. [CrossRef]
94. Goes, A.C.J.; Reis, L.A.B.; Silva, M.B.G.; Kahlow, B.S.; Skare, T.L. Rheumatoid Arthritis and Sleep Quality. *Rev. Bras. De Reumatol.* **2017**, *57*, 294–298. [CrossRef]
95. Son, C.-N.; Choi, G.; Lee, S.-Y.; Lee, J.-M.; Lee, T.-H.; Jeong, H.-J.; Jung, C.-G.; Kim, J.-M.; Cho, Y.-W.; Kim, S.-H. Sleep Quality in Rheumatoid Arthritis, and Its Association with Disease Activity in a Korean Population. *Korean J. Intern. Med.* **2015**, *30*, 384. [CrossRef] [PubMed]
96. Bärebring, L.; Winkvist, A.; Gjertsson, I.; Lindqvist, H. Poor Dietary Quality Is Associated with Increased Inflammation in Swedish Patients with Rheumatoid Arthritis. *Nutrients* **2018**, *10*, 1535. [CrossRef] [PubMed]

97. Zuraikat, F.M.; Wood, R.A.; Barragán, R.; St-Onge, M.P. Sleep and Diet: Mounting Evidence of a Cyclical Relationship. *Annu. Rev. Nutr.* **2021**, *41*, 309–332. [CrossRef] [PubMed]
98. Shi, Z.; McEvoy, M.; Luu, J.; Attia, J. Dietary Fat and Sleep Duration in Chinese Men and Women. *Int. J. Obes.* **2008**, *32*, 1835–1840. [CrossRef] [PubMed]
99. Grandner, M.A.; Jackson, N.; Gerstner, J.R.; Knutson, K.L. Sleep Symptoms Associated with Intake of Specific Dietary Nutrients. *J. Sleep Res.* **2014**, *23*, 22–34. [CrossRef] [PubMed]
100. St-Onge, M.P.; Roberts, A.; Shechter, A.; Choudhury, A.R. Fiber and Saturated Fat Are Associated with Sleep Arousals and Slow Wave Sleep. *J. Clin. Sleep Med.* **2016**, *12*, 19–24. [CrossRef]
101. Spaeth, A.M.; Dinges, D.F.; Goel, N. Objective Measurements of Energy Balance Are Associated with Sleep Architecture in Healthy Adults. *Sleep* **2017**, *40*. [CrossRef] [PubMed]
102. Castro, A.; Gomez-Arbelaez, D.; Crujeiras, A.; Granero, R.; Aguera, Z.; Jimenez-Murcia, S.; Sajoux, I.; Lopez-Jaramillo, P.; Fernandez-Aranda, F.; Casanueva, F. Effect of A Very Low-Calorie Ketogenic Diet on Food and Alcohol Cravings, Physical and Sexual Activity, Sleep Disturbances, and Quality of Life in Obese Patients. *Nutrients* **2018**, *10*, 1348. [CrossRef]
103. Driver, H.S.; Shulman, I.; Baker, F.C.; Buffenstein, R. Energy Content of the Evening Meal Alters Nocturnal Body Temperature but Not Sleep. *Physiol. Behav.* **1999**, *68*, 17–23. [CrossRef]
104. Lieberman, H.R.; Caruso, C.M.; Niro, P.J.; Adam, G.E.; Kellogg, M.D.; Nindl, B.C.; Kramer, F.M. A Double-Blind, Placebo-Controlled Test of 2 d of Calorie Deprivation: Effects on Cognition, Activity, Sleep, and Interstitial Glucose Concentrations. *Am. J. Clin. Nutr.* **2008**, *88*, 667–676. [CrossRef]
105. Irwin, M.R.; Olmstead, R.; Carrillo, C.; Sadeghi, N.; FitzGerald, J.D.; Ranganath, V.K.; Nicassio, P.M. Sleep Loss Exacerbates Fatigue, Depression, and Pain in Rheumatoid Arthritis. *Sleep* **2012**, *35*, 537–543. [CrossRef] [PubMed]
106. Li, Y.; Zhang, S.; Zhu, J.; Du, X.; Huang, F. Sleep Disturbances Are Associated with Increased Pain, Disease Activity, Depression, and Anxiety in Ankylosing Spondylitis: A Case-Control Study. *Arthritis Res. Ther.* **2012**, *14*, R215. [CrossRef] [PubMed]
107. Lopes, F.H.A.; Freitas, M.V.C.; de Bruin, V.M.S.; de Bruin, P.F.C. Depressive Symptoms Are Associated with Impaired Sleep, Fatigue, and Disease Activity in Women with Rheumatoid Arthritis. *Adv. Rheumatol.* **2021**, *61*, 18. [CrossRef] [PubMed]
108. Axford, J.; Butt, A.; Heron, C.; Hammond, J.; Morgan, J.; Alavi, A.; Bolton, J.; Bland, M. Prevalence of Anxiety and Depression in Osteoarthritis: Use of the Hospital Anxiety and Depression Scale as a Screening Tool. *Clin. Rheumatol.* **2010**, *29*, 1277–1283. [CrossRef] [PubMed]
109. McWilliams, L.A.; Cox, B.J.; Enns, M.W. Mood and Anxiety Disorders Associated with Chronic Pain: An Examination in a Nationally Representative Sample. *Pain* **2003**, *106*, 127–133. [CrossRef] [PubMed]
110. Walker, J.G.; Littlejohn, G.O. Measuring Quality of Life in Rheumatic Conditions. *Clin. Rheumatol.* **2007**, *26*, 671–673. [CrossRef] [PubMed]
111. Godos, J.; Currenti, W.; Angelino, D.; Mena, P.; Castellano, S.; Caraci, F.; Galvano, F.; Del Rio, D.; Ferri, R.; Grosso, G. Diet and Mental Health: Review of the Recent Updates on Molecular Mechanisms. *Antioxidants* **2020**, *9*, 346. [CrossRef] [PubMed]
112. Jacka, F.N.; Cherbuin, N.; Anstey, K.J.; Butterworth, P. Dietary Patterns and Depressive Symptoms over Time: Examining the Relationships with Socioeconomic Position, Health Behaviours and Cardiovascular Risk. *PLoS ONE* **2014**, *9*, e87657. [CrossRef]
113. Gomez-Pinilla, F.; Gomez, A.G. The Influence of Dietary Factors in Central Nervous System Plasticity and Injury Recovery. *PMR* **2011**, *3*, S111–S116. [CrossRef]
114. Hsiao, E.Y.; McBride, S.W.; Hsien, S.; Sharon, G.; Hyde, E.R.; McCue, T.; Codelli, J.A.; Chow, J.; Reisman, S.E.; Petrosino, J.F.; et al. Microbiota Modulate Behavioral and Physiological Abnormalities Associated with Neurodevelopmental Disorders. *Cell* **2013**, *155*, 1451–1463. [CrossRef]
115. Maes, M.; Kubera, M.; Leunis, J.-C.; Berk, M. Increased IgA and IgM Responses against Gut Commensals in Chronic Depression: Further Evidence for Increased Bacterial Translocation or Leaky Gut. *J. Affect. Disord.* **2012**, *141*, 55–62. [CrossRef] [PubMed]
116. Sullivan, E.L.; Grayson, B.; Takahashi, D.; Robertson, N.; Maier, A.; Bethea, C.L.; Smith, M.S.; Coleman, K.; Grove, K.L. Chronic Consumption of a High-Fat Diet during Pregnancy Causes Perturbations in the Serotonergic System and Increased Anxiety-Like Behavior in Nonhuman Primate Offspring. *J. Neurosci.* **2010**, *30*, 3826–3830. [CrossRef] [PubMed]
117. Vucetic, Z.; Reyes, T.M. Central Dopaminergic Circuitry Controlling Food Intake and Reward: Implications for the Regulation of Obesity. *Wiley Interdiscip. Rev. Syst. Biol. Med.* **2010**, *2*, 577–593. [CrossRef] [PubMed]
118. Brinkworth, G.D.; Noakes, M.; Buckley, J.D.; Keogh, J.B.; Clifton, P.M. Long-Term Effects of a Very-Low-Carbohydrate Weight Loss Diet Compared with an Isocaloric Low-Fat Diet after 12 Mo. *Am. J. Clin. Nutr.* **2009**, *90*, 23–32. [CrossRef] [PubMed]
119. Papakostas, G.I.; Öngür, D.; Iosifescu, D.V.; Mischoulon, D.; Fava, M. Cholesterol in Mood and Anxiety Disorders: Review of the Literature and New Hypotheses. *Eur. Neuropsychopharmacol.* **2004**, *14*, 135–142. [CrossRef] [PubMed]
120. Kim, E.J.; Hong, J.; Hwang, J.-W. The Association between Depressive Mood and Cholesterol Levels in Korean Adolescents. *Psychiatry Investig.* **2019**, *16*, 737–744. [CrossRef] [PubMed]
121. Nakao, M.; Yano, E. Relationship between Major Depression and High Serum Cholesterol in Japanese Men. *Tohoku J. Exp. Med.* **2004**, *204*, 273–287. [CrossRef]
122. Grundy, S.M. Does Dietary Cholesterol Matter? *Curr. Atheroscler. Rep.* **2016**, *18*, 68. [CrossRef] [PubMed]
123. Ramin, S.; Mysz, M.A.; Meyer, K.; Capistrant, B.; Lazovich, D.; Prizment, A. A Prospective Analysis of Dietary Fiber Intake and Mental Health Quality of Life in the Iowa Women's Health Study. *Maturitas* **2020**, *131*, 1–7. [CrossRef]

124. Saghafian, F.; Sharif, N.; Saneei, P.; Keshteli, A.H.; Hosseinzadeh-Attar, M.J.; Afshar, H.; Esmaillzadeh, A.; Adibi, P. Consumption of Dietary Fiber in Relation to Psychological Disorders in Adults. *Front. Psychiatry* **2021**, *12*, 587468. [CrossRef]
125. Sánchez-Villegas, A.; Bes-Rastrollo, M.; Martínez-González, M.A.; Serra-Majem, L. Adherence to a Mediterranean Dietary Pattern and Weight Gain in a Follow-up Study: The SUN Cohort. *Int. J. Obes.* **2006**, *30*, 350–358. [CrossRef] [PubMed]
126. Woo, J.; Lynn, H.; Lau, W.Y.; Leung, J.; Lau, E.; Wong, S.Y.S.; Kwok, T. Nutrient Intake and Psychological Health in an Elderly Chinese Population. *Int. J. Geriatr. Psychiatry* **2006**, *21*, 1036–1043. [CrossRef] [PubMed]
127. Kelly, J.R.; Keane, V.O.; Cryan, J.F.; Clarke, G.; Dinan, T.G. Mood and Microbes. *Gastroenterol. Clin. N. Am.* **2019**, *48*, 389–405. [CrossRef] [PubMed]
128. Kim, Y.; Hong, M.; Kim, S.; Shin, W.; Kim, J. Inverse Association between Dietary Fiber Intake and Depression in Premenopausal Women: A Nationwide Population-Based Survey. *Menopause* **2021**, *28*, 150–156. [CrossRef] [PubMed]
129. Sandhu, K.V.; Sherwin, E.; Schellekens, H.; Stanton, C.; Dinan, T.G.; Cryan, J.F. Feeding the Microbiota-Gut-Brain Axis: Diet, Microbiome, and Neuropsychiatry. *Transl. Res.* **2017**, *179*, 223–244. [CrossRef] [PubMed]
130. Oishi, J.; Doi, H.; Kawakami, N. Nutrition and Depressive Symptoms in Community-Dwelling Elderly Persons in Japan. *Acta Med. Okayama* **2009**, *63*, 9–17. [CrossRef] [PubMed]
131. Manchishi, S.M.; Cui, R.J.; Zou, X.H.; Cheng, Z.Q.; Li, B. Effect of Caloric Restriction on Depression. *J. Cell. Mol. Med.* **2018**, *22*, 2528–2535. [CrossRef] [PubMed]
132. Martin, C.K.; Bhapkar, M.; Pittas, A.G.; Pieper, C.F.; Das, S.K.; Williamson, D.A.; Scott, T.; Redman, L.M.; Stein, R.; Gilhooly, C.H.; et al. Effect of Calorie Restriction on Mood, Quality of Life, Sleep, and Sexual Function in Healthy Nonobese Adults. *JAMA Intern. Med.* **2016**, *176*, 743. [CrossRef] [PubMed]
133. Kenny, R.; Dinan, T.; Cai, G.; Spencer, S.J. Effects of Mild Calorie Restriction on Anxiety and Hypothalamic-Pituitary-Adrenal Axis Responses to Stress in the Male Rat. *Physiol. Rep.* **2014**, *2*, e00265. [CrossRef]
134. Heliövaara, M.; Aho, K.; Aromaa, A.; Knekt, P.; Reunanen, A. Smoking and Risk of Rheumatoid Arthritis. *J. Rheumatol.* **1993**, *20*, 1830–1835.
135. Nyhäll-Wåhlin, B.-M.; Jacobsson, L.T.H.; Petersson, I.F.; Turesson, C.; BARFOT study group. Smoking Is a Strong Risk Factor for Rheumatoid Nodules in Early Rheumatoid Arthritis. *Ann. Rheum. Dis.* **2006**, *65*, 601–606. [CrossRef] [PubMed]
136. Papadopoulos, N.G.; Alamanos, Y.; Voulgari, P.V.; Epagelis, E.K.; Tsifetaki, N.; Drosos, A.A. Does Cigarette Smoking Influence Disease Expression, Activity and Severity in Early Rheumatoid Arthritis Patients? *Clin. Exp. Rheumatol.* **2005**, *23*, 861–866. [PubMed]
137. Hui, M.; Doherty, M.; Zhang, W. Does Smoking Protect against Osteoarthritis? Meta-Analysis of Observational Studies. *Ann. Rheum. Dis.* **2011**, *70*, 1231–1237. [CrossRef] [PubMed]
138. Kong, L.; Wang, L.; Meng, F.; Cao, J.; Shen, Y. Association between Smoking and Risk of Knee Osteoarthritis: A Systematic Review and Meta-Analysis. *Osteoarthr. Cartil.* **2017**, *25*, 809–816. [CrossRef]
139. McNamara, J.P.H.; Wang, J.; Holiday, D.B.; Warren, J.Y.; Paradoa, M.; Balkhi, A.M.; Fernandez-Baca, J.; McCrae, C.S. Sleep Disturbances Associated with Cigarette Smoking. *Psychol. Health Med.* **2014**, *19*, 410–419. [CrossRef] [PubMed]
140. Morioka, H.; Jike, M.; Kanda, H.; Osaki, Y.; Nakagome, S.; Otsuka, Y.; Kaneita, Y.; Itani, O.; Higuchi, S.; Ohida, T. The Association between Sleep Disturbance and Second-Hand Smoke Exposure: A Large-Scale, Nationwide, Cross-Sectional Study of Adolescents in Japan. *Sleep Med.* **2018**, *50*, 29–35. [CrossRef] [PubMed]
141. Sabanayagam, C.; Shankar, A. The Association between Active Smoking, Smokeless Tobacco, Second-Hand Smoke Exposure and Insufficient Sleep. *Sleep Med.* **2011**, *12*, 7–11. [CrossRef]
142. Safa, F.; Chaiton, M.; Mahmud, I.; Ahmed, S.; Chu, A. The Association between Exposure to Second-Hand Smoke and Sleep Disturbances: A Systematic Review and Meta-Analysis. *Sleep Health* **2020**, *6*, 702–714. [CrossRef] [PubMed]
143. De Bock, G.H.; Kaptein, A.A.; Touw-Otten, F.; Mulder, J.D. Health-Related Quality of Life in Patients with Osteoarthritis in a Family Practice Setting. *Arthritis Care Res.* **1995**, *8*, 88–93. [CrossRef]
144. Devins, G.M.; Edworthy, S.M.; Paul, L.C.; Mandin, H.; Seland, T.P.; Klein, G.; Costello, C.G.; Shapiro, C.M. Restless Sleep, Illness Intrusiveness, and Depressive Symptoms in Three Chronic Illness Conditions: Rheumatoid Arthritis, End-Stage Renal Disease, and Multiple Sclerosis. *J. Psychosom. Res.* **1993**, *37*, 163–170. [CrossRef]
145. Nicassio, P.M.; Wallston, K.A. Longitudinal Relationships among Pain, Sleep Problems, and Depression in Rheumatoid Arthritis. *J. Abnorm. Psychol.* **1992**, *101*, 514–520. [CrossRef] [PubMed]
146. Luqmani, R.; Hennell, S.; Estrach, C.; Birrell, F.; Bosworth, A.; Davenport, G.; Fokke, C.; Goodson, N.; Jeffreson, P.; Lamb, E.; et al. British Society for Rheumatology and British Health Professionals in Rheumatology Guideline for the Management of Rheumatoid Arthritis (the First Two Years). *Rheumatology* **2006**, *45*, 1167–1169. [CrossRef] [PubMed]
147. O'Donnell, S.; Rusu, C.; Hawker, G.A.; Bernatsky, S.; McRae, L.; Canizares, M.; MacKay, C.; Badley, E.M. Arthritis Has an Impact on the Daily Lives of Canadians Young and Old: Results from a Population-Based Survey. *BMC Musculoskelet. Disord.* **2015**, *16*, 230. [CrossRef] [PubMed]
148. Stang, P.E.; Brandenburg, N.A.; Lane, M.C.; Merikangas, K.R.; Von Korff, M.R.; Kessler, R.C. Mental and Physical Comorbid Conditions and Days in Role Among Persons with Arthritis. *Psychosom. Med.* **2006**, *68*, 152–158. [CrossRef] [PubMed]
149. Breslau, N.; Novak, S.P.; Kessler, R.C. Psychiatric Disorders and Stages of Smoking. *Biol. Psychiatry* **2004**, *55*, 69–76. [CrossRef] [PubMed]

150. Brown, R.A.; Lewinsohn, P.M.; Seeley, J.R.; Wagner, E.F. Cigarette Smoking, Major Depression, and Other Psychiatric Disorders among Adolescents. *J. Am. Acad. Child Adolesc. Psychiatry* **1996**, *35*, 1602–1610. [CrossRef]
151. Cuijpers, P.; Smit, F.; ten Have, M.; de Graaf, R. Smoking Is Associated with First-Ever Incidence of Mental Disorders: A Prospective Population-Based Study. *Addiction* **2007**, *102*, 1303–1309. [CrossRef] [PubMed]
152. Mino, Y.; Shigemi, J.; Otsu, T.; Tsuda, T.; Babazono, A. Does Smoking Cessation Improve Mental Health? *Psychiatry Clin. Neurosci.* **2000**, *54*, 169–172. [CrossRef] [PubMed]
153. Taylor, G.; McNeill, A.; Girling, A.; Farley, A.; Lindson-Hawley, N.; Aveyard, P. Change in Mental Health after Smoking Cessation: Systematic Review and Meta-Analysis. *BMJ* **2014**, *348*, g1151. [CrossRef]
154. Mykletun, A.; Overland, S.; Aarø, L.E.; Liabø, H.-M.; Stewart, R. Smoking in Relation to Anxiety and Depression: Evidence from a Large Population Survey: The HUNT Study. *Eur. Psychiatry* **2008**, *23*, 77–84. [CrossRef]
155. Asbridge, M.; Ralph, K.; Stewart, S. Private Space Second-Hand Smoke Exposure and the Mental Health of Non-Smokers: A Cross-Sectional Analysis of Canadian Adults. *Addict. Behav.* **2013**, *38*, 1679–1686. [CrossRef]
156. Wang, R.; Zhang, P.; Lv, X.; Gao, C.; Song, Y.; Li, Z.; Yu, Y.; Li, B. Association between Passive Smoking and Mental Distress in Adult Never-Smokers: A Cross-Sectional Study. *BMJ Open* **2016**, *6*, e011671. [CrossRef] [PubMed]
157. Yolton, K.; Khoury, J.; Hornung, R.; Dietrich, K.; Succop, P.; Lanphear, B. Environmental Tobacco Smoke Exposure and Child Behaviors. *J. Dev. Behav. Pediatr.* **2008**, *29*, 450–457. [CrossRef]
158. de Azevedo Branco, L.G.; Oliveira, S.R.; Corrêa, J.D.; Calderaro, D.C.; Mendonça, S.M.S.; de Queiroz Cunha, F.; Ferreira, G.A.; Teixeira, A.L.; Abreu, L.G.; Silva, T.A. Oral Health–Related Quality of Life among Individuals with Rheumatoid Arthritis. *Clin. Rheumatol.* **2019**, *38*, 2433–2441. [CrossRef] [PubMed]
159. Detert, J.; Pischon, N.; Burmester, G.R.; Buttgereit, F. The Association between Rheumatoid Arthritis and Periodontal Disease. *Arthritis Res. Ther.* **2010**, *12*, 218. [CrossRef] [PubMed]
160. Schmalz, G.; Patschan, S.; Patschan, D.; Ziebolz, D. Oral-Health-Related Quality of Life in Adult Patients with Rheumatic Diseases—A Systematic Review. *J. Clin. Med.* **2020**, *9*, 1172. [CrossRef]
161. Kelsey, J.L.; Lamster, I.B. Influence of Musculoskeletal Conditions on Oral Health among Older Adults. *Am. J. Public Health* **2008**, *98*, 1177–1183. [CrossRef]
162. Pokrajac-Zirojevic, V.; Slack-Smith, L.; Booth, D. Arthritis and Use of Dental Services: A Population Based Study. *Aust. Dent. J.* **2002**, *47*, 208–213. [CrossRef]
163. Millar, W.J.; Locker, D. Smoking and Oral Health Status. *J. Can. Dent. Assoc.* **2007**, *73*, 155.
164. Bergström, J.; Eliasson, S.; Dock, J. A 10-Year Prospective Study of Tobacco Smoking and Periodontal Health. *J. Periodontol.* **2000**, *71*, 1338–1347. [CrossRef]
165. Bloom, B.; Adams, P.F.; Cohen, R.A.; Simile, C. Smoking and Oral Health in Dentate Adults Aged 18–64. *NCHS Data Brief.* **2012**, *85*, 1–8.
166. Torrungruang, K.; Nisapakultorn, K.; Sutdhibhisal, S.; Tamsailom, S.; Rojanasomsith, K.; Vanichjakvong, O.; Prapakamol, S.; Premsirinirund, T.; Pusiri, T.; Jaratkulangkoon, O.; et al. The Effect of Cigarette Smoking on the Severity of Periodontal Disease Among Older Thai Adults. *J. Periodontol.* **2005**, *76*, 566–572. [CrossRef] [PubMed]
167. He, B.; Chen, F.; Yan, L.; Huang, J.; Liu, F.; Qiu, Y.; Lin, L.; Zhang, Z.; Cai, L. Independent and Joint Exposure to Passive Smoking and Cooking Oil Fumes on Oral Cancer in Chinese Women: A Hospital-Based Case-Control Study. *Acta Otolaryngol.* **2016**, *136*, 1074–1078. [CrossRef] [PubMed]
168. Aligne, C.A.; Moss, M.E.; Auinger, P.; Weitzman, M. Association of Pediatric Dental Caries With Passive Smoking. *JAMA* **2003**, *289*, 1258. [CrossRef] [PubMed]
169. Akinkugbe, A.A.; Slade, G.D.; Divaris, K.; Poole, C. Systematic Review and Meta-Analysis of the Association Between Exposure to Environmental Tobacco Smoke and Periodontitis Endpoints Among Nonsmokers. *Nicotine Tob. Res.* **2016**, *18*, 2047–2056. [CrossRef] [PubMed]
170. Preber, H.; Bergström, J. Occurrence of Gingival Bleeding in Smoker and Non-Smoker Patients. *Acta Odontol. Scand.* **1985**, *43*, 315–320. [CrossRef] [PubMed]
171. Nair, P.; Sutherland, G.; Palmer, R.M.; Wilson, R.F.; Scott, D.A. Gingival Bleeding on Probing Increases after Quitting Smoking. *J. Clin. Periodontol.* **2003**, *30*, 435–437. [CrossRef] [PubMed]
172. Bessa-Nogueira, R.V.; do Egito Vasconcelos, B.C.; Duarte, A.P.; Góes, P.S.A.; Bezerra, T.P. Targeted Assessment of the Temporomandibular Joint in Patients with Rheumatoid Arthritis. *J. Oral. Maxillofac. Surg.* **2008**, *66*, 1804–1811. [CrossRef] [PubMed]
173. Martinez-Martinez, R.; Domínguez-Pérez, R.; Sancho-Mata, J.; Abud-Mendoza, C.; Ayala-Herrera, J.; Popoca-Hernandez, E. The Frequency and Severity of Dental Caries, and Counts of Cariogenic Bacteria in Rheumatoid Arthritis Patients. *Dent. Med. Probl.* **2019**, *56*, 137–142. [CrossRef]
174. Monsarrat, P.; Vergnes, J.-N.; Blaizot, A.; Constantin, A.; de Grado, G.F.; Ramambazafy, H.; Sixou, M.; Cantagrel, A.; Nabet, C. Oral Health Status in Outpatients with Rheumatoid Arthritis: The OSARA Study. *Oral. Health Dent. Manag.* **2014**, *13*, 113–119.
175. Riley, J.L.; Tomar, S.L.; Gilbert, G.H. Smoking and Smokeless Tobacco: Increased Risk for Oral Pain. *J. Pain* **2004**, *5*, 218–225. [CrossRef] [PubMed]
176. Kassan, S.S.; Moutsopoulos, H.M. Clinical Manifestations and Early Diagnosis of Sjögren Syndrome. *Arch. Intern. Med.* **2004**, *164*, 1275. [CrossRef] [PubMed]

177. Maeshima, E.; Furukawa, K.; Maeshima, S.; Koshiba, H.; Sakamoto, W. Hyposalivation in Autoimmune Diseases. *Rheumatol. Int.* **2013**, *33*, 3079–3082. [CrossRef] [PubMed]
178. Napeñas, J.J.; Brennan, M.T.; Fox, P.C. Diagnosis and Treatment of Xerostomia (Dry Mouth). *Odontology* **2009**, *97*, 76–83. [CrossRef] [PubMed]
179. Johansson, A.-K.; Johansson, A.; Unell, L.; Ekbäck, G.; Ordell, S.; Carlsson, G.E. Self-Reported Dry Mouth in Swedish Population Samples Aged 50, 65 and 75 Years. *Gerodontology* **2012**, *29*, e107–e115. [CrossRef] [PubMed]
180. Rad, M.; Kakoie, S.; Niliye Brojeni, F.; Pourdamghan, N. Effect of Long-Term Smoking on Whole-Mouth Salivary Flow Rate and Oral Health. *J. Dent. Res. Dent. Clin. Dent. Prospect.* **2010**, *4*, 110–114. [CrossRef]
181. Turner, M.D.; Ship, J.A. Dry Mouth and Its Effects on the Oral Health of Elderly People. *J. Am. Dent. Assoc.* **2007**, *138* (Suppl. S1), S15–S20. [CrossRef] [PubMed]
182. Kulkarni, A.; Demory-Beckler, M.; Kesselman, M.M. The Role of Clock Genes in Maintaining Circadian Rhythm and Rheumatoid Arthritis Pathophysiology. *Cureus* **2023**, *15*, 39104. [CrossRef]
183. Adeola, H.A.; Papagerakis, S.; Papagerakis, P. Systems Biology Approaches and Precision Oral Health: A Circadian Clock Perspective. *Front. Physiol.* **2019**, *10*, 399. [CrossRef]
184. Papagerakis, S.; Zheng, L.; Schnell, S.; Sartor, M.A.; Somers, E.; Marder, W.; McAlpin, B.; Kim, D.; McHugh, J.; Papagerakis, P. The Circadian Clock in Oral Health and Diseases. *J. Dent. Res.* **2014**, *93*, 27–35. [CrossRef]
185. Pundir, M.; Papagerakis, S.; De Rosa, M.C.; Chronis, N.; Kurabayashi, K.; Abdulmawjood, S.; Prince, M.E.P.; Lobanova, L.; Chen, X.; Papagerakis, P. Emerging Biotechnologies for Evaluating Disruption of Stress, Sleep, and Circadian Rhythm Mechanism Using Aptamer-Based Detection of Salivary Biomarkers. *Biotechnol. Adv.* **2022**, *59*, 107961. [CrossRef] [PubMed]
186. Bayon, V.; Berger, M.; Solelhac, G.; Haba-Rubio, J.; Marques-Vidal, P.; Strippoli, M.P.; Preisig, M.; Leger, D.; Heinzer, R. Impact of Night and Shift Work on Metabolic Syndrome and Its Components: A Cross-Sectional Study in an Active Middle-to-Older-Aged Population-Based Sample. *BMJ Open* **2022**, *12*, e053591. [CrossRef] [PubMed]
187. Musiek, E.S.; Holtzman, D.M. Mechanisms Linking Circadian Clocks, Sleep, and Neurodegeneration. *Science* **2016**, *354*, 1004. [CrossRef] [PubMed]
188. Brum, M.C.B.; Filho, F.F.D.; Schnorr, C.C.; Bottega, G.B.; Rodrigues, T.C. Shift Work and Its Association with Metabolic Disorders. *Diabetol. Metab. Syndr.* **2015**, *7*, 45. [CrossRef] [PubMed]

nutrients

Article

Seed Disinfection Treatments Minimized Microbial Load and Enhanced Nutritional Properties of Fenugreek Sprouts Which Alleviated Diabetes-Negative Disorders in Diabetic Rats

Abeer A. Dahab [1,*], Hala M. Bayomy [2], Hemat S. Abd El-Salam [1], Seham E. Almasoudi [2], Nawal A. Ozaybi [2], Gehan A. Mahmoud [3], Amira K. G. Atteya [4] and Rasha S. El-Serafy [5,*]

1 Medicinal and Aromatic Plants Research Department, Horticulture Research Institute, Agricultural Research Center, Giza 12619, Egypt; hemat.sameh@yahoo.com
2 Food Science and Nutrition Department, Faculty of Science, University of Tabuk, Tabuk 71491, Saudi Arabia; hm.mohamed@ut.edu.sa (H.M.B.); salmasoudi@ut.edu.sa (S.E.A.); n.ozaybi@ut.edu.sa (N.A.O.)
3 Fruit Crops Handling Research Department, Horticulture Research Institute, Agricultural Research Center, Giza 12619, Egypt; geh-an.ahmed.11@hotmail.com
4 Horticulture Department, Faculty of Agriculture, Damanhour University, Damanhour 22516, Egypt; amira.khames@agr.dmu.edu.eg
5 Horticulture Department, Faculty of Agriculture, Tanta University, Tanta 31527, Egypt
* Correspondence: dahab.abeer1612@arc.sci.eg (A.A.D.); rasha.elserafi@agr.tanta.edu.eg (R.S.E.-S.)

Abstract: Sprouts are an attractive food product that contains high amounts of nutritional substances and has pro-health features. Sprout consumption has strongly increased despite its potential risk to health due to its microbial load. Both the safety and shelf life of sprouts may be negatively affected by a high microbial load. To reduce the microbial contamination in sprouts before consumption, the initial microbial load on the seeds needs to be controlled. Many herbal sprouts have been recommended for diabetes, and fenugreek is one of these sprouts. Thus, the current experiment aimed at disinfecting fenugreek seeds using microwave (5, 10, and 20 s) and hot water (30, 45, and 60 s) treatments for different durations. The best-disinfected sprouts with the highest nutritional properties were used to evaluate their influence on streptozocin-induced diabetic rats in comparison with fenugreek seed feeding. Microwave treatments showed the highest sprout length, fresh weight, total free amino acids, antioxidants, reducing sugars, and total phenols. Additionally, microwave seed treatments showed the lowest bacteria and mold counts on sprouts produced relative to hot water treatments, and the best seed treatment was a microwave for 20 s, which gave the best values in this respect. Feeding diabetic rats with different fenugreek seeds or sprout rates (0, 5, 7.5, and 10% *w/w*) improved body weight, restricted the growth of glucose levels, lowered total cholesterol and triglycerides, and improved HDL compared with the positive control group, and fenugreek sprouts at higher rates showed the maximum improvements in blood glucose, total cholesterol, and triglycerides. Treating fenugreek seed with microwave radiation for 20 s to disinfect the seeds before sprouting is recommended for lowering the microbial load with optimum nutritional and antioxidant activity, and feeding diabetic rats with these sprouts at the rate of 7.5 and 10% had promising effects on hyperglycemia and associated disorders.

Keywords: disinfection; diabetes; fenugreek sprouts; microwave radiation; antioxidants

Citation: Dahab, A.A.; Bayomy, H.M.; El-Salam, H.S.A.; Almasoudi, S.E.; Ozaybi, N.A.; Mahmoud, G.A.; Atteya, A.K.G.; El-Serafy, R.S. Seed Disinfection Treatments Minimized Microbial Load and Enhanced Nutritional Properties of Fenugreek Sprouts Which Alleviated Diabetes-Negative Disorders in Diabetic Rats. *Nutrients* **2024**, *16*, 2635. https://doi.org/10.3390/nu16162635

Academic Editors: Domenico Tafuri and Francesca Latino

Received: 28 June 2024
Revised: 2 August 2024
Accepted: 5 August 2024
Published: 10 August 2024

1. Introduction

Fenugreek (*Trigonella foenum-graecum* L.) is an important medicinal plant, grows as an annual herb, and is a member of legume seeds. It is cultivated all over the world and originated in Greece. Fenugreek seeds have significant nutritional and therapeutic importance due to their content of nutritional components and bioactive ingredients [1], including carbohydrates (45–60%), proteins (20–30%), fixed oil (5–10%), alkaloids, steroidal saponins, phenolics, and flavonoids, in addition to volatile oils, minerals, and vitamins [2]. Therefore, it is used in the pharmaceutical industry as a steroidal diosgenin source [3], and an antioxidant, antidiabetic, anti-lithogenic, antimicrobial, anticarcinogenic, hypocholesterolemic, and immunological enhancer [4].

Fenugreek has been shown to have a hypoglycemic effect due to its content of trigonelline and fenugrecin alkaloids, as well as the effect of its soluble fibers (e.g., glucomannan) and amino acids (such as 4-hydroxyisoleucine) in stimulating the pancreas to excrete insulin [5]. It also has resulted in a decrease in blood glucose levels in rats that were not insulin dependent as well as in healthy individuals [6]. The addition of seed powder to the diet of diabetic rats induced with alloxan has a positive impact on the levels of glycolytic, gluconeogenic, and lipogenic enzymes, ultimately leading to the restoration of glucose homeostasis [7]. Despite the fact that bioactive substances are found in small quantities in foods, they are vital nutritional components whose consumption promotes numerous health benefits and guards against chronic degenerative diseases [8]. Annually, there is an increase of 10% in bioactive component consumption and their related food products [9]. Several clinical reports state that consuming foods high in antioxidants is associated with a lower risk of degenerative disease occurrence.

Sprouting is a high-impact procedure that increases the amount of available nutrients while lowering the concentration of anti-nutrients at a low cost, with possible health advantages [10]. Sprouts are the byproduct of seeds that are developed in water and collected before true leaves appear. These sprouts are consumed whole, including the seeds [11]. Although consuming different seed sprouts has increased and spread globally due to their high nutritional content and accessibility [11], many consumers are exposed to illness due to the contamination of sprouts. In 1995, the Food and Drug Administration (FDA) in the United States considered sprouts to be a source of foodborne illness and recognized sprouts as a Potentially Hazardous Food [11]. Sprouts can serve as a good food source for bacteria that cause deterioration and pathogenicity. Three distinct stages comprise the germination process: disinfection, steeping, and sprouting. Before seed germination, disinfection is conducted to lower the microbial load of the seeds. Formaldehyde, ethanol, or sodium hypochlorite are the common subjects used for seed disinfection [12]. However, the disinfection phase is skipped because of the possible negative effects and hazards to the food safety of the seeds. The seed sprouting occurred in controlled environments with moist, warm, and nutrient-rich conditions, and this environment is ideal for microbial activity and growth. If the seeds have a microbial load or are not disinfected, the high microbial load can lower the sprouts' shelf life and raise the infection probability [13].

Diabetes is a metabolic defect identified by hyperglycemia due to irregularities in the secretion and action of insulin [14]. Because of the decreased antioxidant levels and increased formation of reactive oxygen species, persistent hyperglycemia causes microvascular problems (such as neuropathy and nephropathy) as well as macrovascular problems (mostly cardiovascular problems) [15]. The worldwide incidence of diabetes mellitus is rising and is predicted to reach 592 million people by 2035, mainly as a result of increasing rates of obesity [16]. Although the etiology, clinical presentation, and disease prevalence of type 1 and type 2 diabetes differ, both kinds of diabetes include secondary consequences such as endothelial dysfunction. Type 2 diabetes is more common in older adults than children, which occurs due to a problem in the way the body uses and regulates glucose as a fuel. Diabetes type 2 has two major issues: (1) the pancreas does not create enough insulin (the hormone that controls the flow of sugar into cells), and (2) cells respond poorly to insulin and consume less sugar. The majority of diabetes problems are incurable, meaning that the health system and society bear a heavy financial cost.

Throughout history, herbal medications have been widely used to cure a variety of illnesses [17–22]. Researchers are attempting to assess both the potential positive and negative effects of these natural compounds on human health because it is still unclear how exactly these products are characterized and how they work [23–26]. Patients prefer to use botanicals over various medications, including insulin analogues, thiazolidine, biguanides, sulphonylureas, dipeptidyl peptidase-4 inhibitors, and α-glucosidase inhibitors, because of the side effects and higher cost of these treatments [15]. Several medicinal plants, including fenugreek, are used as diabetes treatments due to their ability to decrease glucose and lipid levels by inducing insulin secretion, inhibiting glucosidase activity, elevating GLUT4 expression [27], and inhibiting gluconeogenesis [28], in addition to activating the AMP-activated protein pathway [29].

Therefore, to minimize the microbial load of sprouts before consumption, the initial microbial load of the seeds should be controlled [30]. Overlapping this problem can be performed by many strategies, including electromagnetic field applications. Microwave radiation application in food processing to inactivate bacterial growth and different microorganisms is an important technique [31]. Electromagnetic fields induce physiological alteration in plant cell structure, increase nutritional value, and boost crop yields [32]. Physical conditions including light, temperature, disinfection, genotype [33], and seed chemical composition caused a variation in the germination and antioxidant features of seeds [34], including fenugreek. Fenugreek seed sprouts showed a better nutritional profile and a lower fiber level, making them easier to absorb and digest [35], and a higher antioxidant level with more antidiabetic efficiency than their boiled seeds [36]. The release or increased bioavailability of bound antioxidants during sprouting could be the cause of these effects.

Therefore, the objective of this study is to evaluate the most common procedure for fenugreek seed disinfection using microwave and hot water at different durations to minimize the microbial load on the sprouts prior to consumption with optimal nutritional value, and then using these sprouts in the diet of streptozocin-induced diabetic rats to evaluate their effect on serum glucose, total cholesterol, triglycerides, HDL, and kidney markers.

2. Materials and Methods

2.1. Sprout Experiment

2.1.1. Seed Treatments

Fenugreek (*Trigonella foenum-graecum* L.) seeds were provided from the Horticulture Research Institute, Agricultural Research Center, Giza, Egypt. The seeds were washed three times with distilled water, dried, and divided into seven groups. Each group contained 2 g of seeds, which were put in 9 cm glass culture dishes. Three groups of seeds were exposed to irradiation using a microwave (230 volts, 50 Hz, 900–1300 W) for 5 s (M5s), 10 s (M10s), and 20 s (M20s); the other three groups were exposed to hot water (80 ± 1 °C) for 30 s (W30s), 45 s (W45s), and 60 s (W60s), while the last group was not treated, as a control group. After treatment, the seeds for each group were separately soaked in sterilized water for 24 h. During the soaking period, the water was changed one time. After pouring off the soaking water, the seeds were incubated for 4 days at 85% RH and 24 ± 2 °C in a seed germination incubator. Throughout the incubation period, the seeds were rinsed with 8 mL of sterile distilled water every 12 h. The sprouts were promptly removed after germination and stored at −80 °C until use and biochemical analysis.

2.1.2. Microbial Load Determination

Three replicates were performed to assess the number of bacteria and fungi on the fenugreek seeds and on the sprouts produced from the treated seeds. In total, 1 g of samples was mixed with 9 mL of sterile distilled water and homogenized for 1 h using a horizontal shaker (MK201D, Yamato Scientific Co., Ltd., Shanghai, China). After that, 10–4 serial dilutions were made. For uniform distribution, 1 mL of a 10–4 dilution was pipetted into sterile Petri plates along with potato dextrose agar for fungus at 45 °C and molten nutrition agar for total bacteria. Then, Petri plates with molten nutrition agar were incubated for 24 h at 37 °C, and the colonies of bacteria were enumerated. The Petri plates along with potato dextrose agar media were incubated for 5 days at 25 °C, and the colonies of fungus were enumerated [37]. The average number of colonies that developed on the plates was noted. The colony-forming units (log cfu g^{-1} fresh weight) were used to represent the total microbial count.

2.1.3. Biochemical Analysis of Fenugreek

Total free amino acids in seeds and sprouts were estimated according to the procedure outlined by Yemm and Cocking [38] (1955) using the ninhydrin reagent technique. Reducing sugar in seeds and sprouts was extracted as mentioned by Kawamura [39] and determined by the 3,5-dinitrosalicylic acid (DNSA) method, according to Krivorotova and Sereikaite [40]. Antioxidant activity (% as DPPH) was determined based on the radical scavenging ability of reacting with a stable DPPH-free radical, according to Chen et al. [41]. Total phenols were evaluated in dried fenugreek seeds or sprouts, extracted using 80% ethanol and determined according to the colorimetric method described by Boateng et al. [42].

2.2. Animal Experiment

This experiment was performed as given by the standards of animal care and approved by the Ethics Committee of Tanta University, Egypt, during the period of March–April 2022. In total, 48 male Wistar albino rats were acquired from the Faculty of Veterinary Medicine, Cairo University; their weights ranged from 165 to 180 g. The rats were put in cages (22 °C and 45–65% HD) for one week before the beginning of the experiment and were provided with unlimited ordinary rat pellet diets and water.

2.2.1. Diabetes Model

Before beginning streptozocin (STZ) therapy, all rats were weighed, received water as normal, and fasted for 12 h. Then, the rats were injected intraperitoneally with STZ, which was prepared by mixing it with 0.5 mL sodium citrate buffer (pH 4.5) at a rate of 35 mg/kg body weight [43] for all rats except the negative control group, which was injected intraperitoneally with 0.5 mL of citrate buffer (pH 4.5) [44]. After that, the rats were fed and kept on 10% (*w*/*v*) sucrose water for one day, and then tap water was used [44]. One week after STZ injection, fasting glucose levels were estimated for each rat, and rats above 200 mg/dL were considered in the study [45].

2.2.2. Fenugreek Treatments

The rats were divided into eight groups randomly (n = 6): group 1: served as a nondiabetic and untreated control group (Negative control), group 2: functioned as a diabetic control group (Positive control), group 3: diabetic rats receiving 5% (*w*/*w*) fenugreek seeds powder (Diabetic FS1), group 4: diabetic rats receiving 7.5% (*w*/*w*) fenugreek seeds powder (Diabetic FS2), group 5: diabetic rats receiving 10% (*w*/*w*) fenugreek seeds powder (Diabetic FS3), group 6: diabetic rats receiving 5% (*w*/*w*) fenugreek sprouts of microwave treatment of M10s (Diabetic FSP1), group 7: diabetic rats receiving 7.5% (*w*/*w*) fenugreek sprouts of microwave treatment of M10s (Diabetic FSP2), and group 8: diabetic rats receiving 10% (*w*/*w*) fenugreek sprouts of microwave treatment of M10s (Diabetic FSP3). Fenugreek seeds were crushed using an electric grinder, and the powder was added to the standard diet, which was bought from the forage market. The standard diet contains starch (67.0%), casein (15.0%), salts (4.0%), wood fiber (5.0%), corn oil (8.0%), and vitamins (1.0%), and then a little distilled water was added [7], formed into pellets, dried, and kept in plastic containers at 5 °C. The fresh sprouts were crushed and mixed with the standard diet in the same manner as fenugreek seeds.

2.2.3. Blood and Kidney Markers Analysis

After 30 days of the feeding period, the rats' body weight was measured, and blood and kidney marker samples were collected for biochemical analysis. Glucose was analyzed according to Sasaki [46] method using O-toluidine, total cholesterol was estimated according to Richmond [47], triglycerides were assessed using the Trinder [48] method ($mgdL^{-1}$), and high-density lipoprotein (HDL) cholesterol ($mgdL^{-1}$) was measured using the Richmond [47] procedure. Kidney markers of urea, uric acid, and creatinine were estimated by alkaline picrate method.

2.3. Statistical Analysis

Data pooled were statistically analyzed using the COSTAT version 6.4. The Tukey test with a 0.05 probability was used to determine whether there was a significant difference between the mean values [49]. The results were expressed as mean ± SE, n = 6.

3. Results

3.1. Germination and Sprouts Physical Traits

The germinated seeds of all disinfection treatments showed nonsignificant differences regarding germination percentage, as all seeds appeared to have full germination expect for the treatment of W60s, which showed a decrease of about 6.7% compared with other disinfection treatments (Figure 1). Concerning the sprouts' features, seed treatments caused many changes in the sprouts' physical traits as the sprouts' length became taller as affected by hot water and microwave treatments relative to the control seeds, and the highest length was exhibited by the seeds treated with M10s, which was 24.6% taller than the control seeds, followed by M20s and M5s, while the lowest length was given by the control seeds. Hot water treatments showed a significant and gradual increase in sprout length until the W45s treatment, which decreased after that. The fenugreek seeds exposed to microwave treatments showed an increase in the sprouts' fresh weight compared with hot water and control treatments, and the highest fresh weight of 10 sprouts was produced by the M10s treatment, followed by the M5s, but the lowest fresh weight values were given by the hot water treatments, which reached the lowest values by the W45s treatment.

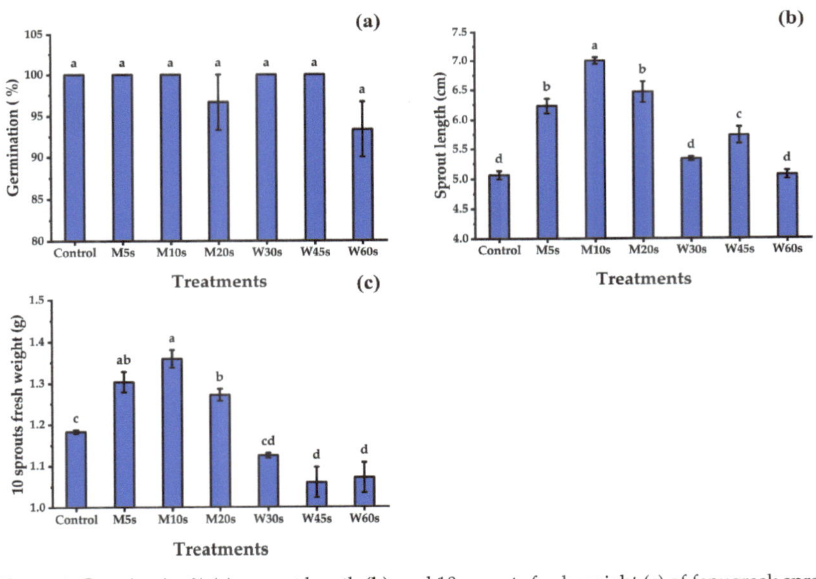

Figure 1. Germination% (**a**), sprout length (**b**), and 10 sprouts fresh weight (**c**) of fenugreek sprouts in response to microwave and hot water seed disinfection treatments. M5s, microwave for 5 s; M10s, microwave for 10 s; M20s, microwave for 20 s; W30s, hot water for 30 s; W45s, hot water for 45 s; and W60s, hot water for 60 s. All data denote mean ± S.E. Means with different letters significantly differed, using Tukey test at $p \leq 0.05$.

3.2. Biochemical Analysis

The data presented in Figure 2 showed the biochemical analysis of fenugreek seeds and sprouts as affected by different disinfection treatments. The fenugreek seed content of total free amino acids and reducing sugar significantly exhibited lower values compared with different fenugreek sprouts. Different disinfection treatments exhibited great improvements in the biochemical analysis of the sprouts produced. Fenugreek seeds subjected to M10s treatment significantly gave the maximum total free amino acid value (0.354 mg 100 mg^{-1}), while the lowest value in this regard was exhibited by the untreated fenugreek seeds (0.129 mg 100 mg^{-1}). Hot water treatments showed an enhancement in total free amino acid values compared with the untreated seeds and control sprouts, where the W45s sprouts presented the highest value in this regard. The presented results in Figure 2b,d show that the M10s treatment recorded the highest antioxidants (58.92%) and total phenols (0.189 mg GAE 100 mg^{-1} DW) values, followed by fenugreek sprouts at W45s for antioxidants and M20s for total phenols. On the other hand, the lowest antioxidants and total phenol content were exhibited by untreated seeds, which means that seed treatments were an effective tool in increasing the antioxidant activity of the sprouts produced. All disinfection treatments led to an enhancement in the reducing sugar content, and increasing the sprout duration from 5 s to 20 s for microwave and from 30 s to 60 s for hot water treatments led to an increase in the reducing sugar content. In this regard, the highest reducing sugar value was given by the M20s treatment, while the lowest reducing sugar content was given by untreated fenugreek seeds.

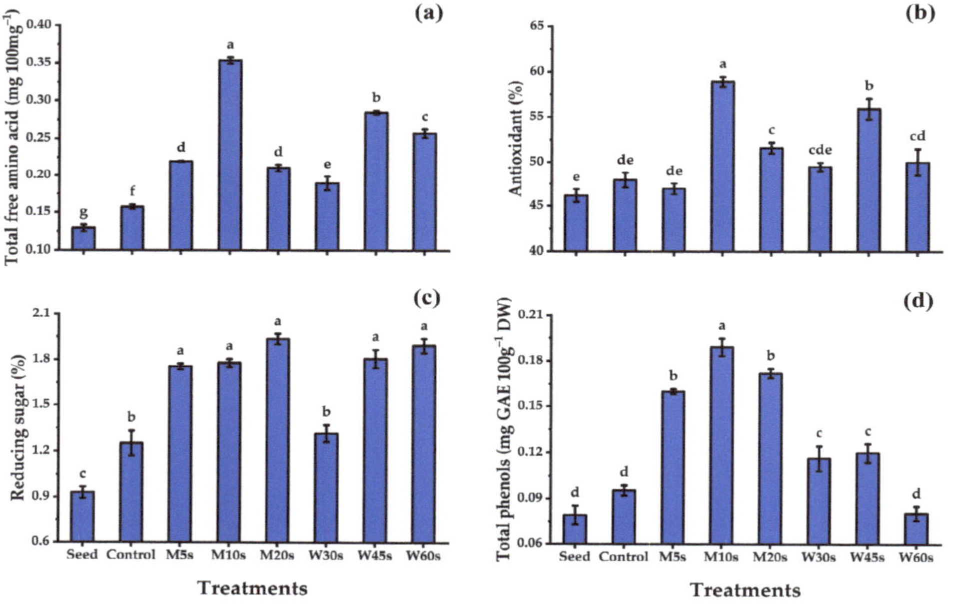

Figure 2. Total free amino acid (**a**), antioxidant (**b**), reducing sugars (**c**), and total phenols (**d**) of fenugreek sprouts in response to microwave and hot water seed disinfection treatments. M5s, microwave for 5 s; M10s, microwave for 10 s; M20s, microwave for 20 s; W30s, hot water for 30 s; W45s, hot water for 45 s; and W60s, hot water for 60 s. All data denote mean ± S.E. Means with different letters significantly differed, using Tukey test at $p \leq 0.05$.

3.3. Microbial Load

The initial total microbial count estimated from the untreated fenugreek seeds was 5.4 log CFU g^{-1} FW for total bacteria and 5.8 log CFU g^{-1} FW for total mold (Figure 3). On the other hand, most of the disinfection applications were effective in substantially decreasing the microbial loads on the fenugreek sprouts. As indicated in Figure 3, the lowest total bacterial count (2.33 log CFU g^{-1} FW) and total mold (1.60 log CFU g^{-1} FW) across all the disinfection treatments were given by M20s-treated seeds, followed by M10s and W60s treatments, with nonsignificant differences among them. The highest microbial load was observed by control sprouts, which gave 6.67 log CFU g^{-1} FW for total bacteria and 6.40 log CFU g^{-1} FW for total mold. These results suggest that the microwave treatment at M20s was the most effective seed disinfection approach compared to other treatments.

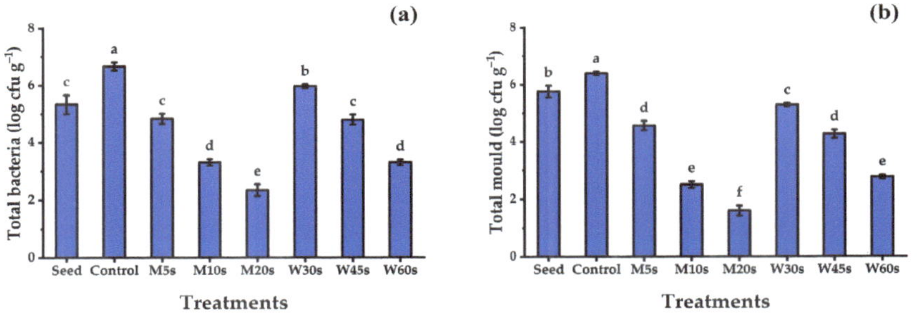

Figure 3. Total bacteria (**a**) and total mold (**b**) of fenugreek sprouts in response to microwave and hot water seed disinfection treatments. M5s, microwave for 5 s; M10, microwave for 10 s; M20s, microwave for 20 s; W30s, hot water for 30 s; W45s, hot water for 45 s; and W60s, hot water for 60 s. All data denote mean ± S.E. Means with different letters significantly differed, using Tukey test at $p \leq 0.05$.

3.4. Rats Body Weight

The body weight performance of negative and positive diabetic rats tabulated in Table 1 revealed that the negative diabetic rats significantly showed the highest body weight gain (37.7 g), while positive diabetic rats exhibited negative body weight changes as their weights decreased by 5.3 g compared with their initial weights. On the other hand, all other experimental groups exhibited an improvement in their body weight performance to different degrees relative to positive diabetic rats. The diabetic rats treated with FS1 exhibited the lowest gain in their body weight, but the diabetic rats treated with FSP2 and FSP3 exhibited the highest improvement in their body weight, but with nonsignificant differences among them.

Table 1. Effect of diabetic rats feeding with fenugreek seed powder and sprouts on the initial weight, final body weight, and changes in body weight of diabetic rats.

Treatments	Initial Body Weight (g)	Final Body Weight (g)	Changes in Body Weight (g)
Negative control	176.0 ± 1.00^{a}	213.67 ± 0.67^{a}	37.7 ± 0.33^{a}
Positive control	173.0 ± 1.15^{a}	167.67 ± 1.45^{g}	-5.3 ± 1.20^{f}
Diabetic FS1	176.7 ± 0.67^{a}	182.33 ± 2.03^{f}	5.7 ± 1.45^{e}
Diabetic FS2	174.6 ± 0.88^{a}	189.33 ± 1.66^{e}	14.7 ± 0.88^{d}
Diabetic FS3	176.0 ± 1.15^{a}	192.33 ± 0.58^{e}	16.0 ± 1.00^{d}
Diabetic FSP1	175.7 ± 1.76^{a}	196.00 ± 1.00^{d}	20.3 ± 1.86^{c}
Diabetic FSP2	173.3 ± 0.88^{a}	200.33 ± 1.33^{c}	27.0 ± 0.57^{b}
Diabetic FSP3	176.3 ± 1.20^{a}	205.00 ± 1.15^{b}	28.7 ± 0.33^{b}

Negative control: nondiabetic and untreated rats, Positive control: diabetic rats, Diabetic FS1: diabetic rats received fenugreek seed powder at 5%, Diabetic FS2: diabetic rats received fenugreek seed powder at 7.5%, Diabetic FS3: diabetic rats received fenugreek seed powder at 10%, Diabetic FSP1: diabetic rats receiving 5% fenugreek sprouts, Diabetic FSP2: diabetic rats receiving 7.5% fenugreek sprouts, and Diabetic FSP3: diabetic rats receiving 10% fenugreek sprouts. Data are presented as mean ± SE (n = 6). Values with different letters in the same column significantly differ at $p \leq 0.05$.

3.5. Blood and Kidney Markers Analysis

The data presented in Table 2 show that the fasting blood glucose levels of different diabetic rats were elevated significantly compared with the negative control rats. The highest elevation in fasting blood glucose level (325.7 mg/dL) was recorded by the positive control rats, but this value was decreased following different fenugreek feeding treatments. The highest reduction in the fasting blood glucose level (117.1 mg/dL) was recorded by the diabetic rats who received fenugreek sprouts at 7.5% in their diet (Diabetic FSP2), followed by diabetic FSP3 rats and FSP1 rats, with no significant difference among them. The results in the same table point out the changes in the lipid profile. Total cholesterol and triglyceride levels of the diabetic rats increased significantly, whereas high-density lipoprotein (HDL) levels decreased significantly in comparison to the negative control rats. Each of the fenugreek seeds or sprouts lowered the lipid profile, but, to some extent, fenugreek sprouts had the best effect on lowering the lipid profile.

Table 2. Effect of diabetic rats feeding with fenugreek seed powder and sprouts on the serum glucose, total cholesterol, triglycerides, and HDL of diabetic rats.

Treatments	Glucose	Total Cholesterol	Triglycerides	HDL
Unit		mg dL^{-1}		
Negative control	94.9 ± 0.17 ^{f}	93.57 ± 0.42 ^{f}	55.96 ± 1.13 ^{c}	60.87 ± 0.64 ^{a}
Positive control	325.7 ± 2.1 ^{a}	114.86 ± 1.06 ^{a}	69.97 ± 0.85 ^{a}	40.08 ± 0.41 ^{f}
Diabetic FS1	278.7 ± 3.8 ^{b}	112.17 ± 0.89 ^{b}	66.89 ± 0.81 ^{ab}	49.16 ± 0.36 ^{e}
Diabetic FS2	227.9 ± 1.6 ^{c}	107.23 ± 0.93 ^{c}	65.26 ± 2.24 ^{b}	50.93 ± 0.25 ^{d}
Diabetic FS3	169.6 ± 4.2 ^{d}	104.79 ± 0.50 ^{d}	66.64 ± 0.51 ^{ab}	52.03 ± 0.73 ^{cd}
Diabetic FSP1	125.15 ± 0.6 ^{e}	99.26 ± 0.39 ^{e}	58.10 ± 0.13 ^{c}	53.62 ± 0.34 ^{bc}
Diabetic FSP2	117.1 ± 0.82 ^{e}	94.08 ± 0.68 ^{e}	56.36 ± 0.40 ^{c}	54.74 ± 0.54 ^{b}
Diabetic FSP3	121.6 ± 0.84 ^{e}	97.43 ± 0.52 ^{f}	56.89 ± 0.28 ^{c}	53.09 ± 0.09 ^{bc}

Negative control: nondiabetic and untreated rats, Positive control: diabetic rats, Diabetic FS1: diabetic rats received fenugreek seeds powder at 5%, Diabetic FS2: diabetic rats received fenugreek seeds powder at 7.5%, Diabetic FS3: diabetic rats received fenugreek seeds powder at 10%, Diabetic FSP1: diabetic rats receiving 5% fenugreek sprouts, Diabetic FSP2: diabetic rats receiving 7.5% fenugreek sprouts, and Diabetic FSP3: diabetic rats receiving 10% fenugreek sprouts. Data are presented as mean ± SE (n = 6). Values with different letters in the same column significantly differ at $p \leq 0.05$.

The administration of the fenugreek sprouts significantly decreased serum total cholesterol and triacylglycerol in diabetic rats compared with nondiabetic control rats, and the lowest values in this respect were recorded by FSP2-rats, who gave 94.1 and 56.4 mg dL^{-1} for total cholesterol and triacylglycerol, respectively. Regarding the nondiabetic control rats group, they had the lowest total cholesterol and triacylglycerol values, but these values were maximized by the positive rats group. It is clearly shown that a significant decrease in HDL values was shown by the diabetic rats and all fenugreek treatments, but the oral administration of FSP2 restricted the reduction of HDL values. The results presented in Table 3 show that uric acid, urea, and creatinine values increased in diabetic control rats when compared with nondiabetic control rats. The fenugreek seeds and sprouts treatments significantly lowered urea, uric acid, and creatinine values compared with the control diabetic rats. The rats group supplemented with fenugreek sprouts of FSP2 significantly recorded the lowest values in this respect relative to the other treatments.

Table 3. Effect of diabetic rats feeding with fenugreek seed powder and sprouts on uric, urea, and creatinine.

Treatments	Uric	Urea	Creatinine
Negative control	2.31 ± 0.08b [c]	34.06 ± 0.36 [f]	0.24 ± 0.008 [e]
Positive control	2.86 ± 0.07 [a]	57.87 ± 0.98 [a]	0.38 ± 0.014 [a]
Diabetic FS1	2.59 ± 0.12 [b]	57.25 ± 0.29 [a]	0.38 ± 0.003 [a]
Diabetic FS2	2.47 ± 0.55 [bc]	52.52 ± 0.80 [b]	0.34 ± 0.01 [b]
Diabetic FS3	2.44 ± 0.02 [bc]	51.48 ± 0.19 [bc]	0.33 ± 0.003 [b]
Diabetic FSP1	2.38 ± 0.02 [bc]	49.65 ± 0.46 [c]	0.27 ± 0.001 [cd]
Diabetic FSP2	2.23 ± 0.02 [c]	45.55 ± 1.0 [d]	0.25 ± 0.005 [c]
Diabetic FSP3	2.29 ± 0.06 [c]	47.54 ± 0.33 [e]	0.28 ± 0.004 [de]

Negative control: nondiabetic and untreated rats, Positive control: diabetic rats, Diabetic FS1: diabetic rats received fenugreek seed powder at 5%, Diabetic FS2: diabetic rats received fenugreek seed powder at 7.5%, Diabetic FS3: diabetic rats received fenugreek seed powder at 10%, Diabetic FSP1: diabetic rats receiving 5% fenugreek sprouts, Diabetic FSP2: diabetic rats receiving 7.5% fenugreek sprouts, and Diabetic FSP3: diabetic rats receiving 10% fenugreek sprouts; HDL: high-density lipoprotein. Data are presented as mean ± SE (n = 6). Values with different letters in the same column significantly differ at $p \leq 0.05$.

4. Discussion

4.1. Effect of Disinfecting Treatments on Germination and Seedling Traits

Seed disinfection has been proposed as an important step in a multi-part technique to decrease the risk of illness caused by polluted sprouts. In this study, all seed treatments showed nonsignificant differences in the germination percentage. The lowest sprout length was recorded by control seeds followed by hot water treatments, but the lowest sprout fresh weight was recorded by hot water treatments. Contrary to our findings, Hermansen et al. [50] found that hot water treatments at 44, 49, and 54 °C on carrot seeds enhanced the germination rate and seedling features. The better crop was obtained by treating okra seeds in hot water for 30 min at 52 °C [51].

4.2. Effect of Disinfecting Treatments on Sprouts' Biochemical Analysis

Microwave treatments showed improvements in the biochemical analysis of total phenols, antioxidants, and total free amino acids; reducing sugar content in the sprouts produced; and increasing the exposure time increased these values. Similar findings were reported by Ragha et al. [52] and Gaurilcikiene et al. [53], who indicated that extending the microwave exposure time increased the antioxidant activities in wheat seedlings. Chen et al. [54] illustrated that lower power and microwave duration decreased antioxidant activities and seedling vigor in *Isatis indigotica* fort. The biologically active ingredient yield of buckwheat has been elevated following microwave treatments [55]. The germination rate and seedling vigor of lentil seeds subjected to microwave radiation (450–730 W) for 30 s were enhanced, but these values decreased with longer exposure times of 60 and 90 s [56]. In this study, seed microwave treatments elicited the bioactive components of fenugreek sprouts. Microwave seed treatments stimulated phenolic production [34]. The nonthermal influences on seeds are induced by the close contact of the microwaves with seed tissue components or molecules as particles try to organize themselves within an electric field and reduce the potential energy [57]. As recorded by Hamada [58], wheat seedlings showed an elevation in amino acid and protein contents as affected by microwave seed treatment. Radish seeds treated with microwave radiation showed an induction in the bioactive component concentration, including a variety of enzymes involved in seed germination, improving the germination rate in addition to boosting the production of specific biological components in the seeds [57]. In plant seedlings, microwave pretreatment increases the expression of genes encoding superoxide dismutase and peroxidase isozymes, which leads to a significant increase in the seeds' germination potential, stem length, and total mass [56,57], in addition to cell membrane integrity, which improves plant resistance to environmental stress [59]. Microwaves can enhance the quality of germinated grains, their nutritional value, and the secondary metabolite content of grains and sprouts [60].

The advantageous effects of microwave frequency on other metabolic processes, including water absorption [61], the degradation of larger phenolic and flavonoid compounds into smaller ones, and phenolic compounds released from glycosidics by irradiation caused an increase in the total phenol content [62] in fenugreek. These findings would also explain the improvement in the phenolic content due to the changes in tissue structure caused by irradiation.

4.3. *Effect of Disinfecting Treatments on the Microbial Load*

Seed treatments with hot water or microwave irradiation decreased the microbial load in fenugreek sprouts. The optimal temperature and length of seed treatment must be determined for each crop and the associated pathogen. This treatment aims to eliminate infections as far as possible without affecting seed germination. As reported by Singh et al. [63], variations in the germination rate of seeds might occur even with a 5 min variation in treatment duration. The ideal combination of temperature and time for a particular plant seed is determined by many variables that interact with the host's heat tolerance, including the state of the outer layers, dormancy, moisture content, age, and vigor [64]. The seeds are more resistant to high temperatures the lower their initial water content is upon heating [63]. Hot water is used to destroy structural components and interfere with vital life processes, such as protein denaturation, which results in the death or inactivation of viruses, bacteria, protozoa, and other pathogens [65]. Hot water treatment for sweet pepper seeds was found to be beneficial in reducing seed-borne infections. Fungal infections were shown to be less common when bell peppers were treated with hot water at 45 °C for 15 min or 53 °C for 4 min before being stored at 8 °C [66]. Hot water treatments for bell pepper seeds led to a significant decline in seed viability, whereas there was no discernible impact on seed vigor [67]. Soaking carrot seeds in 52 °C of hot water for 25 min was effective in eliminating *Xanthomonas hortorum* pv. Carotae bacteria [68].

Microwave radiation heats biological systems dielectrically; this heat is produced when the microwave energy is absorbed by a dielectric substance (usually water) and is then converted into heat by the rotation's internal resistance [69]. The movement of dipolar molecules causes the dissipation of energy received from microwaves due to friction [70]. The thermal theory of microwave radiation states that there is no basis for direct electromagnetic interaction with living systems that is not dependent on these temperature-mediated effects. The bioeffects of microwave radiation are only explained by differences in temperatures or temperature profiles between microwave and conventionally heated systems [70]. The duration of microwave exposure is a significant determinant of the impact of microwaves on living cells, as shown in the current as increasing the exposure period from 5 s to 20 s led to a significant reduction in the microbial count on the fenugreek sprouts. The effects of microwave irradiation are based on the effects of temperature and electromagnetism on organisms. The extent to which microwaves impact microorganism growth is mostly determined by the radiation intensity and total energy assimilated by the organisms. Thus, microwaves are most likely to have a dominant thermal effect and destroy yeast or bacteria when they are applied at high frequencies, with high energy, and over an extended period of time. Generally, seeds have a 3–6 log 10 CFU g/L microbial load [71]. Irradiation can reduce microbial contamination by about 3.18 log CFU/g [72]. According to Danilchuk and Alkhateeb [73], the lethal impact of microwave radiation was significantly reduced in arid environments and manifested itself only after an extended duration of exposure.

4.4. *Effect of Feeding Treatments on the Body Weight and Markers*

Fenugreek is frequently consumed throughout the world. Here, we present a comprehensive examination of the effects of fenugreek seed or sprout feeding on body weight, blood, and kidney markers. Diabetic rats receiving fenugreek seed or sprout exhibited a gain in their body weight, while positive control rats presented a progressive decrease in their weight (Table 1). The gradual rise in body weight following fenugreek treatments indicates that fenugreek, especially when taken in high doses, may be able to reduce the toxicity of STZ [74]. Numerous biochemical changes occur in the seeds during sprouting, such as an increase in the bioavailability of macro- and micro-elements [75]. The highest weight gain was observed with the highest dose of sprout treatment, and this may be explained by the fact that sprout treatments led to better utilization of nutrients in the diet and thus a weight gain [76].

In this study, fenugreek treatments elevated HDL concentration while reducing triglycerides and total cholesterol. The reduction in triglycerides and total cholesterol following fenugreek treatments may be due to the depression in non-esterified fatty acids in diabetic rats, as these acids have an impact on vascular alterations and platelet aggregation through improving the prostacyclin rate in plasma [76]. Fenugreek is rich in bioactive fiber, which has an important role in retarding fat and carbohydrate absorption that participates in the hypolipidemic effect [77]. Additionally, fenugreek has trigonelline, alkaloids, and the saponin compound diasgenin, which has a role in inhibiting intestinal glucose uptake [78]. Also, the amino acids tryptophan and arginine present in fenugreek seeds have hypoglycemic and antidiabetic impacts [5]. Our results showed that treated fenugreek seeds with microwave gave higher antioxidant levels in their sprouts than other treatments, and M10s treatment gave approximately 2.7% higher antioxidant content than fenugreek seeds and control. Similar results were reported by Chand et al. [79], who revealed that fenugreek sprouts have a higher antioxidant content than fenugreek seeds. Khoja et al. [80] revealed that apigenin, one of the isoflavones in fenugreek, is more abundant in the methanolic extract of fenugreek sprouts than seeds. Apigenin has antioxidant activities and can limit or decline oxidative stress by scavenging free radicals and enhancing cellular antioxidant defense activities [81]. A decline in the triglycerides and total cholesterol accumulation in the liver and blood serum of hyperlipidemia rats following apigenin treatment was observed by Muhammed et al. [82]. Apigenin enhances hepatic LDL-C absorption and boosts the transformation of hepatic cholesterol into bile acid [83].

The pathophysiology of diabetic nephropathy is mainly influenced by oxidative stress [84]. The imbalance between the antioxidant defense system and free radicals results in an increase in free radicals [85]. Persistent hyperglycemia exacerbates oxidative stress by boosting the generation of reactive oxygen species (ROS) [86]. Excessive free radicals can cause malfunctioning kidney tissue by oxidizing various cell macromolecules once they have overcome the body's natural antioxidant defense mechanism [87]. The kidneys subjected to hyperglycemia had lower levels of antioxidant indicators, which were restored by apigenin administration [88]. Also, Lin and Yin [89] found a decrease in the endogenous antioxidant levels in diabetic kidney cells. Hyperglycemia can boost renal function biomarkers (uric, creatinine, and urea), which are linked to glomerular atrophic changes, interstitial atrophy, and epithelial necrosis, causing diabetic nephropathy [90]. Apigenin normalizes creatinine, uric, and urea levels significantly, so this antidiabetic compound can maintain renal function in diabetic rats [88].

5. Conclusions

The high microbial load on seeds can decrease the safety and shelf-life of sprouts, so this microbial contamination should be minimized before the consumption of sprouts. In this regard, fenugreek seeds were exposed to different disinfection treatments using microwave and hot water, and the results obtained revealed that microwave treatment for 20 s was the best treatment in improving the sprout antioxidants with 2.7% compared with untreated seeds with minimizing the microbial load of their sprouts. Feeding streptozocin-diabetic rats with these sprouts at 7.5 and 10% enhanced their body weight, restricted the growth of glucose levels, lowered total cholesterol, triglycerides, and improved HDL, as well as enhanced kidney markers.

Author Contributions: Conceptualization, A.A.D., H.S.A.E.-S. and G.A.M.; Methodology, A.A.D., H.S.A.E.-S. and R.S.E.-S.; Software, A.A.D., H.S.A.E.-S. and G.A.M.; Validation, A.A.D., H.S.A.E.-S. and G.A.M.; Formal analysis, A.A.D., H.S.A.E.-S. and G.A.M.; Investigation, A.A.D., H.S.A.E.-S.,

R.S.E.-S. and G.A.M.; Resources, A.A.D., H.S.A.E.-S. and G.A.M.; Data curation, A.A.D., H.S.A.E.-S. and G.A.M.; Writing—Original draft preparation, R.S.E.-S., H.M.B., S.E.A. and N.A.O.; Writing—Review and editing, R.S.E.-S., H.M.B., S.E.A., A.K.G.A. and N.A.O.; Visualization, R.S.E.-S., H.M.B., A.K.G.A., S.E.A. and N.A.O.; Supervision, A.A.D., H.S.A.E.-S. and G.A.M.; Funding acquisition, A.A.D., H.S.A.E.-S. and G.A.M. All authors have read and agreed to the published version of the manuscript.

Funding: This research received no external funding.

Institutional Review Board Statement: The study was approved by the ethical approval committee, Animal Production Department (AY2019-2020/Session 6/2020.01.13), Faculty of Agriculture, Tanta University, Egypt.

Informed Consent Statement: Informed consent was obtained from all subjects involved in the study.

Data Availability Statement: The original contributions presented in the study are included in the article, further inquiries can be directed to the corresponding authors.

Conflicts of Interest: The authors declare no conflicts of interest.

References

1. Kaviarasan, S.; Naik, G.H.; Gangabhagirathi, R.; Anuradha, C.V.; Priyadarsini, K.I. In vitro studies on antiradical and antiox-idant activities of fenugreek (*Trigonella foenum graecum*) seeds. *Food Chem.* **2007**, *103*, 31–37. [CrossRef]
2. Belguith-Hadriche, O.; Bouaziz, M.; Jamoussi, K.; Simmonds, M.S.J.; El-Feki, A.; Makni-Ayedi, F. Comparative study on hypocholesterolemic and antioxidant activities of various extracts of fenugreek seeds. *Food Chem.* **2013**, *138*, 1448–1453. [CrossRef] [PubMed]
3. Petropoulos, G.A. *Fenugreek: The Genus Trigonella. Medicinal and Aromatic Plants—Industrial Profiles*; Taylor & Francis: Abingdon, UK, 2002.
4. Reddy, R.L.R.; Gowda, A.N.S.; Srinivasan, K. Antilithogenic and hypocholesterolemic effect of dietary fenugreek seeds (*Trigonella Foenum-graecum*) in experimental mice. *Med. Plants Int. J. Phytomed. Related Ind.* **2019**, *11*, 145–154. [CrossRef]
5. Eidi, A.; Eidi, M.; Sokhteh, M. Effect of fenugreek (*Trigonella foenum-graecum* L.) seeds on serum parameters in normal and streptozotocin-induced diabetic rats. *Nutr. Res.* **2007**, *27*, 728–733. [CrossRef]
6. Deng, R. A review of the hypoglycemic effects of five commonly used herbal food supplements. *Recent Pat. Food Nutr. Agric.* **2012**, *4*, 50–60. [CrossRef] [PubMed]
7. Raju, J.; Gupta, D.; Rao, A.R.; Yadava, P.K.; Baquer, N.Z. *Trigonella foenum graecum* (fenugreek) seed powder improves glucose homeostasis in alloxan diabetic rat tissues by reversing the altered glycolytic, gluconeogenic and lipogenic enzymes. *Mol. Cell Biochem.* **2001**, *224*, 45–51. [CrossRef] [PubMed]
8. WHO. *Diet, Nutrition, and the Prevention of Chronic Diseases. Report of a WHO Study Group*; WHO Technical Report Series, No. 797; World Health Organization: Geneva, Switzerland, 2003; Volume 916, p. 1990.
9. Milenkovic, D.; Morand, C.; Cassidy, A.; Konic-Ristic, A.; Tomas-Barberan, F.; Ordovas, J.M.; Kroon, P.; De Caterina, R.; Rodriguez-Mateos, A. Interindividual variability in biomarkers of cardiometabolic health after consumption of major plant-food bio-active compounds and the determinants involved. *Adv. Nutr.* **2017**, *8*, 558–570. [CrossRef]
10. Naidu, M.M.; Shyamala, B.N.; Naik, J.P.; Sulochanamma, G.; Srinivas, P. Chemical composition and antioxidant activity of the husk and endosperm of fenugreek seeds. *LWT-Food Sci. Technol.* **2011**, *44*, 451–456. [CrossRef]
11. FDA. FY 2014–2016 Microbiological Sampling Assignment, Summary Report: Sprouts. 2017. Available online: https://www.fda.gov/downloads/Food/ComplianceEnforcement/Sampling/UCM566981.pdf (accessed on 8 July 2024).
12. Limón, R.I.; Peñas, E.; Martínez-Villaluenga, C.; Frias, J. Role of elicitation on the health-promoting properties of kidney bean sprouts. *LWT-Food Sci. Technol.* **2014**, *56*, 328–334. [CrossRef]
13. Taormina, P.J.; Beuchat, L.R.; Slutsker, L. Infections associated with eating seed sprouts: An international concern. *Emerg. Infect. Dis.* **1999**, *5*, 626–634. [CrossRef]
14. Ota, A.; Ulrih, N.P. An overview of herbal products and secondary metabolites used for management of type two diabetes. *Front. Pharmacol.* **2017**, *8*, 436. [CrossRef] [PubMed]
15. Choudhury, H.; Pandey, M.; Hua, C.K.; Mun, C.S.; Jing, J.K.; Kong, L.; Kesharwani, P. An update on natural compounds in the remedy of diabetes mellitus: A systematic review. *J. Tradit. Med. Complement. Ther.* **2018**, *8*, 361–376. [CrossRef] [PubMed]
16. Diabetes UK. *Diabetes: Facts and Stats*; Diabetes UK: London, UK, 2014; Volume 3, pp. 1–21.
17. Alshallash, K.S.; Mohamed, M.F.; Dahab, A.A.; Abd El-Salam, H.S.; El-Serafy, R.S. Biostimulation of *Plectranthus amboinicus* (Lour.) spreng. with different yeast strains: Morphological performance, productivity, phenotypic plasticity, and antioxidant activity. *Horticulturae* **2022**, *8*, 887. [CrossRef]
18. Bahgat, A.-R.; Dahab, A.A.; Elhakem, A.; Gururani, M.A.; El-Serafy, R.S. Integrated action of rhizobacteria with *Aloe vera* and moringa leaf extracts improves defense mechanisms in *Hibiscus sabdariffa* L. cultivated in saline soil. *Plants* **2023**, *12*, 3684. [CrossRef] [PubMed]

19. Youssef, S.M.; El-Serafy, R.S.; Ghanem, K.Z.; Elhakem, A.; Abdel Aal, A.A. Foliar spray or soil drench: Microalgae application impacts on soil microbiology, morpho-physiological and biochemical responses, oil and fatty acid profiles of chia plants under alkaline stress. *Biology* **2022**, *11*, 1844. [CrossRef] [PubMed]
20. Gururani, M.A.; Atteya, A.K.; Elhakem, A.; El-Sheshtawy, A.N.A.; El-Serafy, R.S. Essential oils prolonged the cut carnation longevity by limiting the xylem blockage and enhancing the physiological and biochemical levels. *PLoS ONE* **2023**, *3*, e0281717. [CrossRef] [PubMed]
21. El-Serafy, R.S.; El-Sheshtawy, A.A.; Abd El-Razek, U.A.; Abd El-Hakim, A.F.; Hasham, M.M.A.; Sami, R.; Khojah, E.; Al-Mushhin, A.A.M. Growth, yield, quality, and phytochemical behavior of three cultivars of quinoa in response to moringa and Azollaextracts under organic farming conditions. *Agronomy* **2021**, *11*, 2186. [CrossRef]
22. Atteya, A.K.G.; El-Serafy, R.S.; El-Zabalawy, K.M.; Elhakem, A.; Genaidy, E.A.E. Exogenously supplemented proline and phenylalanine improve growth, productivity, and oil composition of salted moringa by up-regulating osmoprotectants and stimulating antioxidant machinery. *Plants* **2022**, *11*, 1553. [CrossRef] [PubMed]
23. El-Serafy, R.S.; El-Sheshtawy, A.A.; Dahab, A.A.; Al-Ashkar, I. Can yeast extract and chitosan-oligosaccharide improve fruit yield and modify the pharmaceutical active ingredients of organic fennel? *Ind. Crops Prod.* **2021**, *173*, 114130. [CrossRef]
24. El-Serafy, R.S.; Dahab, A.A.; Ghanem, K.Z.; Elhakem, A.; Bahgat, A.R.; Venkatesh, J.; El-Sheshtawy, A.A.; Badawy, A. As a natural antioxidant: *Sesbania grandiflora* Leaf extract enhanced growth and yield performance, active ingredients and tolerance of *Hibiscus sabdariffa* L. under salt-affected soil. *J. Soil Sci. Plant Nutr.* **2024**, *24*, 3406–3420. [CrossRef]
25. Atteya, A.K.G.; El-Serafy, R.S.; El-Zabalawy, K.M.; Elhakem, A.; Genaidy, E.A.E. Brassinolide maximized the fruit and oil yield, induced the secondary metabolites, and stimulated linoleic acid synthesis of *Opuntia ficus-indica* oil. *Horticulturae* **2022**, *8*, 452. [CrossRef]
26. El-Serafy, R.S.; El-Sheshtawy, A.A. Effect of nitrogen fixing bacteria and moringa leaf extract on fruit yield, estragole content and total phenols of organic fennel. *Sci. Hortic.* **2020**, *265*, 109209. [CrossRef]
27. Bi, X.; Lim, J.; Henry, C.J. Spices in the management of diabetes mellitus. *Food Chem.* **2017**, *217*, 281–293. [CrossRef]
28. Adam, S.H.; Giribabu, N.; Kassim, N.; Kumar, K.E.; Brahmayya, M.; Arya, A.; Salleh, N. Protective effect of aqueous seed extract of *Vitis Vinifera* against oxidative stress, inflammation and apoptosis in the pancreas of adult male rats with diabetes mellitus. *Biomed. Pharmacother.* **2016**, *81*, 439–452. [CrossRef] [PubMed]
29. Governa, P.; Baini, G.; Borgonetti, V.; Cettolin, G.; Giachetti, D.; Magnano, A.R.; Miraldi, E.; Biagi, M. Phytotherapy in the management of diabetes: A Review. *Molecules* **2018**, *23*, 105. [CrossRef] [PubMed]
30. Akbas, M.Y.; Olmez, H. Inactivation of *Escherichia coli* and *Listeria monocytogenes* on iceberg lettuce by dip wash treatments with organic acids. *Lett. Appl. Microbiol.* **2007**, *44*, 619–624. [CrossRef]
31. Kumari, N.; Verma, S.; Sharma, V. Manipulating tomato plant electric signaling system by microwave radiation to enhance crop productivity and nutritional value. *Comput. Electron Agric.* **2018**, *154*, 330–340. [CrossRef]
32. Senavirathna, M.D.H.J.; Asaeda, T. Radio-frequency electromagnetic radiation alters the electric potential of *Myriophyllum aquaticum*. *Biol. Plant.* **2014**, *58*, 355–362. [CrossRef]
33. Pour, A.P.; Farahbakhsh, H.; Saffari, M.; Keramat, B. Effects of seed priming on germination and seedling growth under salinity stress in fenugreek. *Int. J. Agric. Crop Sci.* **2012**, *4*, 779–786.
34. Randhir, R.; Shetty, K. Microwave-induced stimulation of L-DOPA, phenolics and antioxidant activity in fava bean (*Vicia faba*) for Parkinson's diet. *Process Biochem.* **2004**, *39*, 1775–1784. [CrossRef]
35. Khan, F.; Negi, K.; Kumar, T. Effect of sprouted fenugreek seeds on various diseases: A review. *J. Diabetes Metab. Disord. Control* **2018**, *5*, 119–125. [CrossRef]
36. Vats, V.; Yadav, S.P.; Grover, J.K. Effect of *Trigonella foenum graecum* on glycogen content of tissues and the key enzymes of carbohydrate metabolism. *J. Ethnopharmacol.* **2003**, *85*, 237–242. [CrossRef] [PubMed]
37. Ejechi, B.; Souzey, J.; Akpomedaye, D. Microbial stability of mango juice preserved by combined application of mild heat and extracts of two tropical spices. *J. Food Protect.* **1998**, *61*, 725–727. [CrossRef] [PubMed]
38. Yemm, E.W.; Cocking, E.C. The determination of amino acids with ninhydrin. *Analyst* **1955**, *80*, 209–213. [CrossRef]
39. Kawamura, S. Quantitative paper chromatography of sugars of the cotyledon, hull, and hypocotyl of soybeans of selected varieties. *Tech. Bull. Fac. Agric. Kagawa Univ.* **1967**, *18*, 117–131.
40. Krivorotova, T.; Sereikaite, J. Determination of fructan exohydrolase activity in the crude extracts of plants. *Electron. J. Biotechnol.* **2014**, *17*, 329–333. [CrossRef]
41. Chen, Y.W.; Wu, S.W.; Ho, K.K.; Lin, S.B.; Huang, C.Y.; Chen, C.N. Characterisation of Taiwanese propolis collected from different locations and seasons. *J. Sci. Food Agric.* **2008**, *88*, 412–419. [CrossRef]
42. Boateng, J.; Verghese, M.; Walker, L.T.; Ogutu, S. Effect of processing on antioxidant contents in selected dry beans (*Phaseolus* spp. L). *LWT-Food Sci. Technol.* **2008**, *41*, 1541–1547. [CrossRef]
43. Srinivasan, K.; Viswanad, B.; Asrat, L.; Kaul, C.L.; Ramarao, P. Combination of high-fat diet-fed and low-dose streptozotocin-treated rat: A model for type 2 diabetes and pharmacological screening. *Pharmacol. Res.* **2005**, *52*, 313–320. [CrossRef]
44. Furman, B.L. Streptozotocin-induced diabetic models in mice and rats. *Curr. Protoc.* **2021**, *1*, e78. [CrossRef]
45. Varghese, R.; Majumdar, A. Current therapies in nephrotic syndrome: HDAC inhibitors, an emerging therapy for kidney dis-eases. *Curr. Res. Biotechnol.* **2021**, 3, 182–194. [CrossRef]

46. Sasaki, T. Effect of acetic acid concentration on the colour reaction in the O-toluidine boric acid method for blood glucose determination. *Rinsho. Kagaku* **1972**, *1*, 346–350.
47. Richmond, W. Preparation and properties of a cholesterol oxidase from *Nocardia* sp. and its application to the enzymatic assay of total cholesterol in serum. *Clin. Chem.* **1973**, *19*, 1350–1356. [CrossRef]
48. Trinder, P. Determination of glucose in blood using glucose oxidase with on alternative oxygen receptor. *Ann. Clin. Biochem.* **1969**, *6*, 24–27. [CrossRef]
49. Steel, R.G.D.; Torrie, J.H. *Principles and Procedures of Statistics*; McGraw-Hill Book: New York, NY, USA, 1960.
50. Hermansen, A.; Brodal, G.; Balvoll, G. Hot water treatments of carrot seeds: Effects on seed-borne fungi, germination, emergence and yield. *Seed Sci. Technol.* **1999**, *27*, 599–613.
51. Begum, M.; Lokesh, S. Effect of hot water and ultra violet radiation on the incidence of seedborne fungi of okra. *Arch. Phytopathol. Plant Prot.* **2012**, *45*, 126–132. [CrossRef]
52. Ragha, L.; Mishra, S.; Ramachandran, V.; Bhatia, M.S. Effects of low-power microwave fields on seed germination and growth rate. *J. Electromagn. Anal. Appl.* **2011**, *3*, 165–171. [CrossRef]
53. Gaurilcikiene, I.; Ramanauskiene, J.; Dagys, M.; Simniskis, R.; Dabkevicius, Z.; Suproniene, S. The effect of strong microwave electric field radiation on: (2) wheat (*Triticum aestivum* L.) seed germination and sanitation. *Zemdirb.-Agric.* **2013**, *100*, 185–190. [CrossRef]
54. Chen, Y.P.; Liu, Y.J.; Wang, X.L.; Ren, Z.Y.; Yue, M. Effect of microwave and He-Ne laser on enzyme activity and biophoton emission of *Isatis indigotica* Fort. *J. Integr. Plant. Biol.* **2005**, *47*, 849–855. [CrossRef]
55. Ji, H.B.; Tang, W.; Zhou, X.L.; Wu, Y. Combined effects of blue and ultraviolet lights on the accumulation of flavonoids in tartary buckwheat sprouts. *Pol. J. Food. Nutr. Sci.* **2016**, *66*, 93–98. [CrossRef]
56. Aladjadjiyan, A. Effect of microwave irradiation on seeds of lentils (*Lens culinaris*, Med.). *Rom. J. Biophys.* **2010**, *20*, 213–221.
57. Radzevičius, A.; Sakalauskienė, S.; Dagys, M.; Simniškis, R.; Karklelienė, R.; Bobinas, Č.; Duchovskis, P. The effect of strong microwave electric field radiation on: (1) vegetable seed germination and seedling growth rate. *Zemdirb.-Agric.* **2013**, *100*, 179–184. [CrossRef]
58. Hamada, E.A.M. Effects of microwave treatment on growth, photosynthetic pigments and some metabolites of wheat. *Biol. Plant.* **2007**, *51*, 343–345. [CrossRef]
59. Chen, W.; Hang, F.; Zhao, J.X.; Tian, F.W.; Zhang, H. Alterations of membrane permeability in *Escherichia coli* and *Staphy-lococcus aureus* under microwave. *Wei Sheng Wu Xue Bao = Acta Microbiol. Sin.* **2007**, *47*, 697–701.
60. Uppal, V.; Bains, K. Effect of germination periods and hydrothermal treatments on in vitro protein and starch digestibility of germinated legumes. *J. Food Sci. Technol.* **2012**, *49*, 184–191. [CrossRef]
61. Nelson, S.O. Review and assessment of radio-frequency and microwave energy for stored-grain insect control. *Trans. Am. Soc. Agric. Eng.* **1996**, *39*, 1475–1484. [CrossRef]
62. Hussain, P.R.; Suradkar, P.; Javaid, S.; Akram, H.; Parvez, S. Influence of postharvest gamma irradiation treatment on the content of bioactive compounds and antioxidant activity of fenugreek (*Trigonella foenum–graceum* L.) and spinach (*Spinacia oleracea* L.) leaves. *Innov. Food Sci. Emerg. Technol.* **2016**, *33*, 268–281. [CrossRef]
63. Singh, S.; Singh, H.; Bharat, N.K. Hot water seed treatment: A review. In *Capsicum*; IntechOpen: London, UK, 2020. [CrossRef]
64. Baker, K.F. Thermotherapy of planting material. *Phytopathology* **1962**, *52*, 1244–1255.
65. Muniz, M.F.B. Control of microorganisms associated with tomato seeds using thermotherapy. *Rev. Brassileira Sementes* **2001**, *23*, 176–280.
66. Aguilar, G.A.; Cruz, R.; Baez, R. Storage quality of bell peppers pretreated with hot water and polyethylene packaging. *J. Food Qual.* **1998**, *22*, 287–299. [CrossRef]
67. Musazura, W.; Bertling, I. Investigations into the effect of multiple hot water treatment of tomato (*Solanum lycopersicum*) and pepper (*Capsicum annuum*) seeds on seed viability and seed vigor. *Acta Hortic.* **2013**, *1007*, 795–799. [CrossRef]
68. Temple, T.N.; Toit, L.J.; Derie, M.L.; Johnson, K.B. Quantitative molecular detection of *Xanthomonas hortorum* pv. *carotae* in carrot seed before and after hot-water treatment. *Plant Dis.* **2013**, *97*, 1585–1592. [CrossRef] [PubMed]
69. Heddleson, R.A.; Doores, S. Factors affecting microwave heating of foods and microwave induced destruction of foodborne pathogens—A review. *J. Food Prot.* **1994**, *57*, 1025–1037. [CrossRef] [PubMed]
70. Shamis, Y.; Croft, R.; Taube, A.; Crawford, R.J.; Ivanova, E.P. Review of the specific effects of microwave radiation on bacterial cells. *Appl. Microbiol. Biotechnol.* **2012**, *96*, 319–325. [CrossRef] [PubMed]
71. Tkalec, V.; Mahnic, A.; Gselman, P.; Rupnik, M. Analysis of seed-associated bacteria and fungi on staple crops using the cultivation and metagenomic approaches. *Folia Microbiol.* **2022**, *67*, 351–361. [CrossRef] [PubMed]
72. Ding, H.; Fu, T.J.; Smith, M.A. Microbial contamination in sprouts: How effective is seed disinfection treatment? *J. Food Sci.* **2013**, *78*, 495–501. [CrossRef] [PubMed]
73. Danilchuk, T.N.; Alkhateeb, M.-K. Effects of microwave radiation on living microorganisms: Effects and mechanisms. *Health Food Biotechnol.* **2021**, *3*, 75–84. [CrossRef]
74. Thakran, S.; Siddiqui, M.R.; Baquer, N.Z. *Trigonella foenum graecum* seed powder protects against histopathological abnormalities in tissues of diabetic rats. *Mol. Cell Biochem.* **2004**, *266*, 151–159. [CrossRef] [PubMed]

75. Swieca, M. Hydrogen peroxide treatment and the phenylpropanoid pathway precursors feeding improve phenolics and antioxidant capacity of quinoa sprouts via an induction of L-Tyrosine and L-Phenylalanine Ammonia-Lyases activities. *J. Chem.* **2016**, *2016*, 1936516. [CrossRef]
76. Xue, W.L.; Li, X.S.; Zhang, J.; Liu, Y.H.; Wang, Z.L.; Zhang, R.J. Effect of *Trigonella foenum-graecum* (fenugreek) extract on blood glucose, blood lipid and hemorheological properties in streptozotocin-induced diabetic rats. *Asia. Pac. J. Clin. Nutr.* **2007**, *16*, 422–426.
77. Hannana, J.M.A.; Rokeya, B.; Faruque, O. Effect of soluble dietary fibre fraction of *Trigonella foenum graecum* on glycemic, insulinemic, lipidemic and platelet aggregation status of Type 2 diabetic model rats. *J. Ethnopharmacol.* **2003**, *88*, 73–77. [CrossRef] [PubMed]
78. Al-Habori, M.; Raman, A.; Lawrence, M.J.; Skett, P. In vitro effect of fenugreek extracts on intestinal sodium-dependent glucose uptake and hepatic glycogen phosphorylase A. *Int. J. Exp. Diabetes Res.* **2001**, *2*, 91–99. [CrossRef] [PubMed]
79. Chand, P.; Singh, K.; Singh, B.N.; Naidu, A. Contribution of fenugreek (*Trigonella foenum graecum* L.) seeds towards the nutritional characterization. *J. Med. Plant Res.* **2013**, *7*, 3052–3058.
80. Khoja, K.K.; Howes, M.J.R.; Hider, R.; Sharp, P.A.; Farrell, I.W.; Latunde-Dada, G.O. Cytotoxicity of fenugreek sprout and seed extracts and their bioactive constituents on MCF-7 breast cancer cells. *Nutrients* **2022**, *14*, 784. [CrossRef] [PubMed]
81. Ahmad, A.; Kumari, P.; Ahmad, M. Apigenin attenuates edifenphos-induced toxicity by modulating ROS-mediated oxidative stress, mitochondrial dysfunction and caspase signal pathway in rat liver and kidney. *Pestic. Biochem. Physiol.* **2019**, *159*, 163–172. [CrossRef] [PubMed]
82. Muhammed, M.T.; Jalil, A.T.; Taher, W.M.; Aminov, Z.; Alsaikhan, F.; Ramz-Coronel, A.A.; Ramaiah, P.; Farhood, B. The effects of apigenin in the treatment of diabetic nephropathy: A systematic review of non-clinical studies. *Mini. Rev. Med. Chem.* **2024**, *24*, 341–354. [CrossRef]
83. Zhang, K.; Song, W.; Li, D.; Jin, X. Apigenin in the regulation of cholesterol metabolism and protection of blood vessels. *Exp. Ther. Med.* **2017**, *13*, 1719–1724. [CrossRef] [PubMed]
84. Vasavada, N.; Agarwal, R. Role of oxidative stress in diabetic nephropathy. *Adv. Chronic Kidney Dis.* **2005**, *12*, 146–154. [CrossRef] [PubMed]
85. Najafi, M.; Mortezaee, K.; Rahimifard, M.; Farhood, B.; Haghi-Aminjan, H. The role of curcumin/curcuminoids during gastric cancer chemotherapy: A systematic review of non-clinical study. *Life Sci.* **2020**, *257*, 118051. [CrossRef]
86. Yan, L.J. Pathogenesis of chronic hyperglycemia: From reductive stress to oxidative stress. *J. Diabetes Res.* **2014**, *2014*, 137919. [CrossRef]
87. Forbes, J.M.; Coughlan, M.T.; Cooper, M.E. Oxidative stress as a major culprit in kidney disease in diabetes. *Diabetes* **2008**, *57*, 1446–1454. [CrossRef] [PubMed]
88. Malik, S.; Suchal, K.; Khan, S.I.; Bhatia, J.; Kishore, K.; Dinda, A.K.; Arya, D.S. Apigenin ameliorates streptozotocin-induced diabetic nephropathy in rats via MAPK-NF-κB-TNF-α and TGF-β1- MAPK-fibronectin pathways. *Am. J. Physiol. Renal Physiol.* **2017**, *313*, F414–F422. [CrossRef] [PubMed]
89. Lin, C.Y.; Yin, M.C. Renal protective effects of extracts from guava fruit (*Psidium guajava* L.) in diabetic mice. *Plant Foods Hum. Nutr.* **2012**, *67*, 303–308. [CrossRef] [PubMed]
90. Hou, Y.; Zhang, Y.; Lin, S.; Yu, Y.; Yang, L.; Li, L.; Wang, W. Protective mechanism of apigenin in diabetic nephropathy is related to its regulation of miR-423-5P-USF2 axis. *Am. J. Transl. Res.* **2021**, *13*, 2006–2020. [PubMed]

nutrients

Article

Patients with Inflammatory Bowel Disease Show Fewer Sex-Related Differences in Their Dietary Behavior Than the General Population: A Qualitative Analysis

Lea Pueschel [1], Fabian Kockelmann [2,3], Momme Kueck [4], Uwe Tegtbur [4], Masoumeh Attaran-Bandarabadi [5], Oliver Bachmann [6], Heiner Wedemeyer [1], Henrike Lenzen [1] and Miriam Wiestler [1,*]

1 Department of Gastroenterology, Hepatology, Infectious Diseases and Endocrinology, Hannover Medical School, 30625 Hannover, Germany; pueschel.lea@mh-hannover.de (L.P.); wedemeyer.heiner@mh-hannover.de (H.W.); lenzen.henrike@mh-hannover.de (H.L.)
2 Department of Surgery, Hospital Dortmund, University Hospital of the University Witten/Herdecke, 44137 Dortmund, Germany; fabian.kockelmann@uni-wh.de
3 Department for Human Medicine, Witten/Herdecke University, 58455 Witten, Germany
4 Clinic for Rehabilitation and Sports Medicine, Hannover Medical School, 30625 Hannover, Germany; kueck.momme@mh-hannover.de (M.K.); tegtbur.uwe@mh-hannover.de (U.T.)
5 Internal Medicine, Private Practice, 30952 Ronnenberg, Germany
6 Department of Internal Medicine I, Siloah St. Trudpert Hospital, 75179 Pforzheim, Germany; obachmann@icloud.com
* Correspondence: wiestler.miriam@mh-hannover.de

Abstract: Background: Since diet is generally recognized as an important factor directly modulating the gut microbiome, it is also considered a potential environmental triggering factor for the pathogenesis and onset of inflammatory bowel disease (IBD). While the habitual and sex-related dietary behavior of the general population has been the subject of extensive study and reporting, data on IBD patients' dietary behavior and especially its sex-related differences are underrepresented. However, as diet is an important factor in the course of IBD, we hypothesized that men and women with IBD have a different dietary profile than the general population. Methods: We performed a cohort analysis of a monocentric, cross-sectional study and compared the sex-related dietary behavior of 82 IBD patients (n = 40 women) to a sex- and age-matched cohort of the general German population [n = 328 (n = 160 women)]. Further on, disease-related quality of life and fecal calprotectin were correlated to the IBD patients' dietary behavior. Results: While sex-related dietary behavior was frequently of statistical difference in the general population within the IBD cohort, only minor numerical differences were observed between the sexes, which were rarely statistically significant. However, correlation analyses of disease-related quality of life (IBDQ) and diet revealed significant differences in male IBD patients but not in female IBD patients (p = 0.007; r = 0.409 for energy intake (kJ/d); p = 0.003, r = 0.449 for adherence to Mediterranean diet). Conclusion: The dietary behavior of IBD patients showed more similarity between the sexes than the general German population. Distinct sex-related trends and differences in correlation with disease parameters demonstrated a significant difference for an adaptive dietary behavior, especially in IBD men.

Keywords: dietary behavior; ultra-processed foods and drinks; sex-related differences; inflammatory bowel disease; diet quality

Citation: Pueschel, L.; Kockelmann, F.; Kueck, M.; Tegtbur, U.; Attaran-Bandarabadi, M.; Bachmann, O.; Wedemeyer, H.; Lenzen, H.; Wiestler, M. Patients with Inflammatory Bowel Disease Show Fewer Sex-Related Differences in Their Dietary Behavior Than the General Population: A Qualitative Analysis. *Nutrients* **2024**, *16*, 2954. https://doi.org/10.3390/nu16172954

Academic Editors: Domenico Tafuri and Francesca Latino

Received: 30 July 2024
Revised: 28 August 2024
Accepted: 29 August 2024
Published: 2 September 2024

1. Introduction

Inflammatory bowel diseases (IBDs), predominantly ulcerative colitis (UC) and Crohn's disease (CD), are now common in the entire developed world, and incidence rates are rising worldwide [1]. IBD and diet appear to be intrinsically governed by bidirectional interactions: while an active disease can affect one's nutrient intake, patients' dietary behavior appears to be a factor that can be either potentially protective or might be contributing to the onset of disease. Recent pilot studies suggest that a westernized diet (e.g., ultra-processed foods) is influencing the risk of developing IBD and exacerbating disease symptoms [2–5]. Indeed, recent data showed an adverse effect of ultra-processed food on health in general, ranging from obesity to cardiovascular diseases but also inflammatory diseases such as IBD [3,6,7]. IBD research has recently shifted to focusing on identifying dietary patterns and their specific influence on the development and disease course, with popular dietary strategies of IBD patients including avoidant (gluten free, lactose free) and restrictive diets (low FODMAP, CDED) as well as the more balanced "Mediterranean diet" [2,8]. In terms of individualized treatment strategies for IBD, one should emphasize diet as part of a holistic patient-centered approach. However, little attention has been paid to sex-related dietary behavior in the field of IBD, even though we know from the general population that there are significant differences in the dietary behavior of men and women, with different health consequences [9,10]. Therefore, investigating sex-related dietary profiles of IBD patients in direct comparison with a sex- and age-matched cohort of the general population as controls seemed advisable.

Objectives

The primary objective was to evaluate the sex-related dietary behavior of IBD patients and to assess the presence of distinct nutritional patterns. The secondary aim was the comparison of dietary behavior between IBD patients and the general German population.

2. Materials and Methods

We performed a cohort analysis of a monocentric, cross-sectional study with the aim to investigate sex-related dietary behavior of IBD patients compared to a sex- and age-matched cohort of the general German population. Physical activity analysis of the IBD cohort has been reported on previously [11].

2.1. Participants and Setting

A total of 94 patients, with a confirmed diagnosis of either ulcerative colitis (UC) or Crohn´s disease (CD), were screened in the IBD outpatient clinic of the Hannover Medical School in the period from January 2017 to October 2017. Of those patients, 91 were recruited into the study. Written informed consent was obtained from each patient prior to screening. Inclusion criteria were a disease duration of at least three months, a confirmed IBD diagnosis of either UC or CD, and disease activity of (a) moderate to severe disease activity or (b) remission. Ostomy, pregnancy, and limitations in everyday life due to cardiovascular or orthopedic diseases were exclusion criteria. The IBD cohort under consideration is ethnically diverse, with the majority of patients identifying as Caucasian.

Control Group

The cohort of the German Health Interview and Examination Survey (DEGS1) [12] was chosen as a control group for food consumption, macronutrient intake, and UPFD consumption. The DEGS1 cohort consists of 7987 subjects (n = 4198 women); data were collected from 2008 to 2011 and made available by the Robert Koch Institute (RKI) as part of a scientific use agreement [13]. This cohort represents a cross-section of the general German population and comprises ethnically diverse and nationally representative survey and measurement data on adults aged 18 to 79 years living in Germany, with the majority of individuals identifying as Caucasian.

2.2. Study Variables and Definitions

2.2.1. Dietary Behavior

In the scope of this study, we understand dietary behavior to mean the composition and quality of daily food intake as well as the determination of daily macronutrient intake. In this work, therefore, dietary behavior must be distinguished from eating habits.

2.2.2. Mediterranean Diet Score

Adherence to a Mediterranean diet was adapted from Trichopoulou et al. [14] based on the sex-specific mean for selected food groups. Food intake was calculated based on the FFQ answers. One point each was awarded whenever the consumption of positive associated foodstuff (vegetables, legumes, fruits and nuts, cereal, fish) was at or above the mean. For negative associated foodstuff (meat, poultry, dairy products), one point each was awarded if the consumption was below the mean. Fat intake was calculated based on the FFQ macronutrient analyses with the ratio of monounsaturated lipids (g) to saturated lipids (g). For ethanol intake, the sex-related values described by Trichopoulou et al. [14] were employed. Failing to reach these thresholds resulted in a score of 0. Overall, the MDS score ranges from 0 to 9, with 9 being maximal adherence to the Mediterranean diet.

2.2.3. Ultra-Processed Foods and Drinks

To calculate ultra-processed food and drink (UPFD) consumption, the NOVA food classification [15] was used to identify mainly Class 4 FFQ items. Tinned/preserved fruits, which are assigned Class 3, were also included due to the frequently added high level of industrial sugars or artificial sweeteners such as cyclamate, as well as additives such as calcium lactate or erythrosine. When calculating UPFD, a distinction was made between foods and drinks. In addition, we calculated the energy content (kJ), the percentage of total energy intake (EN%), and the total weight (g), as so-called "light" or "diet" ultra-processed drinks (UPD) are often consumed in copious quantities but are not high in calories.

2.2.4. Food Groups of the German Nutrition Society

Food groups were created from the FFQ corresponding to the German Nutrition Society (DGE) guideline of food groups [16].

2.2.5. Food Frequency Questionnaire Variables and Macronutrients

The food frequency questionnaire (FFQ) was analyzed by re-coding each question according to the amount and frequency of a single serving. Subsequently, a new variable showing the average daily amount was computed for every question by using the formula for the average daily amount: portion amount × frequency/28 [17,18]. For the subsequent evaluation of the nutrient intake, variables were then calculated using the reference data from the German food nutritional database (Bundeslebensmittelschlüssel [BLS]) [19]. Of the 53 FFQ questions on the frequency of consumption, 52 were converted into variables with the corresponding values of the following nutrients: energy (kJ), protein (g), fat (g), carbohydrates (g), fiber (g), sugar (g), cellulose (mg), lignin (mg), soluble fats (mg), insoluble fats (mg), tyrosine (mg), tryptophan (mg), saturated fats (mg), short-chain fatty acids (mg), medium-chain fatty acids (mg), long-chain fatty acids (mg), omega-3 fatty acids (mg), and omega-6 fatty acids (mg). Energy percentages of the following nutrients were converted using the quantities given here: carbohydrates (17 kJ per gram), proteins (17 kJ per gram), fats (37 kJ per gram), and fiber (8 kJ per gram). To simplify reporting, items were clustered; however, mean comparisons were performed for each item individually.

2.3. Data Sources/Measurement

2.3.1. Questionnaires

Patients were asked to complete various questionnaires, with a demographic survey including data on sex, age, body weight, and height. In addition, and depending on disease entity, disease activity was determined via a physician-guided interview including the Harvey Bradshaw Index (HBI) [20] and the Simple Clinical Colitis Activity Index (SCCAI) [21]. To further measure IBD-related quality of life, the German version of the 36-item IBDQ was employed [22,23]. The Food Frequency Questionnaire (FFQ), developed as part of the "German Health Interview and Examination Survey for Adults" (DEGS11) by the RKI [12], was used as a validated collection tool for nutritional data. The questionnaire retrospectively records the frequency and amounts of food consumption in a total of

53 categories for the last four weeks. In addition, questions are asked about eating and cooking habits in relation to the consumption of animal products, fats, and freshly prepared foods. Subsequent nutrient conversion was based on a recently developed tool [19].

2.3.2. Laboratory Values

Biomaterials (blood and fecal samples) were collected as part of the screening visit for biochemical analysis. Parameters included hemoglobin (g/dL), leukocyte counts (1.000/mL), C-reactive protein (CRP) (mg/L), vitamin D3 (ng/mL), and calprotectin (mg/kg).

2.4. *Study Size*

A priori sample size estimation for the original study specified the inclusion of $\geq$50 patients each in "active disease" and "remission" groups. This group allocation was not used for this cohort analysis; instead, patients were divided according to their biological sex. Due to missing FFQ data, $n = 9$ patients were excluded from the cohort analysis.

2.5. *Statistical Analysis*

Statistical analysis was performed with SPSS Statistics software version 28.0.1.0 (SPSS, IBM, Armonk, NY, USA) and GraphPad PRISM version 10.2.3 (GraphPad Software, Boston, MA, USA). Normal distribution was assessed by the Shapiro–Wilk test. Non-normally distributed data are reported as the median (IQR), except for food data, since reporting the median would result in zero values for low-consumption items. Student's *t*-test was used for comparisons of sex-related within-group and between-group comparisons of food intake, macronutrient intake, UPFD consumption, and adherence to DGE recommendation for the IBD cohort and the DEGS1 cohort. For categorical baseline characteristic variables, the Bonferroni correction was applied to Fisher's exact test. For metric variables, mean difference was assessed using Student's *t*-test. For sex-related distribution of diet quality, the Bonferroni correction was applied to Fisher's exact test. To evaluate sex-related correlations between food data and IBD-related quality of life, we performed precise correlation analyses using Spearman correlation analysis. Food intake, macronutrient intake, UPFD consumption, and mean difference to the DGE reference value are reported as the arithmetic mean, the 95% confidence interval (CI), and the level of significance (p). If not stated otherwise, significance levels are two-sided. Clinical relevance is reported as effect size estimate (d).

2.5.1. Missing Data

Participants who did not complete the FFQ were excluded from the analysis. In the case of individual missing data, it could be assumed that these were either missing completely at random (MCAR) or missing at random (MAR). These cases were omitted analysis by analysis.

2.5.2. Sampling Strategy

To avoid distortion when comparing the IBD cohort with the general population of the DEGS1 cohort, matching was carried out. DEGS1 study participants with IBD were identified and filtered ($n = 102$); additionally, all cases were removed in which daily energy intake (EI) could not be calculated due to missing information in the FFQ. A random sample was then drawn which corresponded to the IBD cohort in a ratio of 4:1, and data were sex- and age-matched.

2.5.3. Bias

Misreporting of energy intake in patient-reported outcomes is a well-documented concern [24]. Therefore, misreporting of energy intake was calculated according to Black's adjustment [25] of Goldberg et al. [26], with the ratio of the reported energy intake (EI_{rep}) to the resting energy expenditure (REE) < individual cut-off for underreporting and

EI_{rep}:REE > 2.4 for overreporting. EI_{rep} was calculated by computing FFQ answers with the reference data from the BLS [19]. While energy expenditure (EE) and total daily energy expenditure (TDEE) were calculated over seven days via the biaxial accelerometer (BodyMedia, Pittsburgh, PA, USA), REE was calculated with the sex-specific Mifflin–St Jeor equations [27], revised from the Harris–Benedict equations [28]. To account for the small study population, the estimated effect size (d) is reported in addition to the statistical significance (*p*), since (d) does not depend on sample size.

3. Results

3.1. Descriptive Data

Of 94 patients screened, 91 were enrolled and reported on in the original monocentric, cross-sectional study. Due to incomplete FFQ data, $n = 9$ had to be excluded. Consequently, the data from 82 patients were stratified by sex ($n = 40$ women) (Figure 1).

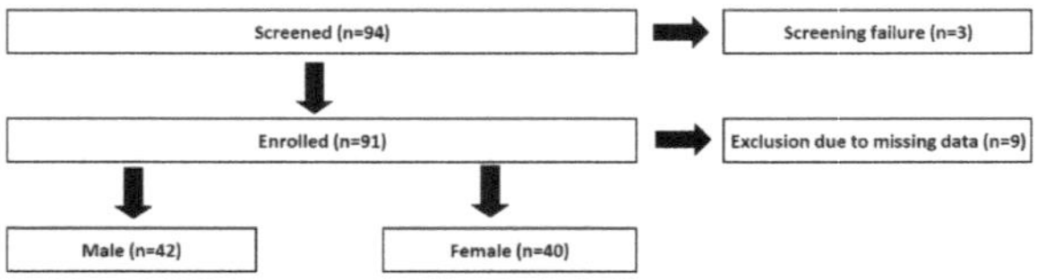

Figure 1. Flowchart of patient enrollment.

Baseline characteristics were well balanced between the sexes, with no significant differences between demographics, laboratory, and IBD-related variables. Median age for men was 38.8 and for women 36.7. Median BMI for men included in this analysis was 26.1 and thus slightly above the national average of 25.5, while median BMI for women was 24.2 and thus slightly below the national average of 25.2 [29]. Median physical activity level (PAL) for men was 1.7, and median PAL for women was 1.6. (Table 1: Baseline characteristics). We saw a discrepancy between TDEE and EI_{rep} in the IBD cohort. The actual PAL-based energy requirement showed differences to EI_{rep} for both sexes of the IBD cohort. Men in particular showed a higher actual need than could be covered by EI_{rep}, with an observed discrepancy of up to 1838 kJ.

Table 1. Baseline characteristics.

		Men (n = 42)	Women (n = 40)	p_{sex}
Entity [n (%)]	CD	19 (45.2%)	25 (62.5%)	0.234
	UC	23 (54.8%)	15 (37.5%)	0.233
Disease duration [median (IQR)] (years)		9 [5, 17]	8 [4, 16]	0.781
HBI [mean (SD)]		6.11 [5.92]	7.84 [5.93]	0.393
SCCAI [mean (SD)]		5.7 [5.05]	3.73 [4.68]	0.404
Location of Crohn's	L1: Ileal	0 (0%)	2 (8%)	0.828
	L2: Colonic	2 (10.5%)	3 (12%)	0.999
	L3: Ileocolonic	12 (63.2%)	12 (48%)	0.999
	L4: Upper GI	5 (26.3%)	8 (32%)	0.999
Crohn's behavior	B1: Inflammatory	5 (26.3%)	11 (44%)	0.681
	B2: Stricturing	10 (52.6%)	10 (40%)	0.999
	B3: Penetrating	4 (21.1%)	4 (16%)	0.999
UC Montreal classification	E1: Proctitis	2 (8.7%)	1 (6.7%)	0.999
	E2: Left-sided colitis	12 (52.2%)	4 (26.7%)	0.358
	E3: Pancolitis	9 (39.1%)	10 (66.7%)	0.999
BMI [median (IQR)] (kg/m^2)		25.70 [23.08, 29.08]	23.15 [20.90, 25.60]	0.066
Age [median (IQR)] (years)		36.50 [27.75, 49.25]	33 [24.50, 47.25]	0.492
Smoker [n (%)]:	yes	38 (90.5%)	5 (12.5%)	0.999
	no	4 (9.5%)	35 (87.5%)	0.999
Calprotectin [median (IQR)] (mg/L)		1081 [233, 1880]	1020 [269, 1556]	0.588
C-reactive protein [median (IQR)] (mg/L)		2.95 [1.08, 11.03]	2.80 [0.90, 10.48]	0.300
Hemoglobin [median (IQR)] (g/dL)		14.10 [11.90, 14.95]	12.65 [12.10, 13.60]	0.164
Leucocytes [median (IQR)] (Tsd/µL)		8.10 [6.15, 11.20]	8.10 [6.50, 10.93]	0.741
Vitamin D3 25-OH [median (IQR)] (µg/L)		21.83 [13.80, 27.80]	22.22 [13.50, 30.95]	0.870

Values as n (%) or median (IQR).

3.2. Main Results

3.2.1. IBD Cohort Versus DEGS1 Cohort

Macronutrient Intake—Mean Daily Amount

The mean daily intake of macronutrients was calculated from food frequency questionnaires (as described above in Section 2.2.4) and showed no significant differences between the sexes in the IBD patient cohort. However, within the general German population, distinguished sex-related differences have been found in the mean daily intake of proteins (g) ($p < 0.001$, d = 0.7), fat (g) ($p < 0.001$, d = 0.7), and sugar (g) ($p = 0.006$, d = 0.3) and in the energy percentage of carbohydrates (%EN) ($p = 0.040$, d = −0.2) and fiber (%EN) ($p < 0.001$, d = −0.5). Further significant differences were observed for the mean daily macronutrient intake of soluble fats (mg) ($p = 0.014$, d = 0.3), tyrosine (mg) ($p < 0.001$, d = 0.6), tryptophane (mg) ($p < 0.001$, d = 0.7), saturated fats (mg) ($p < 0.001$, d = 0.6), short-chain fatty acids (mg) ($p = 0.002$, d = 0.3), medium-chain fatty acids (mg) ($p < 0.001$, d = 0.5), long-chain fatty acids (mg) ($p < 0.001$, d = 0.7), omega-3 fatty acids (mg) ($p < 0.001$, d = 0.5), and omega-6 fatty acids (mg) ($p < 0.001$, d = 0.8). While no significant differences between IBD men and DEGS1 men were detected, compared to DEGS1 women, IBD women had a higher intake of protein (g) with a mean difference of $p = 0.04$, a higher intake of tryptophan (mg) with a mean difference of $p = 0.02$, and a higher intake of omega-6 fatty acids (mg) with a mean difference of $p = 0.04$. Men and women of both cohorts consumed on average more protein (g) daily (IBD men: 93 g; IBD women: 78 g; DEGS1 men: 95 g; DEGS1 women: 68 g) than the recommended sex-related value of the DGE (men: 57–67 g; women: 48–57 g). The recommended maximum daily fat intake of 30%EN was met by both cohorts (IBD men: 30%; IBD women: 29%; DEGS1 men: 29%; DEGS1 women: 28%), while the recommended fiber intake of ≥30 g daily was not reached by both sexes in the IBD (men: 21 g; women: 18 g) and DEGS1 (men: 21 g; women: 19 g) cohorts (Table 2).

Table 2. Macronutrient intake—mean daily amount.

		IBD Cohort					DEGS1-Cohort					IBD vs DEGS1	
Macronutrients—Mean Daily Amount	**Sex**	***n***	**Mean**	**95% CI**	$p_{t\text{-Test(twosided)}}$	***d***	***n***	**Mean**	**95% CI**	$p_{t\text{-Test(twosided)}}$	***d***	$p_{t\text{-Test(twosided)}}$	***d***
Energy (kJ)	Men	42	9358	7658, 11,058	0.205	0.3	168	10,506	9763, 11,249	**<0.001**	0.7	0.184	−0.2
	Women	40	8102	7082, 9122			160	7572	7125, 8019			0.307	0.2
Protein (g)	Men	42	93	72.90, 113.0	0.181	0.3	168	95	87.83, 101.7	**<0.001**	0.7	0.864	0.0
	Women	40	78	67.89, 87.95			160	68	64.31, 72.31			**0.044**	0.4
Protein (%EN)	Men	42	16	15.43, 17.48	0.689	−0.1	168	16	15.15, 16.14	0.731	0.0	0.154	0.2
	Women	40	17	15.46, 18.11			160	16	15.22, 16.33			0.121	0.3
Fat (g)	Men	42	78	59.95, 95.32	0.169	0.3	168	81	74.98, 87.19	**<0.001**	0.7	0.649	−0.1
	Women	40	64	54.30, 73.32			160	57	52.66, 60.77			0.135	0.3
Fat (%EN)	Men	42	30	27.65, 31.72	0.764	0.1	168	29	27.71, 29.74	0.214	0.1	0.400	0.1
	Women	40	29	27.07, 31.42			160	28	26.72, 28.87			0.234	0.2
Carbohydrates (g)	Men	42	272	224.8, 318.4	0.466	0.2	168	323	296.1, 349.9	<0.001	0.5	0.084	−0.3
	Women	40	250	212.8, 286.8			160	242	224.2, 259.1			0.682	0.1
Carbohydrates (%EN)	Men	42	50	47.73, 53.25	0.497	−0.2	168	52	50.17, 52.96	**0.040**	−0.2	0.494	−0.1
	Women	40	52	48.80, 54.97			160	54	52.23, 55.04			0.281	−0.2
Fiber (g)	Men	42	21	15.70, 26.04	0.361	0.2	168	21	19.51, 23.08	0.093	0.2	0.847	0.0
	Women	40	18	15.14, 21.15			160	19	17.61, 20.85			0.546	−0.1
Fiber (%EN)	Men	42	2	1.551, 2.098	0.708	−0.1	168	2	1.590, 1.817	**<0.001**	−0.5	0.367	0.2
	Women	40	2	1.635, 2.155			160	2	1.943, 2.190			0.224	−0.2

Table 2. *Cont.*

			IBD Cohort						DEGS1-Cohort				IBD vs DEGS1	
Macronutrients—Mean Daily Amount	**Sex**	n	**Mean**	**95% CI**	$p_{t\text{-Test(twosided)}}$	d	n	**Mean**	**95% CI**	$p_{t\text{-Test(twosided)}}$	d	$p_{t\text{-Test(twosided)}}$	d
Sugar (g)	Men	42	127	98.83, 155.6	0.883	0.0	168	159	140.1, 178.5	**0.006**	0.3	0.123	−0.3
	Women	40	130	97.47, 163.3			160	126	112.8, 139.6			0.790	0.0
Cellulose (mg)	Men	42	3631	2671, 4591	0.609	0.1	168	3056	2802, 3310	0.124	−0.2	0.249	0.3
	Women	40	3322	2587, 4057			160	3367	3057, 3676			0.901	0.0
Lignin (mg)	Men	42	875	604.4, 1145	0.354	0.2	168	724	647.6, 801.0	0.268	−0.1	0.286	0.3
	Women	40	733	589.1, 877.4			160	794	696.2, 891.2			0.566	−0.1
Soluble fats (mg)	Men	42	6401	4799, 8003	0.308	0.2	168	6803	6247, 7359	**0.014**	0.3	0.559	−0.1
	Women	40	5479	4632, 6326			160	5868	5372, 6364			0.476	−0.1
Insoluble fats (mg)	Men	42	14,189	10,654, 17,724	0.386	0.2	168	14,426	13,225, 15,627	0.138	0.2	0.899	0.0
	Women	40	12,390	10,188, 14,592			160	13,185	12,064, 14,306			0.529	−0.1
Tyrosine (mg)	Men	42	3408	2665, 4152	0.206	0.3	168	3482	3208, 3757	**<0.001**	0.6	0.825	0.0
	Women	40	2871	2458, 3283			160	2535	2370, 2699			0.087	0.3
Tryptophan (mg)	Men	42	1098	861.5, 1335	0.193	0.3	168	1115	1032, 1198	**<0.001**	0.7	0.894	0.0
	Women	40	924	800.7, 1048			160	795	748.5, 842.5			**0.024**	0.4
Saturated fats (mg)	Men	42	35,011	27,309, 42,713	0.231	0.3	168	37,108	34,092, 40,123	**<0.001**	0.6	0.561	−0.1
	Women	40	29,386	24,128, 34,643			160	26,320	24,183, 28,457			0.226	0.2
Short-chain fatty acids (mg)	Men	42	1586	1248, 1924	0.356	0.2	168	1607	1428, 1786	**0.002**	0.3	0.915	0.0
	Women	40	1373	1060, 1687			160	1244	1105, 1383			0.422	0.1
Medium-chain fatty acids (mg)	Men	42	1488	1170, 1805	0.419	0.2	168	1544	1378, 1709	**<0.001**	0.5	0.762	−0.1
	Women	40	1307	989.7, 1624			160	1101	999.6, 1203			0.219	0.3
Long-chain fatty acids (mg)	Men	42	67,875	52,087, 83,664	0.148	0.3	168	71,743	66,309, 77,177	**<0.001**	0.7	0.567	−0.1
	Women	40	54,917	46,519, 63,316			160	49,630	46,035, 53,224			0.208	0.2
Omega-3 fatty acids (mg)	Men	42	1955	1287, 2622	0.513	0.1	168	1792	1658, 1927	**<0.001**	0.5	0.633	0.1
	Women	40	1701	1320, 2082			160	1380	1236, 1524			0.065	0.3
Omega-6 fatty acids (mg)	Men	42	10,755	8355, 13,155	0.134	0.3	168	11,047	10,265, 11,829	**<0.001**	0.8	0.816	−0.1
	Women	40	8742	7551, 9933			160	7550	7043, 8056			**0.045**	0.4

Units of daily intake of macronutrients are reported as kilojoules (kJ), grams (g), or milligrams (mg). Results of Student's t-test between the sexes and the IBD and DEGS1 cohorts are given as the arithmetic mean, the 95% confidence interval (CI), the level of significance (p), and the estimated effect size (d). $P_{t\text{-test}}$ is printed bold when significant. IBD—inflammatory bowel disease; DEGS1—cohort of the German Health Interview and Examination Survey.

Ultra-Processed Foods and Drinks—Mean Daily Amount

Ultra-processed food and drink consumption was calculated as described above (see Section 2.2.3). Minor differences in ultra-processed food and drink consumption between men and women in the IBD cohort were observed. However, those differences were not statistically significant (UPFD p = 0.397, d = 0.2; UPD p = 0.483, d = 0.2; UPF p = 0.262, d = 0.2). In the DEGS1 cohort, medium differences were observed, all of which were statistically significant (UPFD $p \leq 0.001$, d = 0.5; UPD $p \leq 0.001$, d = 0.5; UPF $p \leq 0.001$, d = 0.6). Women of both cohorts on average consumed less ultra-processed foodstuff than their male counterparts, with the highest intake of ultra-processed food (838 g) and ultra-processed drinks (207 g) being reported for healthy men. The lowest intake of UPD was reported for healthy women (366 g), and the lowest intake of UPF was reported for IBD women (146 g) (Table 3).

Table 3. Ultra-processed foods and drinks—mean daily amount.

Ultra-Processed Foods and Drinks—Mean Daily Amount	Sex	IBD Cohort					DEGS1-Cohort					IBD vs DEGS1	
		n	Mean	95% CI	$p_{t\text{-Test(twosided)}}$	*d*	*n*	Mean	95% CI	$p_{t\text{-Test(twosided)}}$	*d*	$p_{t\text{-Test(twosided)}}$	*d*
UPFD (g)	Men	41	871	510.7, 1232	0.397	0.2	150	968	792.7, 1143	**<0.001**	0.5	0.619	−0.1
	Women	40	664	328.9, 999.0			143	522	389.9, 654.0			0.356	0.2
UPFD (kJ)	Men	41	3179	2421, 3938	0.170	0.3	148	3884	3452, 4317	**<0.001**	0.6	0.127	−0.3
	Women	40	2514	1914, 3115			143	2447	2166, 2728			0.830	0.0
UPD (g)	Men	42	685	338.1, 1032	0.483	0.2	156	838	654.7, 1020	**<0.001**	0.5	0.445	−0.1
	Women	40	518	191.4, 845.4			152	366	244.2, 487.3			0.381	0.2
UPD (kJ)	Men	42	987	535.0, 1440	0.233	0.3	156	1266	1008, 1523	**<0.001**	0.6	0.316	−0.3
	Women	40	647	301.5, 991.7			152	504	377.8, 629.4			0.436	0.2
UPF (g)	Men	41	172	131.8, 212.9	0.262	0.3	158	207	187.9, 226.7	**<0.001**	0.6	0.111	−0.3
	Women	40	146	120.3, 170.8			150	148	134.1, 162.0			0.867	0.0
UPF (kJ)	Men	41	2172	1614, 2729	0.37	0.2	156	2683	2416, 2950	**<0.001**	0.5	0.089	−0.3
	Women	40	1867	1479, 2256			150	1924	1709, 2139			0.404	0.0

Units of daily intake of ultra-processed foods and drinks (UPFDs), ultra-processed foods (UPFs), and ultra-processed drinks (UPDs) are reported as kilojoules (kJ) and grams (g). Results of Student's *t*-test between the sexes and the IBD and DEGS1 cohorts are given as the arithmetic mean, the 95% confidence interval (CI), the level of significance (*p*), and the estimated effect size (d). $P_{t\text{-test}}$ is printed bold when significant.

3.2.2. DGE Food Groups—Mean Daily Consumption

Food groups were created from the FFQ corresponding to the German Nutrition Society (DGE) guideline of food groups [16] (see Section 2.2.4).

Fruits and Vegetables

Men in the IBD cohort ate more raw and processed vegetables (65 g) than women within the same cohort (41 g). However, women in the DEGS1 cohort reported a higher intake of raw (69 g) and processed vegetables (43 g) than their male counterparts (56 g and 41 g, respectively). The sexes of both cohorts fell short of the DGE recommendation for fruit and vegetable intake of 550 g daily. The highest mean daily intake of fresh fruits was reported by women in the DEGS1 cohort (238 g), followed by IBD women (193 g). Healthy men reported the lowest intake of fresh fruits (150 g), which showed a significant sex-related difference in comparison ($p \leq 0.001$; d = −0.4), but they had a higher intake of processed fruits (6.4 g) than their female counterparts (5.4 g). The highest intake of processed and canned fruits was reported by IBD women (9.2 g) and the lowest by IBD men (5.4 g) (Table 4).

Table 4. DGE food groups—mean daily consumption.

DGE Food Groups (g/d)		Sex	IBD Cohort					DEGS1-Cohort					IBD vs DEGS1	
			n	Mean	95% CI	$p_{t\text{-Test(twosided)}}$	*d*	*n*	Mean	95% CI	$p_{t\text{-Test(twosided)}}$	*d*	$p_{t\text{-Test(twosided)}}$	*d*
Fruits and vegetables	Fresh Vegetables	Men	42	65.29	27.41, 103.2	0.224	0.3	168	56.47	44.85, 68.09	**0.004**	−0.3	0.656	0.1
		Women	40	40.92	28.01, 53.82			159	81.95	68.90, 95.01			**<0.001**	−0.5
	Processed Vegetables	Men	42	68.26	43.70, 92.81	0.91	0.0	167	40.58	32.93, 48.23	0.053	−0.2	**0.035**	0.5
		Women	40	66.45	45.64, 87.25			159	53.15	42.81, 63.48			0.127	0.2
	Fresh Fruits	Men	42	168.08	104.2, 231.9	0.683	−0.1	167	150.25	121.9, 178.6	**<0.001**	−0.4	0.586	0.1
		Women	40	193.09	86.13, 300.0			160	237.61	197.0, 278.2			0.363	−0.2
	Processed Fruits	Men	42	5.36	1.296, 9.418	0.278	−0.2	166	6.39	3.757, 9.022	0.536	0.1	0.716	−0.1
		Women	40	9.16	3.346, 14.97			160	5.43	3.921, 6.935			0.216	0.3

Table 4. *Cont.*

DGE Food Groups (g/d)			IBD Cohort					DEGS1-Cohort					IBD vs DEGS1	
		Sex	n	Mean	95% CI	$P_{t\text{-Test(twosided)}}$	d	n	Mean	95% CI	$P_{t\text{-Test(twosided)}}$	d	$P_{t\text{-Test(twosided)}}$	d
Juices	Fruit Juice	Men	42	120.96	69.07, 172.9	0.234	−0.3	167	208.25	150.8, 265.7	0.476	−0.1	0.144	−0.3
		Women	40	273.53	23.54, 523.5			159	245.74	158.2, 333.3			0.797	0.0
	Vegetable Juice	Men	42	9.86	−0.4242, 20.15	0.633	0.1	167	2.22	1.000, 3.448	0.145	−0.2	0.144	0.5
		Women	39	6.68	−1.748, 15.12			159	4.2	1.832, 6.569			0.435	0.1
Legumes and pulses		Men	42	19.45	6.867, 32.04	0.272	0.2	168	15.99	11.59, 20.39	**0.024**	0.2	0.525	0.1
		Women	39	11.62	5.613, 17.64			159	10.28	7.993, 12.58			0.314	0.1
Nuts and seeds		Men	42	2.9	1.473, 4.336	0.254	0.3	167	1.99	1.206, 2.767	0.842	0.0	0.290	0.2
		Women	40	1.79	0.4449, 3.126			159	2.09	1.377, 2.811			0.699	−0.1
Potatoes	Boiled Potatoes	Men	42	62.31	39.31, 85.32	0.596	−0.1	167	68.63	57.33, 79.93	0.167	0.2	0.621	−0.1
		Women	40	70.74	48.51, 92.98			160	58.86	50.69, 67.03			0.316	0.2
	Fried Potatoes	Men	42	7.35	4.979, 9.721	0.475	0.2	168	10.08	8.114, 12.06	**0.013**	0.3	0.079	−0.2
		Women	40	6.08	3.372, 8.782			159	6.86	5.267, 8.463			0.653	−0.1
	French fries	Men	42	13.27	5.886, 20.64	0.292	0.2	166	13.44	11.19, 15.69	**<0.001**	0.4	0.953	0.0
		Women	40	8.81	4.852, 12.76			160	7.79	6.192, 9.385			0.591	0.1
Butter and margarine		Men	42	10.05	6.19, 13.91	0.492	0.2	168	9.36	7.77, 10.96	**0.003**	0.3	0.715	0.1
		Women	40	8.28	4.84, 11.72			160	6.36	5.18, 7.53			0.190	0.2
Dairy products	Milk	Men	41	247.17	157.5, 336.8	0.723	0.1	165	311.1	243.7, 378.5	**0.015**	0.3	0.376	−0.2
		Women	40	225.18	138.3, 312.0			158	213.33	172.7, 253.9			0.798	0.0
	Cream Cheese	Men	42	7.23	3.345, 11.12	0.950	0.0	167	5.61	4.202, 7.017	0.635	−0.1	0.433	0.2
		Women	40	7.07	3.485, 10.65			160	6.15	4.377, 7.928			0.648	0.1
	Cheese	Men	42	40.65	24.16, 57.14	0.311	0.2	165	33.67	27.53, 39.80	0.882	0.0	0.346	0.2
		Women	40	30.5	19.30, 41.70			160	32.92	24.97, 40.86			0.777	−0.1
	Yoghurt	Men	42	78.53	39.63, 117.4	0.702	−0.1	164	91.52	68.47, 114.6	0.230	0.1	0.605	−0.1
		Women	40	90.49	40.40, 140.6			159	75.57	63.16, 87.99			0.562	0.1
Fish	Cold fish	Men	42	30.56	15.96, 45.16	0.692	0.1	168	18.51	15.32, 21.70	0.171	0.2	0.111	0.4
		Women	40	26.63	12.96, 40.29			160	15.48	12.51, 18.44			0.114	0.4
	Warm fish	Men	42	12.2	6.782, 17.52	0.675	0.1	167	9.4	7.605, 11.15	0.32	0.1	0.327	0.2
		Women	40	10.7	6.594, 14.86			160	8.8	7.316, 10.27			0.287	0.2
Meat and poultry	Poultry	Men	42	45.5	27.14, 63.87	0.996	0.0	168	32.6	26.02, 39.11	**0.048**	0.2	0.187	0.3
		Women	40	45.6	27.20, 63.94			160	24.5	19.80, 29.16			**0.030**	0.6
	Meat	Men	42	51.8	31.16, 72.51	0.104	0.4	167	53.8	43.86, 63.72	**<0.001**	0.6	0.863	0.0
		Women	40	32.2	20.35, 44.04			160	26.5	22.83, 30.26			0.363	0.2
Cold cuts		Men	42	19.3	10.91, 27.73	0.335	0.2	168	34.0	28.50, 39.54	**<0.001**	0.7	**0.015**	−0.4
		Women	40	14.5	8.994, 19.93			160	14.4	11.55, 17.22			0.982	0.0
Eggs		Men	42	30.56	15.96, 45.16	0.692	0.1	168	18.51	15.32, 21.70	0.171	0.2	0.111	0.4
		Women	40	26.63	12.96, 40.29			160	15.48	12.51, 18.44			0.114	0.4
Cereal products	Breakfast cereals	Men	42	3.3	0.7905, 5.766	0.959	0.0	168	2.9	1.994, 3.881	0.063	0.2	0.765	0.1
		Women	40	3.4	−0.8803, 7.684			159	1.8	0.9100, 2.585			0.448	0.2
	Muesli	Men	42	13.0	−1.690, 27.58	0.372	0.2	166	5.7	3.731, 7.626	0.248	0.1	0.326	0.3
		Women	40	6.1	2.605, 9.605			158	4.3	2.798, 5.693			0.274	0.2
	Wholegrain bread and rolls	Men	42	45.1	12.03, 78.09	0.357	0.2	157	55.7	43.15, 68.23	0.611	−0.1	0.481	−0.1
		Women	40	28.4	15.03, 41.80			160	60.5	46.90, 74.00			**0.001**	−0.4
	Mixed bread	Men	42	44.2	25.83, 62.52	0.897	0.0	166	78.2	60.72, 95.66	**<0.001**	0.5	**0.008**	−0.3
		Women	40	42.6	25.57, 59.56			159	37.0	29.30, 44.65			0.526	0.1
	White bread and rolls	Men	42	48.2	32.88, 63.55	0.955	0.0	166	45.6	37.52, 55.77	**0.006**	0.3	0.875	0.0
		Women	39	49.1	20.11, 78.10			160	30.9	24.19, 37.57			0.222	0.3
	Pasta	Men	42	38.7	20.70, 56.74	0.589	0.1	167	43.6	36.26, 50.84	**0.013**	0.3	0.575	−0.1
		Women	40	33.2	23.79, 42.56			160	32.4	27.48, 37.32			0.888	0.0
	Rice	Men	42	31.6	14.16, 49.04	0.357	0.2	168	18.9	14.77, 22.94	0.759	0.0	0.158	0.4
		Women	40	22.5	13.42, 31.51			160	19.7	16.54, 22.78			0.468	0.1

Food frequency questionnaire (FFQ) variables are sorted by the German Nutrition Society (DGE) guideline of food groups. All units are given as grams per day (g/d). Results of Student's t-test between the sexes and the IBD and DEGS1 cohorts are given as the arithmetic mean, the 95% confidence interval (CI), the level of significance (p), and the estimated effect size (d). $P_{t\text{-test}}$ is printed bold when significant. IBD—inflammatory bowel disease; DEGS1—cohort of the German Health Interview and Examination Survey.

Juices

The mean daily intake of beverages showed numerical differences between the sexes in the IBD cohort, none of which were significant. IBD men reported a higher intake of vegetable juices (9.9 g; women = 6.7 g; $p = 0.633$, d = 0.1), whereas IBD women had a higher intake of fruit juices (273 g; men = 120 g; $p = 0.234$, d = −0.2). DEGS1 women reported a higher intake of vegetable juices (4.2 g; men = 2.2 g; $p = 0.145$, d = −0.2) and fruit juices (245 g; men = 208 g; $p = 0.476$, d = −0.1). The DGE recommendation of 57.14 g per day was exceeded by both sexes in both cohorts (Table 4).

Legumes and Pulses

Men in the IBD cohort ate more raw and processed vegetables (65 g) as well as legumes and pulses (19 g) than women of the same cohort (41 g and 12 g). Women in the DEGS1 cohort reported a mean daily intake of 10 g and men an intake of 12 g. The DGE recommendation for legumes and pulses (a weekly portion size of 125 g computed to 17.85 g daily) was only met by men in the IBD cohort (Table 4).

Nuts and Seeds

However, IBD men had the highest daily intake of nuts and seeds (3 g), followed by DEGS1 women (2.1 g); IBD women reported an intake of 1.8 g and DEGS1 men an intake of 2 g. Sexes of both cohorts, however, did not reach the daily amount for nut and seed intake recommended by the DGE of 25 g daily (Table 4).

Potato Products

On average, men in the DEGS1 cohort consumed the most fried potatoes (10.1 g) and French fries (13.4 g) daily, followed by men with IBD (7.4 g; 13.3 g). Significant sex-related differences were observed in the DEGS1 cohort for fried potatoes ($p = 0.013$, d = 0.3) and French fries ($p \leq 0.001$, d = 0.4). Women with IBD reported the highest daily intake of boiled potatoes (70.7 g) and the lowest intake of fried potatoes (6.1 g), with an additional daily intake of 8.8 g of French fries. Women in the DEGS1 cohort had the lowest daily intake of boiled potatoes (58.9 g) and French fries (7.8 g), with an additional mean daily intake of 6.9 g of fried potatoes. Men with IBD consumed on average 62.3 g of boiled potatoes daily, slightly below the intake of healthy men (68.6 g). The DGE recommendation of 250 g weekly (or 35.7 g daily) for potato intake was therefore exceeded by both sexes of both cohorts (Table 4).

Butter and Margarine

Mean daily intake of spreadable fats (butter and margarine) showed no significant sex-related differences between IBD men and women as well as no significant differences between the cohorts. IBD men and women reported a higher daily intake of spreadable fats (men = 10.05 g; women = 8.28 g) than their same-sex counterparts in the DEGS1 cohort (men = 9.36 g; women = 6.36 g). The differences between the sexes in the DEGS1 cohort were significant ($p = 0.003$; d = 0.3) (Table 4).

Dairy Products

Healthy men reported the highest mean intake of milk (311 g); in comparison with the milk intake of healthy women (213 g), this showed a significant sex-related difference ($p = 0.0015$, d = 0.3). Healthy men further reported the highest intake of yoghurt products (91.5 g), while men with IBD had the highest mean intake of cheese (40.6 g), cream cheese (7.2 g), and eggs (30.6 g). The DGE recommendation for dairy products of 500 g daily was not met by men and women of both cohorts (Table 4).

Fish

The DGE recommends a weekly intake of fish (240 g; 34.3 g daily). While IBD men and women had a higher intake of cold (men = 11.9 g; women = 12.4 g) and warm fish (men = 12.2 g; women = 10.7 g) than healthy men and women (cold fish: men = 5.5 g; women = 7.8 g; warm fish: men = 9.4 g; women = 8.8 g), the recommendation was not met by either sex or cohort. However, the DGE guideline points out that the recommendation for weekly portion sizes for fish and meat also includes the other group (Table 4).

Meat and Poultry

Men and women with IBD reported an almost identical mean daily intake of poultry (men = 45.5 g; women = 45.6 g). Compared to the DEGS1 cohort, a significant difference for poultry consumption was found between IBD women and DEGS1 women (24.5 g; $p = 0.030$, d = 0.6). The sex-related difference in the DEGS1 cohort was also significant (men = 32.6; $p = 0.048$, d = 0.2). DEGS1 men reported the highest intake of meat (53.8 g; women = 26.5 g; $p < 0.001$, d = 0.6). The DGE guideline recommends a weekly maximum of 240 g (34.3 g daily) for the intake of meat products (including beef, pork, and poultry). The recommended intake was on average exceeded by men and women of both cohorts (Table 4).

Cold Cuts

The highest daily intake of cold cuts was reported by DEGS1 men (34 g; women = 14.4; $p < 0.001$, d = 0.7), with a significant difference to IBD men in cold cut intake (19.3 g) ($p = 0.015$, d = −0.4). For women with IBD, a daily intake of 14.5 g was reported. Overall, the recommended intake of 60 g weekly (8.57 g daily) was exceeded (Table 4).

Eggs

Men with IBD reported the highest daily intake of eggs (30.56 g), followed by IBD women (26.63 g). DEGS1 men reported an average intake of 18.51 g per day, followed by DEGS1 women (15.48 g). The recommended intake of a weekly portion size of 60 g of eggs (8.57 g daily) was exceeded by all (Table 4).

Cereal Products and Rice

Men and women with IBD reported a nearly identical intake of breakfast cereals (men = 3.3 g; women = 3.4 g) but showed a numerical yet not significant difference in the intake of muesli (men = 13 g; women = 6.1 g; $p = 0.372$, d = 0.2). The differences in bread and roll consumption are defined by the flour used, with men and women in the DEGS1 cohort reporting a higher mean daily intake of whole grain bread and rolls (men = 55.7 g; women = 60.5 g); while men with IBD had an intake of 45.1 g, women with IBD reported an intake of 28.4 g, with the difference to healthy women being significant ($p = 0.001$, d = −0.4). The difference in consumption of mixed-flour bread was significant between men with IBD (44.2 g) and healthy men (78.2 g; $p = 0.008$, d = −0.3). DEGS1 women reported the lowest intake (37 g), with the sex-related difference to DEGS1 men being significant ($p \leq 0.001$, d = 0.5). The IBD cohort reported the highest mean daily intake of white bread and rolls (men = 48.2 g; women = 49.1 g) and the highest intake of rice (men = 31.6 g; women = 22.5 g). The mean daily intake reported by healthy men was the highest for pasta (43.6 g; women = 32.4 g). The DGE recommends consuming a total of 300 g from the food group "cereals, bread, pasta," and at least one-third of these should be whole grain products. Neither sexes nor both cohorts met this recommendation (Table 4).

3.3. IBD: Sex-Related Trends and Differences in Correlation with Disease Parameters

Food uptake of IBD patients showed significant sex-related trends and differences in correlation with disease-related quality of life (IBDQ). For men, IBDQ positively correlated with daily energy intake (EI) ($p = 0.007$; r = 0.409). Furthermore, for men, IBDQ positively correlated with the Mediterranean Diet Score (MDS) ($p = 0.003$, r = 0.449). Fecal calprotectin as an objective disease activity parameter showed no significant correlation with the EI or the MDS, though distinct negative trends in men are demonstrated. Ultra-processed food and drink intake (UPFD) showed no significant sex-related correlation with IBDQ or fecal calprotectin (mg/kg) (Figure 2).

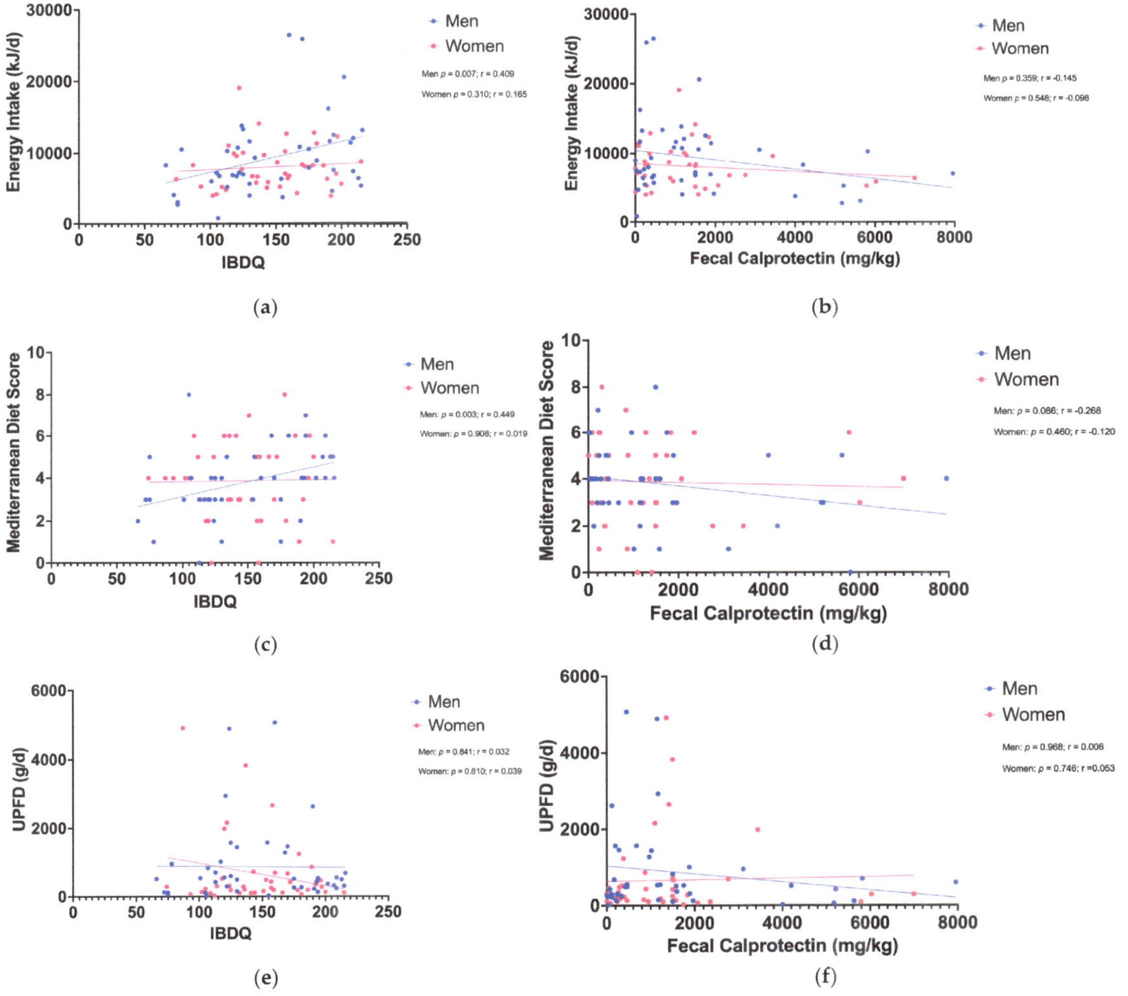

Figure 2. Sex-related trends and differences in correlation with disease parameters: (**a**) Disease-related quality of life positively correlated with daily energy intake (kJ) (p = 0.007; r = 0.409) for men but not for women (p = 3.310, r = 0.165). (**b**) Biochemical disease activity parameter fecal calprotectin (mg/kg) showed no significant correlation with energy intake (kJ) for men and women. (**c**) IBDQ positively correlated with the Mediterranean Diet Score (p = 0.003, r = 0.449) for men but not for women (p = 0.908, r = 0.019). (**d**) MDS showed no correlation with fecal calprotectin (mg/kg) for men and women. (**e**) UPFD intake showed no significant sex-related correlation with IBD-related quality of life (**f**) or fecal calprotectin (mg/kg). IBDQ—Inflammatory Bowel Disease Questionnaire; UPFDs—ultra-processed foods and drinks.

4. Discussion

This analysis shows that the dietary behavior of individuals with IBD exhibits fewer sex-related differences compared to the general German population. This observation may be attributed to the unique challenges and dietary restrictions that IBD patients face, regardless of their sex. The majority of IBD patients modify their diets after diagnosis. One study reported that even 75% of patients modified their dietary habits following their diagnosis [30]. This behavior is observed in both men and women with IBD, as the need to manage symptoms and prevent flares becomes a primary concern, superseding sex-specific dietary preferences that might be more pronounced in the general population. IBD patients frequently modify their diets in response to their disease condition. The majority of IBD patients, irrespective of their sex, consider diet to be a significant factor in their disease management [31]. A study found that 85.4% of IBD patients believed that dietary factors could potentially trigger disease relapses, while 32.9% of the same cohort thought that dietary factors could initiate the disease itself [30]. This shared belief system appears to transcend sex boundaries, resulting in comparable dietary behaviors among male and female IBD patients. One of the most prevalent dietary behaviors observed in individuals with IBD is the avoidance of certain foods. A significant proportion of patients, irrespective of sex, impose dietary restrictions in an attempt to prevent disease relapses. For example, 81.7% of patients in one study indicated that they believed it was necessary to eliminate certain products from their diets [30]. The most commonly avoided foods include spicy and fatty foods, raw fruits and vegetables, alcohol, legumes, cruciferous vegetables, and dairy products. This pattern of avoidance is consistent across both male and female patients, indicating that the disease's impact on dietary choices supersedes typical sex-based differences in food preferences. Interestingly, in our analysis, the consumption of fruits and vegetables did not differ between the IBD patients themselves. However, in comparison to the DEGS1 cohort, female IBD patients especially consumed significantly smaller amounts of fresh vegetables. This may indicate that female IBD patients especially adapt their dietary behavior in this context. When analyzing the general daily energy intake, again no significant differences between IBD patients have been found, whereas in the DEGS1 cohort, men showed a significant higher energy intake compared to women. Moreover, for all macronutrients, we found significant sex-related differences in the general German population, whereas these trends were not significant within the IBD patients. These general sex-related differences in energy uptake and macronutrient consumption are consistent with previously published data. Cross-sectional data from 210.106 men and women of the United Kingdom revealed sex differences in energy intake and distribution. Men had greater intakes of energy and were less likely to have energy intakes above the estimated average requirement. While adherence to recommended dietary guidelines was suboptimal in both sexes, women were significantly more likely than men to exceed recommended intakes of total sugar, total fat, and saturated fat. In contrast, men were more likely to have intakes under the recommended amounts of polyunsaturated fat, carbohydrates, and protein [32]. These differences are partly due to differences in the physiological composition between the sexes [33], but as sex differences in energy and macronutrient intakes vary also by age and socioeconomic status [32], this indicates the necessity for tailored interventions to optimize dietary behavior in men and women across the life course. However, current general dietary recommendations do not incorporate sex-specific factors. The exception to this is during the periods of pregnancy and lactation in women. Despite evidence indicating clear sex-related differences in dietary habits, metabolic processes, hormonal influences, and health outcomes, these factors have not yet been incorporated into dietary recommendations. Dietary recommendations for IBD patients are becoming more and more evident these days. The guidelines from the European Society for Clinical Nutrition and Metabolism (ESPEN) recently recommended a diet rich in fruits and vegetables, rich in n-3 fatty acids, and low in n-6 fatty acids for adult IBD patients. Furthermore, ultra-processed foods and dietary emulsifiers such as carboxymethylcellulose could be associated with an increased risk of IBD and, therefore, generally, such exclusions can be recommended [34]. A more detailed examination of these specific recommendations within our IBD collective indicates a discernible numerical trend among IBD men with respect to an alteration in dietary behavior when compared to the general population of the DEGS1 cohort. To illustrate, with respect to fruit and vegetable consumption, there is a discernible numerical discrepancy, with a proclivity towards greater intake among IBD men in comparison to men in the DEGS1 cohort. With regard to omega-3 fatty acid consumption, both sexes of the IBD cohort demonstrate a numerically higher intake than that observed in the DEGS cohort. With regard to UPFD consumption, IBD men exhibited numerically significant lower consumption compared to DEGS1 men, while IBD women demonstrated higher UPFD consumption compared to DEGS1 women. Correlation analyses between dietary behavior and disease-related quality of life as measured by the IBDQ also demonstrated that caloric intake and healthy dietary patterns as measured by the Mediterranean Diet Score exhibited significant changes with quality of life, particularly in IBD men. A further correlation also demonstrated this trend in relation to fecal calprotectin as an objective measure of IBD activity; however, this was only observed as a numerical trend without reaching statistical significance. Thus, these findings imply that especially male IBD patients tend to adapt their dietary behavior to their disease activity, since there is clear evidence that disease-related quality of life correlates with IBD activity [11,35]. Per se, this seems in line with the overall adaptive eating behavior of IBD patients during different phases of their disease activity. Several studies so far have underlined frequently restricted foods by IBD patients, e.g., dairy products, processed meats, soft drinks, alcoholic beverages, and fast food [36–39]. Furthermore, decreased appetite and reduced enjoyment of eating during IBD relapses appear to impair dietary intake in IBD patients [30,31]. However, it is noteworthy that in our analysis this adaptive dietary behavior was more pronounced in IBD men than in IBD women. Overall, these tendencies of IBD patients to adapt their diet to a concept of "orthorexia nervosa" have been published already. Recently, a prospective cohort of 113 IBD patients and 45 controls showed that the risk of orthorexia among IBD patients is high, at 77%, and significantly higher than that observed in a control group [40]. The prevalence rate of orthorexia is on average 6.9% for the general population [41]. Associations between sex and orthorexia nervosa tendency are not clear yet, but already some studies have found that the orthorexia nervosa tendency was significantly higher in men than in women [41–43]. Thus, this specific effect of an adaptive dietary behavior, in line with the emerging dietary recommendations for IBD patients, might be more pronounced in male IBD patients, leading to an equalization between the sexes.

5. Conclusions

In the present monocentric, cross-sectional study, the dietary behavior of IBD patients displayed a reduced number of sex-related differences when compared to an age- and sex-matched cohort from the general German population (DEGS1 cohort). Our analysis indicates that, in comparison to the distinct sex-related dietary differences observed in the general population, there is a growing convergence in dietary behavior between the sexes in the IBD collective. This phenomenon may be linked to particular dietary recommendations for individuals with IBD. It appears that IBD patients alter their dietary habits in response to the activity of their disease, whereas adherence to recommended dietary guidelines seems suboptimal in both sexes within the general population. Consistent with this, we observed a significant correlation between disease-related factors and the specific dietary behavior; this effect seems to be more pronounced among men with IBD compared to their female counterparts. These findings highlight the importance of providing condition-specific dietary guidance to individuals with IBD, with a particular emphasis on the nutritional challenges that arise from the disease-specific condition of each patient.

Limitations

We acknowledge that our analysis has limitations. First, as this is a post hoc analysis, the quality of data is limited and the analysis exploratory in nature. Second, the original patient cohort was small and included only IBD patients in a single, tertiary center. To address this, we chose the DEGS1 cohort as a healthy control group for all FFQ-based analyses. Although the FFQ is a validated and extensive tool to collect data on dietary behavior, it is limited in some significant areas: first, the recall period covers 4 weeks, and a time span of this length might introduce a recall bias when completing the questionnaire. In addition, some items which have become staple foods in many households are not included, such as processed and ultra-processed foods, especially of the vegetarian or vegan variety.

Author Contributions: Conceptualization, M.W., M.A.-B., U.T. and O.B.; methodology, M.W., M.A.-B., U.T. and O.B.; formal analysis, L.P., F.K. and M.W.; data curation, F.K., L.P., and M.W.; writing—original draft preparation, L.P. and M.W.; writing—review and editing, L.P., F.K., M.W., M.A.-B., U.T., M.K., H.W., H.L. and O.B.; visualization, L.P.; supervision, M.W.; project administration, M.W. All authors have read and agreed to the published version of the manuscript.

Funding: This publication is supported by funds from the Publication Fund NiedersachsenOPEN, funded by zukunft.niedersachsen.

Institutional Review Board Statement: The study was granted ethical approval by the ethics committee of Hannover Medical School on 23 November 2016 (No. 7275/2016), registered at the German Clinical Trials Register (DRKS) as DRKS00011370. The study design is in accordance with the Declaration of Helsinki (2013).

Informed Consent Statement: Written informed consent was obtained from each patient prior to screening.

Data Availability Statement: The original contributions presented in the study are included in the article, further inquiries can be directed to the corresponding author.

Acknowledgments: M.W. was supported by PRACTIS—Clinician Scientist Program of Hannover Medical School, funded by the German Research Foundation (DFG, ME 3696/3-1, RAW). The funders had no influence on study design, data collection and analysis, preparation of the manuscript, or the decision to publish. For the remaining authors, none were declared.

Conflicts of Interest: The authors declare no conflicts of interest.

References

1. Ng, S.C.; Shi, H.Y.; Hamidi, N.; Underwood, F.E.; Tang, W.; Benchimol, E.I.; Panaccione, R.; Ghosh, S.; Wu, J.C.Y.; Chan, F.K.L.; et al. Worldwide incidence and prevalence of inflammatory bowel disease in the 21st century: A systematic review of population-based studies. *Lancet* **2017**, *390*, 2769–2778. [CrossRef]
2. Reznikov, E.A.; Suskind, D.L. Current Nutritional Therapies in Inflammatory Bowel Disease: Improving Clinical Remission Rates and Sustainability of Long-Term Dietary Therapies. *Nutrients* **2023**, *15*, 668. [CrossRef] [PubMed]
3. Vissers, E.; Wellens, J.; Sabino, J. Ultra-processed foods as a possible culprit for the rising prevalence of inflammatory bowel diseases. *Front. Med.* **2022**, *9*, 1058373. [CrossRef] [PubMed]
4. Lo, C.H.; Khandpur, N.; Rossato, S.L.; Lochhead, P.; Lopes, E.W.; Burke, K.E.; Richter, J.M.; Song, M.; Ardisson Korat, A.V.; Sun, Q.; et al. Ultra-processed Foods and Risk of Crohn's Disease and Ulcerative Colitis: A Prospective Cohort Study. *Clin. Gastroenterol. Hepatol.* **2022**, *20*, e1323–e1337. [CrossRef]
5. Narula, N.; Wong, E.C.L.; Dehghan, M.; Mente, A.; Rangarajan, S.; Lanas, F.; Lopez-Jaramillo, P.; Rohatgi, P.; Lakshmi, P.V.M.; Varma, R.P.; et al. Association of ultra-processed food intake with risk of inflammatory bowel disease: Prospective cohort study. *BMJ* **2021**, *374*, n1554. [CrossRef] [PubMed]
6. Elizabeth, L.; Machado, P.; Zinöcker, M.; Baker, P.; Lawrence, M. Ultra-Processed Foods and Health Outcomes: A Narrative Review. *Nutrients* **2020**, *12*, 1955. [CrossRef] [PubMed]
7. Lane, M.M.; Gamage, E.; Du, S.; Ashtree, D.N.; McGuinness, A.J.; Gauci, S.; Baker, P.; Lawrence, M.; Rebholz, C.M.; Srour, B.; et al. Ultra-processed food exposure and adverse health outcomes: Umbrella review of epidemiological meta-analyses. *BMJ* **2024**, *384*, e077310. [CrossRef]
8. Roncoroni, L.; Gori, R.; Elli, L.; Tontini, G.E.; Doneda, L.; Norsa, L.; Cuomo, M.; Lombardo, V.; Scricciolo, A.; Caprioli, F.; et al. Nutrition in Patients with Inflammatory Bowel Diseases: A Narrative Review. *Nutrients* **2022**, *14*, 751. [CrossRef]
9. Grzymisławska, M.; Puch, E.A.; Zawada, A.; Grzymisławski, M. Do nutritional behaviors depend on biological sex and cultural gender? *Adv. Clin. Exp. Med.* **2020**, *29*, 165–172. [CrossRef]
10. Engler, D.; Schnabel, R.B.; Neumann, F.A.; Zyriax, B.C.; Makarova, N. Sex-Specific Dietary Patterns and Social Behaviour in Low-Risk Individuals. *Nutrients* **2023**, *15*, 1832. [CrossRef]
11. Wiestler, M.; Kockelmann, F.; Kück, M.; Kerling, A.; Tegtbur, U.; Manns, M.P.; Attaran-Bandarabadi, M.; Bachmann, O. Quality of Life Is Associated with Wearable-Based Physical Activity in Patients with Inflammatory Bowel Disease: A Prospective, Observational Study. *Clin. Transl. Gastroenterol.* **2019**, *10*, e00094. [CrossRef] [PubMed]

12. Kamtsiuris, P.; Lange, M.; Hoffmann, R.; Schaffrath Rosario, A.; Dahm, S.; Kuhnert, R.; Kurth, B.M. The first wave of the German Health Interview and Examination Survey for Adults (DEGS1): Sample design, response, weighting and representativeness. *Bundesgesundheitsblatt Gesundheitsforschung Gesundheitsschutz* **2013**, *56*, 620–630. [CrossRef] [PubMed]
13. RKI German Health Interview and Examination Survey for Adults (DEGS11). In *Scientific Use File First Version*; Robert Koch Institute. Department of Epidemiology and Health Monitoring: Berlin, Germany, 2015. [CrossRef]
14. Trichopoulou, A.; Costacou, T.; Bamia, C.; Trichopoulos, D. Adherence to a Mediterranean diet and survival in a Greek population. *N. Engl. J. Med.* **2003**, *348*, 2599–2608. [CrossRef] [PubMed]
15. Monteiro, C.A. Nutrition and health. The issue is not food, nor nutrients, so much as processing. *Public Health Nutr.* **2009**, *12*, 729–731. [CrossRef]
16. DGE. *Referenzwerte für die Nährstoffzufuhr*, 2nd ed.; 8th updated edition ed.; Deutsche Gesellschaft für Ernährung, Österreichische Gesellschaft für Ernährung: Bonn, Germany, 2024.
17. Haftenberger, M.; Heuer, T.; Heidemann, C.; Kube, F.; Krems, C.; Mensink, G.B. Relative validation of a food frequency questionnaire for national health and nutrition monitoring. *Nutr. J.* **2010**, *9*, 36. [CrossRef]
18. Mensink, G.B.M.; Truthmann, J.; Rabenberg, M.; Heidemann, C.; Haftenberger, M.; Schienkiewitz, A.; Richter, A. Obst- und Gemüsekonsum in Deutschland: Ergebnisse der Studie zur Gesundheit Erwachsener in Deutschland (DEGS1). *Bundesgesundheitsblatt—Gesundheitsforschung—Gesundheitsschutz* **2013**, *56*, 779–785. [CrossRef]
19. Thieleking, R.; Schneidewind, L.; Kanyamibwa, A.; Hartmann, H.; Horstmann, A.; Witte, A.V.; Medawar, E. Nutrient scoring for the DEGS1-FFQ—From food intake to nutrient intake. *BMC Nutr.* **2023**, *9*, 12. [CrossRef]
20. Harvey, R.F.; Bradshaw, J.M. A simple index of Crohn's-disease activity. *Lancet* **1980**, *1*, 514. [CrossRef]
21. Walmsley, R.S.; Ayres, R.C.S.; Pounder, R.E.; Allan, R.N. A simple clinical colitis activity index. *Gut* **1998**, *43*, 29–32. [CrossRef]
22. Janke, K.-H.; Klump, B.; Steder-Neukamm, U.; Hoffmann, J.; Häuser, W. Validierung der Deutschen Version (Kompetenznetz "Chronisch entzündliche Darmerkrankungen") des Inflammatory Bowel Disease Questionnaire IBDQ-D. *Psychother. Psychosom. Med. Psychol.* **2006**, *56*, 291–298. [CrossRef]
23. Guyatt, G.; Mitchell, A.; Irvine, E.J.; Singer, J.; Williams, N.; Goodacre, R.; Tompkins, C. A new measure of health status for clinical trials in inflammatory bowel disease. *Gastroenterology* **1989**, *96*, 804–810. [CrossRef]
24. Black, A.E.; Goldberg, G.R.; Jebb, S.A.; Livingstone, M.B.; Cole, T.J.; Prentice, A.M. Critical evaluation of energy intake data using fundamental principles of energy physiology: 2. Evaluating the results of published surveys. *Eur. J. Clin. Nutr.* **1991**, *45*, 583–599. [PubMed]
25. Black, A.E. Critical evaluation of energy intake using the Goldberg cut-off for energy intake:basal metabolic rate. A practical guide to its calculation, use and limitations. *Int. J. Obes. Relat. Metab. Disord.* **2000**, *24*, 1119–1130. [CrossRef] [PubMed]
26. Goldberg, G.R.; Black, A.E.; Jebb, S.A.; Cole, T.J.; Murgatroyd, P.R.; Coward, W.A.; Prentice, A.M. Critical evaluation of energy intake data using fundamental principles of energy physiology: 1. Derivation of cut-off limits to identify under-recording. *Eur. J. Clin. Nutr.* **1991**, *45*, 569–581.
27. Mifflin, M.D.; St Jeor, S.T.; Hill, L.A.; Scott, B.J.; Daugherty, S.A.; Koh, Y.O. A new predictive equation for resting energy expenditure in healthy individuals. *Am. J. Clin. Nutr.* **1990**, *51*, 241–247. [CrossRef] [PubMed]
28. Harris, J.A.; Benedict, F.G. A Biometric Study of Human Basal Metabolism. *Proc. Natl. Acad. Sci. USA* **1918**, *4*, 370–373. [CrossRef]
29. Moon, K.; Krems, C.; Heuer, T.; Roth, A.; Hoffmann, I. Predictors of BMI Vary along the BMI Range of German Adults—Results of the German National Nutrition Survey II. *Obes. Facts* **2017**, *10*, 38–49. [CrossRef]
30. Godala, M.; Gaszyńska, E.; Durko, Ł.; Małecka-Wojciesko, E. Dietary Behaviors and Beliefs in Patients with Inflammatory Bowel Disease. *J. Clin. Med.* **2023**, *12*, 3455. [CrossRef]
31. Zallot, C.; Quilliot, D.; Chevaux, J.-B.; Peyrin-Biroulet, C.; Guéant-Rodriguez, R.M.; Freling, E.; Collet-Fenetrier, B.; Williet, N.; Ziegler, O.; Bigard, M.-A.; et al. Dietary Beliefs and Behavior Among Inflammatory Bowel Disease Patients. *Inflamm. Bowel Dis.* **2013**, *19*, 66–72. [CrossRef]
32. Bennett, E.; Peters, S.A.E.; Woodward, M. Sex differences in macronutrient intake and adherence to dietary recommendations: Findings from the UK Biobank. *BMJ Open* **2018**, *8*, e020017. [CrossRef]
33. SACN. *SACN Dietary Reference Values for Energy*; TSO: London, UK, 2011.
34. Bischoff, S.C.; Bager, P.; Escher, J.; Forbes, A.; Hébuterne, X.; Hvas, C.L.; Joly, F.; Klek, S.; Krznaric, Z.; Ockenga, J.; et al. ESPEN guideline on Clinical Nutrition in inflammatory bowel disease. *Clin. Nutr.* **2023**, *42*, 352–379. [CrossRef] [PubMed]
35. Knowles, S.R.; Keefer, L.; Wilding, H.; Hewitt, C.; Graff, L.A.; Mikocka-Walus, A. Quality of Life in Inflammatory Bowel Disease: A Systematic Review and Meta-analyses—Part II. *Inflamm. Bowel Dis.* **2018**, *24*, 966–976. [CrossRef] [PubMed]
36. Murtagh, A.; Cooney, L.; Higginbotham, C.; Heavey, P. Dietary practices, beliefs and behaviours of adults with inflammatory bowel disease: A cross-sectional study. *Ir. J. Med. Sci.* **2023**, *192*, 1115–1124. [CrossRef]
37. Tomar, S.K.; Kedia, S.; Upadhyay, A.D.; Bopanna, S.; Yadav, D.P.; Goyal, S.; Jain, S.; Makharia, G.; Ahuja, V.; Singh, N. Impact of dietary beliefs and practices on patients with inflammatory bowel disease: An observational study from India. *JGH Open* **2017**, *1*, 15–21. [CrossRef] [PubMed]
38. Rajendran, N.; Kumar, D. Food-specific IgG4-guided exclusion diets improve symptoms in Crohn's disease: A pilot study. *Color. Dis.* **2011**, *13*, 1009–1013. [CrossRef]

39. Sigall-Boneh, R.; Pfeffer-Gik, T.; Segal, I.; Zangen, T.; Boaz, M.; Levine, A. Partial enteral nutrition with a Crohn's disease exclusion diet is effective for induction of remission in children and young adults with Crohn's disease. *Inflamm. Bowel Dis.* **2014**, *20*, 1353–1360. [CrossRef]
40. Di Giorgio, F.M.; Modica, S.P.; Saladino, M.; Muscarella, S.; Ciminnisi, S.; Almasio, P.L.; Petta, S.; Cappello, M. Food Beliefs and the Risk of Orthorexia in Patients with Inflammatory Bowel Disease. *Nutrients* **2024**, *16*, 1193. [CrossRef]
41. Donini, L.M.; Marsili, D.; Graziani, M.P.; Imbriale, M.; Cannella, C. Orthorexia nervosa: A preliminary study with a proposal for diagnosis and an attempt to measure the dimension of the phenomenon. *Eat. Weight Disord.* **2004**, *9*, 151–157. [CrossRef]
42. Fidan, T.; Ertekin, V.; Işikay, S.; Kirpinar, I. Prevalence of orthorexia among medical students in Erzurum, Turkey. *Compr. Psychiatry* **2010**, *51*, 49–54. [CrossRef]
43. Donini, L.M.; Marsili, D.; Graziani, M.P.; Imbriale, M.; Cannella, C. Orthorexia nervosa: Validation of a diagnosis questionnaire. *Eat Weight Disord.* **2005**, *10*, e28–e32. [CrossRef]

MDPI AG
Grosspeteranlage 5
4052 Basel
Switzerland
Tel.: +41 61 683 77 34

Nutrients Editorial Office
E-mail: nutrients@mdpi.com
www.mdpi.com/journal/nutrients

www.ingramcontent.com/pod-product-compliance
Lightning Source LLC
Chambersburg PA
CBHW041416210926
51850CB00013B/3289